Marek Pagac · Jiri Hajnys · Tomasz Kozior ·
Hoang-Sy Nguyen · Van Dung Nguyen ·
Akash Nag
Editors

From Smart City to Smart Factory for Sustainable Future: Conceptual Framework, Scenarios, and Multidiscipline Perspectives

SCFF

Springer

Editors
Marek Pagac
Department of Machining, Assembly and Engineering Metrology
Faculty of Mechanical Engineering
Technical University of Ostrava
Ostrava, Czech Republic

Jiri Hajnys
Department of Machining, Assembly and Engineering Metrology, Faculty of Mechanical Engineering
Technical University of Ostrava
Ostrava, Czech Republic

Tomasz Kozior
Department of Metrology and Modern Manufacturing, Faculty of Mechatronics and Mechanical Engineering
Kielce University of Technology
Kielce, Poland

Hoang-Sy Nguyen
Becamex Business School
Eastern International University
Thu Dau Mot, Binh Duong, Vietnam

Van Dung Nguyen
Faculty of Engineering and Technology
Hung Vuong University of Ho Chi Minh City
Ho Chi Minh, Vietnam

Akash Nag
Department of Machining, Assembly and Engineering Metrology
Faculty of Mechanical Engineering
Technical University of Ostrava
Ostrava, Czech Republic

ISSN 2367-3370 ISSN 2367-3389 (electronic)
Lecture Notes in Networks and Systems
ISBN 978-3-031-65655-2 ISBN 978-3-031-65656-9 (eBook)
https://doi.org/10.1007/978-3-031-65656-9

This Springer imprint is published by the registered company Springer Nature Switzerland AG
The registered company address is: Gewerbestrasse 11, 6330 Cham, Switzerland

Lecture Notes in Networks and Systems 1062

The series "Lecture Notes in Networks and Systems" publishes the latest developments in Networks and Systems—quickly, informally and with high quality. Original research reported in proceedings and post-proceedings represents the core of LNNS.

Volumes published in LNNS embrace all aspects and subfields of, as well as new challenges in, Networks and Systems.

The series contains proceedings and edited volumes in systems and networks, spanning the areas of Cyber-Physical Systems, Autonomous Systems, Sensor Networks, Control Systems, Energy Systems, Automotive Systems, Biological Systems, Vehicular Networking and Connected Vehicles, Aerospace Systems, Automation, Manufacturing, Smart Grids, Nonlinear Systems, Power Systems, Robotics, Social Systems, Economic Systems and other. Of particular value to both the contributors and the readership are the short publication timeframe and the world-wide distribution and exposure which enable both a wide and rapid dissemination of research output.

The series covers the theory, applications, and perspectives on the state of the art and future developments relevant to systems and networks, decision making, control, complex processes and related areas, as embedded in the fields of interdisciplinary and applied sciences, engineering, computer science, physics, economics, social, and life sciences, as well as the paradigms and methodologies behind them.

Indexed by SCOPUS, INSPEC, WTI Frankfurt eG, zbMATH, SCImago.

All books published in the series are submitted for consideration in Web of Science.

For proposals from Asia please contact Aninda Bose (aninda.bose@springer.com).

Introduction

The 2nd international conference SCFF24 is organized for researchers, experts, and students to share their vision of the future under the concept of *From Smart City to Smart Factory for Sustainable Future* from multidiscipline perspectives.

The aim of the conference is to provide an opportunity for participants with different expertise to meet up and share their knowledge, disruptive ideas, and solutions to existing problems that hinder us from moving forward to a more modern and sustainable life using technologies. The conference facilitates the mutual exchange of experiences and the establishment of long-lasting international cooperation. The conference scope includes discussions about innovative and sustainable technologies, solutions, and frameworks for applications of state-of-the-art technologies to Smart Cities and Smart Factories.

Scope

The conference scope includes but is not limited to the below topics, emphasizing the use of technologies in Smart Cities and Smart Factories for a Sustainable Future. The objective is to promote innovative research and work in the field of applied sciences, technologies, engineering, and management. Participants can submit to the following categories:

Track 1
Smart Factory, Non-conventional Manufacturing, Materials Science, Simulation, Robotics, Automation, Autonomous Driving, Digital Twin, Smart Electronics, Internet of Things, Wireless Communication.

Track 2
Artificial Intelligence, Machine Learning, Cloud Computing, Cyber Security, Network Analysis, Big Data, Data Visualization, Blockchain, Metaverse.

Track 3
Smart Economy, Industrial Management, Mathematical and Simulation Model, Digital Transformation, E-Commerce, Business Intelligence, Economics, Political Economy, Management, Logistics.

Organizers

VSB—Technical University of Ostrava, VSB—TUO, Ostrava, Czech Republic
Hanoi University of Science and Technology, HUST, Ha Noi, Vietnam
University of Finance - Marketing, UFM, Ho Chi Minh, Vietnam
Hung Vuong University of HCM city, DHV, Ho Chi Minh, Vietnam
Eastern International University, EIU, Thu Dau Mot, Vietnam
Thu Dau Mot University, TDMU, Thu Dau Mot, Vietnam
Ton Duc Thang University, TDTU, Ho Chi Minh, Vietnam
Kielce University of Technology, KUT, Kielce, Poland
Indian Institute of Technology (Indian School of Mines) Dhanbad, IIT (ISM), Dhanbad, India
Research Construction Technology Institute, RCTI, Thu Dau Mot, Vietnam

Conference Board

Honorary Chair

Prof. Robert Cep, VSB—TUO

General Co-chairs

Prof. Jana Petru, VSB—TUO
Assoc. Prof. Zdenek Poruba, VSB—TUO
Assoc. Prof. Marek Pagac, VSB—TUO
Prof. Huynh Trung Hai, HUST
Dr. Truong Thanh Cong, UFM
Assoc. Prof. Duong Hoai Nghia, EIU
Dr. Kang Seung Won, EIU
Dr. Narayan Chandra Debnath, EIU
Dr. Hoang-Sy Nguyen, EIU
Dr. Van Dung Nguyen, DHV
Dr. Huynh Van Van, TDTU
Dr. Nguyen Quoc Cuong, TDMU

Organizing Co-chairs

Dr. Jiri Hajnys, VSB—TUO
Dr. Jakub Mesicek, VSB—TUO
Prof. Nguyen Duc Hoa, HUST
Dr. Nguyen Trung Nghiep, EIU
Dr. Lam-Thanh Tu, TDTU
Assoc. Prof. Tomasz Kozior, KUT

International Coordinator

M.Sc. Quoc-Phu Ma, VSB—TUO
Dr. Dang Quoc Khanh, HUST
M.Sc. Vu Thi Thanh Huong, UFM
Ms. La Viet Ha, EIU
Dr. Le Truong An, TDMU
M.Sc. Khanh-Duy Ho Hoang, RCTI

Technical Committee

International Committee

Scientific Board

Keynote Speakers

Dr. Luc Phan Tan

Thu Dau Mot University, TDMU, Vietnam

Green Advertising: A Review Based on Bibliographic Coupling Analysis

Dr. Luc Phan Tan is a program director at the Thu Dau Mot University, Binh Duong Province, Vietnam. His research interests include entrepreneurship, social entrepreneurship, consumer behavior, corporate social responsibility, and green marketing. His research has been published in the *Journal of Social Entrepreneurship, Journal of Workplace Behavioral Health, The Journal of Entrepreneurship, Journal of Marketing for Higher Education, Journal of Marketing Communications, Journal of Business-to-Business Marketing*, and *Marketing Intelligence and Planning.*

Dr. Van Dung Nguyen

Hung Vuong University of HCM city, DHV, Vietnam

The Cloud-to-Edge-to-IoT Continuum Systems: Survey and Challenges

Dr. Van Dung Nguyen received the B.Eng. and M.Eng. degrees in telecommunications engineering from the Ho Chi Minh City University of Technology, Vietnam, in 2008 and 2012, respectively. He had a Ph.D. degree in computer science and engineering from Kyung Hee University in 2018. He was a post-doctoral researcher with the ICNS Laboratory, Kyung Hee University, from 2018 to 2021. From 2009 to 2010, he was a developer with TOSADENSHI, Vietnam. From 2021–2023, he was a lecturer of electronics and telecommunications, Ton Duc Thang University. Now he is the dean of Faculty of Engineering and Technology, Hung Vuong University of Ho Chi Minh City. His research interests include designing the MAC protocols in vehicular ad hoc networks and wireless sensor networks, cloud computing, mobile edge computing, Internet of Things, AI, and Mobile Edge Orchestrator.

Prof. Sergej Hloch

VSB—Technical University of Ostrava, VSB—TUO, Czech Republic

Water Droplet as a Universal Tool

Prof. Sergej Hloch received his Ph.D. from Technical University of Košice. He is currently the full professor of Advanced Manufacturing Technologies at Technical University of Košice and a senior researcher at Institute of Geonics of the Czech Academy of Sciences. His research covers research on water jetting focusing on ultrasonic pulsating water jet, studying erosion wear phenomena on different materials, including surface treatment of biocompatible materials and erosion resistance of materials and focuses on the transfer of basic research results to the applications of surface treatment of materials and to the development of universal technology with interdisciplinary overlapping. He has 117 publications and h-index of 26 on Web of Science.

Contents

Smart Factory, Non-conventional Manufacturing, Materials Science, Simulation, Robotics, Automation, Autonomous Driving, Digital Twin, Smart Electronics, Internet of Things, Wireless Communication

Smart Economy, Industrial Management, Mathematical and Simulation Model, Digital Transformation, E-Commerce, Business Intelligence, Economics, Political Economy, Management, Logistics

Smart Factory, Non-conventional Manufacturing, Materials Science, Simulation, Robotics, Automation, Autonomous Driving, Digital Twin, Smart Electronics, Internet of Things, Wireless Communication

3D Print of Clamping Tools Used in Electric Motor Manufacturing

Michal Prusa(✉), Petr Nemecek, Jiri Hajnys, Jakub Mesicek, Marek Pagac, Jana Petru, and Robert Cep

Department of Machining Assembly and Engineering Metrology, Faculty of Mechanical Engineering, VSB - Technical University of Ostrava, 70800 Ostrava, Czech Republic
michalprusa91@gmail.com

Abstract. The article deals with the testing of special clamps that were produced with the help of additive 3D printing technology. The primary purpose of the entire test was to test the clamps and to reduce the costs of manufacturing the prototype of the clamping devices. As part of the experiment itself, a special composite of ONYX plastic with carbon fiber was used. The primary purpose was to find out whether the clamps made by 3D printing from composite will be able to replace conventionally made metal clamps and, due to the reduction in the weight of the entire rotary system, whether it will be possible to increase the production parameters and thus obtain a higher quality of machining. A total of two clamp tests were performed. The clamp was given the working name of the external clamp. The design was completely taken from the original metal clamps, which were made of steel and used for everyday production. The main idea was only to change the material, so that the clamps for subsequent and possible future conventional production no longer had to change in terms of design, but only to replace the material used. As part of prototype production, it happens that the clamps are used only once, mostly during development and possibly the zero series, because the further development of the design of the electric motor shield requires a change in the design of the clamps as well. The usability of this study was mainly conceived for the prototype production of shields during the introduction of new types of electric motors, or their facelifts, and thus for the subsequent machining of all important surfaces.

Keywords: Additive manufacturing · Clamps · 3D print · Plastic · Markforged · Clamp printed · Machining · Turning · Composite

1 Introduction

As part of the application of additive technologies, the emphasis is mainly on changing the design in order to make the most of the potential of 3D printing. The design can be maximally adapted to the needs of the given application and is therefore the main advantage of the entire technology. A change in design can further lead to a reduction in weight and thus the resulting cost of production. Last but not least, 3D printing technology can be combined with conventional machining, milling, drilling or grinding

M. Pagac et al. (Eds.): SCFF 2024, LNNS 1062, pp. 3–13, 2024.
https://doi.org/10.1007/978-3-031-65656-9_1

technology. Conventional technologies are primarily used in additive technologies as finishing operations to ensure the necessary surface quality and part accuracy. Topological optimization [1] can not only reduce weight, but possibly also increase strength for load. The main idea is to know the forces and their distribution and direction of action on the test element. This idea of changing the design was also one of the possible directions of clamps printed on a 3D printer, with the fact that the weak point is the subsequent transfer of clamp production to conventional production, and therefore another optimal design would have to be created and the entire time requirement of the project would again be she extended.

An exemplary example of the use of additive technology in the application of clamps is not use clamps only for machining or other conventional operations in engineering but is used in biomedicine [2, 3] and to ensure the stability of the spine sample. Clamps made with additive 3D printing technology ensure in such a case an exact copy of the grip of the counterpart as well as perfect fixation stability compared to the conventional method (pouring), which is used in biomedicine, especially in connection with the spine.

Topological optimization and its use in practice in the form of a 3D printed scooter is dealt [4]. 3D printing achieves maximum success and use in conjunction with design changes and the application of various internal structures, either to reinforce the entire system or to create functional internal cooling channels.

On the other hand, shows the flexibility of the ONYX composite under load [5], using an experimental method including mathematical verification. The mechanical properties of ONYX composite still represent a weakness in the use of plastic composites, even though it is still a marked improvement in strength compared to conventional plastics. The completes the further processing of composites and their testing [6–8]. The processing of polymer composites is dealt [9], which describes and examines the 3D printing technique for polymer composites and, last but not least, applications in biomedicine, electronic or aerospace industries. Another study focused on the bending behavior of thin-walled samples with square hollow cross-sections produced using fused deposition modeling (FDM) [10, 11].

As already mentioned, the primary reason for this experiment was mainly to reduce the production costs of clamps, which, due to the high number of productions, are renewed quite often. Replacing metal with a plastic carbon composite [12–16] brings several advantages. A secondary reason was also the reduction of the weight of the entire rotary system and thereby ensuring more stable, more accurate and faster production, or more precisely, adjusting (increasing) the machining parameters.

2 Materials and Method

2.1 Selection of the Type of Machined Part

The priority criterion for the correct selection was above all the type and shape of the machined material. It was very important to think about making the machined part as small, light and as simple as possible. These factors are important above all for reasons of safety and possible damage to the machine or machine equipment if the component becomes loose. Any damage to the machine or machine equipment may affect the maintenance of the necessary production quality in the future.

Based on the decision criteria mentioned above, an aluminum electric motor shield was chosen, used for an axial height of 63 mm (Figs. 1 and 2). It was therefore the smallest and at the same time the lightest part. As part of the operations, the simplest and shortest possible machining cycle was chosen. So that, if necessary, the component can be detached and the component can be visually inspected at shorter intervals.

Fig. 1. Machined part of the electric motor - "end shield".

Fig. 2. Machined part of the electric motor - "end shield".

2.2 Selection of the Appropriate Machine Tool and Chuck

Based on the selected part, it was subsequently possible to select the machinery on which the given part of the electric motor can be manufactured. Due to the diverse nature of production, the clamping jaws that belong to the given machine and thus to the associated part can only be selected after selecting the machine. In this case, however, it was a relatively new machine tool marked EMAG VL5. This machine stands out for its excellent parameters and high production quality.

The chuck was purchased from Forward Deutschland and was a type 3QLC 200 Z6.

2.3 Choose of Clamps

The experiment itself was performed on two different samples (3 + 3 pieces). Both sets of clamps were not different from each other at first glance. However, the difference was in the direction and density of laying the base material and the composite fiber. The work

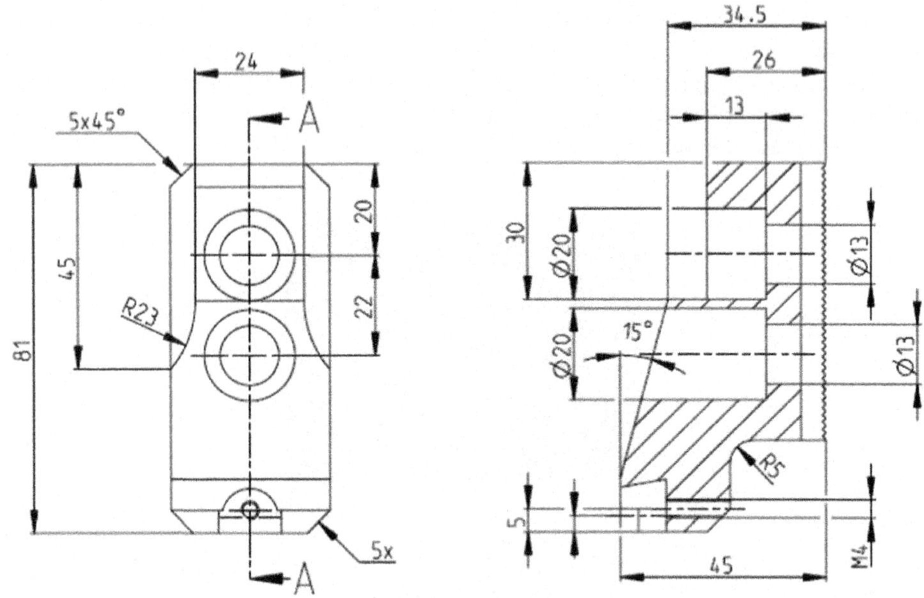

Fig. 3. Sketch of the external clamp.

clamp was designated as "external" because it clamped the workpiece on its outside. A detailed sketch of the outer clamp is shown in Fig. 3.

The assembly of clamps together with the machined part (Fig. 4) then looks as follows.

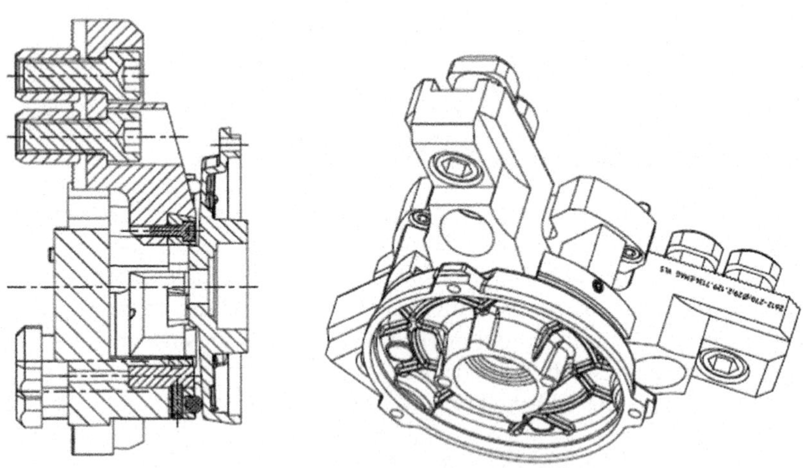

Fig. 4. Sketch of the outer clamp assembly.

2.4 Selection of 3D Printing Technology to Produce Clamps

After selecting the part, the machine, and the clamps, it was important to ensure the right production technology and the material from which to make the clamps by 3D printing. The main criteria were aimed at high strength, medium-low weight, and low flexibility. In the framework of a simple analysis, a technology called FDM or FFF was clearly offered, but in this case, it was accompanied by high-quality materials that compensated for the disadvantage of the chosen technology.

2.5 Selection of Equipment for 3D Printing and Choice of Material

As part of the 3D printing of the clamping jaws, a 3D printer marked X7 from the Markforged Company was used, mainly due to the possibility of printing special materials - composites, which were used for this machining test. It was a composition of ONYX material with carbon fiber. This material should be stronger than ONYX with Kevlar fiber, but its negative aspects include, for example, less shear strength [17].

2.6 Finalization and Testing of the Clamp

The printed clamps had to be fitted with special metal stones (Fig. 5), which firmly but at the same time sensitively "bite" into the aluminum shield, and the machining therefore takes place with maximum precision and high quality.

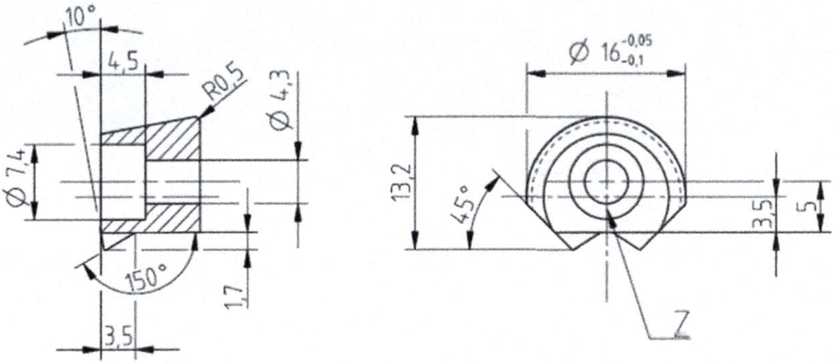

Fig. 5. Sketch of a clamping stone.

Clamping pressures for aluminum parts are set to 10 bar as standard. The minimum pressure for attaching an aluminum part is 8 bar.

3 Conclusion

3.1 Testing the External Clamp

Test Nr. 1

A clamp with a smaller leverage effect, and therefore with a reduced demand for flexibility, was defined. The clamp was designated as external (due to the external clamping of the workpiece).

The first test was conducted with ONYX material with carbon fiber and based on the mechanical properties there was a high probability of success of the entire test.

During the first clamping of the part, there was no plastic deformation, as was predicted by some, but there was a direct breakage of the clamp (Fig. 6). The fracture occurred at the junction of both ONYX materials + composite fiber.

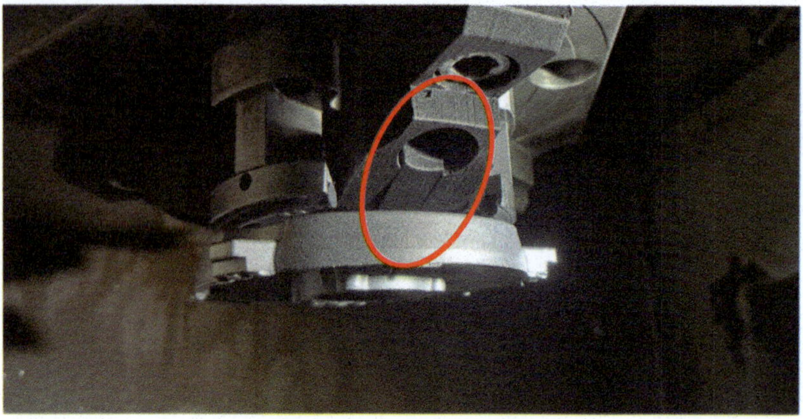

Fig. 6. Defect on the clamp – detail.

Based on the results of this first experiment, the test was evaluated as unsuccessful, with the need to change the laying of the inner fibers, so that the load of the entire system is uniform and primarily resistant to shear deformation.

Due to the nature of the crack, the clamp was subsequently divided by a direct cut and both parts were examined in detail. The clamps showed no over-tolerance inaccuracies.

Test Nr. 2

Due to the results of test no. 1, it was necessary to change the direction of laying the carbon fibers by 90°. According to theoretical reasoning, the clamp should already be much more reinforced.

However, even this test result turned out to be unsuccessful. The clamp broke again in the place under the most stress (Fig. 7), but this time at the point of transition to the bolted part – the base.

Fig. 7. Deformation of the clamp under load - from the right.

Since only one of the three clamps broke, it was important to carry out a detailed inspection and find out whether it is a printing defect and, if necessary, rule out this cause. A cut was made in the company's laboratory, followed by a detailed examination of the cut structure.

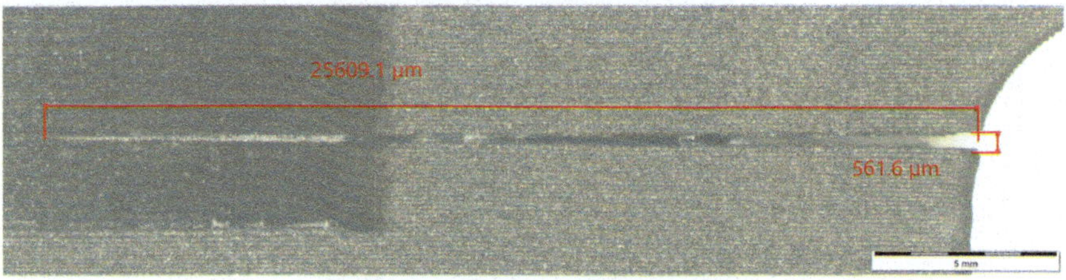

Fig. 8. Detail of a clamp break.

The crack reached almost 26 mm (Fig. 8). After cutting it open, we found that the method of laying the inner threads contained defects, but according to the subsequent discussion and further investigation, these were not the primary reason for the crack of the clamp. In the pictures below, the color change of the fibers is visible, as well as their visible fatigue from the load. It is therefore evident that the composite together with the base material could not withstand the clamping and material fatigue occurred in the weakest point (Figs. 9, 10, 11, 12 and 13). The clamps showed no over-tolerance inaccuracies.

Fig. 9. Fatigue from loading ONYX base material.

Fig. 10. Detail of critical point and threads.

4 Discussion

Unfortunately, the whole experiment with plastic clamps turned out to be unsatisfactory for practical implementation in industrial use.

The outer clamp could not withstand the clamping pressure and at the weakest point the layers of the composite broke off, breaking the base material. The quality of the print and the discovered deformation - lunkers - probably played a role. However, due to the 3D printing technology used, the print quality was still at a very good level.

However, this result has opened up some other possibilities that can be explored and further developed.

Fig. 11. Detail of critical point and threads.

Fig. 12. The thickness of the layers of the fiber grouping.

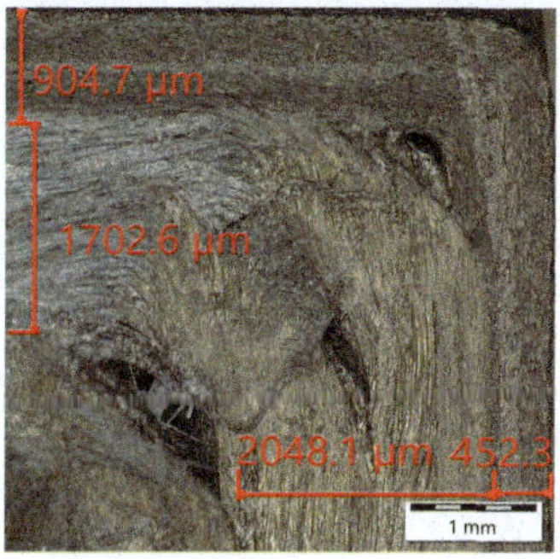

Fig. 13. The thickness of the layers of the fiber grouping.

Tips for possible future development and development:

a) measure the real load of the clamps and use topological optimization to focus on the most stressed places and reinforce them with fibers
b) simulation based on real forces during machining and design clamps based on it

c) a study based on the direction of laying the fibers of the composite, their density and ratio with the base material in relation to the direction and magnitude of the load
d) the creation of a study that will further deal with plastic materials, their composites, testing and, last but not least, comparison with the above tests/results.

References

1. Sotola, M., Marsalek, P., Rybansky, D., Fusek, M., Gabriel, D.: Sensitivity analysis of key formulations of topology optimization on an example of cantilever bending beam. Symmetry **13**, 712 (2021). https://doi.org/10.3390/sym13040712
2. Cornaz, F., et al.: 3D printed clamps for fixation of spinal segments in biomechanical testing. J. Biomech. **125**, 110577 (2021). ISSN 0021-9290. https://doi.org/10.1016/j.jbiomech.2021.110577
3. Marsalek, P., et al.: Modeling and testing of flexible structures with selected planar patterns used in biomedical applications. Materials **14**, 140 (2021). https://doi.org/10.3390/ma14010140
4. Mesicek, J., et al.: Comprehensive view of topological optimization scooter frame design and manufacturing. Symmetry **13**, 1201 (2021). https://doi.org/10.3390/sym13071201
5. Krzikalla, D., et al.: On flexural properties of additive manufactured composites: experimental, and numerical study. Compos. Sci. Technol. **218**, 109182 (2022). ISSN 0266-3538. https://doi.org/10.1016/j.compscitech.2021.109182
6. Ignazio, B.: The use of composite materials in 3D printing. J. Compos. Sci. **4**, 42 (2020). https://doi.org/10.3390/jcs4020042
7. Wickramasinghe, S., Do, T., Tran, P.: FDM-based 3D printing of polymer and associated composite: a review on mechanical properties, defects and treatments. Polymers **12**, 1529 (2020). https://doi.org/10.3390/polym12071529
8. Al Abadi, H., Thai, H.-T., Paton-Cole, V., Patel, V.I.: Elastic properties of 3D printed fibre-reinforced structures. Compos. Struct. **10** (2018). ISSN 0263-8223. https://doi.org/10.1016/j.compstruct.2018.03.051
9. Wang, X., Jiang, M., Zhou, Z., Gou, J., Hui, D.: 3D printing of polymer matrix composites: a review and prospective. Compos. Part B Eng. **110** (2017). ISSN 1359-8368. https://doi.org/10.1016/j.compositesb.2016.11.034
10. Jerzy, B., Kozior, T., Musialek, M.: Flexural properties of thin-walled specimens with square hollow sections 3D printed from ABS reinforced with aramid fibers. Fibers **11**, 77 (2023). https://doi.org/10.3390/fib11090077
11. Jerzy, B., Blasiak, M., Kozior, T.: A comparative study of the mechanical properties of FDM 3D prints made of PLA and carbon fiber-reinforced PLA for thin-walled applications. Materials **14**, 7062 (2021). https://doi.org/10.3390/ma14227062
12. Ning, F., Cong, W., Qiu, J., Wei, J., Wang, S.: Additive manufacturing of carbon fiber reinforced thermoplastic composites using fused deposition modeling. Compos. Part B Eng. **9** (2015). ISSN 1359-8368. https://doi.org/10.1016/j.compositesb.2015.06.013
13. Tekinalp, H.L., et al.: Highly oriented carbon fiber–polymer composites via additive manufacturing. Compos. Sci. Technol. **6** (2014). ISSN 0266-3538. https://doi.org/10.1016/j.compscitech.2014.10.009
14. Goh, G.D., et al.: Characterization of mechanical properties and fracture mode of additively manufactured carbon fiber and glass fiber reinforced thermoplastics. Mater. Des. **10** (2018). ISSN 0264-1275. https://doi.org/10.1016/j.matdes.2017.10.021
15. Sanei, S.H.R., Popescu, D.: 3D-printed carbon fiber reinforced polymer composites: a systematic review. J. Compos. Sci. **4**, 98 (2020). https://doi.org/10.3390/jcs4030098

16. van de Werken, N., Tekinalp, H., Khanbolouki, P., Ozcan, S., Williams, A., Tehrani, M.: Additively manufactured carbon fiber-reinforced composites: state of the art and perspective. Addit. Manuf. (2020). ISSN 2214-8604. https://doi.org/10.1016/j.addma.2019.100962
17. MARKFORGED: 3D Printing Materials (2024). Online. https://markforged.com/materials. [Cited 2024-04-01]

Analysis of Bi-directional Functionally Graded Porous Beams Vibrations Utilizing Ritz Solutions and Higher-Order Shear Deformation Theory

Quoc-Cuong Le[1] and Ba-Duy Nguyen[2](✉)

[1] Institute of Engineering and Technology, Thu Dau Mot University, 06 Tran Van On Street, Thu Dau Mot City, Vietnam

[2] Faculty of Architecture, Thu Dau Mot University, 06 Tran Van On Street, Thu Dau Mot City, Vietnam

duynb@tdmu.edu.vn

Abstract. This paper presents a novel hybrid function for analyzing the free vibration of bi-directional functionally graded porous (2D-FGP) beams. The higher-order shear deformation theory (HSDT) serves as the foundation for the development of the displacement field of the 2D-FGP beams. The variation in material along the x- and z-axes within the beam volume is determined by a power-law relationship. The porosity of the functionally graded material is represented by three different distributions. By using Lagrange's equation, the governing equations can be inferred. To define the displacement field, a hybrid function that combines an exponential-trigonometric function is devised. Different boundary conditions (simply-supported, clamped-clamped, and clamped-free), gradation exponents in the x and z directions, the porosity coefficient, the porosity distribution types (FGP-1, FGP-2, and FGP-3), and the slenderness ratio are utilized to calculate the frequency. A study of finite element method results shows that the suggested method is correct and works well.

Keywords: Ritz solution · Bi-directional functionally graded beam · Vibration

1 Introduction

The utilization of functionally graded materials (FGMs) in engineering applications is attributed to their capacity to effectively reduce delamination at the interfaces of structural layers. FGMs are able to efficiently satisfy customer requirements for constructions that are lightweight while still preserving stiffness and strength. FGMs consist of two distinct materials, and their characteristics exhibit continuous changes in a certain direction, adhering to specific distribution principles such as the rule of the mixture or the exponential law. Research attention has been directed towards bi-directional functionally graded materials due to the increasing demand for materials that can endure harsh operating conditions and offer strength in lighter constructions, as outlined in the aforementioned

M. Pagac et al. (Eds.): SCFF 2024, LNNS 1062, pp. 14–24, 2024.
https://doi.org/10.1007/978-3-031-65656-9_2

study [1]. Nevertheless, various obstacles encountered in manufacturing operations, such as substandard quality protocols, technical complications, insufficient expertise, inadequate upkeep, and logistical difficulties, give rise to the presence of porosities. These porosities are observed as a network of empty spaces distributed throughout the material structure in bi-directional functionally graded configurations [2, 3].

Some mechanical behavior of structures comprised of bi-directional functionally graded porous (2D-FGP) materials are also precisely analyzed. Numerous theories have been postulated to explore the behaviours of 2D-FGP beams, with the equivalent single-layer theory gaining traction owing to its simplicity and reduced computational demands [4]. Sari et al. [5] investigated the vibration response of 2D-FGP beams with the EBT and Galerkin techniques.

The exclusion of shear deformation in the mathematical model results in mistakes in the predictions generated by the EBT, leading to an overestimation of the beam's frequency. As a result, the Timoshenko Beam Theory (TBT), which is acknowledged for its ability to incorporate shear effects, has been extensively developed and applied in numerous research activities. The analysis conducted by Simsek [6] focused on the examination of the free and forced vibration characteristics of a bi-directional functionally graded (BDFG) Timoshenko beam subjected to the movement of a load.

Using TBT, nonlocal elasticity theory, and GDQM, Mirjavadi et al. [7] looked into the thermomechanical motions of 2D-FGP nanobeams.

The TBT has been utilized in structural analysis to include shear deformation in beams of different dimensions, accommodating both slender and thick configurations. Nevertheless, in order to precisely measure real displacements, TBT requires the inclusion of a shear correction factor. However, the task of establishing a suitable value for this adjustment factor presents a significant issue. As a result, the field of beam analysis has employed higher-order beam theories (HBT) as a means to tackle this particular concern. Fouda et al. [8] conducted a finite element analysis on the bending, buckling, and vibration of a functionally graded porous beam. Turan et al. [9] used a higher-order finite element model to investigate the free vibration and buckling of porous two-directional functionally graded beams. Adiyaman [10] investigated the free vibration analysis of a porous bi-directional (2D) functionally graded (FG) beam using a higher-order shear deformation theory (HSDT). In their study, Mellal et al. [11] proposed a three-variable HSDT based analytical method to investigate the free vibration and stability of defective and perfect functionally graded (FG) beams supported by variable elastic foundations (VEFs).

Despite extensive study in this field, this literature review reveals a scarcity of papers examining the free vibration characteristics of 2D-FGP beams. As far as current knowledge is concerned, there is a lack of research that investigates the integration of higher-order beam theory with the Ritz methods for vibration analysis in the context of 2D-FGP beams. Further investigation is required to have a full understanding of the behavior of 2D-FGP beams in order to address this research gap.

This study aims to analyze the free vibration behaviors of 2D-FGP beams comprehensively, taking into account the existence of porosities. This analysis utilizes higher-order beam theory and the Ritz method, representing the first application of this approach. The study concentrates on 2D-FGP beams to investigate the impact of functionally graded material throughout the thickness and length dimensions. We have created a Ritz approach that combines exponential-trigonometric function to address this issue. The governing equations have been derived using Lagrange's equation. The numerical findings we got using the proposed method are consistent with the current body of literature. The study thoroughly investigates how various factors such as boundary conditions, porosity coefficient, porosity distribution, growing exponent, and fundamental frequency affect the system.

2 Theoretical Formulation

2.1 Material Properties of 2D-FGB

The FGP beam under consideration has the dimensions $b \times h$ and L, as illustrated in Fig. 1. The constituent volume fraction of the ceramic and metal materials comprising the FGP beams causes a continuous variation in their properties. The power law describes the effective properties of FGP beams as follows:

$$P(x, z) = (P_c - P_m)\left(\frac{x}{L}\right)\left(\frac{z}{h} + \frac{1}{2}\right)^{p_z} + P_m - \frac{e}{2}f(z)(P_c + P_m) \quad (1)$$

where $P(x, z)$ is material elastic moduli as Young modulus $E(x, z)$, mass density ρ and Poisson's ratio ν; P_c, P_m are material elastic properties of ceramic and metal, respectively; $f(z)$ is a function showing the distribution of the void along the beam's thickness; e represents the porosity ratio. In this paper, three porosity distributions (FGP-1, FGP-2 and FGP-3) shown in Fig. 2 are considered as follows:

$$\begin{aligned} &\text{FGP-1:} \quad f(z) = 1 \\ &\text{FGP-2:} \quad f(z) = 1 - \frac{2|z|}{h} \\ &\text{FGP-3:} \quad f(z) = \sin\left(\frac{|z|\pi}{h}\right) \end{aligned} \quad (2)$$

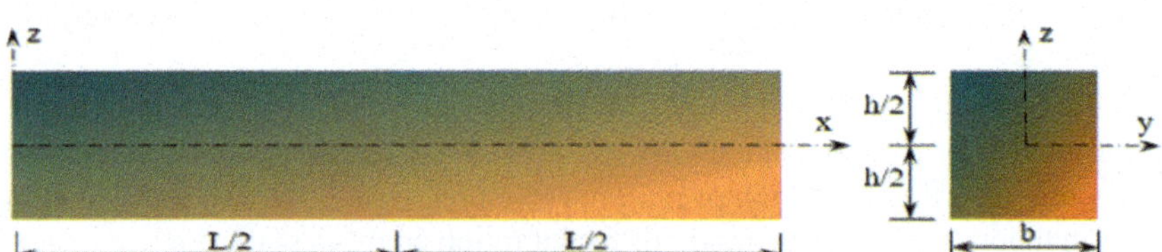

Fig. 1. Geometry of FGP beams.

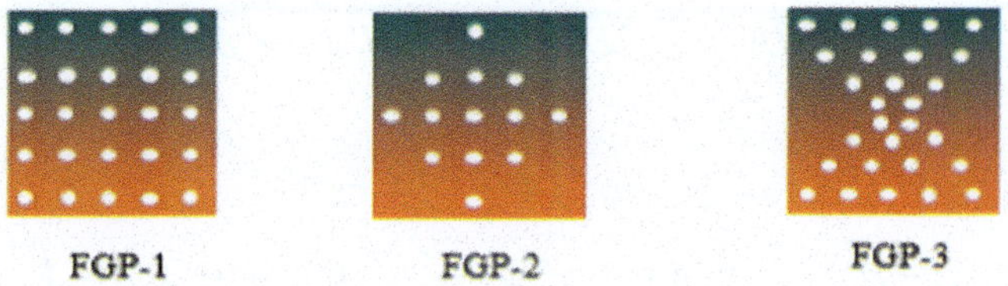

Fig. 2. The utilization of cross-sectional beams in diverse distribution functions.

2.2 Higher-Order Shear Deformation Theory

The displacement fields of 2D-FGP beams is chosen from previous study [12, 13]:

$$u_1(x, z, t) = u(x, t) - zw_{,x}(x, t) + \kappa(z)\phi(x, t); \quad u_3(x, z, t) = w(x, t) \tag{3}$$

where $u(x, t)$ and $w(x, t)$ are displacement components in the x- and z-directions of a point at z = 0, respectively. $\phi(x, t)$ is the rotation of a transverse normal about the y-axis. $\kappa(z) = \sinh^{-1}\left(\frac{rz}{h}\right) - \frac{8z^3}{3h^3\sqrt{r^2+4}}$ is the shear function. In this notation, the comma indicates partial differentiation concerning the coordinate subscript that follows.

The nonzero strain of 2D-FGP beams is written:

$$\varepsilon_x = u_{1,x} = u_{,x} - zw_{,xx} + \kappa\phi_{,x}; \quad \gamma_{xz} = u_{1,z} + u_{3,z} = \kappa_{,x}\phi \tag{4}$$

2.3 Constitutive Equations

The stress and strain relation of 2D-FGP beams are elucidated in the ensuing section [9]:

$$\begin{Bmatrix} \sigma_{xx} \\ \sigma_{xz} \end{Bmatrix} = \begin{bmatrix} Q_{11} & 0 \\ 0 & Q_{55} \end{bmatrix} \begin{Bmatrix} \varepsilon_{xx} \\ \gamma_{xz} \end{Bmatrix} \tag{5}$$

where $Q_{11} = E(x, z)$ and $Q_{55} = \frac{E(x,z)}{2[1+\nu(x,z)]}$.

2.4 Equations of Motion

The equations of motion are derived using a Lagrangian function.

$$\Pi = U - K \tag{6}$$

where U represents the strain energy, while K represents the kinetic energy.

The system's strain energy U can be expressed as [14]:

$$U = \frac{1}{2}\int_V (\sigma_{xx}\varepsilon_{xx} + \sigma_{xz}\gamma_{xz})dV$$
$$= \frac{1}{2}\int_0^L \Big[Au_{,x}^2 - 2Bu_{,x}w_{,xx} + 2B^s u_{,x}\phi_{,x}$$

$$+Dw_{,xx}^2 - 2D^s w_{,xx}\phi_{,x} + H^s\phi_{,x}^2 + A^s\phi^2\Big]dx \tag{7}$$

where

$$\left(A,\ B,\ D,\ B^s,\ D^s,\ H^s\right) = \int_{-h/2}^{h/2} \left(1;\ z;\ z^2,\ \kappa,\ z\kappa,\ \kappa^2\right) Q_{11} b dz;\ A^s = \int_{-h/2}^{h/2} \left(\kappa_{,z}\right)^2 Q_{55} b dz \tag{8}$$

The formulation of the kinetic energy for the 2D-FGP beam is expressed as follows [14].

$$K = \frac{1}{2}\int_V \rho(x, z)\left[\dot{u}_1^2 + \dot{u}_3^2\right] dV$$
$$= \frac{1}{2}\int_0^L \left[I_0\dot{u}^2 - 2I_1\dot{u}\dot{w}_{,x} + I_2\dot{w}_{,x}^2 + 2J_1\dot{\phi}\dot{u} - 2J_2\dot{\phi}\dot{w}_{,x} + J_3\dot{u}^2 + I_0\dot{w}^2\right] dx \tag{9}$$

where the notation of a dot superscript denotes differentiation with respect to time in accordance with established conventions. The determination of inertia coefficients $(I_0,\ I_1,\ I_2,\ J_1,\ J_2,\ J_3)$ is reliant upon the following equations:

$$(I_0,\ I_1,\ I_2,\ J_1,\ J_2,\ J_3) = \int_{-h/2}^{h/2} \rho(x, z)\left(1;\ z;\ z^2,\ \kappa,\ z\kappa,\ \kappa^2\right) b dz \tag{10}$$

By substituting Eqs. (7) and (9) into Eq. (6), Lagrangian function is explicitly expressed as:

$$\Pi = \frac{1}{2}\int_0^L \Big[Au_{,x}^2 - 2Bu_{,x}w_{,xx} + 2B^s u_{,x}\phi_{,x}$$
$$+Dw_{,xx}^2 - 2D^s w_{,xx}\phi_{,x} + H^s\phi_{,x}^2 + A^s\phi^2\Big]dx$$
$$-\frac{1}{2}\int_0^L \left[I_0\dot{u}^2 - 2I_1\dot{u}\dot{w}_{,x} + I_2\dot{w}_{,x}^2 + 2J_1\dot{\phi}\dot{u} - 2J_2\dot{\phi}\dot{w}_{,x} + K_2\dot{u}^2 + I_0\dot{w}^2\right] dx \tag{11}$$

2.5 Ritz Solution

Based on the Ritz method, the axial and transverse displacements at location (x, z) of the 2D-FGP beam can be generally approximated in the following forms:

$$u(x, t) = \sum_{j=1}^{m} \psi_j(x) u_j e^{i\omega t};\ w(x, t) = \sum_{j=1}^{m} \varphi_j(x) w_j e^{i\omega t};\ \phi(x, t) = \sum_{j=1}^{m} \psi_j(x)\phi_j e^{i\omega t} \tag{12}$$

where u_j, w_j, ϕ_j are unknown values to be determined; $i^2 = -1$; ω is natural frequency; $\psi_j(x)$ and $\varphi_j(x)$ are the shape functions which are proposed in Table 1 for simply-supported (SS), clamped-clamped (CC) and clamped-free (CF) boundary conditions (BCs).

Substituting Eq. (12) into Eq. (11), the following characteristic equation is obtained:

$$\left\{\begin{bmatrix} \mathbf{K}^{11} & \mathbf{K}^{12} & \mathbf{K}^{13} \\ T\mathbf{K}^{12} & \mathbf{K}^{22} & \mathbf{K}^{23} \\ T\mathbf{K}^{13} & T\mathbf{K}^{23} & \mathbf{K}^{33} \end{bmatrix} - \omega^2 \begin{bmatrix} \mathbf{M}^{11} & \mathbf{M}^{12} & \mathbf{M}^{13} \\ T\mathbf{M}^{12} & \mathbf{M}^{22} & \mathbf{M}^{23} \\ T\mathbf{M}^{13} & T\mathbf{M}^{23} & \mathbf{M}^{33} \end{bmatrix}\right\} \begin{Bmatrix} \mathbf{u} \\ \mathbf{w} \\ \boldsymbol{\phi} \end{Bmatrix} = \begin{Bmatrix} \mathbf{0} \\ \mathbf{0} \\ \mathbf{0} \end{Bmatrix} \tag{13}$$

where the components of the stiffness matrix $\mathbf{K}$ and the mass matrix $\mathbf{M}$ are given by:

$$K_{ij}^{11} = A\int_0^L \psi_{i,x}\psi_{j,x}dx,\ K_{ij}^{12} = -B\int_0^L \psi_{i,x}\varphi_{j,xx}dx,\ K_{ij}^{13} = B^s\int_0^L \psi_{i,x}\psi_{j,x}dx$$

$$K_{ij}^{22} = D\int_0^L \varphi_{i,xx}\varphi_{j,xx}dx,\ K_{ij}^{23} = -D^s\int_0^L \varphi_{i,xx}\psi_{j,x}dx,$$

$$K_{ij}^{33} = H^s\int_0^L \psi_{i,x}\psi_{j,x}dx + A^s\int_0^L \psi_i\psi_j dx$$

$$M_{ij}^{11} = I_0\int_0^L \psi_i\psi_j dx,\ M_{ij}^{12} = -I_1\int_0^L \psi_i\varphi_{j,x}dx,\ M_{ij}^{13} = -J_1\int_0^L \psi_i\psi_j dx$$

$$M_{ij}^{22} = I_0\int_0^L \varphi_i\varphi_j dx + I_2\int_0^L \varphi_{i,x}\varphi_{j,x}dx,\ M_{ij}^{23} = -J_2\int_0^L \varphi_{i,x}\psi_j dx,\ M_{ij}^{33} = J_3\int_0^L \psi_i\psi_j dx \tag{14}$$

Finally, the dimensionless fundamental frequency (DFF) of FGB can be determined by solving Eq. (13).

Table 1. Functions and fundamentals of BC beams.

BCs	$\psi_j(x, z)$	$\varphi_j(x, z)$	$x = 0$	$x = L$
S-S	$\cos(\frac{\pi x}{L})e^{-jx/L}$	$\sin(\frac{\pi x}{L})e^{-jx/L}$	$w = 0$	$w = 0$
C-F	$\sin(\frac{\pi x}{2L})e^{-jx/L}$	$(1 - \cos(\frac{\pi x}{2L}))e^{-jx/L}$	$u = 0,\ w = 0,$ $w_{,x} = 0$	
C-C	$\sin(\frac{\pi x}{L})e^{-jx/L}$	$\sin^2(\frac{\pi x}{L})e^{-jx/L}$	$u = 0,\ w = 0,$ $w_{,x} = 0$	$u = 0,\ w = 0,$ $w_{,x} = 0$

3 Numerical Results

This section is dedicated to the presentation of numerical exemplifications aimed at validating the proposed methodology. Additionally, it seeks to scrutinise the impact of parameters such as slenderness, boundary conditions, power-law index, porosity ratio, and distribution type on the DFF, within the context of 2D-FGP beams. The computations are carried out in an Intel® Core™ i7 with 4 cores at 3.4 GHz, 8 GB of RAM, and calculations in MATLAB R2015a. It is postulated that the 2D-FGP beam is constituted of both metallic and ceramic constituents, the specific material attributes of which are delineated in Table 2.

$$\overline{\omega} = \frac{\omega L^2}{h}\sqrt{\frac{\rho_m}{E_m}} \tag{15}$$

where subscript m donates properties related to metal.

Table 2. Material property

Material	E (GPa)	ρ (kg/m^3)	ν
Ceramic (Al_2O_3)	380	3960	0.3
Metal (Al)	70	2702	0.3

3.1 Convergence Study

The study investigates the convergence properties of a FGP-2 beam with different boundary conditions, where $L/h = 20$, $p_x = p_z = 1$, and $e = 0.1$. Table 3 presents the DFF of 2D-FGP beams. The current findings reach a consensus at $m = 10$ for the DFF. This particular number acts as the convergence criterion for the numerical examples in the next publication.

Table 3. Convergence analyses of the DFF of FGP beams ($L/h = 20, p_x = p_z = 1, e = 0.1$)

BCs	m						
	2	4	6	8	10	12	14
S-S	3.6781	3.6669	3.6661	3.6661	3.6661	3.6661	3.6661
C-F	1.9464	1.0848	1.0620	1.0618	1.0618	1.0618	1.0618
C-C	8.3935	8.0810	8.0653	8.0618	8.0601	8.0591	8.0591

3.2 Free Vibration Analysis

A variety of 2D-FGP beams are analyzed for validation, each with different power-law indexes, porosity ratios, slenderness, porosity distribution patterns, and boundary conditions. Table 4 displays the DFF of clamped-clamped (CC) beams. The results are compared with those produced by Turan et al. [9] using higher-order beam theory (HBT) and finite element method (FEM). The current results are consistent with those documented in the reference cited.

Figure 3 shows the changes in the DFF for FGP-1 SS beams ($L/h = 20$, $e = 0.2$) in relation to the power-law increase in p_x and p_z. The simultaneous rise in the metal volume in the beam resulted in a decrease in both stiffness and inertial mass. Despite a decrease in both stiffness and inertial mass, the rate of decline in stiffness exceeds that of inertial mass, leading to a significant decrease in the DFF.

Figure 4 displays the DFF fluctuations of FGP-2 beams with $p_x = p_z = 1$ and $e = 0.2$ in respect to the L/h ratio under different boundary conditions. Increasing the L/h ratio results in a notable rise in the DFF for CC beams, but the increase is minimal for SS and CF beams. CC beams have the highest DFF, while CF beams have the lowest values. Examining how different porosity distribution patterns affect the DFF involves analyzing a CF beam ($L/h = 20$, $p_z = 1$, and $e = 0.2$) from the FGP-1, FGP-2, and FGP-3 series.

Figure 5a illustrate the DFF of FGP-1, FGP-2, and FGP-3 beams in relation to p_x. The DFF of all beams decreases significantly for $0 \leq p_x < 2$, and this decline becomes minimal for p_x values greater than 2. FGP-2 beams have the highest DFF, whereas FGP-1 beams have the lowest values.

Figure 5b explore how the porosity ratio affects the DFF of CC beams with $L/h = 20$ and $p_x = p_z = 1$. Increasing the porosity ratio leads to a large decrease in the DFF for FGP-1 and FGP-3 beams, while FGP-2 beams show minimal change. The tendency is explained by the drop in stiffness and inertial mass due to a higher porosity ratio, with the loss in stiffness being more significant than that in inertial mass for FGP-1 and FGP-3 beams. However, this difference is insignificant for FGP-2 beams.

4 Conclusions

The study employs a hybrid method that integrates an exponential-trigonometric functions to examine the free vibration properties of 2D-FGP beams with various boundary conditions, utilizing the higher-order beam theory. The material properties of the beams change exponentially in both the axial and thickness dimensions. The paper analyzes three distinct porosity distribution patterns. The governing equation is obtained through the use of Lagrange's equation. A formulation utilizing exponential-trigonometric function is developed to represent the unexplained displacement field. The Ritz method is then employed to computationally solve the problem. Numerical experiments are conducted to confirm the suggested methodology and gain a more profound understanding of the behavior of 2D-FGP beams. Numerical investigations show the efficiency of the approach and examine the impact of several parameters, including porosity ratio, porosity distribution pattern, slenderness ratio, and boundary condition, on the fundamental

Table 4. DFF of 2D-FGP beams with CC boundary condition

p_z	e	Reference	$L/h = 5$				$L/h = 20$			
			$p_x = 0$	$p_x = 1$	$p_x = 2$	$p_x = 5$	$p_x = 0$	$p_x = 1$	$p_x = 2$	$p_x = 5$
FGP-1										
1	0.1	Present	7.8234	6.3382	5.9226	5.5021	9.2271	7.6221	7.1656	6.6667
		Turan [9]	7.7919	6.2999	5.8941	5.4806	9.2250	7.6192	7.1634	6.6651
	0.2	Present	7.6324	5.7742	5.2035	4.6569	8.9315	6.9295	6.3154	5.6494
		Turan [9]	7.6033	5.7301	5.1750	4.6381	8.9296	6.9260	6.3132	5.6480
5	0.1	Present	6.0317	5.4254	5.2447	5.0449	7.5425	6.7575	6.4901	6.2003
		Turan [9]	6.0020	5.3975	5.2201	5.0242	7.5392	6.7549	6.4879	6.1986
	0.2	Present	5.2476	4.5683	4.3486	4.1054	6.4536	5.6872	5.3855	5.0485
		Turan [9]	5.2217	4.5427	4.3272	4.0891	6.4506	5.6846	5.3835	5.0472
FGP-2										
1	0.1	Present	7.9653	6.6140	6.2467	5.8774	9.4803	8.0601	7.6741	7.2511
		Turan [9]	7.9325	6.5757	6.2173	5.8543	9.4780	8.0569	7.6716	7.2492
	0.2	Present	7.9738	6.4946	6.0701	5.6683	9.5275	8.0469	7.6401	7.1958
		Turan [9]	7.9412	6.4534	6.0385	5.6428	9.5252	8.0430	7.6367	7.1930
5	0.1	Present	6.3322	5.7740	5.6172	5.4442	8.0643	7.3249	7.0874	6.8314
		Turan [9]	6.2970	5.7424	5.5894	5.4205	8.0600	7.3215	7.0847	6.8291
	0.2	Present	6.1107	5.5249	5.3620	5.1833	7.9401	7.2286	6.9932	6.7364
		Turan	6.0665	5.4860	5.3285	5.1552	7.9336	7.2232	6.9887	6.7328
FGP-3										
1	0.1	Present	7.8019	6.3918	6.0046	5.6001	9.1646	7.5948	7.1491	6.6626
		Turan [9]	7.7698	6.3541	5.9746	5.5765	9.1625	7.5921	7.1469	6.6610
	0.2	Present	7.6053	5.9665	5.4892	4.9849	8.8357	6.9626	6.3937	5.7762
		Turan [9]	7.5747	5.9278	5.4572	4.9600	8.8337	6.9599	6.3914	5.7745
5	0.1	Present	6.1694	5.5618	5.3789	5.1772	7.6046	6.7878	6.5159	6.2240
		Turan [9]	6.1421	5.5343	5.3533	5.1542	7.6021	6.7856	6.5139	6.2223
	0.2	Present	5.7013	5.0087	4.7786	4.5232	6.8523	5.9528	5.6260	5.2733
		Turan [9]	5.6788	4.9830	4.7529	4.4998	6.8506	5.9509	5.6241	5.2716

frequency of 2D-FGP beams. A concise summary of the most significant findings derived from the numerical results is as follows:

- By employing a combination of exponential-trigonometric functions, the mechanical properties of 2D-FGP beams can be examined in a straightforward and practicable manner.
- An increase in the power-law indexes leads to a decrease in the DFF of the beams.

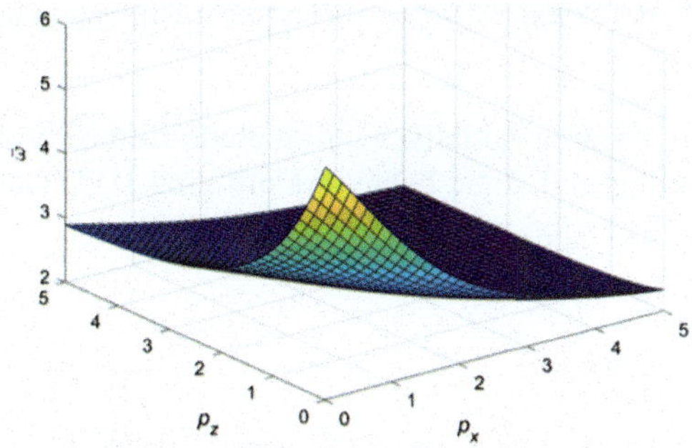

Fig. 3. Variation of DFF of FGP-1 SS beams with respect to power-law index

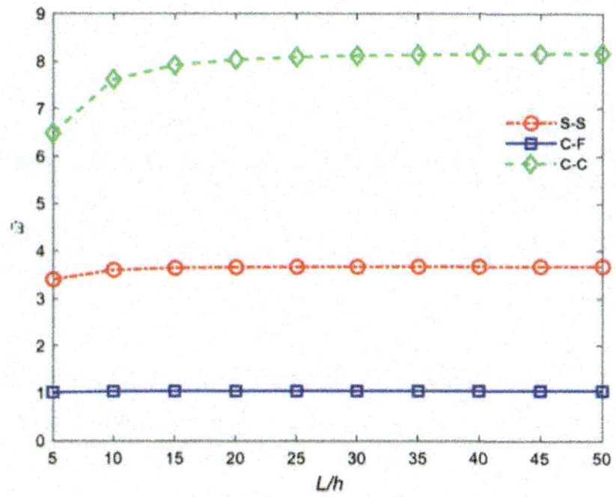

Fig. 4. Variation of DFF of FGB-2 beams with respect to slenderness

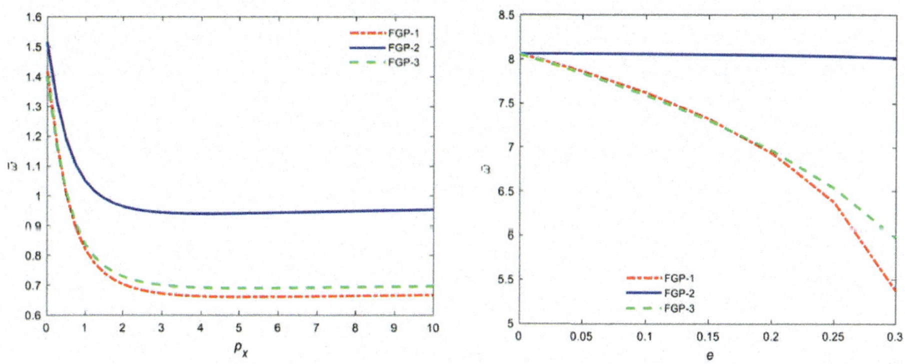

a. CF b. CC

Fig. 5. DFF with various porosity distribution types

- Changes in the porosity ratio have a big effect on how FGP-1 and FGP-3 beams behave mechanically, but not as much on FGP-2 beams.
- Contrarily, CF beams demonstrate the least DFF, whereas CC beams exhibit the greatest DFF.

- An augmentation in the slenderness of the beams results in an equivalent rise in their DFF.

Our group's future research initiatives will include further in-depth development of 2D-FGP, such as buckling and static behaviors utilizing HSDT or an improved FSDT for static bending, and free vibration analysis of 2D-FGP.

References

1. Nemat-Alla, M.: Reduction of thermal stresses by developing two-dimensional functionally graded materials. Int. J. Solids Struct. **40**(26), 7339–7356 (2003)
2. Ghatage, P.S., Kar, V.R., Sudhagar, P.E.: On the numerical modelling and analysis of multi-directional functionally graded composite structures: a review. Compos. Struct. **236**, 111837 (2020)
3. Chen, D., et al.: Functionally graded porous structures: analyses, performances, and applications – a review. Thin-Walled Struct. **191**, 111046 (2023)
4. Garg, A., et al.: A review of the analysis of sandwich FGM structures. Compos. Struct. **258**, 113427 (2021)
5. Sari, M.E., Al-Dahidi, S., Hammad, B.: Free and forced nonlinear vibrations of bi-directional functionally graded Euler–Bernoulli porous beams. J. Vib. Control **29**, 107754632210843 (2022)
6. Simsek, M.: Bi-directional functionally graded materials (BDFGMs) for free and forced vibration of Timoshenko beams with various boundary conditions. Compos. Struct. **133**, 968–978 (2015)
7. Mirjavadi, S.S., et al.: Thermal vibration of two-dimensional functionally graded (2D-FG) porous Timoshenko nanobeams. Steel Compos. Struct. **25** (2017)
8. Fouda, N., El-Midany, T., Sadoun, A.: Bending, buckling and vibration of a functionally graded porous beam using finite elements. J. Appl. Comput. Mech. **3**(4), 274–282 (2017)
9. Turan, M., Adiyaman, G.: Free vibration and buckling analysis of porous two-directional functionally graded beams using a higher-order finite element model. J. Vib. Eng. Technol. **12**(1), 1133–1152 (2024)
10. Adiyaman, G.: Free vibration analysis of a porous 2D functionally graded beam using a high-order shear deformation theory. J. Vib. Eng. Technol. **12**(2), 2499–2516 (2024)
11. Mellal, F., et al.: On the vibration and buckling behaviors of porous FG beams resting on variable elastic foundation utilizing higher-order shear deformation theory. Acta Mech. **234**(9), 3955–3977 (2023)
12. Nguyen, T.-K., et al.: Vibration and buckling analysis of functionally graded sandwich beams by a new higher-order shear deformation theory. Compos. Part B Eng. **76** (2015)
13. Nguyen, T.-K., et al.: Hygro-thermal effects on vibration and thermal buckling behaviours of functionally graded beams. Compos. Struct. **176**, 1050–1060 (2017)
14. Reddy, J.N.: Mechanics of Laminated Composite Plates and Shells: Theory and Analysis. CRC Press (2003)

Analysis of the Tensile Strength of PA 2200 Material Depending on the Direction of SLS Additive Technology Printing

Paweł Szczygieł(✉), Jerzy Bochnia, Tomasz Kozior, Wiktor Szot, Mateusz Rudnik, and Natalia Kowalska

Kielce University of Technology, Kielce, Poland
pszczygiel@tu.kielce.pl

Abstract. This paper presents results from static tensile tests of ISO 527 standardized samples (type 1BA). The samples were made of PA 2200 material in white powder form using Selective Laser Sintering technology. The tests additionally included measuring the weight of the sample, the surface roughness, and checking the actual dimensions of the samples. The results of this study demonstrated that a number of parameters are impacted by the printing direction. Based on the lowest surface roughness and highest tensile strength as criterion, the printing direction X should be recommended as the most appropriate.

Keywords: PA 2200 · Tensile strength · SLS

1 Introduction

Polymers are a group of materials very often used in additive technologies due to their versatility, affordability and ease of use [1]. Their application is wide and includes industrial section such as aerospace, electronics industry [2], and environment sectors, specifically in the areas of wastewater management, gas separation, and the development of batteries [3]. Materials classified as biocompatible for medical use must meet normative requirements (ISO 10993) for: cytotoxicity, irritation, delayed-type hypersensitivity, genotoxicity, chemical characterization. Polymeric biomaterials are used to manufacture implants including craniofacial implants, dental molds, prosthetic parts, on-demand medical equipment, surgical models, scaffolds for tissue regeneration, organ printing, implantable biosensors, and tissue models for drug discovery [4]. One of the most popular polymers used in additive technologies is Polyamide 12 (as PA 2200). It is a great choice for functioning prototypes or end-use parts because of its mechanical properties, flexibility, and heat resistance. Printing the most complex designs is possible since laser printing PA 12 doesn't require support structures. Among the major polyamides, Nylon 12 ($C_{12}H_{23}NO$) has the lowest melting point, ranging from 178 to 180 °C. The mechanical properties of PA 12 (hardness, tensile strength, and abrasion resistance) are similar to those of Nylon 6 and Nylon 66. Its relatively long hydrocarbon chain length, which results in its dimensional stability and almost paraffin-like structure, also contributes to its low water absorption and density of 1.01 g/cm^3. Nylon 12 is not sensitive to stress cracking and is resistant to chemicals [5].

M. Pagac et al. (Eds.): SCFF 2024, LNNS 1062, pp. 25–35, 2024.
https://doi.org/10.1007/978-3-031-65656-9_3

The additive manufacturing process makes it possible to produce customized objects in moderate quantities. Additive manufacturing technologies create new opportunities for manufacturing and fabrication paradigms by introducing the ability to produce components with complex shapes. It results in a significant reduction in production time, shorter time to market for new designs and faster satisfaction of customer needs [6]. However, additive technologies are not without their drawbacks, one of which is the anisotropy of mechanical properties, whereby the direction of printing has a significant impact on the strength of the manufactured part [7], moreover, also on dimensional accuracy and geometric tolerances [8]. Along with other technological parameters, e.g. layer height, these materials are subjected to continuous testing. The tests carry valuable information for designers of printed parts, so that the best possible strength properties can be obtained using the appropriate technological parameters.

In this article, the PA 2200 material used in powder form by Selective Laser Sintering additive technology was tested to determine the tensile strength depending on the direction of printing. In addition, the dimensions of the samples were compared with the nominal dimensions, and the surface roughness and weight of the samples were measured.

2 Experimental Part

2.1 Materials

The material used in the study was PA 2200 in powder form (EOS GmbH, Krailling, Germany). Features of this material include high strength and stiffness, good chemical resistance, excellent long-term constant behavior, high selectivity and detail resolution, and a variety of finishing options (metallization, stove enameling, vibratory grinding, tub coloring, bonding, powder coating, flocking, etc.). It is also biocompatible per EN ISO 10993-1 and USP/level VI/121 °C, and it is approved for food contact in accordance with EU Plastics Directive 2002/72/EC. High-quality, fully functional plastic pieces are typical uses for this material. The material is frequently utilized in place of standard injection-molded plastics because of its superior mechanical qualities. Because of its great abrasion resistance and biocompatibility, it can be used in prosthetics and to realize moving part connections, among other things [9]. Table 1 displays a selection of mechanical properties.

2.2 Methods

The study used Selective Laser Sintering technology belongs to the Powder Bed Fusion additive manufacturing category (ISO 52900). The samples were designed in the CAD (computer aided design) software SolidWorks (Dassault Systemes SolidWorks Corp., Waltham, MA, USA) and fabricated using a Formiga P100 3D printer (EOS GmbH, Krailling, Germany).

The samples were fabricated in accordance with ISO 527 (Fig. 1) in three print directions X (flat), Y (on edge) and Z (vertical). A total of 36 samples were made (12 for each series). After the fabrication process, they were cleaned of residual untreated

Table 1. Mechanical properties of PA 2200 [9].

Mechanical properties	Value	Unit	Test standard
Tensile modulus (X direction)	1700	MPa	ISO 527-1/-2
Tensile modulus (Y direction)	1700	MPa	ISO 527-1/-2
Tensile modulus (Z direction)	1650	MPa	ISO 527-2/-2
Tensile strength (X direction)	48	MPa	ISO 527-2/-2
Tensile strength (Y direction)	48	MPa	ISO 527-2/-2
Tensile strength (Z direction)	47	MPa	ISO 527-2/-2
Strain at break (X direction)	24	%	ISO 527-2/-2

powder, weighed using a JS-100xV scale (Jennings Scale, China) and measured in cross-section using a Mitutoyo digital micrometer (Mitutoyo, Kanagawa, Japan). Then, using a surface roughness tester (INSIZE CO., Suzhou New District, China), surface roughness measurements were made on the upper surface of the sample - in the area of the measurement base, i.e. along the X axis of the X samples, along the Y axis of the Y samples, and along the Z axis of the Z samples. Measuring condition settings are shown in Table 2. The tensile tests were carried out on an Inspekt Mini strength testing machine using LabMaster (Hegewald and Peschke, Nossen, Germany).

Table 2. Roughness measuring condition settings.

Parameter	Value/setting
Cut off length λc	2.50 mm (according to PN-ISO 4288)
Number of sampling lengths (xn)	5
Filter	Gauss

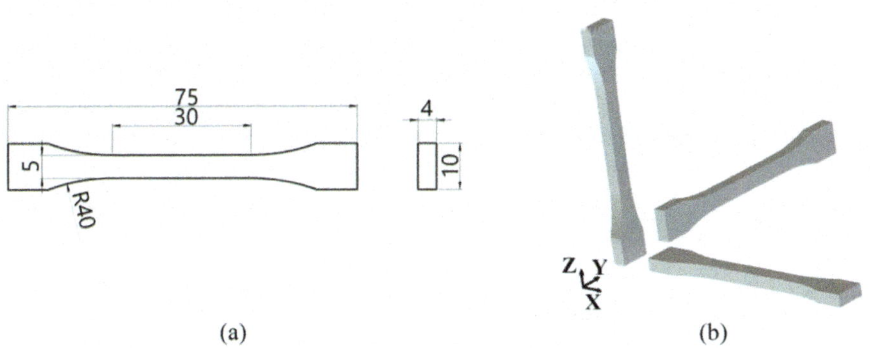

Fig. 1. Sample: (a) dimensions, (b) print directions.

In addition, virgin powder was used (it was not mixed with the powder already used in the manufacturing process). Density of laser-sintered part was 930 kg/m^3 [9]. The technological parameters that were used in the manufacturing process are shown in Table 3. The energy density was calculated with Eq. 1, and was ED = 0.056 J/mm^2. According to the manufacturer: fine resolution, very high surface quality, and detail resolution are achieved with a layer thickness choice of 0.1 mm [9].

$$ED = \frac{P}{vh}x, \ \mathrm{J/mm^2} \tag{1}$$

Table 3. Printing processing parameters.

Parameter	Layer height	Laser power	Laser scanning speed	Hatch distance	Beam diameter	Overlay ratio $x = \frac{d}{h}$
	Lh [mm]	P [W]	v [mm/s]	h [mm]	d [mm]	x
Values	0.1	21	2500	0.25	0.42	1.68

3 Results and Discussion

3.1 Samples Weight

Each sample was weighed three times, then the mean overall weight of each of the X, Y and Z sample series was calculated. The results are shown along with the standard deviation in Fig. 2. Samples X had the highest weight and takes an mean value of 1.94 g. Samples Y had a lower mean weight of about 0.5%, and samples Z had a lower mean weight of almost 5%. It is worth noting that the nominal volume of the sample determined in SolidWorks is 2024.30 mm^3 and, according to the manufacturer, the laser-sintered part density is 0.00093 g/mm^3, so it follows that the nominal mass should be 1.88 g. The determined average masses and the density given by the manufacturer allow to calculate the actual volumes of samples X, Y and Z, which are respectively: 2086.02 mm^3; 2075.27 mm^3 and 1989.25 mm^3. Thus, it can be seen that the closest to the nominal value were Z samples, and the largest discrepancy occurs for X samples. Each of the adopted printing directions was characterized by different number of layers: Z > Y > X, so the conclusion arises that the more layers the print has, the closer the mass of the sample is to the nominal one.

3.2 Cross-Section Measurement

The samples were measured in cross-section to check the actual dimensions. The nominal dimensions were A = 4 mm and B = 5 mm (Fig. 1a). Each measurement was taken three times, and Fig. 3 shows the overall means for each series along with the standard

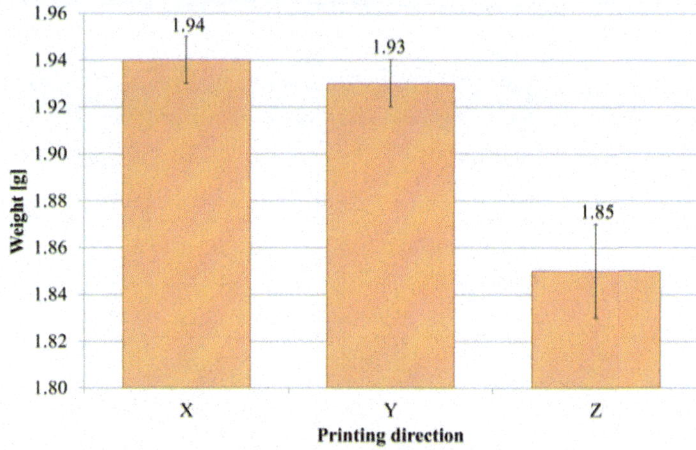

Fig. 2. Weight of samples depending on the direction of printing.

deviations. Comparing the results with nominal dimensions, in the case of dimension A, the smallest deviations were obtained for the Y series of samples, while in the case of dimension B for the X series of samples. The deviations of dimensions A and B were positive for X samples, negative for Z samples, and in the case of the Y series, they were negative for dimension A and positive for dimension B. The negative deviations of the Z series justify the lower weight of these samples. The mean A-dimension of the Y samples was about 1% smaller than the mean A-dimension of the X samples, and about 3% smaller for the Z series compared to the X series. In the case of dimension B, comparing the mean values to X samples, Y samples were larger by 3% and Z samples were smaller by about 1%. Z samples had the smallest sample cross-section, which definitely influenced the smallest weight of the samples. The nominal cross-sectional area of the sample is 20 mm^2, thus X samples were closest to this value. In view of the differences, it should be noted that the direction of printing has a significant impact on the actual dimensions of the printed part. The results also indicate that components printed in the Z-axis and having more layers may require the assumption of an allowance to correct the actual dimensions obtained after the additive manufacturing process.

It is worth noting that in the article [10], cylindrical samples were made in three different printing directions (vertical, horizontal, 45°). It turned out that in each case negative dimensional deviations were obtained compared to the nominal diameter dimension.

3.3 Roughness

Surface roughness measurements were performed on each sample three times and the overall mean values are shown in Fig. 4. Ra is the arithmetic mean deviation of the measured profile from the centerline of the evaluation length, and Rz is the average distance between the highest peak and lowest valley in each sampling area. The lowest values of Ra and Rz parameters were obtained for the X series of samples. Values higher in Ra parameter by about 24% were recorded for Y and Z samples. As for the Rz parameter values, they were 23% larger for Y samples and 25% larger for Z samples. In

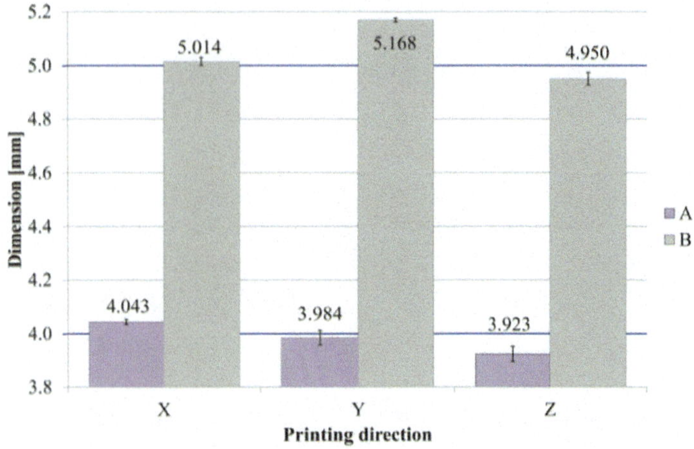

Fig. 3. Actual mean dimensions of the samples.

the article [11], Mitutoyo SJ-410 was used for roughness measurements and an average Ra parameter of 10.75 was obtained. The material used in the study was also PA 2200 in the form of virgin powder.

In the case of surface roughness, the differences are not as clear as in the case of technologies from the MEX (Material Extrusion) group, where "stair-stepping," a phenomenon where the layers of the printed object are visibly and tangibly noticeable, giving the surface of the object a stepped or stair-like appearance. The results indicate that in SLS technology this effect does not occur, moreover, the difference between the surface roughness made in the Y and Z directions is insignificant (less than 1% for the Ra parameter and less than 2% for the Rz parameter). It should be pointed out, however, that the surface roughness was the lowest for samples fabricated in the X direction.

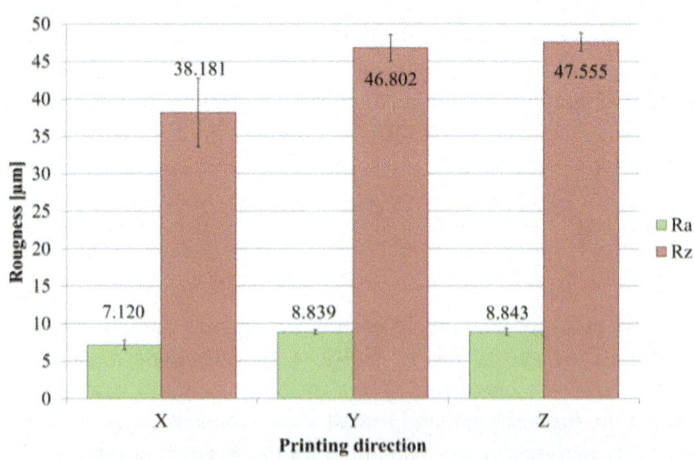

Fig. 4. Mean values of the roughness parameters Ra and Rz.

3.4 Tensile Strength

The figure shows the results of the static tensile tests. Both X-direction and Y-direction printed samples achieved strain of about 50% (Fig. 5a, b). The Z samples cracked much faster with a strain of about 17% (Fig. 5c). The highest mean tensile strength value was obtained for X-direction samples (Fig. 5d). Samples printed in the Y-direction showed a lower mean tensile strength of about 2%, while a lower one of about 23% was registered for samples made in the Z direction. The highest standard deviation was also obtained for the Z-series of samples. A comparison of the results with other articles is shown in Table 4.

Analyzing Table 3, it can be seen that technological parameters have a huge impact on tensile strength. In this study, the results were lower than those reported by the manufacturer [9] by: 7% for sample X, 9% for sample Y and 27% for sample Z. The biggest difference is noticeable in the study [12], where samples made with identical parameters obtained an average tensile strength that was 84% lower for the Y direction and 78% lower for the Z samples. In this study, it was shown that higher values of Laser Power and lower values of Laser Scanning Speed result in increased tensile strength. In the study [13], the team obtained used a different sample standard but also results most similar to the manufacturer's reported values. The paper [14] examined the effect of different Hatch Distance values. It was found that increasing the value of this parameter resulted in lower tensile strength. In the study [15], a larger Layer Height was used (0.12 mm), but the results showed small difference in tensile strength depending on the direction of printing and, in addition, samples made in the Z direction obtained the highest strength. The article [16] analyzed two parameters: Laser Power and Laser Scanning Speed. The recorded results showed that an increase in Laser Power results in an increase in tensile strength, and for the second parameter, the best results were obtained for a value of 1400 mm/s. There were also large differences in the tensile strength values depending on the direction of printing, where the largest took samples made in the X direction.

4 Conclusions

The results showed that the printing direction affects many parameters, taking the lowest surface roughness and the highest tensile strength as the criteria, X printing direction should be indicated as the most suitable. The recorded differences in the studied parameters depending on the direction of printing are due to the layered nature of additive technologies. Depending on the direction of printing, the measured parameters were as follows:

- samples produced in the X direction had the highest average weight; in the case of actual dimensions, they had the smallest deviation from the nominal dimension B (+ 0.014 mm) and in the case of dimension A deviation of + 0.043 mm; the surface roughness of these samples was the smallest, these samples also achieved the highest tensile strength;

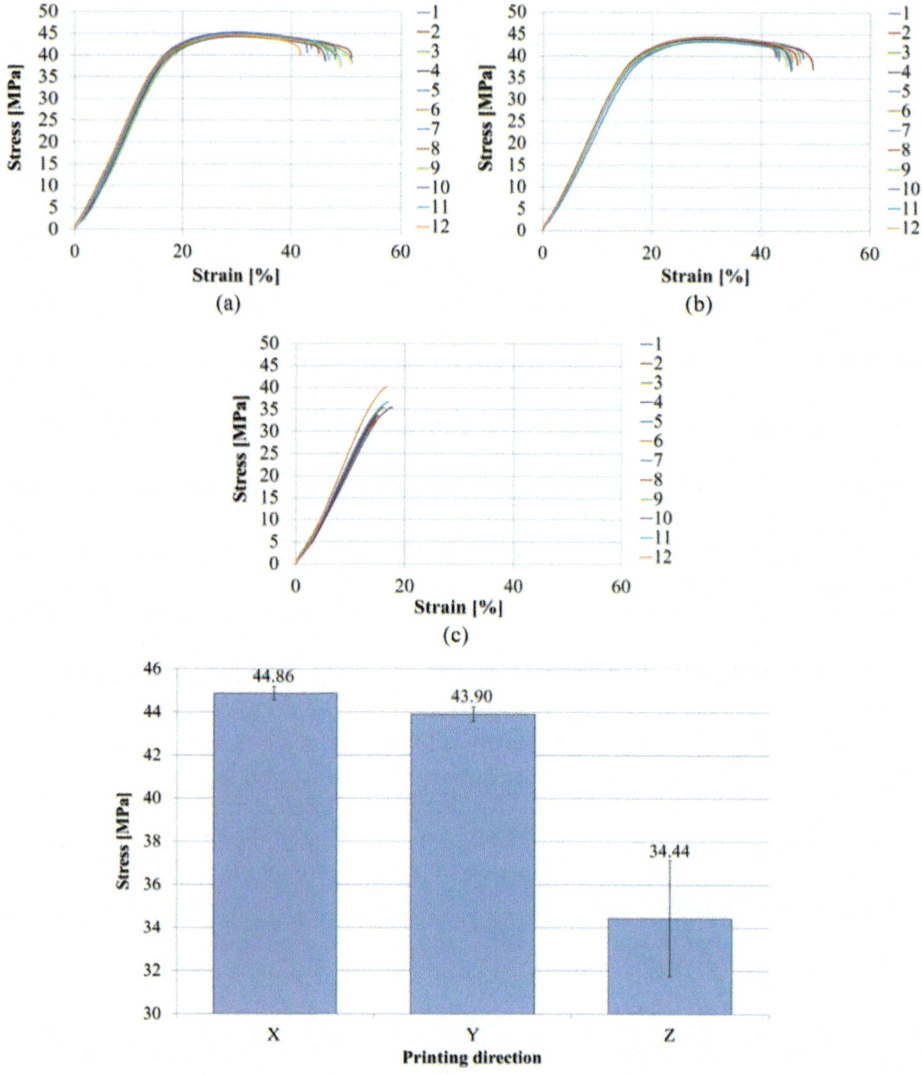

Fig. 5. Tensile strength: (a) X series, (b) Y series, (c) Z series, (d) average values.

- samples produced in the Y direction had a lower average weight than X samples by 0.01 g; for actual dimensions, the determined averages showed a deviation from the A dimension of – 0.016 mm and + 0.168 mm for the B dimension; the surface roughness of these samples was higher than X samples by about 24% Ra; the average tensile strength was lower by about 2% compared to X samples;
- samples made in the Z direction had the smallest average weight of 1.85 g; in terms of actual dimensions, they had negative deviations from nominal dimensions of – 0.077 mm and – 0.05 mm; the surface roughness of these samples was the highest,

but the difference with respect to Y samples was small; during the tensile test, the lowest strength was registered for Z samples.

Samples produced by laser beam sintering of PA 2200 powder indicate that the direction of printing has a significant effect on the weight and actual dimensions, and thus on the volume of the parts. In the case of positive dimensional deviations, it is possible to use post-process machining, such as milling, to get the desired dimension.

Table 4. Mean tensile strength obtained for PA 2200 samples in comparison with results obtained in other works at different parameters rates [9, 12–16].

	Printing direction	Mean tensile strength [MPa]	Sample standard	Layer height [mm]	Laser power [W]	Laser scanning speed [mm/s]	Hatch distance [mm]	Beam diameter [mm]
This work	X	44.86	ISO 527	0.1	21	2500	0.25	0.42
	Y	43.90						
	Z	34.44						
Manufacturer [9]	X	48	ISO 527	Unknown				
	Y	48						
	Z	47						
Stoia et al. [12]	Y	27.06	ISO 527	0.1	25	1500	0.25	0.15 (beam offset)
	Z	27.87						
	Y	15.53			23	2000		
	Z	15.28						
	Y	7.08			21	2500		
	Z	7.51						
Morano et al. [13]	Y	47.10	ASTM-D 638	0.1	21 (inner) 16 (outer)	2500 (inner) 1500 (outer)	0.25	Unk.
Pilipović et al. [14]	X	46.27	ISO 527	Unk.	16 (contour) 21 (hatching)	1500 (contour) 2500 (hatching)	0.23	Unk.
		30.62					0.42	
		8.16					0.6	
Zakręcki et al. [15]	X	42.4	ISO 527	0.12	Unknown			
	Y	42.08						
	Z	43.73						
Kundera et al. [16]	X	1.00	ISO 527	Unk.	7	3000	Unknown	
	Z	0.20						
	X	15.00			14	3000		
	Z	0.40						
	X	37.00			25	3000		
	Z	5.00						
	X	40.00			21	2500		

(continued)

Table 4. (*continued*)

	Printing direction	Mean tensile strength [MPa]	Sample standard	Layer height [mm]	Laser power [W]	Laser scanning speed [mm/s]	Hatch distance [mm]	Beam diameter [mm]
	Z	3.50						
	X	46.00			22	1400		
	Z	33.00						
	X	41.50			22	1000		
	Z	39.00						

However, the problem arises in the case of negative deviations, so depending on the direction of the print, it is good practice to put on an appropriate allowance.

References

1. Iftekar, S.F., Aabid, A., Amir, A., et al.: Advancements and limitations in 3D printing materials and technologies: a critical review. Polymers **15**(11), 1–23 (2023)
2. Ortiz-Acosta, D., Moore, T.: Functional 3D Printed Polymeric Material, pp. 1–15. IntechOpen (2018)
3. Siddiqui, S., Surananai, S., Sainath, K., et al.: Emerging trends in development and application of 3D printed nanocomposite polymers for sustainable environmental solutions. Eur. Polym. J. **196** (2023)
4. Pugliese, R., Beltrami, B., Regondi, S., et al.: Polymeric biomaterials for 3D printing in medicine: an overview. Ann. 3D Print. Med. **2**, 1–10 (2021)
5. Martynková, S.G., Slíva, A., Kratošová, G., et al.: Polyamide 12 materials study of morpho-structural changes during laser sintering of 3D printing. Polymers **13**(5), 1–14 (2021)
6. Mohsen, A.: The rise of 3-D printing: the advantages of additive manufacturing over traditional manufacturing. Bus. Horiz. **60**(5), 1–12 (2017)
7. Bochnia, J., Kozior, T., Szot, W., et al.: Selected mechanical and rheological properties of medical resin MED610 in PolyJet matrix three-dimensional printing technology in quality aspects. 3D Print. Addit. Manuf. 1–15 (2022)
8. Świderski, J., Makieła, W., Dobrowolski, T., et al.: The study of the roundness and cylindricity deviations of parts produced with the use of the additive manufacturing. Int. J. Adv. Manuf. Technol. **121**, 7427–7437 (2022)
9. EOS PA 2200 Product Information Data Sheet. https://drukarki3d.pl/wp-content/uploads/2015/09/karta-materia%C5%82owa-PA-2200-ENG.pdf. Accessed 20 Feb 2024
10. Bernaczek, J., Dębski, M., Gontarz, M., et al.: Analysis of torsional strength of PA2200 material shape additively. Adv. Sci. Technol. Res. J. **17**(2), 12–24 (2023)
11. Omar, M.R., Abdullah, M.I.H.C., Alkahar, M.R., et al.: Effect of polyamide-12 material compositions on mechanical properties and surface morphology of SLS 3D printed part. J. Mech. Eng. **19**(1), 57–70 (2022)
12. Stoia, D.I., Marşavina, L., Linul, E.: Correlations between process parameters and outcome properties of laser-sintered polyamide. Polymers **11**(11), 1–15 (2019)
13. Morano, C., Alfano, M., Pagnotta, L.: Effect of strain rates and heat exposure on polyamide (PA12) processed via selective laser sintering. Materials **16**(13), 1–13 (2023)

14. Pilipović, A., Ilinčić, P., Tujmer, M., et al.: Impact of part positioning along chamber Z-axis and processing parameters in selective laser sintering on polyamide properties. Appl. Sci. **14**(13), 1–21 (2024)
15. Zakręcki, A., Cieślik, J., Bazan, A., Turek, P.: Innovative approaches to 3D printing of PA12 forearm orthoses: a comprehensive analysis of mechanical properties and production efficiency. Materials **17**(3), 1–27 (2024)
16. Kundera, C., Kozior, T.: Influence of printing parameters on the mechanical properties of polyamide in SLS technology. Mechanika 31–37 (2016)

Comparative Analysis of Cell Structure Models with Standardized Models Manufactured by PolyJet Matrix Additive Technology Based on Compressive Relaxation Tests

Mateusz Rudnik(✉), Jerzy Bochnia, Tomasz Kozior, Wiktor Szot, and Paweł Szczygieł

Kielce University of Technology, Kielce, Poland
mrudnik@tu.kielce.pl

Abstract. From the following tests, one of the rheological properties of the RGD720 material can be assessed in the form of a photo-curable resin and a molding silicone. Then, one of the characteristics which is the percentage stress drop for comparison with joints of inhomogeneous materials can be compiled, which will enable the information to be used to determine the effect of material volume on the characteristics of the models. It was indicated that the percentage decrease in force at constant load was greatest for the hexagonal cell structure. For the other two structures, a situation occurred where the percentage decrease and maximum force values are greater for the QSS structure, relative to the helical structure. The results may serve in the development of the design and manufacturing of further cell structures in larger models.

Keywords: Cell structures · Additive manufacturing · Stress relaxation

1 Introduction

Modern industry is undergoing a revolution that is changing not only the way products are made, but the entire manufacturing ecosystem. Industry 4.0 is a typical industrial approach that combines advanced information technology, the Internet of Things (IoT), artificial intelligence (AI) and data analytics, opening the door to intelligent manufacturing systems [1]. In this context, 3D printing has a major role to fill, and cell structures are gaining ground, revolutionizing the printed model construction sector [2]. The innovation of 3D printing cellular structures is based on the ability to add material layer by layer to create complex structures [3]. However, what makes this technology even more compelling is the use of a cellular structure that allows the production of material-optimized, lightweight yet durable objects [4]. The cellular structures used in 3D printing resemble those found in nature, and their use is bringing revolutionary changes in many different areas. One important aspect is the lightness of these structures, which results from the optimization of the amount of material [5]. This makes 3D printed objects more material-intensive, which is important in the context of sustainability. This is why

M. Pagac et al. (Eds.): SCFF 2024, LNNS 1062, pp. 36–46, 2024.
https://doi.org/10.1007/978-3-031-65656-9_4

there is an increasing discussion of the concept of metamaterials, which are characterized by unusual properties and specific structures, rather than the properties of the materials themselves used [6]. Consequently, contemporary research into the rheological characteristics of solids is now both necessary and practical. Recently, there has been an increasing discussion on the concept of metamaterials, i.e. materials with unique properties and specific structures, which are becoming the focus of researchers.

Metamaterials, as opposed to conventional materials, shape their properties based on precisely designed structures, opening up new horizons in materials science [7, 8]. This phenomenon highlights the importance of understanding the rheological properties of solids, both in the context of traditional materials and contemporary developments such as metamaterials [9]. This dynamic combination of rheological research and innovative metamaterials concepts represents a significant advance in materials engineering, enabling the creation of materials with unusual properties beyond the traditional definition of solid materials [10, 11]. The study of the rheological properties of solids is a key area that focuses on the analysis of the stress-strain behavior of solid materials [12, 13]. A fundamental rheological characteristic of solids is their elasticity, which means that these materials can return to their original form once stress is applied and terminated, as opposed to liquids, which persist in their altered form once external forces cease. Concurrently, the growing interest in metamaterials emphasizes the importance of studying the rheological properties of solids, thus opening up new horizons in the field of materials and mechanical engineering [14].

1.1 Materials and Methods

PolyJet Matrix Technology

Photo-curable liquid polymer resin (PJM) technology is a method in which a liquid resin is injected onto the printer's working platform, in cross-section of the model. This is followed by a UV irradiation step, which initiates the polymerization process and the transition of the polymer from the liquid state to the solid state, thus making a 3D model [15, 16]. The minimum layer thickness for this technology is 16 μm. The gel-support material of this technology can be removed easily with a water jet, but in this process, it is necessary to design a so-called open structure. All models for the resulting research were manufactured by using Stratasys' Objet350 Connex 3D printer. The samples were manufactured by using the model material RGD720 (Stratasys Corp, Minnesota, Minneapolis) and the support material SUP705 (Stratasys Corp, Minnesota, Minneapolis). The RGD720 material was defined by the supplier as a stiff material. The main applications for models manufactured from this resin include functional prototyping or application to fluid flow testing.

Samples Preparation

As part of the tests conducted to determine the rheological properties, a compression relaxation test was carried out on cylindrical samples. These samples had dimensions of $d = 13$ mm and $h = 6.3$ mm in accordance with ISO-3384 and were manufactured from the materials RGD720 and molding silicone. As shown in Fig. 1, 30 samples with print directions of 0°, 45°, 90° were manufactured from RGD720 material [17]. The 10 samples were created from molding silicone by using a mould manufactured by Fused Deposition Modeling technology on a Makerbot Sketch printer.

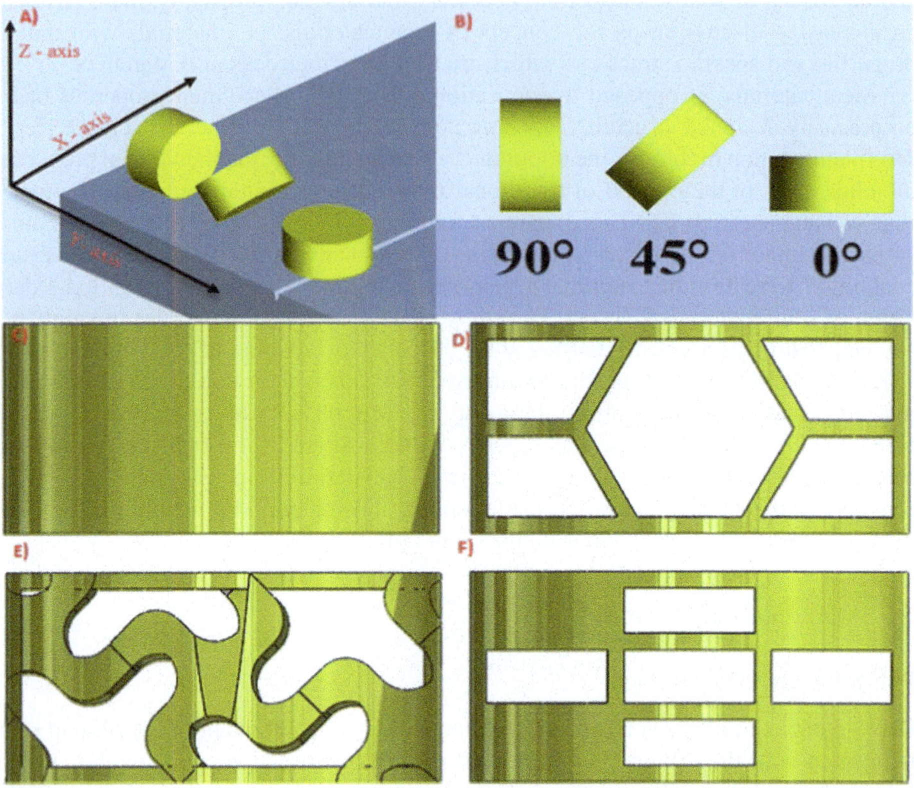

Fig. 1. Arrangement of samples on the build platform and type of samples: A) isometric view; B) angles of samples, C) solid sample, D) hexagonal sample, E) spiral sample, F) quasi-self-similar sample

Then, based on the compression relaxation tests performed, samples with cell structures were manufactured in one print direction - 0°, where RGD720 resin was also used as the model material and molding silicone served as the infill material for the cell structures shown in Fig. 1, where the hexagonal, spiral and quasi-self-similar structures are presented respectively. While the concept of the hexagonal and spiral structure is known and generated recursively, the quasi-self-similar cells themselves were loosely based on one of the main issues of fractals, which concerns self-similarity [18]. The cell structure was generated according to the assumptions on the rectangular section of the ISO-3384 standardized sample, where the self-similarity was based on the corners of the section alone, which were filled, the other 5 holes indicated in Fig. 2 as subsection D represent the free zone, they are separated by a lattice connecting the infilled parts of the model depicting the corner connections of each infilled part. In addition, to maintain the uniformity of the model, a shell was made around the sample. These elements are

0.5 mm in thickness. The stress relaxation tests were carried out on an Inspekt Mini 3 kN strength testing machine using LabMaster (Hegewald and Peschke, Nossen, Germany). The duration of the test carried out was 600 s with a displacement stroke of 0.63 mm.

Maxwell-Wiechert Model

The five-parameter Maxwell-Wiechert model was used to describe stress relaxation curves in the simulation part. The five parameter Maxwell-Wiechert model (M-W) is a combination of Hooke and Newton models. More specifically, these are two basic Maxwell models combined in parallel with the Hooke model.

Equation (1) presents a five-parameter Maxwell-Wiechert model:

$$\sigma(t) = \sigma_0 + \sigma_1 e^{\frac{-t}{t_1}} + \sigma_2 e^{\frac{-t}{t_2}} \tag{1}$$

σ - stress, t_1, t_2 – relaxation times described by the formula (2):

$$t_1 = \frac{\mu_1}{E_1}, t_2 = \frac{\mu_2}{E_2} \tag{2}$$

μ - viscosity, E_1, E_2 - moduli of elasticity.

Equation (1) has been transformed into Eq. (3):

$$\sigma(t) = \varepsilon_0\left(E_0 + E_1 e^{\frac{1}{t_1}} + E_2 e^{\frac{1}{t_2}}\right) \tag{3}$$

ε_0 – initial singular strain.

Equations (1) and (3) respectively were used to derive the values of the elastic modulus (E_0, E_1, E_2) and dynamic viscosity coefficients (μ_1, μ_2). Rheological models were fitted to the relaxation curves by the Levenberg-Marquardt method in OriginPro software. In addition to the calculated coefficients of the five-parameter model, the authors also obtained: the coefficient of determination R^2 and the coefficient of agreement Chi^2. The fit was estimated from these two factors.

The equivalent modulus was calculated from the formula (4):

$$E_s = E_0 + E_1 + E_2 \tag{4}$$

E_0, E_1, E_2 – moduli of elasticity of the M-W model.

Furthermore, the mean value, standard deviation was determined for each of the partial stresses and relaxation times. In addition, the uncertainty of the type A measurement was also determined by using Eq. (5):

$$u_A = \sqrt{\frac{1}{n(n-1)}\sum\nolimits_{i=1}^{n}(x_i - \overline{x})^2} \tag{5}$$

x_i – value of the i-th measurement, $\overline{x}$ – mean value, n – number of measurements.

The percentage decrease in force $F_{\%}$ was determined from Eq. 6, which is defined as the quotient of the difference in maximum force F_{max} and minimum force F_{min} with the maximum force and is expressed as a percentage.

$$F_{\%} = \frac{F_{max} - F_{min}}{F_{max}} \cdot 100\% \tag{6}$$

2 Results

The aim of the investigation was to perform compression relaxation tests on selected cell structures on the basis of the results obtained using a standardized approach derived from ISO-3384. Through this carefully planned research process, the results obtained in this experiment were not only accurate, but can also be compared with other samples tested. Verifying and comparing standardized samples with different configurations of cellular structures allows us to explore different scenarios and potential changes in cellular structures, which contributes to our knowledge of these systems. In a scientific perspective, this research procedure is an essential step in the quest to discover novelty and improve existing configurations of cellular structures. In this study, the Maxwell-Wiechert rheological model was used to investigate the rheological properties of samples made of RGD720 material and molding silicone, and then part of the results of these tests served as a benchmark for the comparative analysis of cellular structure tests to characterize the behavior of cellular structures under constant loading over time.

Figures 2 and 3 showed example fittings for samples of 10 from each test sample to illustrate how the five-parameter model fitting is performed. Based on Table 1, where the statistical values for the determined fit factors are presented, the following characteristics stand out. The highest standard deviations for the relaxation times t_1, t_2 have been observed for a print orientation of 45° and amounted to $t_1 = 1.41$ s and $t_2 = 16.48$ s, respectively. The lowest values of the mentioned relaxation times were observed for $t_1 = 0.63$ s for the 90° print orientation and $t_2 = 9.49$ s for the 0° print orientation, respectively. In addition, it can be assessed that for the coefficients t_1, t_2 corresponding to the relaxation times, the uncertainty was lowest for time t_1 in the 90° print orientation and was $u_A = 0.2$ s, for time t_2 the lowest uncertainty was in the 0° print orientation and was $u_A = 3$ s.

Based on Table 2 showing the determined elastic moduli and dynamic viscosity coefficients, the highest equivalent modulus was determined for a print orientation of 90° and was $E_z = 103.83$ MPa, and the lowest for a print orientation of 45° and was $E_z = 102.13$ MPa. The lowest dynamic viscosity coefficient was determined for the 0° print orientation and was $\eta_1 = 40.24$ MPa·s, and the highest dynamic viscosity coefficient was $\eta_1 = 57.87$ MPa·s for a print orientation of 45°. For the dynamic viscosity coefficient η_2 the lowest value was determined for a print orientation of 90° and was $\eta_2 = 1160.90$ MPa·s, and the highest value was recorded for the 45° print direction and was $\eta_2 = 1464.76$ MPa·s. Referring to the χ^2 and R^2 coefficients determined, a high accuracy fit of the five-parameter model should be found for each configuration of the RGD720 material print and for the samples manufactured from silicone. The lowest χ^2 values closest to 0 and the values closest to 1 for R^2 occurred for a print angle of 0°.

According to Table 3, it can be determined that, for samples made of homogeneous materials, the lowest mean value of percentage decrease was recorded for samples printed in 0° orientation and was 6.63% with a standard deviation of 1.08 percentage points. The highest average maximum force value was recorded for the 45° and 90° print orientations and was 1394.60 N, and the lowest average minimum force value was recorded for the 45° print orientation and was 1246.48 N. For the molding silicone, a significant percentage decrease was recorded with correspondingly low average maximum and minimum force values. Comparing these values with the results of the cellular structures, it should

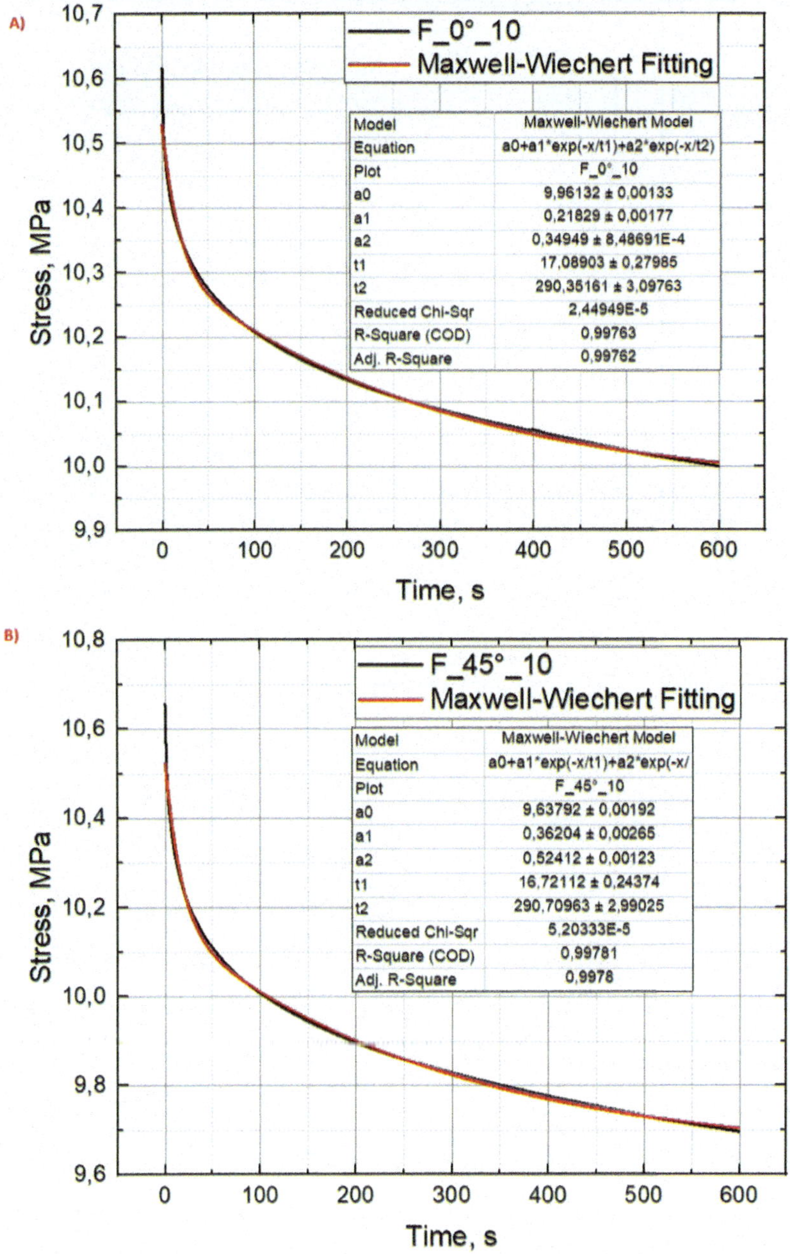

Fig. 2. Fitting the Maxwell-Wiechert model for selected ISO-3384 samples from RGD720 material manufactured: A) at an angle of 0°; and B) at an angle of 45°

be noted that the lowest average maximum force decrease was recorded for the spiral structure and was 57.17%, which is slightly more than 10 percentage points lower on

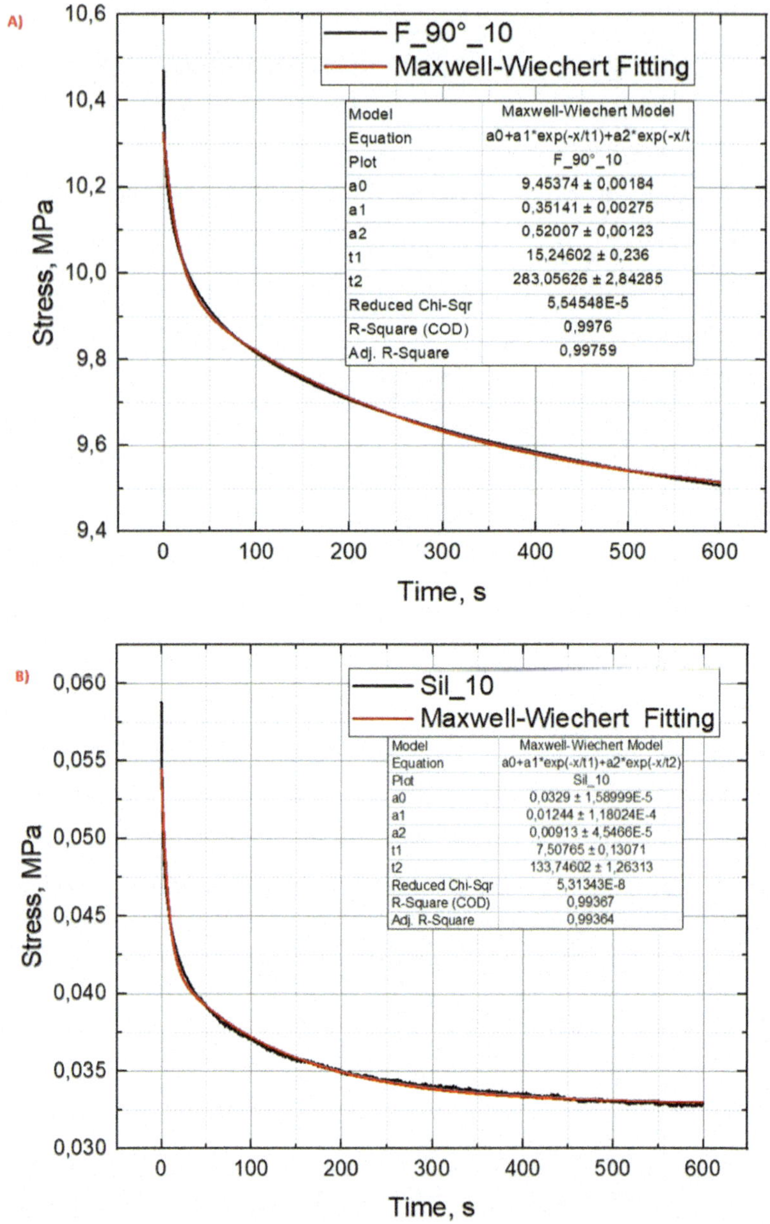

Fig. 3. Fitting the Maxwell-Wiechert model for selected ISO-3384 samples from RGD720 material manufactured: A) at an angle of 90°; and B) from molding silicone.

average than for the hexagonal structure. In addition, the hexagonal structure recorded diametrical differences in each of the statistics presented in relation to the other two types of cell structure due to the fact that the predominant material in this cell structure

Table 1. Results of matching relaxation test curves to the approximation curve

	σ_0, MPa	σ_1, MPa	σ_2, MPa	t_1, s	t_2, s	χ^2	R^2
Statistics	*F_0_RGD*720						
$\overline{X}$	9.61	0.24	0.40	16.98	301.68	0.00003	0.998
SD	0.42	0.03	0.08	1.41	9.49	0.00001	0.001
u_A	0.13	0.01	0.02	0.45	3.00		
u_C	0.30	0.02	0.05	1.01	6.79		
Statistics	*F_45_RGD*720						
$\overline{X}$	9.34	0.36	0.52	16.09	283.75	0.000057	0.998
SD	0.40	0.04	0.08	1.42	16.48	0.000013	0.0004
u_A	0.13	0.01	0.02	0.45	5.21		
u_C	0.29	0.03	0.06	1.01	11.79		
Statistics	*F_90_RGD*720						
$\overline{X}$	9.67	0.30	0.41	15.10	280.64	0.00004	0.997
SD	0.35	0.07	0.11	0.63	15.99	0.00002	0.0003
u_A	0.11	0.02	0.03	0.20	5.06		
u_C	0.25	0.05	0.08	0.45	11.44		
Statistics	*Molding silicone*						
$\overline{X}$	0.027	0.007	0.005	7.18	140.46	0.00000003	0.988
SD	0.005	0.003	0.002	1.22	16.143	0.00000002	0.004
u_A	0.002	0.001	0.001	0.387	5.105		
u_C	0.004	0.002	0.002	0.876	11.548		

Table 2. Calculated moduli of elasticity, equivalent modulus and dynamic viscosity coefficients

	E_0, MPa	E_1, MPa	E_2, MPa	η_1, MPa·s	η_2, MPa·s	E_Z, MPa
F_0_RGD720	96.12	2.37	3.99	40.24	1203.28	102.48
F_45_RGD720	93.37	3.60	5.16	57.87	1464.76	102.13
F_90_RGD720	96.67	3.02	4.14	45.61	1160.90	103.83
Molding silicone	0.33	0.10	0.08	0.97	12.03	0.51

is molding silicone, which accounts for almost 71% of the total sample. Comparing the other two cellular structures, it can be determined that the QSS structure, despite a much higher percentage decrease in force, achieved on average almost twice the maximum force values and almost one and a half times the minimum force values as a result of the compression relaxation test.

Table 3. Percentage decrease in force during the compression relaxation test for all samples tested and volumetric share of materials.

	$\overline{X}_{F\%}$, %	$s_{F\%}$, %	$F_{\%max}$, %	$F_{\%min}$, %	$\overline{X}_{F_{max}}$, N	$s_{F_{max}}$, N	$\overline{X}_{F_{min}}$, N	$s_{F_{min}}$, N
QSS	65.25	8.81	75.67	49.38	892.66	106.83	315.94	107.16
Hexagonal	67.86	23.74	86.28	19.15	121.51	9.62	39.69	31.43
Spiral	57.17	10.65	73.27	44.49	512.72	53.58	220.05	60.34
F_0_RGD720	6.63	1.08	8.64	5.30	1372.15	50.28	1281.32	54.91
F_45_RGD720	9.31	1.16	10.72	7.58	1394.60	48.63	1246.48	53.74
F_90_RGD720	7.62	1.91	11.24	5.16	1394.60	27.83	1288.61	45.45
Silicone	33.22	3.20	38.03	28.77	4.77	1.37	3.16	0.79

Percentage and volumetric share of materials

	V_S, mm^3	V_{RGD}, mm^3	V_S, %	V_{RGD}, %
QSS	31996	59216	38.26	61.74
Hexagonal	59216	24405	70.81	29.18
Spiral	31999	51622	38.27	61.73

3 Conclusions

The presented results of stress relaxation tests on PolyJet Matrix (PJM) samples allow conclusions to be drawn:

1. According to the presented results, very good and reliable fits of the five-parameter Maxwell-Wiechert model were obtained, which makes it possible in the future to apply these test results in the aspect of engineering calculations, mainly in computer simulations for practical application. As a result of the values obtained for the compressive relaxation test, the direction of printing of the normalized models with included cell structures was selected for comparative analysis.
2. The examination of the cellular structures made it possible to determine, on the basis of the corresponding series, the percentages of maximum slump, which disqualified the silicone-filled hexagonal structures due to their significant differences from the other cellular structures in terms of structure, volume and applied initial force.
3. In the case of the other cellular structures, particular attention should be paid to the construction of both structures, as, based on the so-called quasi-similarity, the initial force is almost twice as high and the minimum force is 1.5 times higher, while the fastest drop is almost 8 times higher.
4. On the basis of the tests carried out, it was indicated that the use of molding silicone as a structure filler allowed tests to be carried out for each of the cell structures under the same test conditions for a force stroke of 0.63 s and for a test execution time of 10 min. In addition, in the light of the tests carried out, it was indicated that the structures manufactured from the RGD720 material were highly susceptible to constant loading over time, and that the molding silicone, whose main function was

to dampen the impact force, had an impact on the performance of the test even though it was not a homogeneous part of the cellular structure, but this did not change the fact that the cellular structure deformed significantly after the test.

The conclusions of this study will also be important for further research in the field of rheology and will provide a detailed understanding of the mechanical properties of the materials and cell structures made from the RGD720 material. Possible next steps in the research are the combination of two materials on the Connex 350 printer in Digital Material mode, where 2 materials can be combined to study further aspects in the field of cell structures, or the use of epoxy resins to fill the printed cell structure constructions.

References

1. Nguyen, H.-S., Ma, Q.-P., Hajnys, J., Mesicek, J., Pagac, M.: Research trend in the field of additive manufacturing with bibliometrics study. MM Sci. J. **2023** (2023). https://doi.org/10.17973/MMSJ.2023_06_2023032
2. Gao, C., Yu, H., Gu, B., Lin, Y.: Modeling and analysis of assembly variation with non-uniform stiffness condensation for large thin-walled structures. Thin-Walled Struct. **191**, 111042 (2023). https://doi.org/10.1016/j.tws.2023.111042
3. Liang, Y., et al.: Energy absorption of gradient triply periodic minimal surface structure manufactured by stereolithography. Rev. Adv. Mater. Sci. **63** (2024). https://doi.org/10.1515/rams-2023-0185
4. Pascual-Francisco, J.B., Farfan-Cabrera, L.I., Cuan-Urquizo, E., Álvarez-Trejo, A., Roman-Flores, A.: Additive manufacturing and viscoelasticity evaluation of Bézier metamaterials fabricated via vat photopolymerization. Addit. Manuf. **60**, 103281 (2022). https://doi.org/10.1016/j.addma.2022.103281
5. Mahesh, V., Maladkar, P.G., Sadaram, G.S., Joseph, A., Mahesh, V., Harursampath, D.: Experimental investigation of the in-plane quasi-static mechanical behaviour of additively-manufactured polyethylene terephthalate/organically modified montmorillonite nanoclay composite auxetic structures. J. Thermoplast. Compos. Mater. **36**, 4021–4041 (2023). https://doi.org/10.1177/08927057221147826
6. Yang, H., et al.: 4D printed zero Poisson's ratio metamaterials with vibration isolation properties for magnetic response. Smart Mater. Struct. **33**, 025015 (2024). https://doi.org/10.1088/1361-665X/ad1c3c
7. Fatıma Faidallah, R., Hanon, M.M., Szakál, Z., Oldal, I.: Study of the mechanical characteristics of sandwich structures FDM 3D-printed. Acta Polytech. Hung. **20**, 7–26 (2023). https://doi.org/10.12700/APH.20.6.2023.6.1
8. Li, X., et al.: 3D-printed lattice structures for sound absorption: current progress, mechanisms and models, structural-property relationships, and future outlook. Adv. Sci. **11** (2024). https://doi.org/10.1002/advs.202305232
9. Zhang, Y., Huang, Z., Wang, F., Li, J., Wang, H.: Design of bioinspired highly aligned bamboo-mimetic metamaterials with structural and functional anisotropy. IEEE Trans. Dielectr. Electr. Insul. **30**, 1170–1177 (2023). https://doi.org/10.1109/TDEI.2023.3264964
10. Hu, Q., Lu, G., Tse, K.M.: Tensile mechanical behaviors of re-entrant and Kelvin cell lattice structures. JOM **76**, 387–396 (2024). https://doi.org/10.1007/s11837-023-06193-8
11. Feng, K., et al.: Effect of geometric structure and fiber orientation on crashworthiness of honeycomb-inspired composite thin-walled tubes. Polym. Compos. (2024). https://doi.org/10.1002/pc.28430

12. Szot, W.: Rheological analysis of 3D printed elements of acrylonitrile butadiene and styrene material using multiparameter ideal body models. 3D Print. Addit. Manuf. (2023). https://doi.org/10.1089/3dp.2022.0298
13. Kozior, T., Kundera, C.: Rheological properties of cellular structures manufactured by additive PJM technology. Tehn. Vjesn. Techn. Gaz. **28** (2021). https://doi.org/10.17559/TV-20191007145545
14. Sha, Y., Zhao, W., Tang, X., Zhao, F.: Acoustic and vibration response and fatigue life analysis of thin-walled connection structures under heat flow conditions. Aerospace **11**, 287 (2024). https://doi.org/10.3390/aerospace11040287
15. Kozior, T., Ehrmann, A.: First proof-of-principle of PolyJet 3D printing on textile fabrics. Polymers (Basel) **15**, 3536 (2023). https://doi.org/10.3390/polym15173536
16. Bochnia, J.: A study of the mechanical properties of naturally aged photopolymers printed using the PJM technology. Materials **16**, 400 (2023). https://doi.org/10.3390/ma16010400
17. ISO 3384-1:2019: Rubber, Vulcanized or Thermoplastic Determination of Stress Relaxation in Compression Part 1: Testing at Constant Temperature
18. Cao, N.T.N.K., Mayol, C., Salmachi, A., Labrinidis, A.: Modelling and 3D printing of a random porous media using iterative function system fractals. APPEA J. **63**, 47–55 (2023). https://doi.org/10.1071/AJ22144

Comparison of Selected Methods and Strategies for Measuring the Surface Topography of Models Manufactured by Additive SLS Technology

Dominik Malara[1,2](✉), Tomasz Kozior[1], Aleksandra Bochenek[2], and Michał Nawotka[2]

[1] Department of Metrology and Unconventional Manufacturing Methods, Faculty of Mechatronics and Mechanical Engineering, Kielce University of Technology, 25-314 Kielce, Poland

dmalara@tu.kielce.pl

[2] Metrological IT Laboratory, Digital Technology Metrology Department, Central Office of Measures, 00-139 Warsaw, Poland

Abstract. The following article presents the results of a study of the measurement of the surface topography of components manufactured by the additive Selective Laser Sintering (SLS) method, using polyamide powder PA 2200. The measurement of surface topography was carried out using three different optical measurement methods. The optical microscopy methods used in the study were compared: interferometric, confocal and focus variation. A Leica DCM8 microscope was used to carry out optical measurements. The conducted tests showed significant differences in the number of non-measured points in terms of selecting the appropriate measurement system for the elements produced by the SLS method, which may affect the reliability of the tests.

Keywords: Metrology · Additive technology · Surface texture · SLS · PA 2200

1 Introduction

The continuous development of additive technologies brings with it increasing dimensional and shape requirements from those using these manufacturing methods in various industries. The constantly expanding range of materials available for these technologies, requires the definition of guidelines indicating the selection of an appropriate measurement strategy for a particular method and material. The evaluation of surface topography fits closely into these requirements. There are no appropriate guidelines for selecting a suitable measurement system for evaluating the surface layer of 3D printed models. Contact and optical systems have their limitations, which in many cases disqualify them as suitable tools for reliable evaluation.

The study of surface topography of models produced by 3D printing is a point of interest for many researchers, and many articles have been written on this topic [1–8].

M. Pagac et al. (Eds.): SCFF 2024, LNNS 1062, pp. 47–56, 2024.
https://doi.org/10.1007/978-3-031-65656-9_5

In the article [1], a study was conducted to determine the 2D and 3D surface texture of polylactic acid (PLA) polymer and Al-Si10Mg metal parts produced by fused deposition modelling (FDM) and direct metal laser sintering (DMLS), respectively. The authors conducted surface topography studies using optical and contact methods. Samples produced by the FDM method, tested on the surface opposite to the table contact surface during manufacturing, had lower values for the parameters: Ra, Rq, Rz, Sa, Sq, Sz, than for samples produced by the DMLS method. The parameters Ra, Rq, Rz in all analysed cases had lower values than the parameters Sa, Sq, Sz for the same samples.

The authors [2] conducted a study of the surface topography of models produced by 3D printing technologies such as PolyJet Matrix (PJM), Fused Deposition Modelling (FDM), Selective Laser Sintering (SLS) and Selective Laser Melting (SLM), for materials such as: FullCure 720 (PJM), ABS P430 (FDM), PA 2200 (SLS) and 316L for (SLM). Samples were produced for each method in three different orientations, regardless of the location on the working plane, models produced with SLS technology for optical measurement showed the highest number of non-measured points, which may affect the accuracy of the obtained results.

The authors of this paper [3] conducted a study of the surface topography of models produced by 3D printing using laser powder bed fusion (PBF-LB) and electron beam powder bed fusion (PBF-EB) technologies. The results showed that surface measurements with metal additives are highly susceptible to changes in measurement control parameters for Sa, with deviations within 5% of the average parameter value, for the same surface and measured area. The number of non-measured points and measurement noise were more variable and depended on the choice of measurement parameters.

The selection of a suitable method for measuring surface topography is a key issue in terms of conducting a reliable assessment of the surface layer. Optical and contact methods have their limitations, such as scratching with a blade mapping the surface being tested [4], as well as the inability to measure soft materials, such as the Tango Black material used in the Poly Jet Matrix (PJM) method, and non-measured points when using optical methods [9, 10].

The main aim of the conducted research was to compare the results of surface topography of SLS-produced models for interferometric, confocal and focus variation microscopy. Preliminary studies conducted by the authors in paper [2] showed a large number of non-measured points for models produced by SLS and PA 2200 polyamide powder, tested using a Talysurf CCI Lite device.

2 Materials and Methods

The tests were carried out on samples manufactured by the additive Selective Laser Sintering (SLS) method.

To perform optical measurements, a DCM8 microscope (Leica, Wetzlar, Germany) was used, giving the possibility of optical measurement by three methods: confocal, interferometric and focus variation. The MountainsLab (V10, Digital Surf, Besançon, France) program was used to analyse the conducted tests.

2.1 Samples Manufacturing

The sample model was designed using the 3D CAD program, SolidWorks (2022, Dassault Systèmes, Vélizy-Villacoublay, France). The designed model had a cuboid shape with dimensions: 20 mm length, 20 mm width, 2.5 mm height. The model designed in this way was saved in STL format, which creates a solid model using triangles. The model saved in this format was created with a combination of 12 triangles.

The Selective Laser Sintering (SLS) method is based on the sintering of a powdered material (usually plastics such as polyamide) by a laser at predefined locations, successively the distributor arm applies another layer of material, and the working table of the device is lowered by a dimension equal to the thickness of the layer of the manufactured model [11, 12].

A Formiga P100 (EOS, Krailling, Germany) was used to produce the samples. The samples were manufactured from PA 2200 material.

The samples were made with a preset thickness of the built-up layer of 0.1 mm and the following laser operating parameters: power of 21 W, scanning speed of 2500 mm/s and energy density of 0.056 J/mm^2 calculated from Eq. (1) [13].

$$E_d = \frac{P}{vh}x \tag{1}$$

where: E_d - Energy density, J/mm^2; P – laser power, W; v – laser speed, mm/s; h – hatch distance, 0.25 mm; x – beam overlay ratio, 1.68.

2.2 Measurement Methodology

A Leica DCM8 was used for optical measurements. The microscope gives the possibility of optical measurement by three methods: confocal with magnification ×5, ×20, ×50, ×100; interferometric ×10, ×50; focus variation ×5, ×20, ×50, ×100 [14, 15].

A model of the sample with the marked measurement area is shown in Fig. 1. The analysed area was 0.95 mm × 0.71 mm.

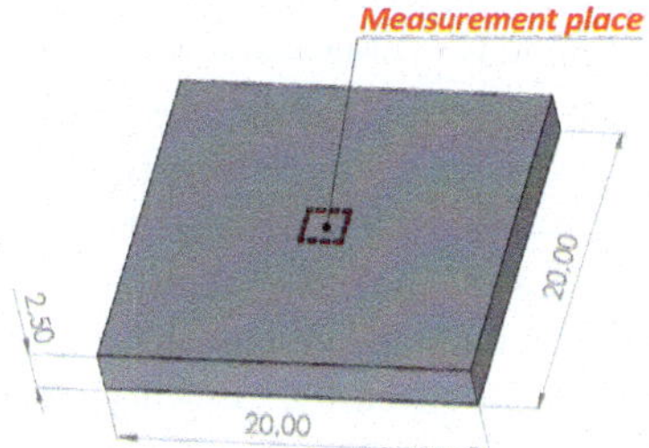

Fig. 1. A model of the designed sample in SolidWorks software with the measurement location marked.

Figure 2 shows a view of the Leica DCM8 during the measurements of the presented tests.

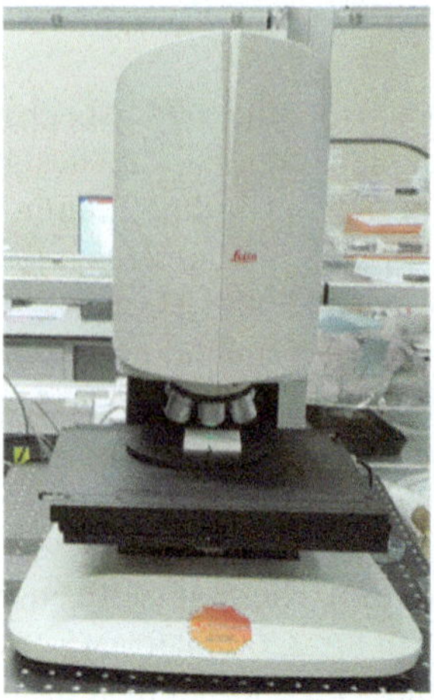

Fig. 2. Realization of the measurement with the Leica DCM8.

In confocal microscopy, light passing through an aperture in the aperture falls on the surface to be examined and the resulting image is directed through a light-dividing system into the aperture to a detector, which analyses the resulting light beam. This method is distinguished by better horizontal resolution and contrast than that of ordinary optical microscopes, and also offers the possibility of measuring greater slopes of the surface [14, 16].

Focus variation microscopy involves illuminating with light of appropriate modulation and transmitting and focusing through the optical system to the surface to be measured, the reflected focused light goes to the detector. While the method allows for the measurement of steep slopes, it exhibits poorer resolution than confocal microscopy [16].

Interferometry involves splitting a light source into two separate beams in a light splitting system, one falls on a reference surface, while the other falls on the test surface, after reflection both beams fall on the light splitting system and are transmitted to the detector, where they interfere with each other. The advantage of this method is high resolution, while the disadvantage is problems with measuring slopes [16].

The tests were carried out at ×50 magnification for three measurement methods (only this is common to all): confocal, interferometric and focus variation for an area of 0.95 mm × 0.71 mm. Each measurement consisted of 9 measured areas, which were stitched together with 15% coverage. The area had an x-y axis resolution of 3672 × 2765 pixels, 1 pixel = 0.26 μm. It is noteworthy that the sample was measured at the same location without changing its position on the working plane of the measuring device. The measurement was taken at the central point of the top surface of the sample because the sample was characterized by a small dispersion of results on its surface.

3 Result

Table 1 shows the percentage of measured points for each measurement method at ×50 magnification. In order to obtain the best value of measured points, the light setting was selected for each measurement method, which was at a different level for each method.

Table 1. Number of points measured for each method.

Measurement method	Number of points measured, %
Confocal	90.65
Interferometry	94.7
Focus variation	21.12

Figures 3, 4 and 5 present topographic images showing the measurement of the sample after levelling for each of the methods used.

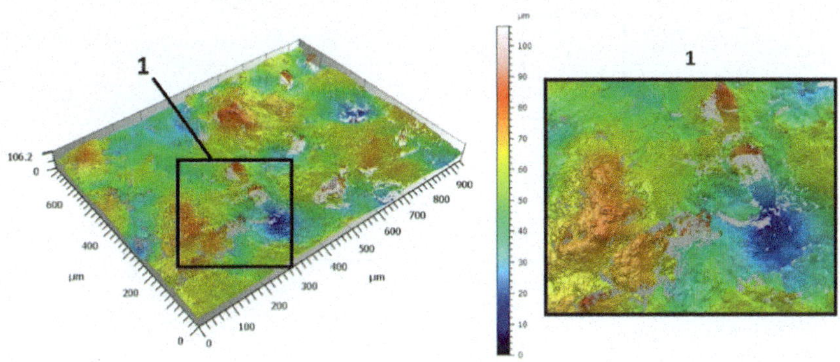

Fig. 3. Surface topography after confocal measurement before filling in non-measured points.

Figure 6 presents topographic images showing the measurement of the sample after levelling and filling in non-measured points (using the least squares method of neighbouring points) for: (a) confocal, (b) interferometric and (c) focus variation methods.

Table 2 presents a comparison of the results of 3D parameter measurements according to ISO 25178 [17].

Table 3 presents a comparison of the results of 3D parameter measurements according to ISO 25178 [17], using a 0.25 mm Gaussian filter.

Table 4 shows a summary of parameter results for profile, roughness and waviness according to ISO 21920 [18], using a Gaussian filter for roughness and waviness of 0.25 mm.

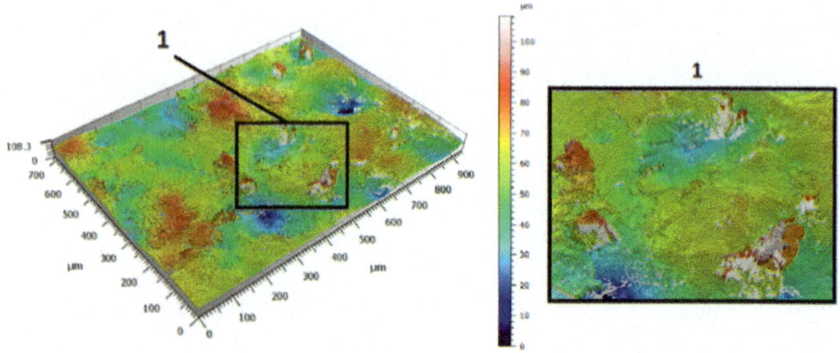

Fig. 4. Surface topography after interferometric measurement before filling in non-measured points.

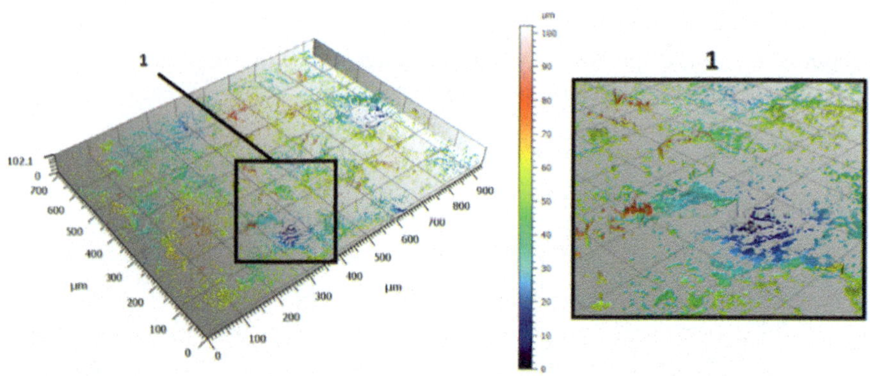

Fig. 5. Surface topography after focus variation measurement before filling in non-measured points.

4 Discussion

Analysing the obtained results shown in Table 1, it can be seen that the interferometric method showed the highest percentage of measured points. The value of the obtained points in this method is higher than the number of points measured in the confocal method by about 4% and is higher than for the focus variation method by about 78%.

Comparing the results shown in Figs. 3, 4 and 5, it can be concluded that the non-measured points in interferometric microscopy (Fig. 4) occur at the locations of very steep slopes of peaks and valleys, in confocal microscopy (Fig. 3) the locations of non-measured points are the same as in the interferometric method (locations of steep slopes) and appear on the slopes of peaks and valleys with lower slope angles. Analysing Fig. 5 where the measurements were made using the focus variation microscopy method, it can be observed that most of the measured points occur on peaks, valleys and on slopes with a low angle to the measured surface.

Comparing the obtained results of the spatial parameters shown in Table 2 without filtering and in Table 3 with a Gaussian filter of 0.25 mm, for the parameter Sq - the mean

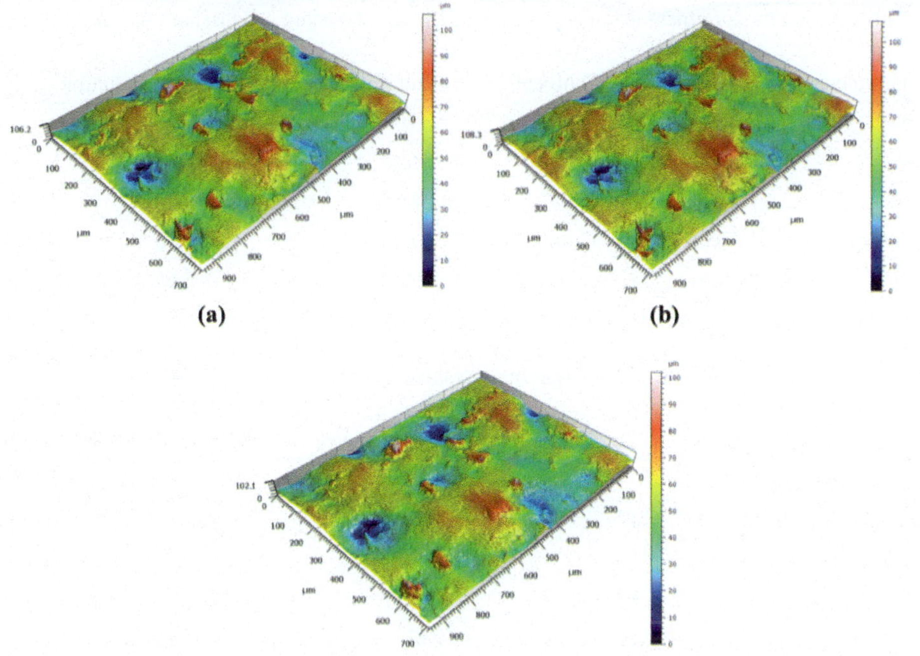

Fig. 6. Sample surfaces after levelling and filling in non-measured points: (a) confocal method; (b) interferometric method; (c) focus variation.

Table 2. 3D measurement results according to ISO 25178.

Parameter	Unit	Confocal	Interferometry	Focus variation
Sq	μm	12.98	12.79	13.03
Sp	μm	54.31	53.34	53.31
Sv	μm	51.70	53.31	48.58
Sz	μm	106.01	106.66	101.89
Sa	μm	10.22	10.07	10.26

square height of the area after filtering is lower by about 40% for all three methods tested. The parameter Sp, describing the maximum peak height, increases by about 4% after filtration in the three methods. The value of the parameter Sv describing the maximum depth of pit, after filtration, is lower by about 30% for all three methods tested. The parameter Sz, describing the maximum height for all measurement methods, decreases by approximately 14% after filtration. The value of the parameter Sa describing the arithmetic mean height retains a tendency to decrease in value after filtration by about 45%.

Table 3. 3D measurement results according to ISO 25178 using a Gaussian filter of 0.25 mm.

Parameter	Unit	Confocal	Interferometry	Focus variation
Sq	µm	7.82	7.59	7.84
Sp	µm	56.29	55.31	55.67
Sv	µm	35.77	36.40	31.48
Sz	µm	92.06	91.71	87.15
Sa	µm	5.66	5.51	5.65

Table 4. 2D measurement results according to ISO 21920.

	Unit	Confocal	Interferometry	Focus variation
Primary profile				
Pq	µm	12.09	11.83	12.06
Pp	µm	11.94	13.06	12.05
Pv	µm	10.70	11.42	10.44
Pz	µm	17.14	19.02	17.59
Pa	µm	9.78	9.59	9.74
Roughness profile				
Rq	µm	5.49	5.35	5.45
Rp	µm	12.13	12.43	11.68
Rv	µm	9.85	10.74	9.62
Rz	µm	21.91	23.16	21.27
Ra	µm	4.10	3.95	4.05
Waviness profile				
Wq	µm	8.94	8.79	8.86
Wp	µm	6.66	6.63	6.62
Wv	µm	5.69	5.17	5.68
Wz	µm	3.73	3.41	3.73
Wa	µm	7.63	7.50	7.55

Comparing the results in Tables 2 and 3, it can be observed that the results obtained between the individual methods differ slightly and the difference is no greater than 6% only the Sv parameter for the focus variation method after filtration is lower than the others by about 13%.

The results shown in Table 4, describing the primary profile with the parameters Pp, Pv, Pz, Pc, Pa, Pq for all three tested methods, remain at a similar level and the difference is no greater than 11%. The values of the parameters describing surface roughness Rp, Rv,

Rz, Ra, Rq also show no significant differences between the methods and the maximum difference is less than 10%. The values of the parameters describing ripple Wp, Wv, Wz, Wa, Wq differ by less than 10%.

Comparing the results of Tables 3 and 4 for the parameters Sq, Sp, Sv, Sz, Sa with the parameters Rq Rp, Rv, Rz, Ra, it can be observed that the parameters S have in each case higher values than the parameters R. The value of the parameter Sq is greater than the parameter Rq by about 43%, the values of the parameters Sp, Sv and Sz are greater than the parameters Rp, Rv, Rz by between 260 and 380%, the parameter Sa is greater than the parameter Ra by about 40%.

5 Conclusions

The development of additive technologies determines the need to define measurement guidelines for the surface layer. The research carried out showed significant differences in the number of non-measured points, depending on the measurement method used. In the study presented here, the model complementing the non-measured points worked by approximating neighbouring points. Analysing the results obtained, it can be observed that for each of the tested methods, the values of the parameters of the geometric structure of 2D and 3D surfaces, regardless of the number of non-measured points, were at a similar level. A very valuable conclusion is that the focus variation method at a magnification of ×50 showed the lowest number of measured points, which raises further questions, e.g. what will be the results for the other magnifications for this method and for the confocal and interferometric methods. The authors of the study see a need for further research related to the selection of suitable methods for the evaluation of surface topography for different methods and materials of parts manufactured with 3D printing technologies.

Another conclusion is that 2D and 3D parameter values are significantly different and should not be substituted for each other.

References

1. Sagbas, B., Durakbasa, N.M.: Profile and areal surface characterization of additive manufacturing polymer and metal parts. In: Proceedings of the 12th International Conference on Measurement and Quality Control-Cyber Physical Issue: IMEKO TC 14 2019, vol. 12, pp. 240–246 (2019)
2. Zmarzły, P., Kozior, T., Gogolewski, D.: The effect of non-measured points on the accuracy of the surface topography assessment of elements 3D printed using selected additive technologies. Materials (Basel) **16**, 460 (2023). https://doi.org/10.3390/ma16010460
3. Newton, L., et al.: Optimisation of imaging confocal microscopy for topography measurements of metal additive surfaces. Metrology **3**, 186–221 (2023). https://doi.org/10.3390/metrology3020011
4. Senin, N., Thompson, A., Leach, R.K.: Characterisation of the topography of metal additive surface features with different measurement technologies. Meas. Sci. Technol. **28**, 95003 (2017). https://doi.org/10.1088/1361-6501/aa7ce2
5. Caminero, M.Á., Chacón, J.M., Garcia-Plaza, E., Núñez, P.J., Reverte, J.M., Becar, J.P.: Additive manufacturing of PLA-based composites using fused filament fabrication: effect of graphene nanoplatelet reinforcement on mechanical properties, dimensional accuracy and texture. Polymers (Basel) **11**, 799 (2019). https://doi.org/10.3390/polym11050799

6. Cabanettes, F., et al.: Topography of as built surfaces generated in metal additive manufacturing: a multi scale analysis from form to roughness. Precis. Eng. **52**, 249–265 (2018). https://doi.org/10.1016/j.precisioneng.2018.01.00
7. Tato, W., Blunt, L., Llavari, I., Aginagalde, A., Townsed, A., Zabala, A.: Surface integrity of additive manufacturing parts: a comparison between optical topography measuring techniques. Procedia CIRP **87**, 403–408 (2020). https://doi.org/10.1016/j.procir.2020.02.082
8. Everton, S.K., Hirsch, M., Stravroulakis, P., Leach, R.K., Clare, A.T.: Review of in-situ process monitoring and in-situ metrology for metal additive manufacturing. Mater. Des. **95**, 431–445 (2016). https://doi.org/10.1016/j.matdes.2016.01.099
9. Pawlus, P., Reizer, R., Wieczorowski, M.: Problem of non-measured points in surface texture measurements. Metrol. Meas. Syst. **24**, 525–536 (2017). https://doi.org/10.1515/mms-2017-0046
10. Wieczorowski, M.: Teoretyczne podstawy przestrzennej analizy nierówności powierzchni. Inżynieria Masz. **18**, 7–34 (2013)
11. Siemiński, P., Budzik, G.: Techniki przyrostowe: druk drukarki 3D. Oficyna Wydawnicza Politechniki Warszawskiej, Warszawa (2015)
12. Rokicki, P., et al.: Manufacturing of aircraft engine transmission gear with SLS (DMLS) method. Aircr. Eng. Aerosp. Technol. Int. J. **88**, 397–403 (2016)
13. Kozior, T.: The influence of selected selective laser sintering technology process parameters on stress relaxation, mass of models, and their surface texture quality. 3D Print. Addit. Manuf. **7**, 126–138 (2020). https://doi.org/10.1089/3dp.2019.0036
14. Niemczewska-Wojcik, M., Madej, M., Kowalczyk, J., Piotrowska, K.: A comparative study of the surface topography in dry and wet turning using the confocal and interferometric modes. Measurement **204**, 112144 (2022). https://doi.org/10.1016/j.measurement.2022.112144
15. Get the Full Picture for Surface Metrology. https://www.leica-microsystems.com/dcm8. Accessed 2024/02/10
16. Wieczorowski, M.: Metrologia nierówności powierzchni-metody i systemy. ZAPOL, Szczecin (2013)
17. ISO 25178-2 Geometrical Product Specifications (GPS) - Surface Texture: Areal - Part 2: Terms, Definitions and Surface Texture Parameters (2012)
18. ISO 21920-2 Geometrical Product Specifications (GPS)—Surface Texture: Profile—Part 2: Terms, Definitions and Surface Texture Parameters (2021)

Effect of Carbon Fiber Content in Polyethylene Terephthalate Glycol on Tribological Behavior of 3D-Printed Parts

Hoang Tung Nguyen, Jiri Prochazka(✉), David Dobrocky, and Zbynek Studeny

Department of Mechanical Engineering, Faculty of Military Technology, University of Defence, Brno, Czech Republic
{hoangtung.nguyen,jiri.prochazka}@unob.cz

Abstract. This study explores the impact of carbon fiber reinforcement on the friction properties of PETG-based composite materials fabricated using FFF/FDM printing technology. The effect of carbon fiber content and orientation on mechanical properties are investigated, specifically focusing on tribological behavior. Results indicate a significant correlation between the content of carbon fiber and friction coefficient. Moreover, the arrangement of carbon fibers introduces anisotropic friction properties in the PETG composites. Additionally, the application of ironing treatment substantially enhances surface quality without significant alteration to internal material properties. These findings contribute to the understanding of the interplay between carbon fiber reinforcement, printing parameters, and friction characteristics, thereby facilitating the optimization of PETG-based composite materials for various applications.

Keywords: Tribological · FFF/FDM · PETG · Carbon fiber reinforced

1 Introduction

In the realm of digital fabrication technology, 3D printing, also known as additive manufacturing, has emerged as a revolutionary tool. This technology creates physical objects from a geometrical representation, enabling the production of complex designs and shapes that were previously difficult or impossible to achieve with traditional manufacturing methods. The potential of new materials in 3D printing has led to an exponential increase in the technology's applications across various industries [1–3].

One of the most common additive manufacturing processes is material extrusion (MEX) according to ISO/ASTM 52900:2021 which includes technology such as Fused Deposition Modelling (FDM) or similar Fused Filament Fabrication (FFF) respectively. FFF technology is an efficient and cost-effective choice for rapid prototyping and other applications. It involves the process of making physical objects by building up successive layers of material with an extruded thermoplastic filament. The key advantages of FFF include ease of use, speed, expandability, and a wide range of materials developed specifically for the purpose of this technology [4–6].

M. Pagac et al. (Eds.): SCFF 2024, LNNS 1062, pp. 57–66, 2024.
https://doi.org/10.1007/978-3-031-65656-9_6

Each material offers unique properties that make them suitable for specific applications [7–9]. Polymers like PLA (Polylactic Acid), ABS (Acrylonitrile Butadiene Styrene), PETG (Polyethylene Terephthalate Glycol), PA (Polyamide), PC (Polycarbonate), and many others as well as their composite derivatives containing admixtures like carbon fiber, glass fiber etc. are commercially used filaments in 3D printing [10]. Recently, using carbon fiber to reinforce these materials has gained attention. Carbon fiber reinforcement in 3D printing materials leads to an increase in the strength, stiffness, heat resistance, and durability of the printed parts [11–16].

The components built through this technology find their applications in automotive, medical, electrical, military technology, and other applications [4, 7, 17, 18]. In many cases, the printed components can be exposed to wear and friction, especially in the prototyping of mechanical devices.

The article concentrates on analyzing the impact of carbon fiber content on the frictional characteristics of selected materials, along with exploring how printing parameters influence the properties of the printed part.

2 Materials and Methods

2.1 Materials

One of the most commonly used materials in this printing technology is PETG which stands out as a prominent polymer in the realm of 3D printing materials due to its versatile properties. PETG offers a unique combination of attributes, including compatibility with lower printing temperatures, low tendency to shrinkage, and resistance to humidity and chemicals [19], which enhances its suitability for a wide range of applications.

In the present study, PETG was employed, along with their carbon fiber-reinforced counterparts, denoted as PETG/CF, respectively. These materials were procured from Devil Design, Spectrum, and Recreus Company and used as filaments with a uniform diameter of 1.75 mm. For detailed information regarding the specific characteristics of these materials, please refer to Table 1.

2.2 Sample Preparation

In this current work, the models were designed using SolidWorks 2022 software with a size (30 × 30 × 8) mm of square plat. These models were then converted into the .STL file format, which is commonly used in 3D printing, and sliced using Bambu Studio software.

For the printing process, three materials were focused on: Polyethylene Terephthalate Glycol (PETG), carbon fiber-reinforced Polyethylene Terephthalate Glycol (CF/PETG) with 10% carbon fiber, and another PETG/CF whose carbon fiber content is kept secret by the manufacturer, and is noted PETG/CF with unknown carbon fiber content. The extrusion temperature and bed temperature of all samples were set at 250 °C and 80 °C. Other printing parameters, such as layer thickness were set to 0.2 mm, ironing on top surfaces with rectilinear pattern, speed 30 mm/s, and ironing line spacing 0.15 mm were used. The infill density was set at 100%, the infill direction 0°, and the printing speed was set at 100 mm/s. The resulting G-code was directly imported into the 3D printer for the printing of the specimens. In this study, the Bambu Lab X1-Carbon 3D printer (Bambu Lab, Shenzhen, China) was employed to print the test samples.

For further details on the specific characteristics of the materials and their printing parameters, please refer to Table 1.

Table 1. Material list and process parameters chosen for printing samples

Sample	Material	Carbon fiber content (%)	Density (g/cm^3)	Non-ironing	Ironing
A1	PETG-Devil	0	1.23	X	
A2	PETG-Devil	0	1.23		X
B1	PETG/CF-Spectrum	10	1.28	X	
B2	PETG/CF-Spectrum	10	1.28		X
C1	PETG/CF-Recreus	Unknown	1.32	X	
C2	PETG/CF-Recreus	Unknown	1.32		X

2.3 Hardness Measurement

The surface hardness of the samples was assessed using the Shore D method with DIGI-Test II hardness tester (Bareiss, Oberdischingen, Germany) in accordance with ASTM D2240. The Shore D method, utilized for measuring the surface hardness of materials such as polymers, hard synthetic plastics, thermoplastics, and vinyl coatings, is a crucial tool in various industries. A specified impact force to an indenter is applied, which then penetrates the material, causing surface deformation to a specific depth.

2.4 Surface Morphology

The surface of the specimens was characterized using the LEXT™ OLS5100 3D Laser Scanning Microscope (Keyence Corporation, Nagoya, Japan). This method allowed for a detailed examination of the roughness of the sample surface, the Ra and Rz parameters were evaluated as part of this process. The 20 times magnification lens was utilized, scanning an area of (3.5 $\times$ 3.5) mm, equivalent to (5500 $\times$ 5500) pixels, and scanning with 3D (Fine + color) mode.

The volume and depth of the tribological wear track were also measured under identical conditions, but within a range of (1.8 $\times$ 7.0) mm, ensuring the complete observation of the entire friction trace.

2.5 Tribological Measurement

The friction coefficient is a crucial parameter in the field of tribology, which is defined as the ratio of the tangential force to the normal force [20]. In the experiment, the friction coefficient was measured by Bruker/UMT-3 Tribolab universal tester (Bruker Corporation, Massachusetts, USA). The tests were performed under a ball-on-flat dry sliding condition, following the guidelines outlined in the ASTM G133-22 standard. An indenter-type ball (94% WC + 6% C) with a diameter of 6.35 mm was used for the tests.

During the wear tests, a normal constant load of 20 N was applied to the specimens. The stroke length was set at 10 mm, with an oscillation frequency of 5 Hz, period 500 s. Friction testing was performed in 2 directions: one direction parallel to the 3D printing direction (0°) and the other direction perpendicular to the printing direction (90°). Each sliding condition was repeated three times to ensure the reliability of the results.

3 Results and Discussion

3.1 Hardness

The change in hardness of materials, including PETG and PETG/CF with different carbon fiber compositions after 3D printing, is a significant area of interest. Adding carbon fibers to PETG can lead to changes in the mechanical properties of the resulting composite materials, including hardness [21].

The surface hardness of the samples was measured on the DIGI-Test II hardness tester by using the Shore D method. In this study, each material sample underwent five measurements, and the average values are presented in Table 2.

Table 2. Surface hardness of samples

Number of measurement	PETG Devil Design	PETG/10%CF Spectrum	PETG/unknown %CF Recreus
1.	72.4	68.8	67.2
2.	73.1	71.5	67.7
3.	73.1	71.1	66.8
4.	72.7	71.9	68.3
5.	73.4	69.8	66.9
Average value	72.9 ± 0.32	70.6 ± 1.04	67.5 ± 0.5

Figure 1 shows that the highest surface hardness value is achieved by PETG, the addition of 10% carbon fiber to PETG results in a decrease in hardness from 72.9 ± 0.32 to 70.6 ± 1.04. Furthermore, for PETG/CF material with an unspecified carbon fiber content, the lowest hardness observed is 67.5 ± 0.5. This suggests that the carbon fiber content of this material is likely to exceed 10%, as the carbon fiber content exerts a significant influence on the hardness of PETG [22].

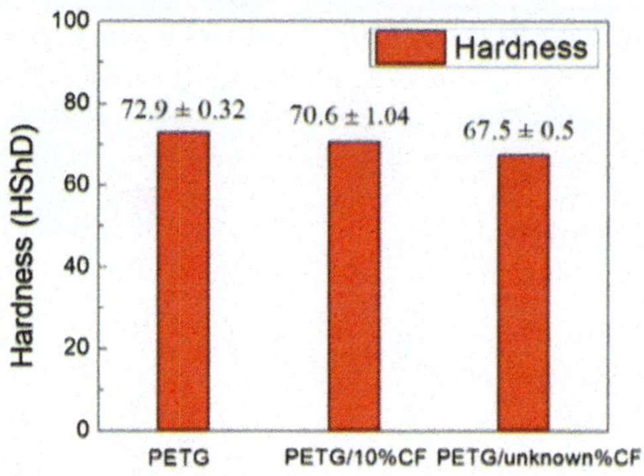

Fig. 1. Surface hardness

3.2 Surface Morphology Assessment

Figure 2 shows the comparison of the PETG surface texture with ironing and without ironing treatment.

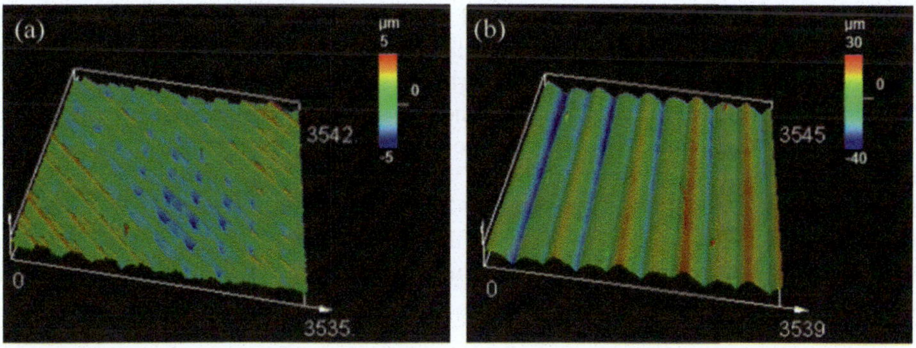

Fig. 2. Surface texture of PETG: (a) with ironing; (b) without ironing; measured by LEXT™ OLS5100

The presence of carbon fiber reinforcement in PETG material is associated with a significant increase in the roughness of the sample surface. This is evident in the impact of carbon fiber-reinforced PETG materials on the surface quality and mechanical properties of 3D-printed parts.

The use of ironing treatment has been observed to significantly reduce the surface roughness of PETG and PETG/CF materials after 3D printing, as shown in Fig. 3. There has been a considerable reduction in both the Rz and Ra parameters of the product samples. Specifically, for PETG samples, the Rz parameter plummeted from 41.2 µm to 4.4 µm, while parameter Ra also indicated a sharp decline from 7.8 µm to 0.7 µm. Similarly, for PETG/CF Recreus, the Rz value dropped from 102.1 µm to 17.1 µm, and Ra decreased from 15.8 µm to 2.2 µm. The effect of ironing treatment on PETG/10% CF Spectrum samples is shown at least in terms of changes in Rz and Ra values. The values of parameter Rz decreased from 50.5 μm to 30.9 μm, and from 9.6 μm to 3.9 μm with Ra parameter.

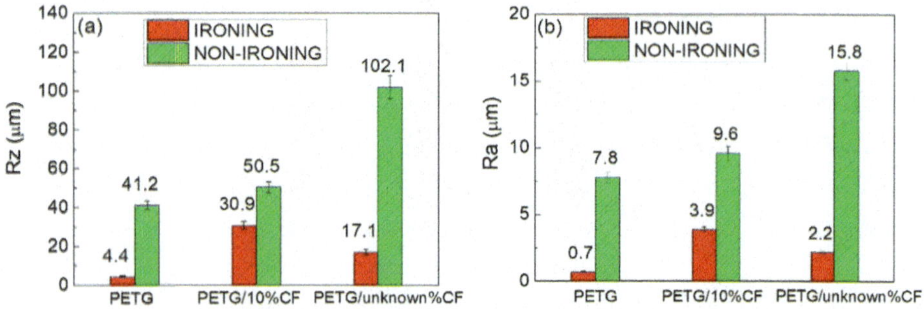

Fig. 3. Roughness variation in PETG and PETG/CF

This improvement in surface roughness is attributed to the ironing process, which involves the nozzle performing a pass over the top surface of the 3D print, the surface peaks are melted and ignited into valleys, and make the surface smoother.

Additionally, it can be easily observed that, in samples without ironing treatment, the influence of carbon fiber content on surface roughness is clearly demonstrated. With increasing carbon content, the surface roughness of the sample is also seen to increase. The smallest Rz and Ra values of 41.2 μm and 7.8 μm are reached with the original PETG sample. When the carbon fiber content increases by 10%, the Rz and Ra increase by about 20% to reach a value of 50.5 μm and 9.6 μm, separately. Furthermore, the highest Rz value of 102.1 μm 150% higher than PETG, is achieved with the PETG/unknown % CF sample, which was identified as having the highest carbon fiber content (greater than 10%). Regarding the Ra parameter, the value increases 2 times and reaches 15.8 μm in this sample.

3.3 Tribological Behavior

Effect of ironing treatment on tribological behavior

The friction coefficient of 3D-printed PETG samples is shown in Fig. 4. The COF (coefficient of friction) of the samples surveyed, the ironing process in 3D printing does not significantly affect the friction coefficient of the materials. The COF shows almost no change in both directions (0° - parallel to the 3D printing direction, and 90° - perpendicular to the 3D printing direction).

The COF of 3D printed materials can vary significantly depending on material composition, and printing parameters (print speed, layer height, infill density, etc.) [23]. While the ironing process is effective in improving surface roughness, its impact on the friction coefficient appears to be minimal, particularly for PETG.

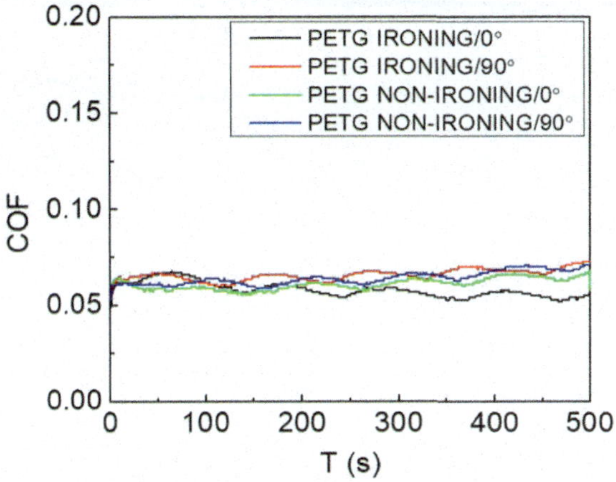

Fig. 4. The effect of ironing on COF of PETG samples

The observation that the ironing process in 3D printing results in "unclear" friction tracks on test samples (see Fig. 5). Specifically, for PETG printed samples without ironing, discontinuous "jumps" were observed during the friction test. This further confirms the benefits of the ironing process in enhancing the surface quality of printed parts.

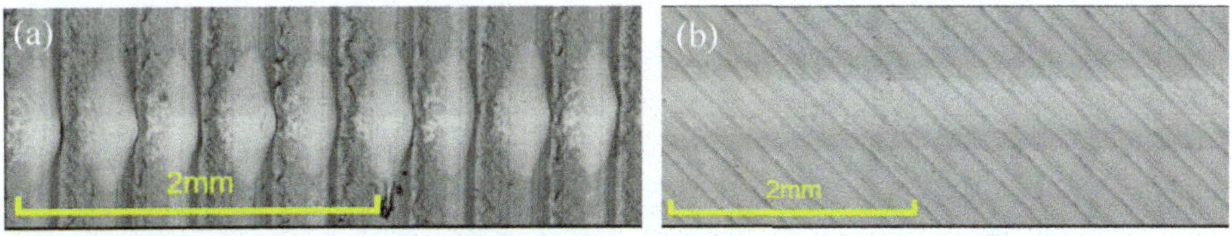

Fig. 5. The wear track profile of PETG: (a) without ironing; (b) with ironing

Effect of carbon fibers content on tribological behavior

The information provided in Fig. 6 shows the influence of carbon fiber content on the friction coefficient of PETG/CF composite material. As the carbon fiber content increases, the friction coefficient of the material gradually rises. Specifically, when conducting a friction test at 0° direction (parallel to the printing direction), the coefficient of friction (COF) value for the original PETG is approximately 0.07, while with 10% carbon fiber (CF) content, the COF value increases to about 0.12. Moreover, with another carbon fiber content of the sample PETG/CF Recreus, it has been expected greater than 10%, and the COF value further rises to 0.18 (see Fig. 6a).

This increase in friction coefficient can be attributed to several factors. On the one hand, the addition of carbon fibers can lead to increased surface roughness which can increase the friction coefficient. Carbon fibers can change the surface characteristics of PETG, potentially affecting how it interacts with other materials. On the other hand, the tensile strength of PETG is altered with the addition of carbon fiber [14], which can influence the frictional properties.

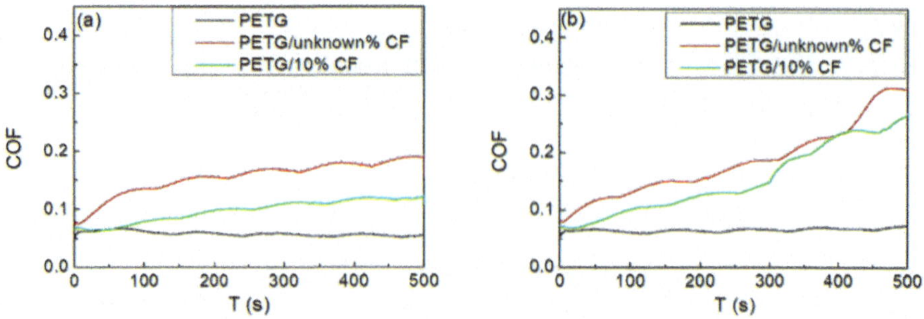

Fig. 6. The effect of carbon fiber on tribological behavior: (a) ironing/0°; (b) ironing/90°

In Fig. 6 (b), the sudden increase in the coefficient of friction (COF) was observed between 300 and 500 s in samples PETG/CF with ironing treatment and tested 90° perpendicular to the 3D printing direction. This phenomenon can indeed be attributed to the presence of carbon fiber reinforcement and its distribution within the composite material.

In composite materials, the orientation of reinforcement carbon fibers plays a significant role in determining mechanical properties [24, 25]. Due to the directional distribution of carbon fibers, the friction properties of the composite material can become anisotropic, meaning they vary with the direction of applied force or motion. Friction tests conducted perpendicular to the 3D printing direction may expose different fiber orientations and interfacial interactions compared to tests conducted along the printing direction.

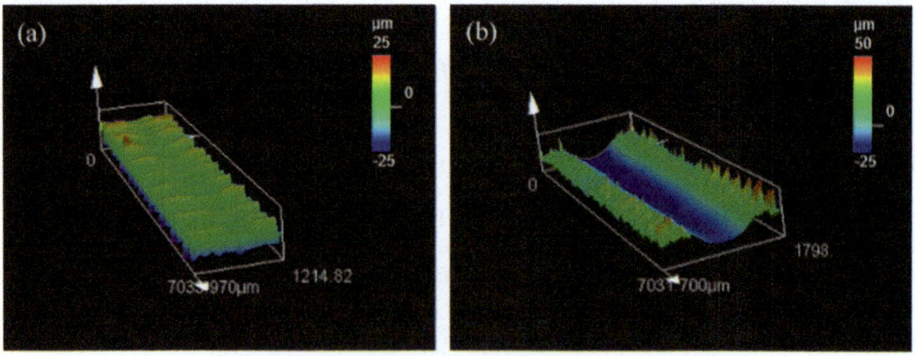

Fig. 7. Tribological wear track: (a) PETG; (b) PETG/CF.

The influence of carbon fiber on tribological behavior is also in wear tracks. The friction track on PETG/CF shows a larger depth and volume when compared to plain PETG in Fig. 7. Along with the material becoming less hard, the ball tip in a friction test could penetrate deeper into the material under the same impact load, increasing the contact area and potentially the friction and wear.

The impact of carbon fiber content becomes evident when measuring the volume and depth of wear tracks. With a higher carbon fiber content, there is a notable increase in both the depth and volume of wear tracks, as illustrated in Fig. 8.

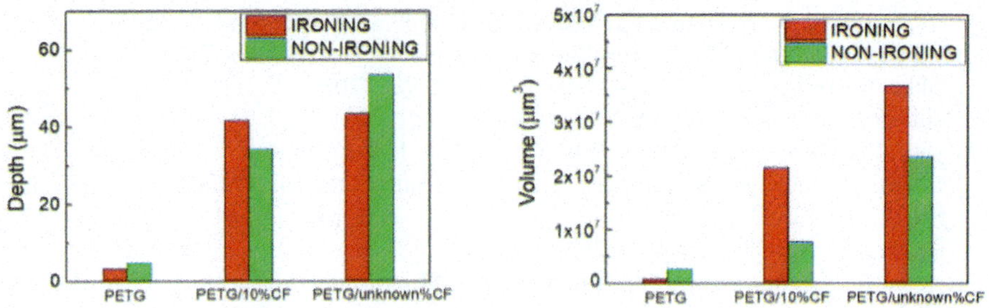

Fig. 8. The volume and depth of wear tracks

4 Conclusions

The ironing process has played a significant role in enhancing the surface quality of printed materials. Specifically, the PETG sample exhibited a decrease in parameter Rz from 41.2 µm to 4.4 µm. Similarly, the PETG/10%CF Spectrum showed reductions with the Rz value decreasing from 50.5 µm to 30.9 µm. Notably, the PETG/CF Recreus sample demonstrated the most substantial improvements, with the parameter Rz reducing nearly 5 times, from 102.1 µm to 17.1 µm. A similar trend in results was found regarding the Ra parameter, with a noticeable decline of values when ironing was employed.

The presence of carbon fiber content in PETG materials affects their friction behavior noticeably. In particular, during a friction test conducted in the 0° direction (parallel to the printing direction), the coefficient of friction (COF) for the base PETG material is approximately 0.07. However, when incorporating 10% carbon fiber (CF), the COF value increases to around 0.12. Additionally, in the PETG/CF Recreus sample, where the concentration exceeds 10%, the COF value escalates further to 0.18.

In composite materials PETG/CF where PETG is reinforced with carbon fibers, the content and orientation of these fibers during synthesis play a crucial role in determining the internal properties of the material. In both PETG/10%CF Spectrum and PETG/CF Recreus samples, the COF rose by approximately 50% when subjected to friction testing at 90° (perpendicular to the printing direction) in comparison to 0° (parallel to the printing direction).

Acknowledgment. The research outlined in this paper received support from two specific research projects: "SV23-216" and "SV24-216" at the Department of Mechanical Engineering, University of Defence, Brno, and from the Project for the Development of the Organization "VAROPS (DZRO VAROPS) Military autonomous and robotic assets" by the Ministry of Defence of the Czech Republic.

References

1. Kanishka, K., Acherjee, B.: J. Manuf. Process. **107**, 574–619 (2023)
2. Karakurt, I., Lin, L.: Curr. Opin. Chem. Eng. **28**, 134–143 (2020)
3. Shahrubudin, N., Lee, T.C., Ramlan, R.: Procedia Manuf. 1286–1296 (2019)
4. Park, S., Shou, W., Makatura, L., Matusik, W., Fu, K.K.: Matter **5**, 43–76 (2022)
5. Bakhtiari, H., Aamir, M., Tolouei-Rad, M.: Appl. Sci. (Switzerland) **13** (2023)
6. Prabhu, R., Devaraju, A.: Mater. Today Proc. 781–788 (2020)
7. Rajib Hossain, K., et al.: J. Chem. Lett. **4**, 103–116 (2023)
8. Ryan, K.R., Down, M.P., Banks, C.E.: Chem. Eng. J. **403** (2021)
9. Velu, R., Raspall, F., Singamneni, S.: Green Composites for Automotive Applications, pp. 171–196. Elsevier (2018)
10. Yilmaz, S.: Eur. J. Sci. Technol. (2022)
11. Batista, M., Lagomazzini, J.M., Ramirez-Peña, M., Vazquez-Martinez, J.M.: Appl. Sci. **13**, 12701 (2023)
12. Li, S., Wang, K., Zhu, W., Peng, Y., Ahzi, S., Chinesta, F.: Constr. Build. Mater. **340** (2022)
13. Saeed, K., et al.: Compos. Struct. **282** (2022)
14. Kichloo, A.F., Raina, A., Haq, M.I.U., Wani, M.S.: J. Mater. Eng. Perform. **31**, 1021–1038 (2022)
15. Wang, P., Zou, B., Ding, S., Li, L., Huang, C.: Chin. J. Aeronaut. **34**, 236–246 (2021)
16. Wang, P., et al.: Compos. B Eng. **198** (2020)
17. Shahrubudin, N., Lee, T.C., Ramlan, R.: Procedia Manuf. 1286–1296 (2019)
18. Colorado, H.A., Cardenas, C.A., Gutierrez-Velazquez, E.I., Escobedo, J.P., Monteiro, S.N.: J. Mater. Res. Technol. **27**, 3900–3913 (2023)
19. Sedlak, J., et al.: Materials **16** (2023)
20. Prochazka, J., Vilis, J., Dobrocky, D., Sperka, P.: Manuf. Technol. **22**, 724–732 (2022)
21. Mansour, M., Tsongas, K., Tzetzis, D., Antoniadis, A.: Polym. Plast. Technol. Eng. **57**, 1715–1725 (2018)
22. Kumar, M.A., Khan, M.S., Mishra, S.B.: Mater. Today Proc. 975–983 (2020)
23. Ranganathan, S., Kumar, K.S., Gopal, S., Palanivelu, R.: SAE Techn. Pap. SAE International (2020)
24. Zhou, W.D., Chen, J.S.: Mater. Sci. Forum 558–563 (2018)
25. Peng, X., Zhang, M., Guo, Z., Sang, L., Hou, W.: Compos. Commun. **22** (2020)

Effect of Heat Treatment on Surface Integrity and Tribological Properties of LPBF Printed AlSi10Mg

Pratik Kumar Shaw[1], Suryank Dwivedi[1], Preeti Gautam[2], Amit Rai Dixit[1](✉), and Alokesh Pramanik[3]

[1] Indian Institute of Technology (Indian School of Mines), Dhanbad, Jharkhand 826004, India
amitraidixit@iitism.ac.in
[2] VSB - Technical University of Ostrava, Ostrava, Poruba 70800, Czech Republic
[3] Curtin University, Kent Street, Perth, Bentley, WA 6102, Australia

Abstract. This study investigates the influence of heat treatment on the surface integrity and tribological properties of LPBF printed AlSi10Mg alloy. The AlSi10Mg samples underwent various heat treatment procedures, including stress relieving, heating the sample to 285 °C for 2 h, and direct ageing processes involving heating the sample to 170 °C for 6 h followed by furnace cooling. Comprehensive characterization was conducted using surface profilometry, scanning electron microscopy, and tribological testing. The results demonstrate significant enhancements in surface roughness and tribological properties following heat treatment. Specifically, the surface roughness of the sample treated at 170 °C for 6 h showed Sa and Sq values of 7.388 μm and 9.374 μm, respectively, compared to untreated samples which exhibited values of 8.338 μm and 10.197 μm, respectively. Heat treatment at 285 °C for 2 h led to substantial improvements in Sa and Sq values, measuring 5.523 μm and 7.000 μm, respectively. The microhardness of samples subjected to direct ageing increased by 15%, whereas stress relieving resulted in a 35% reduction in microhardness due to stress relief from the surface. Furthermore, tribological assessments under lubricating oil revealed enhanced wear resistance and reduced friction coefficients for heat-treated specimens subjected to direct ageing compared to their as printed and stress reliving counterparts. The findings underscore the efficacy of post-processing heat treatments in enhancing the surface integrity and tribological performance of LPBF printed AlSi10Mg components.

Keywords: LPBF · Heat treatment · Surface morphology · Tribology

1 Introduction

Additive manufacturing (AM), particularly Laser Powder Bed Fusion (LPBF), is gaining traction for its ability to create intricate shapes efficiently [1–5]. Aluminum alloys like AlSi10Mg are notable for their mechanical properties and lightweight nature, ideal for aerospace and automotive uses [6–9]. However, LPBF-produced parts often lack surface quality and tribological properties compared to traditionally made parts due to rapid solidification and thermal cycling during printing [10].

M. Pagac et al. (Eds.): SCFF 2024, LNNS 1062, pp. 67–77, 2024.
https://doi.org/10.1007/978-3-031-65656-9_7

Heat treatment involves controlled heating and cooling to modify material microstructure and properties, mitigating LPBF-related issues. Understanding its impact on LPBF-printed AlSi10Mg parts is crucial for optimizing performance. Surface integrity, including roughness, microstructure, residual stresses, and porosity, affects mechanical and tribological properties [11–13]. LPBF's thermal gradients and rapid solidification create complex microstructures and stresses, affecting part quality. Heat treatment refines microstructures, relieves stress, and heals defects, improving surface integrity. Tribological properties, like wear resistance and friction, are vital for functional components [14–16]. LPBF-printed AlSi10Mg parts often underperform due to surface roughness, microstructural heterogeneity, and residual stresses. Heat treatment can enhance tribological behavior by optimizing microstructure, surface finish, and mechanical properties.

While studies on conventionally manufactured AlSi10Mg alloys exist, research on their LPBF-printed counterparts is limited, highlighting the need for comprehensive investigations. This study aims to understand how heat treatment affects the surface integrity and tribological properties of LPBF-printed AlSi10Mg alloys. Through experimental techniques like surface roughness analysis and tribological testing, this explores how heat treatment influences material behavior and establishes correlations between processing parameters, surface roughness changes, and tribological performance. This knowledge will aid in optimizing the heat treatment process for LPBF-produced AlSi10Mg components. This research contributes to understanding post-processing techniques for enhancing additively manufactured metal alloys, facilitating their adoption in various industries.

2 Materials and Methods

2.1 Sample Fabrication

The AlSi10Mg samples, measuring 20 mm × 20 mm × 3 mm, were fabricated using DMP Flex 200 LBPF machines. The AlSi10Mg powder utilized ranged from 15 μm to 40 μm in size, with its chemical composition detailed in Table 1. OEM printing parameters were applied for the production of the parts, as outlined in Table 2.

Table 1. Chemical composition of AlSi10Mg powder

Element	Al	Si	Fe	Cu	Mn	Mg
Wt.%	Bal.	9–11	0.55	0.05	0.45	0.20–0.45

2.2 Heat Treatment of AlSi10Mg Samples

A Nabertherm GmbH muffle furnace was utilized for the heat treatment of printed AlSi10Mg samples. The furnace boasts a maximum temperature range of 3000 °C and

Table 2. Printing parameters used for AlSi10Mg samples preparation.

Printing parameters	Laser power	Laser beam diameter	Scanning speed	Layer thickness	Scanning strategy	Strip width	Hatch distance
Set values	325 W	0.06 mm	1400 mm/s	0.03 mm	Hexagon	10 mm	0.09 mm

was filled with argon gas during operation. Two distinct temperatures were chosen for the heat treatment of LPBF printed AlSi10Mg samples, 285 °C for a duration of 2 h, which is OEM parameter of manufacturer aimed at reducing tensile residual stresses, and 170 °C for a period of 6 h obtained from the literature [17]. The names of all samples were used in this study has been indicated in Table 3.

Table 3. Samples with designations

Designations	UT	HT170	HT285
Samples	Untreated samples	Heat treated at 170 °C, 6 h	Heat treated at 285 °C, 2 h

2.3 Surface Roughness and Morphology

Zygo NewView TM 9000, 3D optical profilometer was used for measuring the surface roughness and 3D surface image of the as printed and heat-treated AlSi10Mg samples considering the scan area of 1.563 mm × 1.563 mm. The parameters used to measure surface roughness are indicated with S, according to ISO 25178 with SL-filter, while the waviness measured for L-filter surfaces only 1.5 mm cut-off was considered for the roughness measurement. Sa (arithmetical mean height) and Sq (Root mean square height) both values were measured for this study as these metrics are typically reported in relation to friction, and wear.

2.4 Hardness Measurement

HM220A microhardness testing machine (Manufacturer: MITUTOYO, Japan), was used to measure the Vickers microhardness, conducted utilizing the applying a 0.1 kg load to both the as printed and heat-treated AlSi10Mg samples. Measurements were obtained at 8 specific points for each sample to determine the hardness value along the build direction. The interval between each measurement was 1.5 mm.

2.5 Tribology Test

Tribology tests were conducted on ball on disc reciprocating tribometer in lubricating oil. Tribometer uses 4 mm steel ball and reciprocates at a frequency of 3 Hz for 5 min for 3 N, 6 N and 9 N loads. Wear tracks for all the samples has been recorded using 3D profilometer. The wear volume was determined by combining the width and depth profile of the wear track, which was then multiplied by the wear track length. To calculate the specific wear rate Eq. 1 was used.

$$W_s = \frac{\Delta V}{L \times TSD}, \tag{1}$$

ΔV = wear volume, L = applied load and TSD = total sliding distance.

3 Result and Discussion

3.1 Effect of Heat Treatment on Surface Roughness and Surface Morphology

Figure 1 illustrates the 3D surface morphology and waviness for five designated slices under both as printed and heat-treated conditions. Specifically, Fig. 1(a), (c), and (e) display the 3D surface profiles of the UT, HT170, and HT285 samples, respectively, while Fig. 1(b), (d), and (f) depict the average waviness of the five marked slices on the 3D surfaces. Upon comparison of the 3D surface profiles across all samples, it was observed that the UT sample exhibits marginally higher surface waviness compared to the HT170 sample. However, the HT285 sample demonstrates superior surface morphology relative to both UT and HT170 samples. This observation is supported by Fig. 1, wherein the average waviness for the slices exhibits a wider range for UT and HT170, whereas the waviness for the HT285 sample demonstrates a more uniform surface compared to the preceding samples. Figure 2 represents the Sa and Sq values for all the samples, showing nearly a 30% improvement in both Sa and Sq for the HT285 sample and 12% and 7% for HT170 when compared with untreated samples. Thermal expansion and contraction during heat treatment causes to fill the voids obtained after the fabrication process and creates a uniform surface which is evident from the FESEM surface morphology depicted in Fig. 3, improving surface roughness and waviness. High temperature heat treatment of AlSi10Mg sample causes a reduction in residual thermal stresses from the surface and produces improved surface morphology as shown in Fig. 1(e).

Figure 3 shows the FESEM images of surface morphology for UT, HT170, and HT285 samples, clearly indicating improvement in surface roughness with the heat treatment for the residual stress relieving cycle HT285. However, pores have been observed in the samples due to the presence of unfused powder particles during melting and their removal during the sandblasting process. Pores weaken materials, causing premature failure under stress. Porosity level and characteristics affect wear. Small pores trap debris, while larger ones cause delamination. Increased porosity boosts wear resistance but lowers hardness [18].

3.2 Change in Hardness Value Along the Build Direction with Heat Treatment

Figure 4 illustrates the Vickers microhardness values for the UT (untreated), HT170 (heat-treated at 170 °C), and HT285 (heat-treated at 285 °C) samples. Figure clearly demonstrates that the microhardness values for the untreated sample ranged from 134.4 to

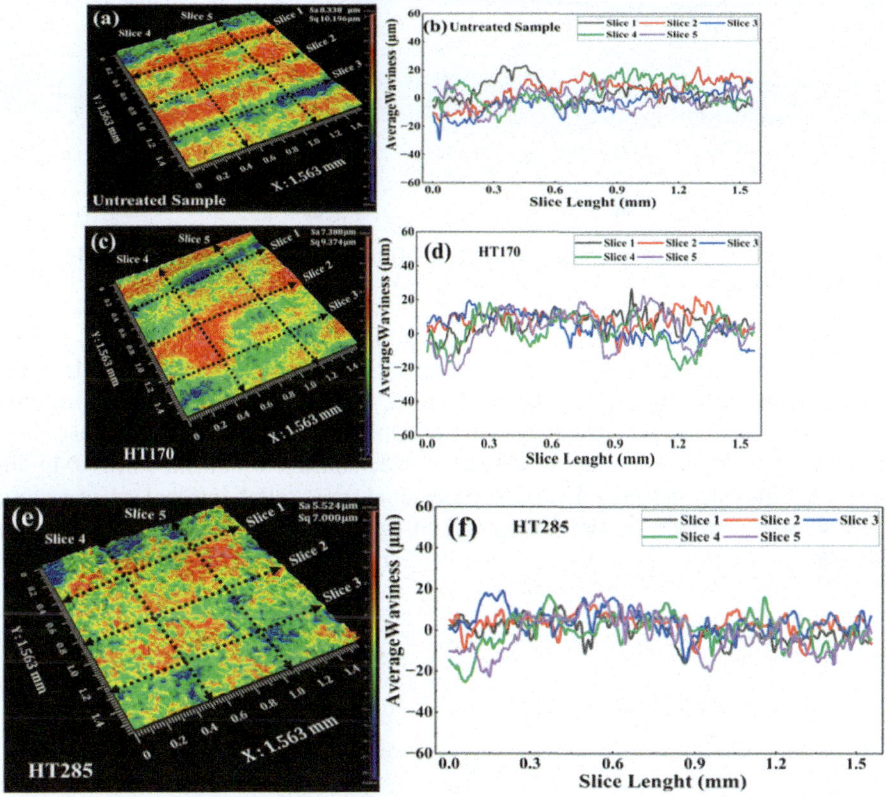

Fig. 1. (a), (c), and (e) 3D surface roughness profile and (b), (d), and (f) waviness for the five designated slices under both as printed and heat-treated samples.

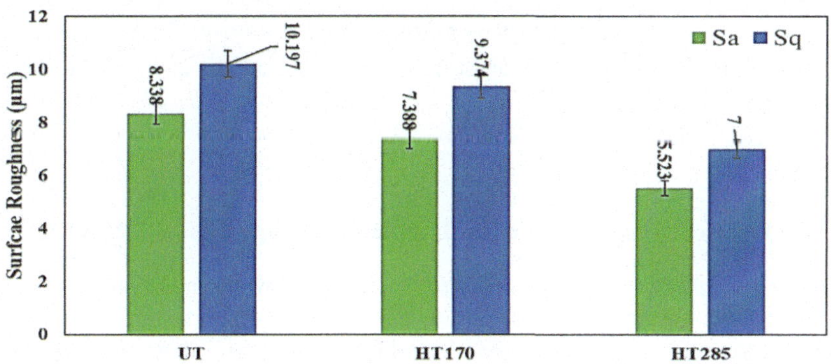

Fig. 2. Sa and Sq values of all AlSi10Mg samples

148.6. Upon subjecting the sample to heat treatment at 170 °C for 6 h, the microhardness value increased to a range of 148.6–165.5. Conversely, when the sample underwent heating at 285 °C for 2 h, the hardness decreased to a range of 79.7–90.8 along the build

Fig. 3. SEM micrograph of surface morphology of (a) untreated sample, (b) HT170 and (c) HT285 AlSi10Mg samples

direction. The observed increase in microhardness from the untreated (UT) to the HT170 samples suggests that the heat treatment at 170 °C for 6 h induced a strengthening effect. This enhancement can be attributed to the precipitation of strengthening phases or the improvement of solid solution strengthening mechanisms within the AlSi10Mg alloy matrix. Additionally, during the process of residual stress relief, internal stresses present on the sample's surface are diminished, resulting in reduced resistance towards plastic deformation and consequently a decrease in microhardness.

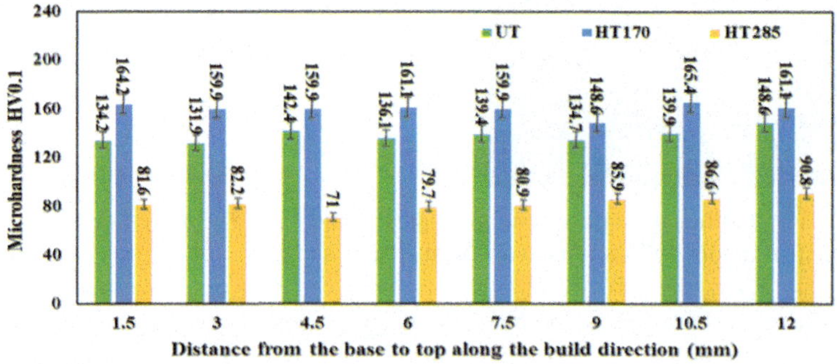

Fig. 4. Microhardness of UT HT170 and HT285 AlSi10Mg samples along the build direction from bottom to top.

3.3 Effect of Heat Treatment on the Tribological Properties of AlSi10Mg

Figure 5 (a–c) depict the 3D wear track profiles of AlSi10Mg samples subjected to loads of 3 N, 6 N, and 9 N, respectively. Figure 5(d) illustrates the average wear width and depth for all samples. The wear test, conducted with lubricating oil, reveals that untreated samples exhibit greater wear track width compared to HT170 samples under all load conditions. Conversely, the wear track of HT285 samples displays a wider range of wear track dimensions compared to untreated samples. The wear width and depth values for all wear tracks are detailed in Table 4, indicating that wear for HT285 was more pronounced than for other samples. Also, the effect of varying loads has been analysed in the 3D profile of wear track where minimal surface contact characterizes low load of

3 N. Adhesive wear, caused by direct asperity contact, and abrasive wear from lubricant contaminants prevail, inducing surface damage [19]. While at 6 N load, increase contact area, elevating abrasive, and adhesive wear severity. Surface deformation and fatigue, especially in high-stress regions, become evident. At 9 N load, significant pressures at high loads induce material deformation and microstructural changes. Surface fatigue intensifies with cyclic loading, leading to crack formation [20]. Increased adhesive wear and delamination occur, aggravated by abrasive contaminants. Also, the microhardness trends across samples correlate as anticipated: wear resistance decreases with material softness. Hardness influences a material's ability to withstand wear-induced deformation, with softer materials more susceptible to abrasion.

Table 4. Width and depth values of wear track of UT, HT170 and HT285 samples

Load	3 N		6 N		9 N	
Sample(s)	Width (μm)	Depth (mm)	Width (μm)	Depth (mm)	Width (μm)	Depth (mm)
UT	207.67	1.478	239.133	1.88	261.159	2.35
HT170	197.99	1.15	235.7	1.6	248.2	2.2
HT285	254.936	2.4	365.094	5.405	459.515	9.169

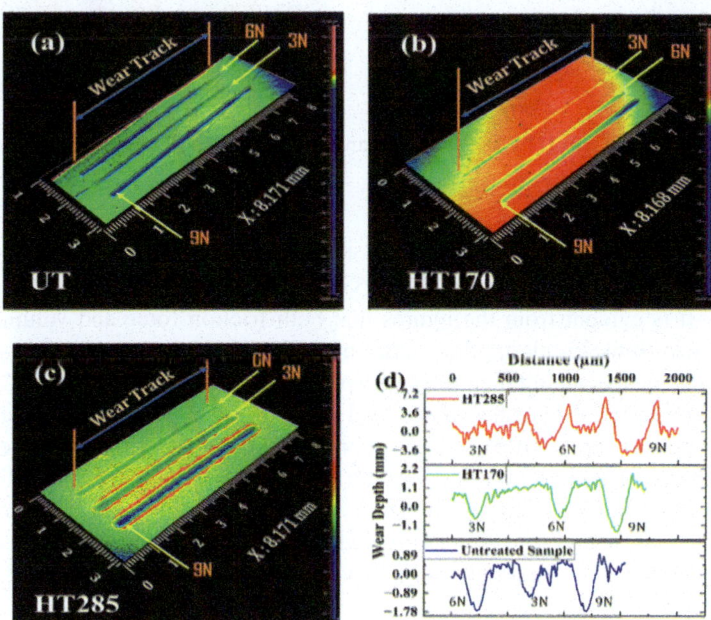

Fig. 5. 3D optical profile of wear track, (a) UT, (b) HT170, (c) HT285 and (d) 2D wear track of all AlSi10Mg samples at 3 N, 6 N and 9 N load.

Figure 6 (a–c) depicts the coefficient of friction (COF) of all AlSi10Mg samples across varying load conditions, while (d) presents the average COF values. The data illustrates a direct correlation between load increase and COF elevation across all samples, attributable to heightened adhesion strength within the layer.

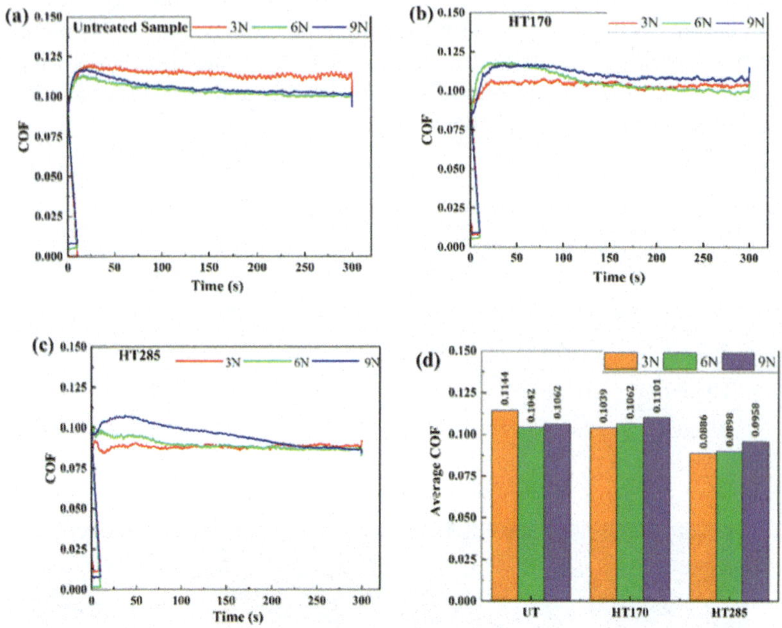

Fig. 6. COF of (a) UT, (b) HT170, (c) HT285 and (d) average COF of all AlSi10Mg samples at 3 N, 6 N and 9 N load.

Figure 7(a, b, c) illustrates the friction force, volumetric wear rate, and specific wear rate respectively for UT, HT170, and HT285 samples subjected to 3 N, 6 N, and 9 N load conditions. It is evident from the figures that both friction force and volumetric wear rate increase with the escalating load, attributed to the augmented adhesion strength as the load intensifies, reaching their peak values under the 9 N load condition. Conversely, the specific wear rate diminishes as the load escalates, exhibiting the highest wear rate under the 3 N load condition. His observation implies that under lower load conditions, such as the 3 N load condition, the specific wear rate reaches its peak, indicating the highest wear per unit load. The trends observed in Fig. 7(a), (b), and (c) underscore the intricate interplay between load conditions, frictional forces, and wear characteristics, elucidating the dynamic nature of wear behavior in the UT, HT170, and HT285 samples.

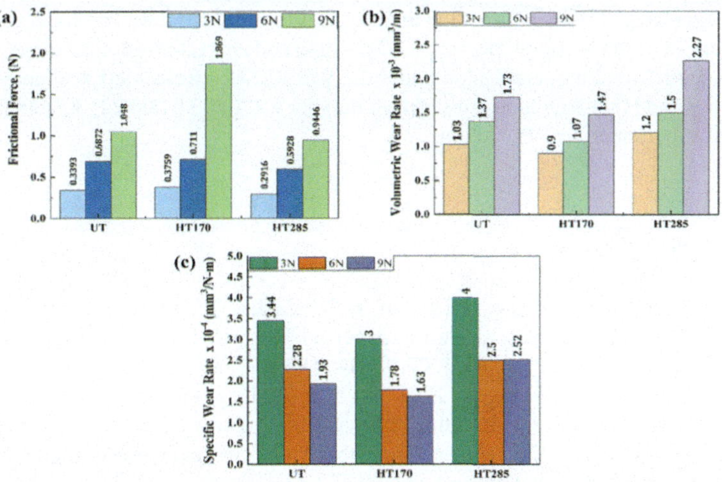

Fig. 7. (a) Frictional force, (b) volumetric wear rate and (c) specific wear rate of all AlSi10Mg samples at 3 N, 6 N and 9 N load.

4 Conclusion

The AlSi10Mg samples were examined with respect to surface morphology, roughness analysis, microhardness, and wear behavior. Upon analysis of the obtained results, the following conclusions can be drawn:

1. Heat treatment at 285 °C significantly improves surface morphology and reduces surface waviness compared to both untreated and HT170 samples, as evidenced by 3D surface profiles and average waviness measurements.
2. Heat treatment induces a substantial enhancement in microhardness, particularly at 170 °C, indicating strengthened material properties attributed to precipitation of strengthening phases or solid solution mechanisms within the alloy matrix. However, a decrease in microhardness is observed at 285 °C due to the reduction in residual stress relief, leading to decreased resistance towards plastic deformation.
3. Wear resistance inversely correlates with material hardness, as demonstrated by wear track analysis. HT170 samples exhibit superior wear resistance compared to untreated samples under all load conditions, while HT285 samples show increased wear track dimensions, indicating diminished wear resistance.
4. Load increase directly correlates with the coefficient of friction (COF) elevation across all samples, reflecting heightened adhesion strength within the layer. Additionally, friction force and volumetric wear rate increase with escalating load conditions, reaching peak values under the 9 N load condition, while the specific wear rate diminishes as the load intensifies, highlighting the dynamic nature of wear behavior influenced by load conditions.

It is recommended that the combined effect of residual stress followed by direct aging of the sample will yield a more promising surface finish, along with enhanced wear resistance properties in the AlSi10Mg samples.

Acknowledgement. The authors extend their regards to the Central Research Facility of the IIT (ISM), Dhanbad, India for providing metal 3D printing and characterization facilities. This article was co-funded by the European Union under the REFRESH – Research Excellence for Region Sustainability and High-tech Industries project number CZ.10.03.01/00/22_003/0000048 via the Operational Programme Just Transition.

References

1. Mezghani, A., Nassar, A.R., Dickman, C.J., Valdes, E., Alvarado, R.: Laser powder bed fusion additive manufacturing of copper wicking structures: fabrication and capillary characterization. Rapid Prototyp. J. **27**, 1181–1188 (2021)
2. Nagarajan, B., Hu, Z., Song, X., Zhai, W., Wei, J.: Development of micro selective laser melting: the state of the art and future perspectives. Engineering **5**, 702–720 (2019)
3. Tur, E.: A comprehensive review on sustainability and environmental impact of laser powder bed fusion additively manufactured as-built Ti-6Al-4V parts. El-Cezeri **10**, 612–644 (2023)
4. Srivastava, A.K., Kumar, A., Kumar, P., Gautam, P., Dogra, N.: Research progress in metal additive manufacturing: challenges and opportunities. Int. J. Interact. Des. Manuf. (IJIDeM) (2023)
5. Dwivedi, S., Rai Dixit, A., Kumar Das, A.: Wetting behavior of selective laser melted (SLM) bio-medical grade stainless steel 316L. Mater. Today Proc. **56**, 46–50 (2022)
6. Gite, R.E., Wakchaure, V.D.: A review on process parameters, microstructure and mechanical properties of additively manufactured AlSi10Mg alloy. Mater. Today Proc. **72**, 966–986 (2023)
7. Han, Q., Jiao, Y.: Effect of heat treatment and laser surface remelting on AlSi10Mg alloy fabricated by selective laser melting. Int. J. Adv. Manuf. Technol. **102**, 3315–3324 (2019)
8. Ghio, E., Cerri, E.: Additive manufacturing of AlSi10Mg and Ti6Al4V lightweight alloys via laser powder bed fusion: a review of heat treatments effects. Materials **15** (2022)
9. Ponnusamy, P., Rahman Rashid, R.A., Masood, S.H., Ruan, D., Palanisamy, S.: Mechanical properties of SLM-printed aluminium alloys: a review. Materials **13** (2020)
10. Khan, H.M., Karabulut, Y., Kitay, O., Kaynak, Y., Jawahir, I.S.: Influence of the post-processing operations on surface integrity of metal components produced by laser powder bed fusion additive manufacturing: a review. Mach. Sci. Technol. **25**, 118–176 (2021)
11. Li, W., et al.: Effect of heat treatment on AlSi10Mg alloy fabricated by selective laser melting: microstructure evolution, mechanical properties and fracture mechanism. Mater. Sci. Eng. A **663**, 116–125 (2016)
12. Alghamdi, F., Song, X., Hadadzadeh, A., Shalchi-Amirkhiz, B., Mohammadi, M., Haghshenas, M.: Post heat treatment of additive manufactured AlSi10Mg: on silicon morphology, texture and small-scale properties. Mater. Sci. Eng. A **783**, 139296 (2020)
13. Ramoni, M., Shanmugam, R., Ross, N.S., Gupta, M.K.: An experimental investigation of hybrid manufactured SLM based Al-Si10-Mg alloy under mist cooling conditions. J. Manuf. Process. **70**, 225–235 (2021)
14. Amanov, A.: Effect of local treatment temperature of ultrasonic nanocrystalline surface modification on tribological behavior and corrosion resistance of stainless steel 316L produced by selective laser melting. Surf. Coat. Technol. **398**, 126080 (2020)
15. Unune, D.R., Brown, G.R., Reilly, G.C.: Thermal based surface modification techniques for enhancing the corrosion and wear resistance of metallic implants: a review. Vacuum **203**, 111298 (2022)
16. Prashar, G., Vasudev, H., Thakur, L.: Influence of heat treatment on surface properties of HVOF deposited WC and Ni-based powder coatings: a review. Surf. Topogr. **9**, 43002 (2021)

17. Di Egidio, G., Ceschini, L., Morri, A., Martini, C., Merlin, M.: A novel T6 rapid heat treatment for AlSi10Mg alloy produced by laser-based powder bed fusion: comparison with T5 and conventional T6 heat treatments. Metall. Mater. Trans. B **53**, 284–303 (2022)
18. Islam, M.A., Farhat, Z.N.: Effect of porosity on dry sliding wear of Al–Si alloys. Tribol. Int. **44**, 498–504 (2011)
19. Gahr, K.-H.Z. (ed.): Sliding wear, chap. 6. In: Microstructure and Wear of Materials, pp. 351–495. Elsevier (1987)
20. Stachowiak, G.W., Batchelor, A.W. (eds.): 11 abrasive, erosive and cavitation wear. In: Engineering Tribology, pp. 557–612. Elsevier (1993)

Implementation of Advanced Techniques for Quality Assessment of Metal Parts Produced by SLM Technology

Miroslav Matuš[1](✉), Vladimír Bechný[1], Michal Šajgalík[1], Jaromír Markovič[1], Mário Drbúl[1], Miroslav Cedzo[1], Richard Joch[1], and Dominik Krišák[2]

[1] Faculty of Mechanical Engineering, University of Zilina, Univerzitná 8215/1, 010 26 Žilina, Slovakia
miroslav.matus@fstroj.uniza.sk

[2] Department of Machining, Assembly and Engineering Technology, Faculty of Mechanical Engineering, Center of 3D Printing Protolab, VSB-TU Ostrava, 17. Listopadu 2172/15, 708 00 Ostrava, Czech Republic

Abstract. Over time, additive manufacturing has found a place in almost every sector of industry. This is helped by the fact that it is possible to process an increasing amount of material, especially metals, plastics and composite materials. As with any technology, additive manufacturing includes many advantages and disadvantages, which we are continuously striving to optimize through research and development. In the study, deviations between scanned parts and basic CAD models were evaluated. The individual parts were produced by additive SLM technology, which is a metal-based additive manufacturing technique, and the technology uses the principle of powdered material processing, where deformations occur during the manufacturing process due to the influence of heat on the part. The aim of the present study is the design and implementation of optimization processes based on advanced quality assessment methods using simulations, by scanning models and then comparing the deviations against nominal values of the models. The study identifies individual undesirable phenomena that arise with each technology and seeks to minimize them. These are the deformations produced during the manufacturing process and the deviations from nominal values caused by this.

Keywords: Additive manufacturing · SLM · Simulation · Optimization

1 Introduction

In recent years, the additive technology industry has experienced a revolutionary development, bringing not only new opportunities in production but also challenges in maintaining and improving the quality of the resulting products [1, 5]. Additive manufacturing has become a key element in both industrial and research environments. Due to its ability to create complex and geometrically intricate objects with high precision and efficiency.

M. Pagac et al. (Eds.): SCFF 2024, LNNS 1062, pp. 78–88, 2024.
https://doi.org/10.1007/978-3-031-65656-9_8

However, despite these advantages, the quality of products from additive manufacturing remains a constant subject of investigation and improvement. The implementation of advanced techniques is becoming an essential part of the process of evaluating and ensuring product quality in this dynamic field [3, 4, 6].

Selective laser melting (SLM) additive technology is just the right technology for the job. Currently, SLM is an innovative technology that is rapidly gaining popularity in a variety of industries including aerospace, medical, and automotive. Its ability to create complex metal components with high precision and efficiency provides engineers with new opportunities in design and innovation [7, 8]. The process of the technology itself involves the sequential deposition of layer after layer using high-power laser equipment. This additive technology allows for today's modern manufacturing to create parts with high precision and density, optimize the design itself, and finally, reduce the weight of the manufactured parts [9–14].

It can be said that simulations are now an integral part of additive manufacturing. They help us to predict and analyze the results of individual production processes under certain production conditions in real time. Simulations can be carried out prior to the creation of the support structures themselves, helping us to identify critical areas of significant deformation or internal stresses during production. Thanks to the simulations, it is possible to apply customized support structures to minimize deformations, to change the orientation of the part due to changes in the heat accumulation area and, finally, to change the geometry in order to improve the quality of the final part. On the other hand, it must be said that the simulations after the generation of the supports ensure that the final dimensions of the part are placed within a specific range of tolerances. This means that they evaluate the influence of different parameters of the manufacturing process [15–21].

The aim of the present study is to provide an overview of SLM technology, its development, principles, and importance in the current industrial environment. The study also includes the design and implementation of optimization processes based on advanced quality evaluation methods, which consist of pre-production simulations, scanning of final samples and their subsequent comparison of deviations against baseline models, together with comparative interpretation of the results obtained.

2 Materials and Methods

The material used in the present study was 316L stainless steel. The steel type in question belongs to austenitic steels with high corrosion and heat resistance. For application in the additive manufacturing industry, it exhibits relatively good strength, ductility, and high hardness. It is these properties that make it one of the steels used in various industries. Table 1 shows the chemical composition of the material [22]. The equipment used for the SLM technology was a Renishaw AM400 printer, which offers high accuracy and productivity. The laser power for printing was 400 W, the layer thickness was 50 μm and the defined print orientation was in the Z-axis.

Table 1. Chemical composition of 316L material

Element	Cr	Ni	Mo	Mn	Si	N	P	C	S
Composition (%)	16–18	10–14	2–3	≤ 2	≤ 1	≤ 0.1	≤ 0.05	≤ 0.03	≤ 0.03

2.1 Design of Models

The design of the hollow structure in joining of the individual walls causes the concentration of higher stresses and heat, which leads to significant deviations in the manufacturing process, which are caused by the higher deformations in the additive manufacturing process. The aim of this study is to analyze the critical areas that arise in the additive process, ensuring the repeatability of the manufacturing process and the proposal to improve the quality of the manufactured parts. The research was carried out on 3 samples, namely needle, cube, and gear (Fig. 1). All the above-mentioned models had 316L material assigned in the simulations themselves.

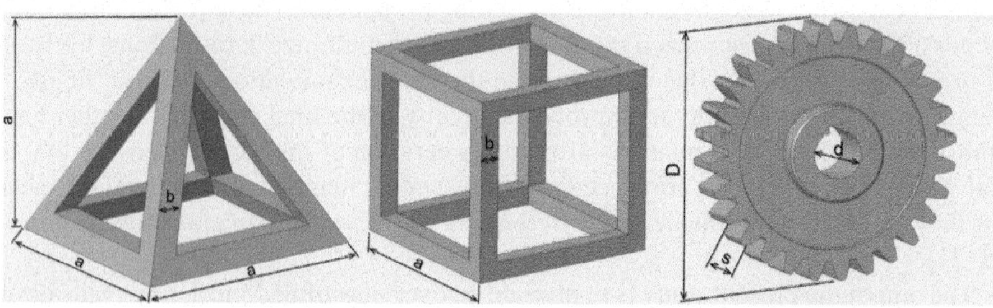

Fig. 1. Design model of a needle, cube, and gear wheel

The spur gear is designed with a spur gear, a hole for the shaft and a groove for the tongue. The choice of this model was mainly due to its more complex design and the need to maintain high precision in its manufacture, as it is applied in various industrial structures.

2.2 Scanning with Ein Scan-SP

The present study, parts were scanned using an advanced desktop 3D scanner with high accuracy and scanning speed. The scanner in question is used to capture the geometry of small to medium sized objects. It works on the principle of structured white light and uses a rotary table with markers to capture objects through 360°. The scanner scans to an accuracy of 0.03 mm. Before the actual scanning, it is necessary to calibrate the device to ensure sufficient accuracy and quality of the desired model. For the present study, multiple scans of the parts were necessary. For the needle and cube models, 2–3 scans were sufficient and for the gear wheel, 3–4 scans were needed. When scanning the models no marker was used but a spray was used to prevent surface glossiness.

2.3 ZEISS GOM Inspect

The software is used to analyze deviations of real models from their original CAD design models. For our study, an analysis based on superimposing two models evaluating the deviations over color maps was chosen. The principle itself consisted in the interleaving of the scanned model and the base CAD model. Before the actual analysis, it was necessary to ensure that the two models were accurately constrained to each other for the most reliable results. The results of the comparison are presented in a color map (scale on the right), which shows the deformation regions according to the magnitude of their values (red color represents the highest deformations in the positive direction and blue color the highest de-deformations in the negative direction). The scale values were chosen equally in all cases, according to the maximum deviations of the model, for better qualitative resolution. In this way, the locations that are most deviated from the base CAD model can be pinpointed.

3 Results

3.1 Analysis of Needles by SLM Technology

The first model evaluated was a needle that had not been modified or filmed. The support structures that were tied to the model base were not considered in the evaluation. From Fig. 2 the model is made relatively accurately, but the problem occurs at the nodal points and at the points where multiple walls are joined and at the top surface of the walls of the substructure. In these areas there is sagging and overlapping of the planar surface where the largest deformations are also present.

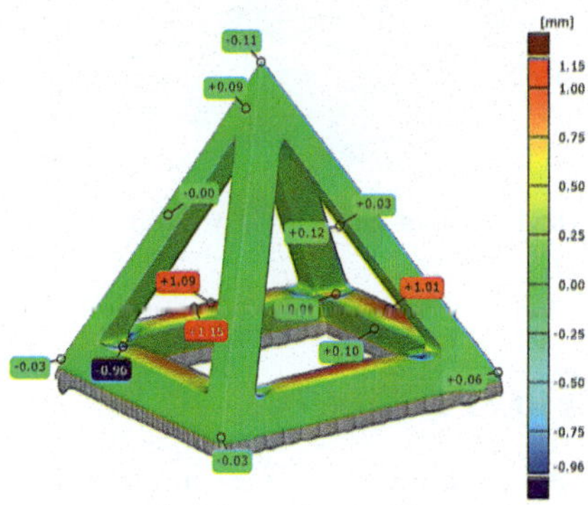

Fig. 2. Variations of the basic model of needle

The second model evaluated was a compensated model of a needle, where deformations from Autodesk Fusion 360 were also considered. The supporting structures were also not considered in the evaluation. From Fig. 3 we can determine that the deviations for this model are almost twice smaller compared to the first one. The area of the largest deformations is again at the nodal points of the system. The difference at the nodal points compared to the basic model is the absence of deflection at the corners.

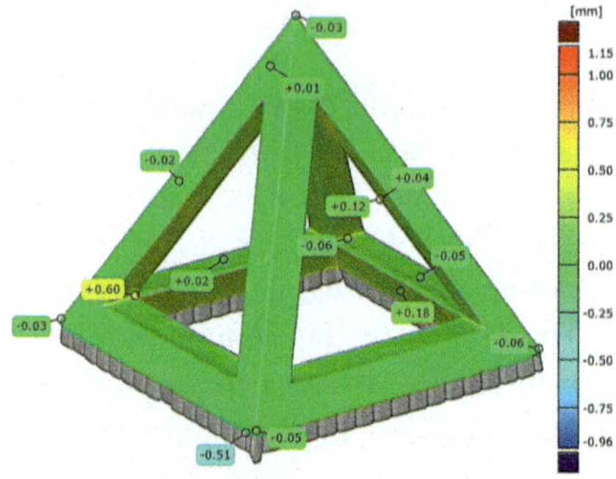

Fig. 3. Deviations of the compensated needle model

The last model analyzed was the optimally oriented model of the needle, where rotation occurred in the X and Y axes. The design itself was generated by Inspire Print 3D software. For this specimen, it was possible to remove all the supports, and from Fig. 4 the source of the largest deformations are the bottom edges of the needle, the nodal points and the top tip of the needle.

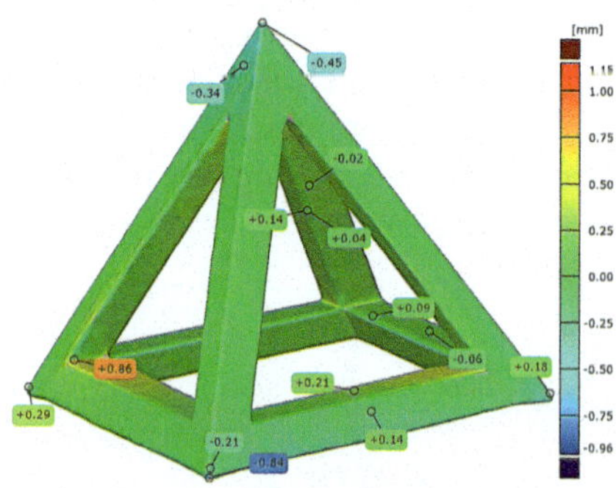

Fig. 4. Deviations of the optimally oriented needle model

3.2 Cube Deviations by SLM Technology

Similarly to the needles, the cube models were supported from the bottom edge of the base. The base cube model was evaluated first and was not modified in any way. It can be seen in Fig. 5 that the de-formations have an increasing tendency in the positive direction of the Z-axis. The area of concentration of the largest de-deformations is the upper nodal points, where the de-deformations reached values of 1.44–1.45 mm. In this case, there is a deflection of the parallel shape and a significant deflection, which is manifested by a narrowing of the shape below the upper walls.

The second cube model evaluated was a compensated model (Fig. 6) in which deformations were also considered. Also in this case, it can be noted that the de-formations took on an increasing tendency in the positive Z-axis direction, also at the upper nodal

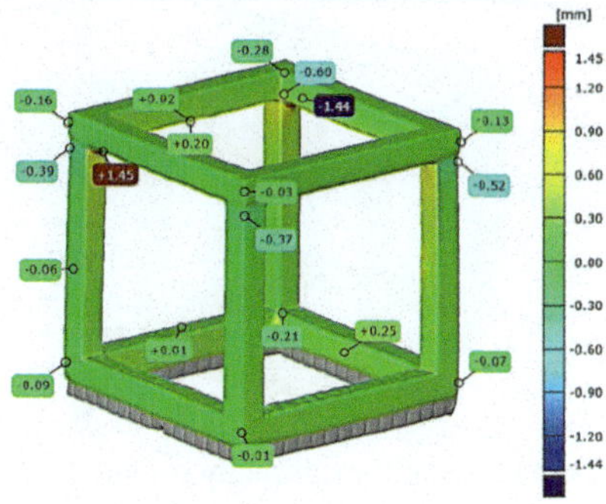

Fig. 5. Deviations of the basic cube model

points, but compared to the first cube model, the maximum values of the deviations reached a limit of 0.98 mm. In the same way, in this model it can be stated that there was a deflection of the parallel shape and the appearance of the deflection. The deviations for this part reached values of 0.3 mm.

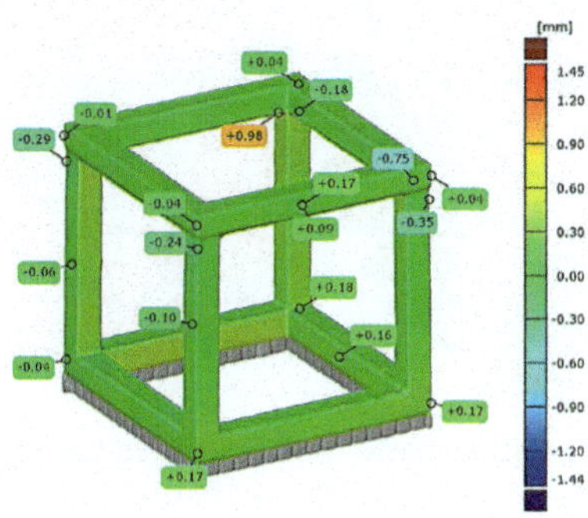

Fig. 6. Deviations of the compensated cube model

The last model evaluated was an optimally oriented cube model, whose rotations in the X and Y axes were provided by Inspire Print3D software. In Fig. 7 we can see that the model shows the least de-formation among the other models. Compared to the compensated model, the values of the maximum deflections are not significantly different, but there is no deflection of the parallel state of the walls at the top. Also, the deflections do not acquire an increasing tendency in the positive Z-axis direction but are concentrated only at the nodal points of the upper part.

3.3 Gear Wheel Analysis by SLM Technology

The last part analyzed was the gear wheel. A basic model in the horizontal position, which was not modified or oriented in any way, was evaluated first. One of the main reasons that began to influence the value of the maximum deflection was the remains of the structural supports on the back of the wheel. From Fig. 8 higher deflections were

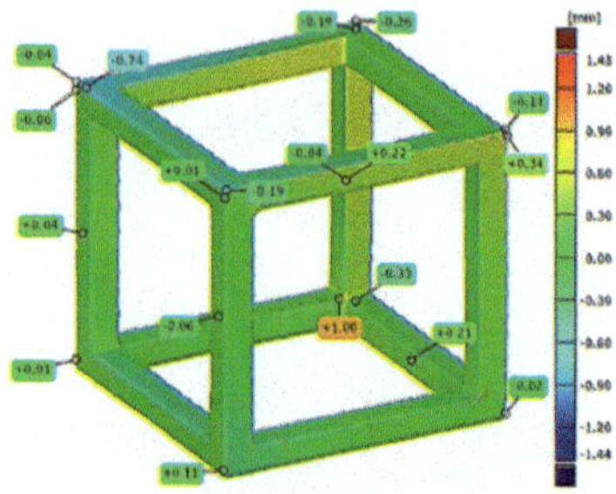

Fig. 7. Deviations of the optimally oriented cube model

also on the side where the structural supports were not. This means that the model is deformed even by the influence of the supports alone.

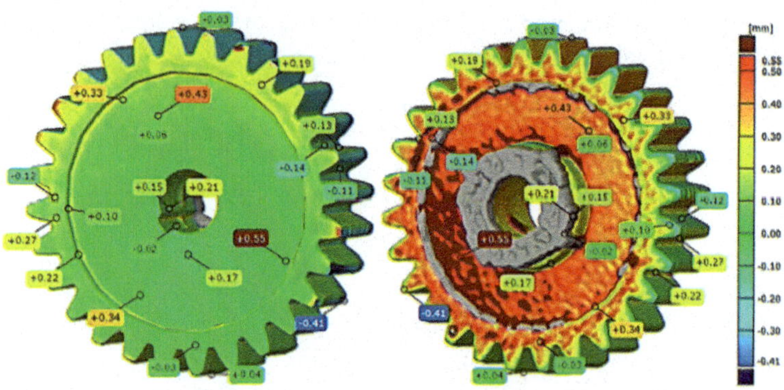

Fig. 8. Deviations of the basic wheel model in the horizontal position

The second model analyzed was the upright model (Fig. 9), which was also not modified or oriented in any way. In this case, it was possible to eliminate the structural supports from the software evaluation. The maximum deviation values were 0.49 mm and – 0.33 mm lower compared to the previous model. From the front side it is possible to see the critical areas, which are colored yellow. In view of how the orientations are fitted to the supports, it is not appropriate to produce the model. It is mainly a matter of subsequently removing the supports between the teeth, which is very inefficient, with a very high risk of affecting the shape of the gearing.

The third model analyzed was a compensated gear model in the horizontal position (Fig. 10). For this case, the deformations from Autodesk Fusion 360 software were included. For the given model, the back side of the model was covered with structural supports, which manual removal was not possible. The deflections were 0.42 mm and – 0.25 mm. Compared to its base model they show a partial improvement. The overall maximum deflections were reduced, as well as if the groove area for the tongue where the deflections were reduced to 0.09 mm and 0.16 mm. The tooth spacing and tooth profiles show comparable deviations with the previous two cases. It is also considered

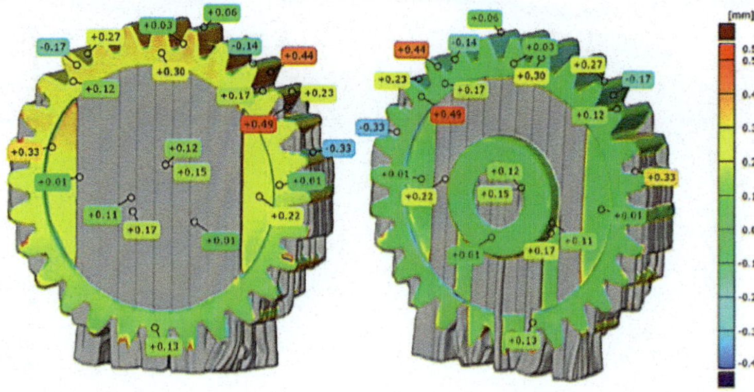

Fig. 9. Deviations of the basic wheel model in the vertical position

to be an advantage in this variant that the functional part of the gear is not affected in any way by the removal of the supports. It can therefore be stated with certainty that it shows a better suitability for use compared to its basic variant.

Fig. 10. Deviations of the compensated wheel model in the horizontal position

The fourth model analyzed was a compensated gear model in the horizontal position (Fig. 11). The deformations were obtained from Autodesk Fusion 360 software. The given one has very similar deformation values compared to its basic variant. There was a slight improvement in the maximum deflections of 0.48 mm and – 0.17 mm and an improvement in the geometry of its front side. Again, the presence of supports appears to be a negative, as they significantly affect the functional parts of the model, and their removal is very risky.

The last model analyzed was the optimally oriented model (Fig. 12). The rotation was performed in the X y Y axis and was generated by Inspire Print3D software. This model shows the lowest deformation among all the models. The maximum rank - the deviation in the positive direction is 0.47 mm. It is caused by the remnant of the support.

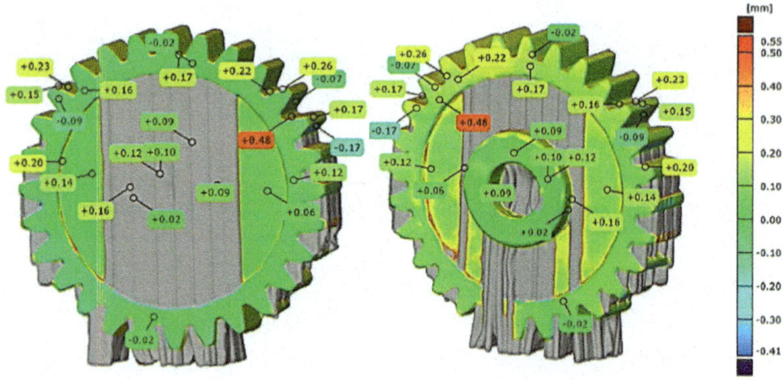

Fig. 11. Deviations of the compensated wheel model in the vertical position

In the negative direction, the maximum value of the deflection is the most acceptable at − 0.23 mm. On the other hand, the largest deflections are on the rear side of the model colored yellow. On the positive side, the removal of the supports in this case does not appear to be difficult and the supports have the least effect on the outer and inner surfaces of the model.

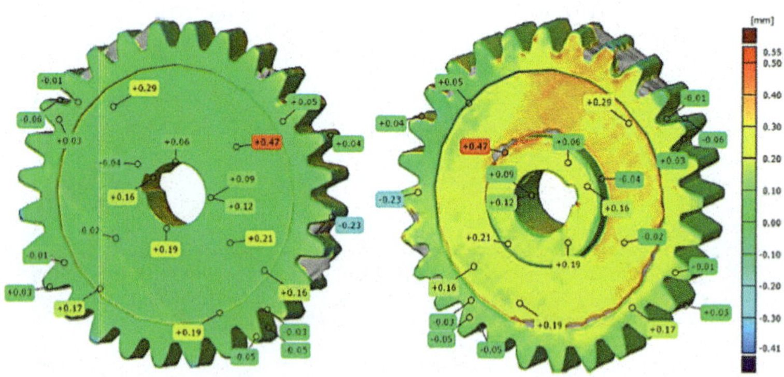

Fig. 12. Deviations of the optimally oriented wheel shape

4 Conclusion

From the present study it can be confirmed that additive manufacturing is currently an innovative production method, but it has some shortcomings. Based on our findings, it can be said that parts produced by SLM technology have a great influence on the resulting functionality and the actual quality of the manufactured parts due to the presence of structural supports. It can also be evaluated that not only the presence of support structures but also the actual area of support placement on a given model can account for the difficulty and time of support removal. The research results show that the models which are not modified in any way show the highest values of deviations in both directions. The most suitable models produced by the SLM technology are the compensated needle model, the optimally oriented cube model, and the gear model. Based on the research, it can be clearly determined that the modification of the models, either by compensation or orientation, can significantly improve the quality of the resulting model. On the other hand, it can be said that the modification of the base model does not always guarantee the smallest possible deviations on the whole component. In the future, consideration can be given in more easily identifying the most optimal manufacturing method for a given component based on the identified deviations. This makes it possible to improve the expected outcome of the production process.

Acknowledgement. The article was supported by the grant scheme KEGA 017/ŽU-4/2022: "Implementation of digital technologies and simulations into the teaching process of machining technology", APVV-20-0216: "Research on the implementation of high-impact surface technologies for precision automotive structural components", APVV-22-0328: "Design of a Methodology and its Verification for the Measurement of Selected Parameters of Ti Implants in the Manufacturing Process". This research was funded by the University of Žilina project APVV 20-0561: "Research on the implementation of new measurement methods for the calibration of measurement systems for industrial metrology practice", and by the grant scheme UNIZA: "Identification of dynamic properties of samples produced by additive technologies".

References

1. Ngo, T.D., Kashani, A., Imbalzano, G., Nquyen, K.T., Hui, D.: Additive manufacturing (3D printing): a review of materials, methods, applications and challenges. Compos. B Eng. **143**, 172–196 (2018)
2. Bechný, V., Czán, A., Holubják, J., Czánová, T., Cedzo, M., Timko, P.: Analysis and comparison of production parameters of additive technologies in the construction of objects based on metals using the ADAM and SLM methods. AIP Conf. Proc. **2976**(1) (2023)
3. Gibson, I., et al.: Addit. Manuf. Technol. **17**, 160–186 (2021)
4. Bourell, D., et al.: Materials for additive manufacturing. CIRP Ann. **66**(2), 659–681 (2017)
5. Joch, R., et al.: The application of additive composites technologies for clamping and manipulation devices in the production process. Materials **16**(10) (2023)
6. Wu, Y., Fang, J., Wu, C., Li, C., Sun, G., Li, Q.: Additively manufactured materials and structures: a state-of-the-art review on their mechanical characteristics and energy absorption. Int. J. Mech. Sci. **246** (2023)
7. Hajnys, J., Pagac, M., Mesicek, J., Petru, J., Spalek, F.: Research of 316L metallic powder for use in SLM 3D printing. Adv. Mater. Sci. **20**(1), 5–15 (2020)
8. Paraschiv, A., Matache, G., Condruz, M.R., Frigioescu, T.F., Ionică, I.: The influence of laser defocusing in selective laser melted in 625. Materials **14**(13), 3447 (2021)
9. Das, P., et al.: Constitutive modelling and damage prediction of AlSi10Mg alloy manufactured by SLM technology with emphasis on ratcheting in LCF regime. Int. J. Fatigue **181**, 108115 (2024)
10. Song, X., Zhai, W., Huang, R., Fu, J., Fu, M.W., Li, F.: Metal-based 3D-printed micro parts & structures. In: Encyclopedia of Materials: Metals and Alloys, pp. 448–461 (2021)
11. Sternadelova, K., Krupova, H., Hajnys, J., Mesicek, J., Mohyla, P.: Surface pretreatment of selectively laser melted 316L stainless steel substrate and its effect on the quality of the vitreous enamel coating. MM Sci. J. (2024)

12. Kong, D., et al.: Bio-functional and anti-corrosive 3D printing 316L stainless steel fabricated by selective laser melting. Mater. Des. **152**, 88–101 (2018)
13. Gao, B., Zhao, H., Peng, L., Sun, Z.: A review of research progress in selective laser melting (SLM). Micromachines **14**(1), 57 (2022)
14. Nancharaiah, T.: A review paper on metal 3D printing technology. In: Advances in Materials Processing and Manufacturing Applications: Proceedings of iCADMA 2020, pp. 251–259 (2021)
15. Song, X., et al.: Advances in additive manufacturing process simulation: residual stresses and distortion predictions in complex metallic components. Mater. Des. **193** (2020)
16. Oelschläger, F., Klein, D., Müller, S., Roth, J.: Molecular dynamics simulation of selective laser melting. In: International Conference on High Performance Computing in Science and Engineering, pp. 65–79 (2021)
17. Khan, S.A., Koç, M.: Numerical modelling and simulation for extrusion-based 3D concrete printing: the underlying physics, potential, and challenges. Results Mater. **16** (2022)
18. Singh, P., Singari, R.M., Mishra, R.S.: A review of study on modeling and simulation of additive manufacturing processes. Mater. Today Proc. **56**, 3594–3603 (2022)
19. Baiges, J., Chiumenti, M., Moreira, C.A., Cervera, M., Codina, R.: An adaptive finite element strategy for the numerical simulation of additive manufacturing processes. Addit. Manuf. **37** (2021)
20. Chen, S.G., Gao, H.J., Wu, Q., Gao, Z.H., Zhou, X.: Review on residual stresses in metal additive manufacturing: formation mechanisms, parameter dependencies, prediction and control approaches. J. Mater. Res. Technol. **17**, 2950–2974 (2022)
21. Xu, S., Liu, J., Ma, Y.: Residual stress constrained self-support topology optimization for metal additive manufacturing. Comput. Methods Appl. Mech. Eng. **389** (2022)
22. Renishaw. https://www.renishaw.com/resourcecentre/en/details/data-sheet-ss-316l-0407-powder-for-additive-manufacturing--90802. Accessed 28 Feb 2024

Induction Motor Drive Using Model Prediction Torque Control

Nguyen Quang Dung and Hau Huu Vo(✉)

Faculty of Electrical and Electronics Engineering, Ton Duc Thang University, Ho Chi Minh City, Vietnam
{nguyenquangdung,vohuuhau}@tdtu.edu.vn

Abstract. The paper addresses the prediction of stator flux and motor torque through the application of an induction machine model for the drive system. Initially, the model is presented, followed by its application to approximate the state vector at the $(k + 1)$th instant. A cost function is formulated based on the absolute errors in stator flux magnitude and motor torque. Subsequently, simulations are conducted, and comprehensive comparisons are drawn in conjunction with pulse-width modulation direct torque-controlled induction motor drive.

Keywords: Induction motor model · Induction motor drive · Predictive torque control

1 Introduction

Induction Motors (IMs) are widely utilized across various industrial sectors [1]. Among control methods for achieving high-performance of IM drives, Field-Oriented Control (FOC) [2], Direct Torque Control (DTC) [3], and Predictive Control [4–10] emerge prominently. The predictive control techniques enumerated for IMs are fundamentally rooted in IM models.

Predictive control methods have been extensively applied to IMs [4]. A novel approach employing flux prediction, as opposed to torque prediction, was developed [5]. Additionally, predictive control has been augmented with delta-sigma modulation to realize high-performance IM drives [6]. Furthermore, strategies for disturbance rejection have been integrated into predictive control for IMs [7]. Notably, Model Predictive Torque Control (PTC) has been successfully implemented in sensorless IM drives [8, 9], and it has even been extended to dual IM drive systems [10]. The subsequent section provides a detailed explanation of Model Predictive Torque Control (PTC) for IM drives, followed by simulation results in Sect. 3 and concluding remarks in the final section. In this paper, a PTC algorithm based on estimation is proposed. The proposed algorithm is further validated through simulation and compared with the traditional DTC algorithm.

This diagram employs a feedback control strategy that integrates model-based estimation and prediction to enhance the precision of motor speed regulation.

The inputs for Model-based Estimation and Prediction are derived from voltage reconstruction and the Clarke transformation block of currents. The Clarke transform

M. Pagac et al. (Eds.): SCFF 2024, LNNS 1062, pp. 89–96, 2024.
https://doi.org/10.1007/978-3-031-65656-9_9

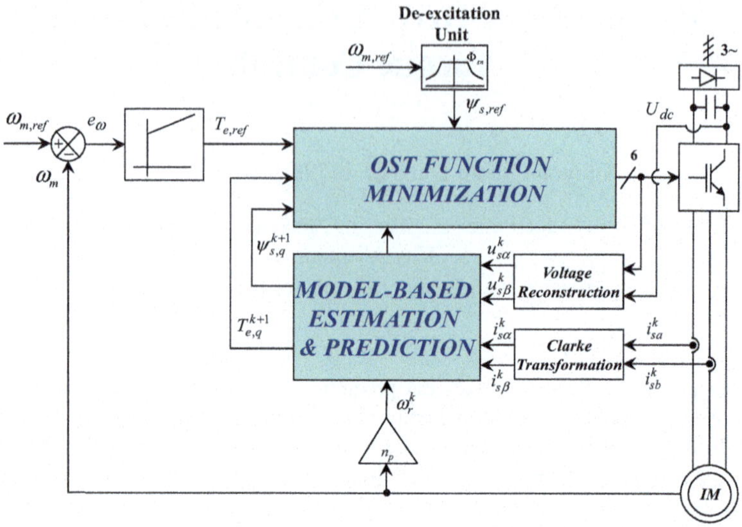

Fig. 1. Predictive torque controlled drive using induction motor model

uses three-phase stator currents i_{sa}, i_{sb}, and i_{sc} to calculate currents in the two-phase orthogonal stator axis: $i_{s\alpha}$ and $i_{s\beta}$. The voltage reconstruction block calculates voltages by considering the DC link voltage as a constant DC voltage and incorporating the states of the switching table.

2 Model Predictive Torque Control for IM Drive

Model prediction torque control for induction motor drive is shown in Fig. 1. The linear state-space model of an induction drive is described by Eq. (1):

$$\dot{X} = \mathbf{A}X + \boldsymbol{B}U \tag{1}$$

where:

$$X = \left[\, i_{s\alpha}\ i_{s\beta}\ \psi_{r\alpha}\ \psi_{r\beta} \,\right]^T \tag{2}$$

$$U = \left[\, u_{s\alpha}\ u_{s\beta} \,\right]^T \tag{3}$$

$$\mathbf{A} = \begin{bmatrix} c_1 & 0 & c_2 & c_3\omega_r \\ 0 & c_1 & -c_3\omega_r & c_2 \\ \frac{L_m R_r}{L_r} & 0 & -\frac{R_r}{L_r} & -\omega_r \\ 0 & \frac{L_m R_r}{L_r} & \omega_r & -\frac{R_r}{L_r} \end{bmatrix} \tag{4}$$

$$\boldsymbol{B} = \begin{bmatrix} \frac{1}{\sigma L_s}\, 0\, 0\, 0 \\ 0\, \frac{1}{\sigma L_s}\, 0\, 0 \end{bmatrix}^T \tag{5}$$

$$c_1 = -\frac{L_m^2 R_r + L_r^2 R_s}{\sigma L_s L_r^2},\ c_2 = \frac{L_m R_r}{\sigma L_s L_r^2},\ c_3 = \frac{L_m}{\sigma L_s L_r},\ \sigma = 1 - \frac{L_m^2}{L_s L_r} \tag{6}$$

The electrical torque generating by the IM can be determined by Eq. (7).

$$T_e = \frac{3}{2}\frac{L_m}{L_r} n_p \left(\psi_{r\alpha} i_{s\beta} - \psi_{r\beta} i_{s\alpha}\right) \tag{7}$$

Motor speed can be simply calculated by Eq. (11)

$$J_m \frac{d\omega_m}{dt} = T_e - T_L \tag{8}$$

Model-Based Estimation and Prediction (MBEP) block estimates components of rotor flux vector at *kth* instant according to Eqs. (9)–(12):

$$\psi_{s\alpha}^k = \psi_{s\alpha}^{k-1} + t_c\left(u_{s\alpha}^k - R_s i_{s\alpha}^k\right) \tag{9}$$

$$\psi_{s\beta}^k = \psi_{s\beta}^{k-1} + t_c\left(u_{s\beta}^k - R_s i_{s\beta}^k\right) \tag{10}$$

$$\psi_{r\alpha}^k = \frac{L_r}{L_m}\psi_{s\alpha}^k - \frac{L_s L_r - L_m^2}{L_m} i_{s\alpha}^k \tag{11}$$

$$\psi_{r\beta}^k = \frac{L_r}{L_m}\psi_{s\beta}^k - \frac{L_s L_r - L_m^2}{L_m} i_{s\beta}^k \tag{12}$$

where t_c is control period. It also predicts stator flux and motor torque at $(k+1)$th instant as follows [5]:

$$\boldsymbol{X}_{p,q}^{k+1} = t_c\left(\mathbf{A}\boldsymbol{X}^k + \boldsymbol{B}\boldsymbol{U}_q^{k+1}\right) + \boldsymbol{X}^k \tag{13}$$

$$\boldsymbol{X}_q^{k+1} = \frac{t_c}{2}\mathbf{A}\left(\boldsymbol{X}_{p,q}^{k+1} - \boldsymbol{X}^k\right) + \boldsymbol{X}_{p,q}^{k+1} \tag{14}$$

where:

$$\boldsymbol{U}_q^{k+1} = \left[u_{s\alpha,q}^{k+1}\ u_{s\beta,q}^{k+1}\right]^T \tag{15}$$

$$\psi_{s\alpha,q}^{k+1} = \frac{L_m}{L_r}\psi_{r\alpha,q}^{k+1} + \frac{L_s L_r - L_m^2}{L_r} i_{s\alpha,q}^{k+1} \tag{16}$$

$$\psi_{s\beta,q}^{k+1} = \frac{L_m}{L_r}\psi_{r\beta,q}^{k+1} + \frac{L_s L_r - L_m^2}{L_r} i_{s\beta,q}^{k+1} \tag{17}$$

$$\psi_{s,q}^{k+1} = \sqrt{\left(\psi_{s\alpha,q}^{k+1}\right)^2 + \left(\psi_{s\beta,q}^{k+1}\right)^2} \tag{18}$$

$$T_{e,q}^{k+1} = \frac{3}{2} n_p \left(\psi_{s\alpha,q}^{k+1} i_{s\beta,q}^{k+1} - i_{s\alpha,q}^{k+1}\psi_{s\beta,q}^{k+1}\right) \tag{19}$$

where $q = 1, \ldots, 7$ denotes for one of seven voltage vectors for two-level inverter voltage vector utilized for prediction process. Cost Function Minimization block seeks voltage vector that minimizes cost function:

$$CF = \left|T_e^{ref} - T_{e,q}^{k+1}\right| + \lambda_\psi \frac{T_n}{\Phi_{sn}} \left|\psi_s^{ref} - \psi_{s,q}^{k+1}\right| \tag{20}$$

where $\lambda_\psi > 0$ is chosen by trial-and-error method with many simulations. If it is too large, the motor torque is not guaranteed, while accuracy of control stator flux is not obtained if it is too small.

3 Simulations

Parameters of induction motor [3] are listed in Table 1. Simulations are implemented at $\omega_{m,ref} = \{250\,rpm, 50\,rpm\}$, $T_L = \{2\,N\,m, 5\,N\,m, 8\,N\,m\}$. . Responses of the PTC (Figs. 2, 3, 4, 5, 6, 7, 8 and 9) are shown and compared to the PWM-DTC [3].

Table 1. Parameters of induction motor

Symbol	Value	Symbol	Value
n_N	1420 rpm	J_m	0.0047 kg·m^2
T_N	14.8 N·m	R_s	3.179 Ω
Φ_{sn}	0.78 Wb	$L_s = L_r$	0.209 H
n_p	2	L_m	0.192 H
		R_r	2.118 Ω

Figures 2 and 3 indicate that the PTC achieves speed-control performance comparable to that of the DTC. However, the responses of the stator flux and stator currents for the PTC exhibit less smoothness compared to the DTC (see durations 0.6–1.0 s in Figs. 4, 5, 6, 7 and Figures 14–19); nevertheless, the stator flux and stator current magnitude is ensured. The response of motor torque for the PTC is rapid and robust, similar to that of the DTC (see Figs. 8 and 9).

To evaluate the performance of PTC relative to DTC, the Integral of Time multiplied by the Absolute Error (ITAE) of speed is employed.

$$ITAE = \int_0^1 t|e_\omega|dt \tag{21}$$

The simulation results are shown in Figs. 2, 3, 4, 5, 6, 7, 8 and 9.

The motor speeds at the reference speed of 250 rpm, when subjected to two different load torques introduced at 0.3 and 0.4 s, are illustrated in Fig. 2a. The speed responses of the proposed PTC and DTC algorithms are closely resemble each other.

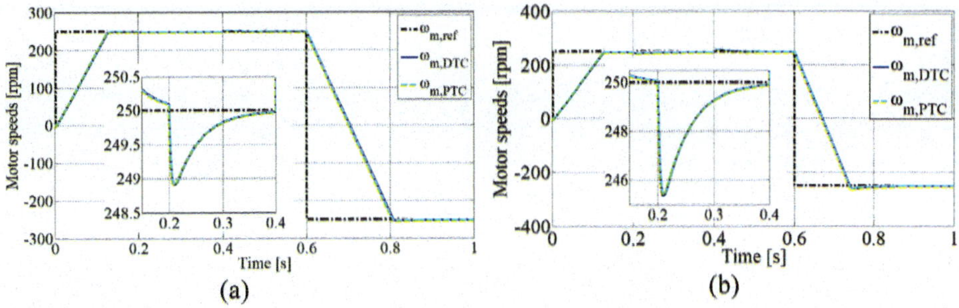

Fig. 2. Motor speeds at (a) $\omega_{m,ref}$ = 250 rpm, T_L = 2 N·m and at (b) $\omega_{m,ref}$ = 250 rpm, T_L = 5 N·m

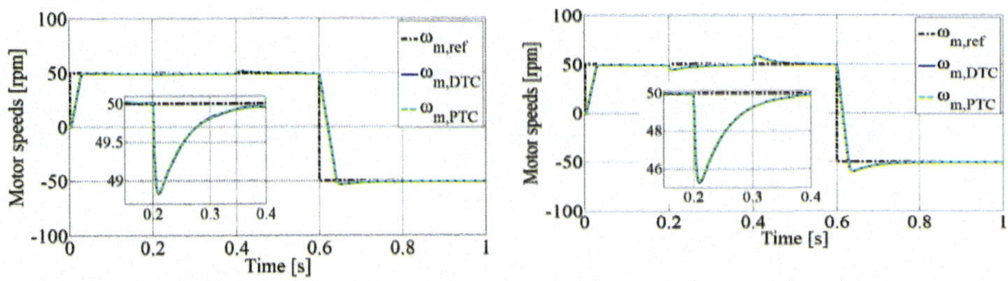

Fig. 3. Speeds at $\omega_{m,ref}$ = 50 rpm, T_L = 2 N·m. SPEEDS at $\omega_{m,ref}$ = 50 rpm, T_L = 8 N·m

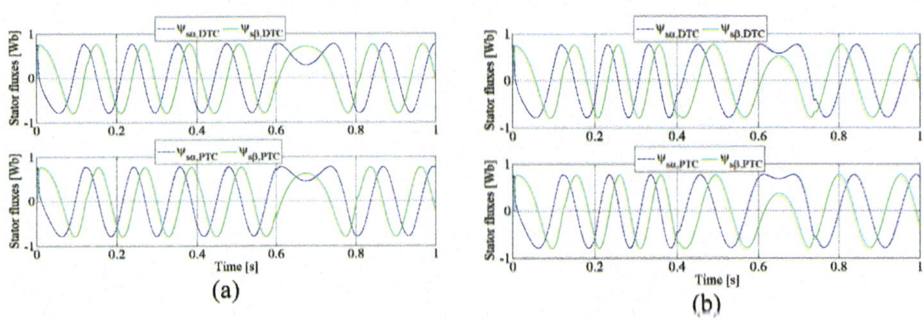

Fig. 4. Stator flux at $\omega_{m,ref}$ = 250 rpm, T_L = 2 N·m and $\omega_{m,ref}$ = 250 rpm, T_L = 8 N·m

Similarly, the stator flux of both PTC and DTC demonstrate similar characteristic patterns under different speeds and loads as shown in Figs. 4, 5, 6 and 7.

According to Figs. 6 and 7, the stator currents in the orthogonal stator axis domain at the two investigated different speeds of 250 rpm and 50 rpm for both methods are closely aligned.

Figures 8 and 9 illustrate that the torque responses of the induction motor model employing PTC and DTC algorithms show closely each other under various investigating speeds and loads.

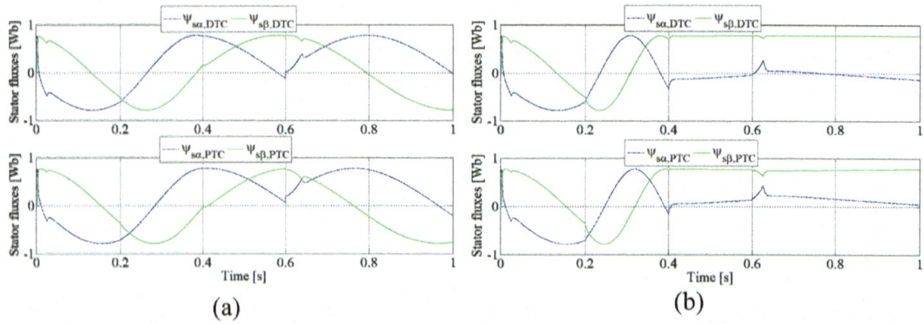

Fig. 5. Stator flux at $\omega_{m,ref} = 50$ rpm, $T_L = 2$ N·m and at $\omega_{m,ref} = 50$ rpm, $T_L = 8$ N·m

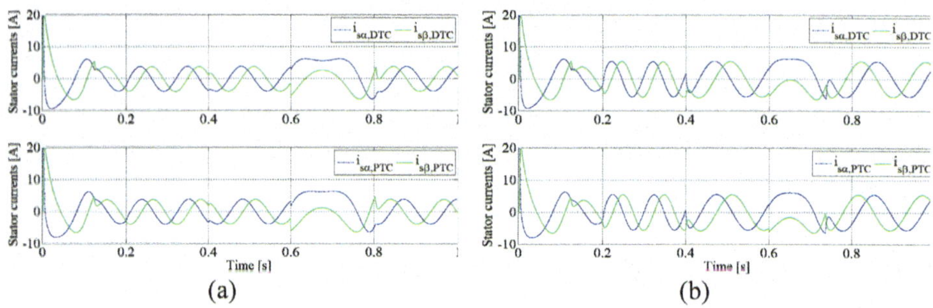

Fig. 6. Stator currents at $\omega_{m,ref} = 250$ rpm, $T_L = 2$ N·m and at $\omega_{m,ref} = 250$ rpm, $T_L = 8$ N·m

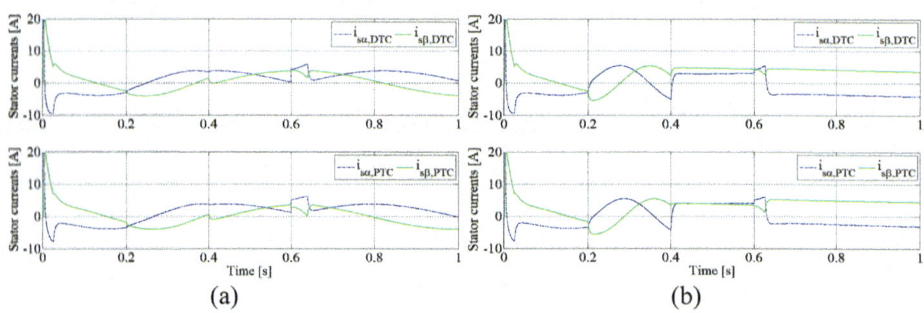

Fig. 7. Stator currents at $\omega_{m,ref} = 50$ rpm, $T_L = 2$ N·m and at $\omega_{m,ref} = 50$ rpm, $T_L = 8$ N·m

According to Figs. 2, 3, 4, 5, 6, 7, 8 and 9, the responses of the proposed method PTC are closed to the results employed by traditional DTC method. In closed zooms, the proposed PTC method presented a lightly better results compared to the DTC method.

ITAE values were calculated for both DTC and PTC methods and listed in Table 2.

As shown in Table 2, the ITAE performance of the PTC algorithm is slightly lower than that of the DTC algorithm at both investigated speeds (50 rpm and 250 rpm) across three different load torques (2 N·m, 5 N·m and 8 N·m).

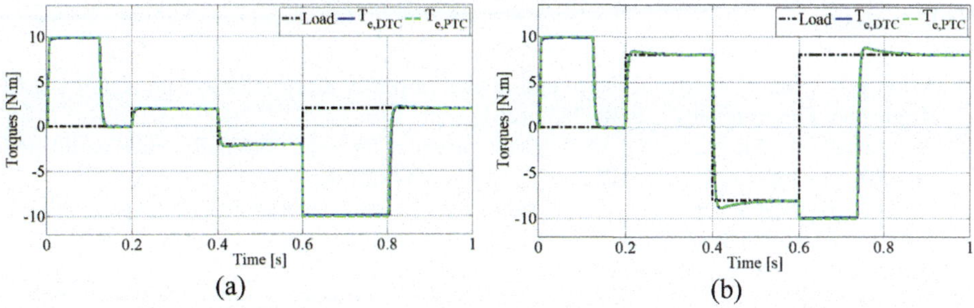

Fig. 8. Torques response at $\omega_{m,ref}$ = 250 rpm, T_L = 2 N·m and at $\omega_{m,ref}$ = 250 rpm, T_L = 8 N·m

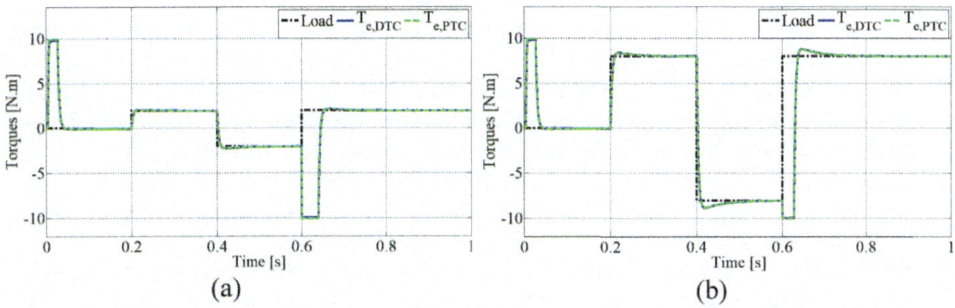

Fig. 9. Torques response at $\omega_{m,ref}$ = 50 rpm, T_L = 2 N·m and at $\omega_{m,ref}$ = 50 rpm, T_L = 8 N·m

Table 2. ITAE value [rpm·s^2]

$\omega_{m,ref}$ [rpm]	$T_L = 2$ N·m		$T_L = 5$ N·m		$T_L = 8$ N·m	
	DTC	PTC	DTC	PTC	DTC	PTC
250	35.56	35.24	28.24	28.05	23.64	23.51
50	1.461	1.451	1.435	1.431	1.503	1.502

4 Conclusions

The model predictive torque control using IM model was presented in the paper. The PTC is comparable to the DTC in terms of ITAE. However, selection of the coefficient λ_ψ in the cost function requires a large number of simulations. Predictive method based on flux [5] can be deployed to overcome the problem. Besides that, control performance of stator flux is ensured. Estimation techniques [3, 6, 8, 9] could be employed to obtain IM parameters and motor speed. The ITAE performance of the proposed method illustrates slightly better than the traditional DTC method.

References

1. Bose, B.K.: Global energy scenario and impact of power electronics in 21st century. IEEE Trans. Ind. Electron. **60**(7), 2638–2651 (2013)
2. Brandstetter, P., Skotnica, M.: ANN speed controller for induction motor drive with vector control. Int. Rev. Electr. Eng.-IREE **6**(7), 2947–2954 (2011)
3. Vo, H.H., Brandstetter, P., Tran, T.C., Dong, C.S.T.: An implementation of rotor speed observer for sensorless induction motor drive in case of machine parameter uncertainty. Adv. Electr. Electron. Eng. **16**(4), 426–434 (2018)
4. Rodriguez, J., Cortes, P.: Predictive Control of Power Converters and Electrical Drives. Wiley, New York (2012)
5. Zhang, Y., Yang, H., Xia, B.: Model-predictive control of induction motor drives: torque control versus flux control. IEEE Trans. Ind. Appl. **52**(5), 4050–4060 (2016)
6. Mir, T.N., Singh, B., Bhat, A.H.: Predictive speed estimation for encoderless control of an induction motor fed by delta sigma modulated matrix converter. IEEE Trans. Ind. Appl. **56**(6), 6477–6485 (2020)
7. Mousavi, M.S., Davari, S.A., Nekoukar, V., Garcia, C., Rodriguez, J.: A robust torque and flux prediction model by a modified disturbance rejection method for finite-set model-predictive control of induction motor. IEEE Trans. Power Electron. **36**(8), 9322–9333 (2021)
8. Boyar, A., Kabalci, E., Kabalci, Y.: MRAS based model predictive torque control of induction motor drive for electric vehicles. In: Proceedings of the 2022 4th Global Power, Energy and Communication Conference (GPECOM), Nevsehir, Turkey, pp. 51–56 (2022)
9. Nguyen, N.D., Nam, N.N., Yoon, C., Lee, Y.I.: Speed sensorless model predictive torque control of induction motors using a modified adaptive full-order observer. IEEE Trans. Ind. Electron. **69**(6), 6162–6172 (2022)
10. Choi, D., Lee, J.S., Lim, Y.S., Lee, K.B.: Priority-based model predictive control method for driving dual induction motors fed by five-leg inverter. IEEE Trans. Power Electron. **38**(1), 887–900 (2023)

Influence of Machining on Selected Tensile Properties of MEX-Produced PETG Materials

Natalia Kowalska(✉), Paweł Szczygieł, Michał Skrzyniarz, and Sławomir Błasiak

Kielce University of Technology, Kielce, Poland
nkowalska@tu.kielce.pl

Abstract. The subject of this study is to determine the effect of machining on selected properties of specimens made from glycol-doped polyethylene terephthalate (PETG) material. The specimens to be tested were prepared in accordance with ISO527 type 1BA with allowance and fixture holders. Two infills of the specimens were selected for testing: rectilinear and triangles. The specimens were printed using Material Extrusion Technology (MEX) in the X-direction (flat) of the printer. The specimens were then measured and machined. The results of the tensile test show that, in the case of filled triangles, the difference between the average tensile strengths of the reference and machined specimens is about 3.31%, while in the case of rectilinear-filled specimens the difference was about 0.03%.

Keywords: PETG · MEX · Tensile strength · Machining finishing

1 Introduction

The term Additive Technologies refers to a group of technologies characterised by the building of a model layer by layer. The first mention of additive technologies dates back to 1986, when stereolithography - SLA - was developed [1]. Subsequent years of research led to the development of further additive technologies such as Fused Deposition Modelling (FDM), selective laser sintering (SLS) or PolyJet Matrix (PJM) [2].

The unquestionable advantage of these technologies is the ability to use materials with different, often extreme, properties. The most commonly used materials include all polymers, some metal alloys, ceramics and even concrete. Therefore, additive technologies are used to produce prototypes, scientific aids, and also to repair, for example, replacing damaged machine parts [3]. Furthermore, the current level of additive technologies makes it possible to make functional components and even tools, e.g. drills, with applications to plastics [4]. Due to the widespread application of these technologies, it is necessary to conduct research into the strength of materials, e.g. static tensile testing [5] and research into rheological properties [6].

The current state of advancement of additive technologies has contributed to work on hybrid technology, which is a combination of two and more technologies. An example of hybrid technologies is the combination of additive technologies and CNC machining

M. Pagac et al. (Eds.): SCFF 2024, LNNS 1062, pp. 97–105, 2024.
https://doi.org/10.1007/978-3-031-65656-9_10

[7]. The use of machining aims to reduce roughness and improve dimensional and shape accuracy. In addition, the use of hybrid machining results in less waste material [8].

The article [9] describes an example of a finishing process using machining and specimens produced using FDM technology. An important aspect of the work cited was the determination of the depth of cut depending on the inclination of the model surface, in the case of additive manufactured models, there is a stepped effect in the case of sloping walls or those consisting of an arc. This effect has a significant impact on the amount of material to be collected. Analysis of the surface enabled the authors to use the developed method to provide a variable depth of cut in the case of a design based on a Pelton wheel. In other cases, in order to remove the stair-step effect, a cutting tool equipped with a spherical-tipped heating plate has been developed to enable the shearing of the tops of the 'stair-steps', as well as the fusing by means of high temperature of the waste fragments in the voids between the layers [10]. With FDM technology, the average surface roughness can be improved by up to 70% by machining [11].

Although there has been extensive research into the change in roughness of specimens under investigation, few articles address the subject of the change in shape deviations or the change in mechanical properties after finishing. These issues are critical for the operation of models produced with additive technologies, as finishing is unavoidable, especially to improve the roughness of the working surfaces of the model. Complementary research into the effects of machining on the mechanical properties of different materials is therefore crucial. Therefore, the aim of this article is to determine the effect of machining with a burr on selected mechanical properties.

2 Experimental Part

2.1 Materials

The material used in this study is Polyethylene terephthalate glycol PETG (Bambu Lab, Shenzhen, China), which is a derivative of PET (used, for example, in bottle production) enriched with the addition of glycol. As a result, the material is characterised by its high resistance to mechanical as well as chemical and atmospheric influences. Selected mechanical properties developed by the manufacturer [12] are shown in Table 1.

Table 1. Mechanical properties of PETG [12]

Mechanical properties	Value	Unit	Test standard
Young's modulus (X-Y direction)	1460 ± 190	MPa	ISO 527, GB/T 1040
Young's modulus (Z direction)	1120 ± 130	MPa	ISO 527, GB/T 1040
Tensile strength (X-Y direction)	32 ± 4	MPa	ISO 527, GB/T 1040
Tensile strength (Z direction)	28 ± 4	MPa	ISO 527, GB/T 1040
(X-Y direction)	11.2 ± 0.8	%	ISO 527, GB/T 1040
Breaking elongation rate (Z direction)	5.7 ± 0.6	%	ISO 527, GB/T 1040

The manufacturing parameters adopted by the manufacturer were nozzle temperature of 255 °C, bed temperature of 70 °C, print speed of 150 mm/s and infill density of 100%, respectively.

2.2 Methods

A Bambu Lab P1P printer (Bambu Lab, Shenzhen, China) belonging to the MEX layered material extrusion technology group (ISO 52900) was used in the study. An example of how the MEX technology operates, based on the above-mentioned printer, is shown in Fig. 1. The Bambu Lab P1P printer has a single print head, a nozzle with a diameter of 0.4 mm and a heated bed. The material in the form of a filament is fed into the extruder, where the material is heated to a set temperature. As a result of the high temperature, it is possible to extrude the material through the nozzle and apply a single line on the platform. After the model has been fully sectioned, the table is lowered (in the case of a kinematic system in which the table moves in the Z axis and the head in the X and Y axes).

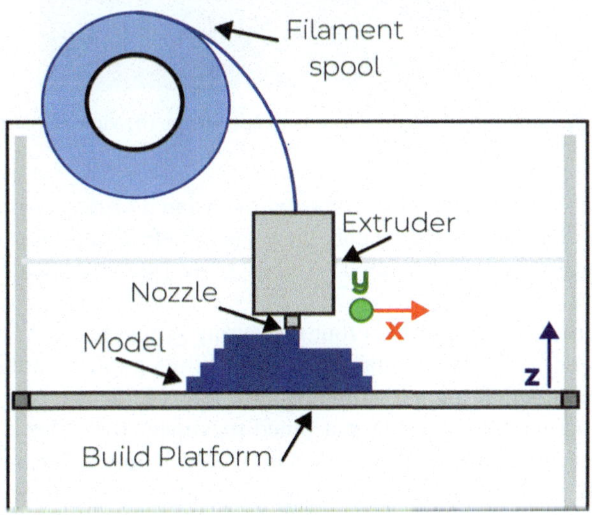

Fig. 1. MEX technology

The specimens for testing were prepared according to ISO 527 1BA, due to the need to fix the specimen in the holder, additional blades with holes were designed (Fig. 2a) and a 0.2 mm material allowance was added. The prepared specimens were printed in the position corresponding to the X axis (Fig. 2b).

The specimens were printed with the following parameters: nozzle temperature 255 °C, bed temperature 70 °C, filament application speed 150 mm/s, infill density 95% and layer thickness 0.2 mm. In the study, an infill density different from that specified by the manufacturer was used (Table 1), as it is not possible to print with 100% density for some of the available infill patterns (in the case of the study - triangles). Two infills were also used: rectilinear and triangles.

a)

b)

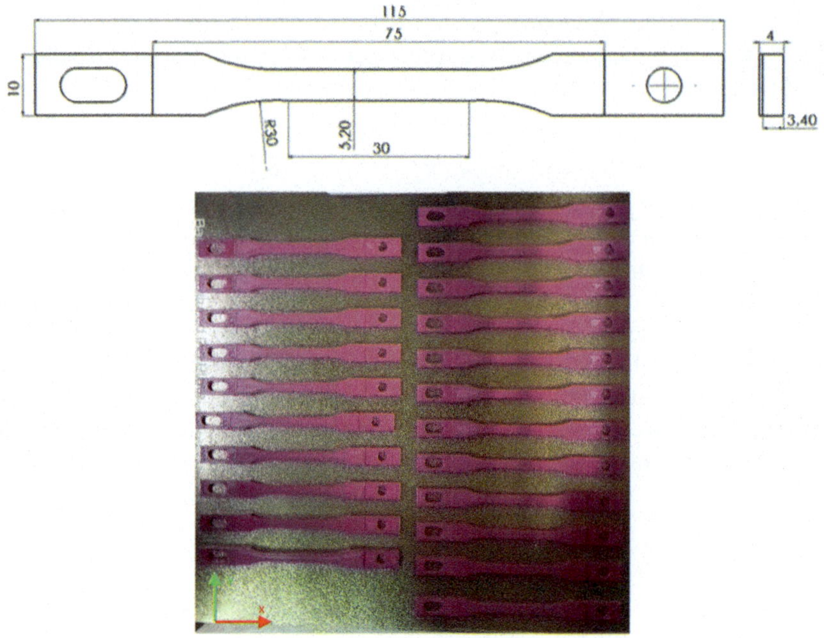

Fig. 2. Specimens: (a) dimensions of specimens (b) view of platform

All specimens were measured in cross-section using a Mitutoyo micrometer (Mitutoyo, Kanagawa, Japan). Measurements were taken before and after machining. The specimens were also weighed using JS-100xV scale (Jennings Scale, China) before machining.

The specimens were divided into four measuring series: series A, being a group of unfinished reference specimens with a rectilinear filling pattern, series B, a reference group of specimens with a triangle filling pattern, series C, machined specimens (rectilinear pattern) and the final series D, machined specimens (triangles pattern). Material was only removed from two sides corresponding to the nominal dimension (in the standard) $a = 5$ mm and $b = 4$ mm. A Hermle B 300 machining centre (HERMLE AG, Gosheim, Germany) equipped with a three-edge cutter and ZrCN Ultra-N coating by Atorn, shown in Fig. 3 (HAHN + KOLB, Ludwigsburg, Germany), was used for the machining. To ensure that the specimens were equally positioned, a handle was fabricated (Fig. 3) to which the specimens were mounted. This made it possible to process both walls indicated in one process.

Machining was carried out according to the following parameters: cutting speed 80 m/min, feed rate per blade 0.02 mm/blade in addition to the use of coolant to eliminate material deformation and excessive chip length.

A static tensile test was then performed on an Inspekt Mini machine from LabMaster (Hegewald and Peschke, Nossen, Germany). A tensile speed of 2 mm/min was used for each series of specimens.

(a) (b)

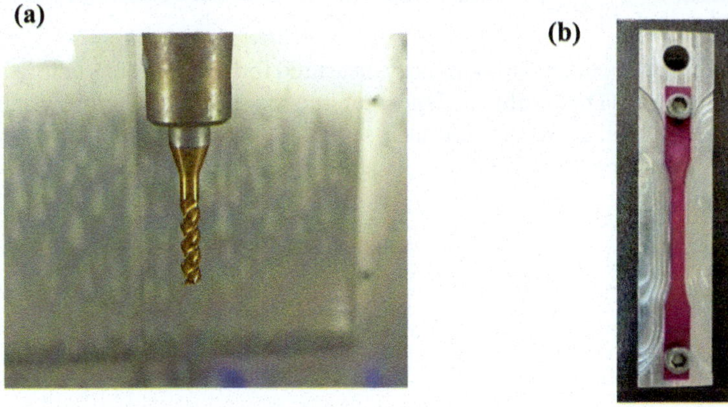

Fig. 3. Machining: a) tool cutter b) holder

3 Results and Discussion

3.1 Cross-Section Measurement

All specimens were measured at various points within the specimen measurement base before machining. Table 2 shows the average values of dimensions *a* and *b*. In the case of the PETG material, shrinkage of the material was observed for dimension *a* - with an added 0.2 mm allowance. In the case of the height of the specimen, a shrinkage of the material can be observed because no allowance was applied for dimension *b*.

Table 2. Dimensions of specimens

	Unit	Series A	Series B	Series C	Series D
$\bar{a}$	mm	5.167	5.155	5.189	5.159
$\bar{b}$	mm	4.088	4.163	4.079	4.154

After machining, cross-sectional measurements were again taken for the C and D series only. The results are shown in Table 3. The use of machining resulted in dimensions close to the nominal dimensions. Proper dimensional matching is particularly important for assemblies with a specific fit tolerance.

Table 3. Dimensions of specimens after machining

	Unit	Series C	Series D
$\bar{a}$	mm	5.016	5.007
$\bar{b}$	mm	4.079	4.154

3.2 Samples Weight

The weight of the specimens was similarly measured for all specimens before machining and after machining. The results in Table 4 refer to the specimens before machining and show the average weight results for each series. The study shows that the specimens made with filled triangles have a lower weight despite the identical degree of filling.

Table 4. Weight of specimens

	Unit	Series A	Series B	Series C	Series D
g̅	g	3.68	3.62	3.68	3.61

After treatment, samples from series C and D were reweighed and the results are shown in Table 5.

Table 5. Weight of specimens

	Unit	Series C	Series D
g̅	g	3.62	3.59

When comparing the results obtained, a slight decrease in weight was observed, corresponding to the material being cut.

3.3 Tensile Strength

The results of the static tensile test are shown in Fig. 4 for series A and B, the average tensile strength values were 39.39 MPa and 29.28 MPa, respectively. Similar values were obtained in the article [13] in which the tensile strength was 48.3 MPa for 100% filled specimens. The difference in the obtained tensile strength between series A and B is about 25.5%, a confirmation of the influence of the chosen filling on the tensile properties. According to research carried out in the article [14], the highest tensile strength values were obtained for grid (36.34 MPa), honeycomb (34.26 MPa) and rectilinear (33.35 MPa) fillings. On the other hand, the lowest values for the same specimens were obtained for the fillings: concentric (13.54 MPa), triangles (18.73 MPa) and rectangular (19.72 MPa). The results described were obtained with the following printing parameters: 80% infill destiny and 0.1 mm layer thickness. The difference between series A and B is therefore only due to the infill pattern used.

The difference between the A and C series specimens is 0.03%, while the difference between the B and D series specimens is 3.31%. It can be concluded that, in the case of PETG material, the application of machining does not significantly affect the strength properties. It is important to note that in the case of the triangles infill pattern, the difference between the series (B and D) is much greater than for the rectilinear infill series.

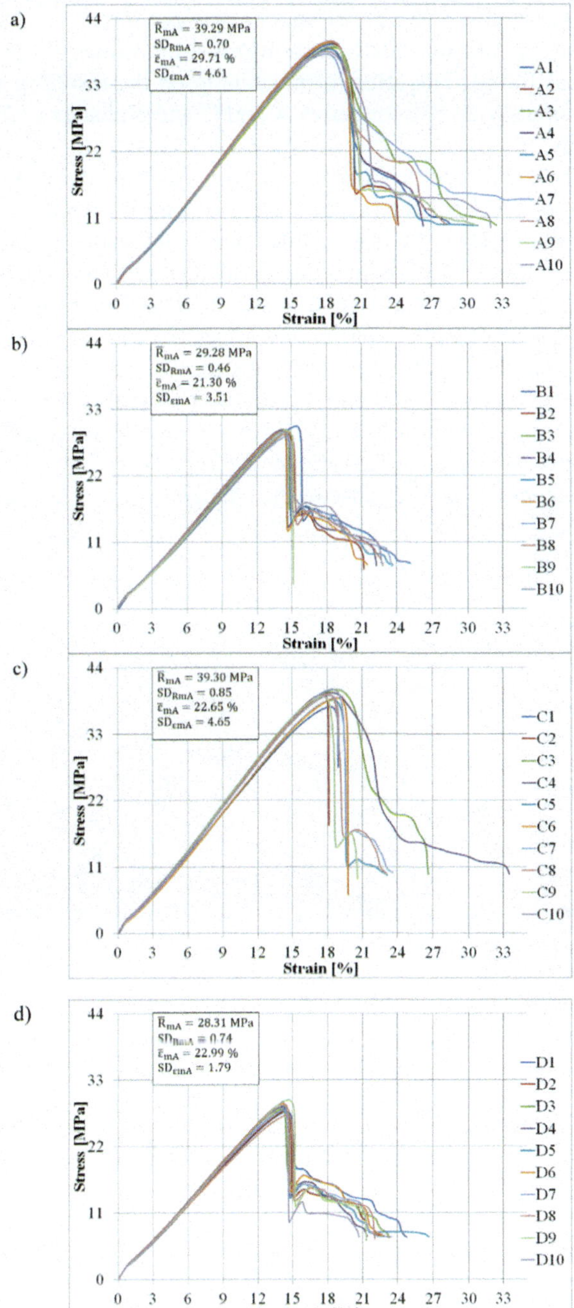

Fig. 4. Tensile strength: a) series A, b) series B, c) series C, d) series D

In the case of maximum strain, the difference between series A and B is about 28.3%, and according to the results presented, much higher strain values were observed in series A - with rectilinear filling. The effect of machining on the resulting strain values is a change of approximately 23.8% for series A and C and a change of 7.9% for series B and D, respectively.

Analysis of the tensile curve indicates uneven crack of the different parts of the specimen. During the test, the internal structures at the neck opening gradually broke, as can be seen in the graphs (Fig. 4). The recorded phenomenon demonstrates the plasticity of the PETG material. It is worth noting that the B-series specimen B9 and the C-series specimens C1, C2, C6 and C10 broke much faster and no neck formation was observed during the testing. This phenomenon mainly affects rectilinear filled specimens after machining. Although, the tensile strength for the mentioned specimens did not change drastically the specimens cracked without previously forming a neck. The exact reason for the change in the behaviour of the specimens is unknown, as the other specimens underwent similar deformation to the reference group. Therefore, the use of machining cannot be clearly identified as the cause of the change.

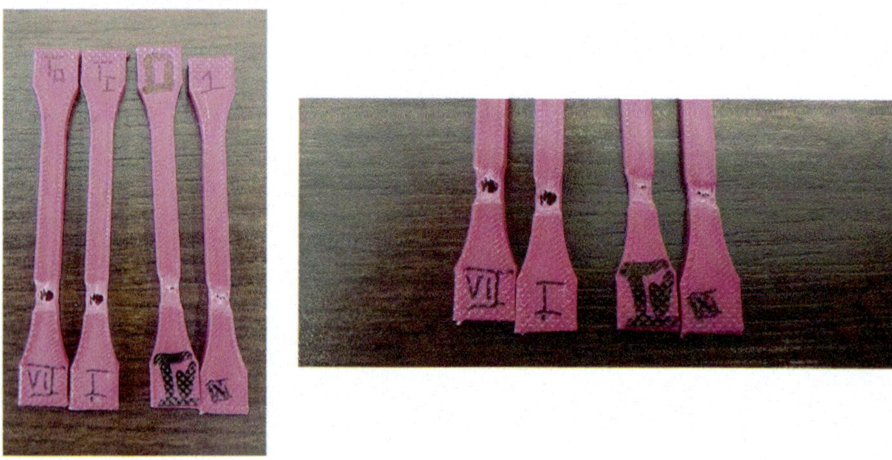

Fig. 5. Specimens after breaking

The static tensile test continued until a 75% load drop was reached. The specimens shown in Fig. 5 did not rupture, but neck formation could be observed during the test.

4 Conclusions

Analysis of the results allows to draw the following conclusions:

- The difference between the reference series (series A - 39.39 MPa, series B - 29.28 MPa) is due to the use of different infills. As with the cited articles, the rectilinear infill pattern achieved a higher tensile strength.

- Machining has no significant effect on the strength properties of specimens made of PETG material. The largest difference between the reference specimens was obtained in the case of infill triangles. However, the difference remains minimal at only 3.31%.
- Due to the addition of glycol, PETG specimens behave plastically; in the case of the tested specimens, the neck formed and the infill corresponding to the tested filling broke first.
- Some specimens broke completely without neck formation. Unfortunately, no direct cause for the phenomenon can be identified. It is suggested to carry out more tests with a detailed analysis of the crack propagation of the specimens after machining.
- The results of the tests indicate a much greater influence of machining on the strain of the specimens.

References

1. Fink, J.K.: 3D Industrial Printing with Polymers. Wiley (2018)
2. Kabir, S.M.F., Mathur, K., Seyam, A.-F.M.: A critical review on 3D printed continuous fiber-reinforced composites: history, mechanism, materials and properties. Compos. Struct. **232**, 111476 (2020)
3. Kim, H., Cha, M., Kim, B.C., Lee, I., Mun, D.: Maintenance framework for repairing partially damaged parts using 3D printing. Int. J. Precis. Eng. Manuf. **20**, 1451–1464 (2019)
4. Nowakowski, L., Skrzyniarz, M., Blasiak, S., Rolek, J., Vasileva, D., Avramova, T.: Analyzing the potential of drill bits 3D printed using the direct metal laser melting (DMLM) technology to drill holes in polyamide 6 (PA6). Materials **16**(8), 1–14 (2023)
5. Bochnia, J., et al.: Selected mechanical and rheological properties of medical resin MED610 in PolyJet matrix three-dimensional printing technology in quality aspects. 3D Print. Addit. Manuf. **11**, 299–313 (2024)
6. Szot, W.: Rheological analysis of 3D printed elements of acrylonitrile butadiene and styrene material using multiparameter ideal body models. 3D Print. Addit. Manuf. (2023)
7. Kale, A., Lakshman Kumar, A., Murali Krishna Kumar, M., Prakasah, M.: Optimization of hybrid manufacturing process parameters by using FDM in CNC machine. IOP Conf. Ser. Mater. Sci. Eng. **402**, 12088 (2018)
8. Dvorak, J., et al.: A machining digital twin for hybrid manufacturing. Manuf. Lett. **33**, 786–793 (2022)
9. Boschetto, A., Bottini, L., Veniali, F.: Finishing of fused deposition modeling parts by CNC machining. Robot. Comput.-Integr. Manuf. **41**, 92–101 (2016)
10. Taufik, M., Jain, P.K.: CNC-assisted selective melting for improved surface finish of FDM parts. Virtual Phys. Prototyp. **11**, 319–341 (2016)
11. Parandoush, P., Deines, T., Lin, D., Zhang, H., Ye, C.: Mechanical finishing of 3D printed continuous carbon fiber reinforced polymer composites via CNC machining. In: Volume 1: Additive Manufacturing; Manufacturing Equipment and Systems; Bio and Sustainable Manufacturing. American Society of Mechanical Engineers (2019)
12. PETG Product Information Data Sheet
13. Kumaresan, R., Samykano, M., Kadirgama, K., Pandey, A.K., Rahman, M.M.: Effects of printing parameters on the mechanical characteristics and mathematical modeling of FDM-printed PETG. Int. J. Adv. Manuf. Technol. **128**, 3471–3489 (2023)
14. Srinivasan, R., Nirmal Kumar, K., Jenish Ibrahim, A., Anandu, K.V., Gurudhevan, R.: Impact of fused deposition process parameter (infill pattern) on the strength of PETG part. Mater. Today Proc. **27**(2), 1801–1805 (2020)

Inspection Planning Based on DOProC Method for Steel Structures Exposed to Fatigue

Martin Krejsa(✉), Jiří Brožovský, and Petr Lehner

Faculty of Civil Engineering, Department of Structural Mechanics, VSB–Technical University of Ostrava, Ludvika Podeste 1875/17, 708 00 Ostrava-Poruba, Czech Republic
martin.krejsa@vsb.cz

Abstract. The paper describes the fundamentals of the methodology for probabilistic assessment of cyclically loaded steel building structures with attention to fatigue. The phenomenon of the formation and propagation of fatigue cracks is stochastically modeled to design a control system during the operation of the structure. A real composite steel-concrete highway bridge was chosen as a demonstration example of the proposed methodology, the reliability of which was assessed in a detail susceptible to fatigue damage—at the point where the cross member connects to the stringer. The calculation procedure is based on the fully probabilistic DOProC method and has also been extended to investigate load-bearing elements made of HSS steels.

Keywords: Probability of failure · fatigue · steel structure · HSS · DOProC

1 Introduction

The reliability of building structures [1] that are subjected to repeated loading is affected by degradation processes caused by fatigue of the material. There are several feasible approaches to address this phenomenon. One of the approaches is use of the linear elastic fracture mechanics (LEFM) [2]. However, there still are some challenges. Importantly, input data are random and there are uncertainties. Therefore, approaches based on reliability theory using stochastic methods should be considered. For example, probabilistic prediction of fatigue damage can be operationally solved with a detailed analysis of the speed of propagation of fatigue phenomena in load-bearing elements susceptible to this type of damage [3].

In this paper, the fully probabilistic DOProC method [4] is used to calculate the failure probability. The resulting reliability indicator is then compared with the design value, which is defined by valid standard regulations. The DOProC is a method which was originally developed with the goal to omit the use of any simulation techniques based on crude Monte Carlo method principles. The DOProC stochastic calculation method can be used similarly to other probabilistic approaches for solving problems where the input variables are affected by randomness. A significant advantage of this procedure is the high efficiency for appropriately solved engineering problems and the achieved estimation of the resulting degree of reliability with the maximum possible achievable

M. Pagac et al. (Eds.): SCFF 2024, LNNS 1062, pp. 106–115, 2024.
https://doi.org/10.1007/978-3-031-65656-9_11

accuracy. There are no imprecisions caused by working with random generators. There are only rounding errors and imprecisions caused by the discretization of input and output quantities.

The aforementioned computational stochastic method has been successfully applied in solving many technical problems in the construction industry. Currently, the advantage of its use in the issue focused on assessing the reliability of load-bearing structures that are exposed to loads acting in cycles is becoming apparent. The resulting reliability of the analyzed structure is also related to the occurrence of fatigue phenomena, which are also taken into account in the described calculation model.

There are procedures for a possible assessment of fatigue-resistant buildings, which pay particular attention to edge and surface cracks. This approach can be used as a basis for the design of a system for inspection of cyclical load constructions, e.g. [5–7]. The goal is to assess, based on numerical modeling, the real behavior of the element with regard to fatigue damage [8]. If it is found that the expected damage to the supporting structure did not occur, the subsequent behavior of the analyzed detail can be predicted by effective procedures based on conditional probability [9, 10].

To use the full potential of the DOProC a special software FCProbCalc was developed.

2 The Essence of Stochastic Modeling

Fundamentals of proposed DOProC calculation approach were summarized in [4]. Unlike stochastic methods based on simulations, where the random aspect of the input variables is solved by using pseudo-random number generators, in DOProC the calculation procedure is determined by an algorithm using a specially developed procedure for numerical integration of the safety function. In this aspect the DOProC is not influenced by possible deficiencies of random number generation and the results do not depend on the number of simulations. This approach comes at a cost because the number of computations in the DOProC depends on other factors. At the same time, the randomness and uncertainty of the input data is expressed in the same way as in other similarly focused procedures using histograms, which can be defined by the results of physical experiments or on data obtained by observation and monitoring, which can take over a longer period of structural operation [11].

In the DOProC method, the number of calculations needed to compute the resulting probability of failure, p_f, is determined by the number of inputs, taking into account their random nature. It is worth noting that if there are too many random variables, the calculation can be quite time-consuming, even with the use of high-performance computing on supercomputers. An important parameter for a satisfactory calculation time is also the number of subintervals in the used input histograms, which can ultimately affect the achieved accuracy of the result. However, there are approaches that can be used to optimize calculation procedures in order to reduce the computational effort without any noticeable impact on the quality of the calculation results.

3 Probabilistic Analysis of Solved Fatigue Task

The phenomenon of the initiation and subsequent propagation of fatigue cracks in the details of steel structures represents a long-term engineering problem that has been solved for many decades. It is also important to note that this process is greatly influenced by the method and technology of manufacturing steel structures, which necessarily involve the creation of details prone to crack initiation. During the process of sizing and subsequent production of structural details, it is very unusual to pay attention to random occurrences of initiation cracks from the surface (see Fig. 1).

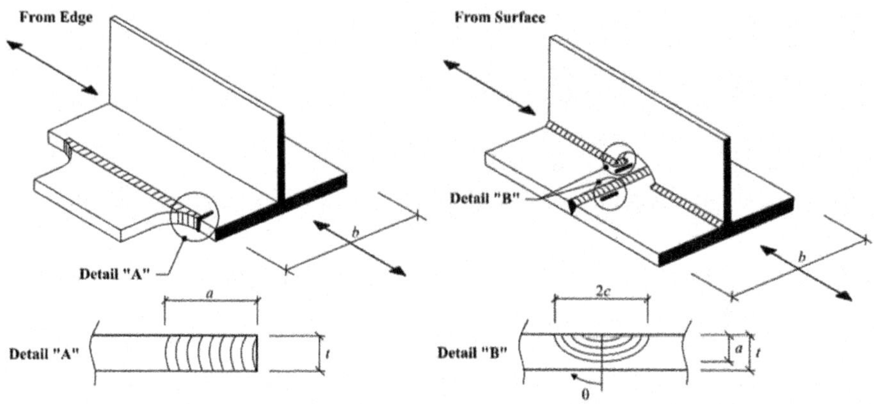

Fig. 1. Examples of typical fatigue damage in a steel structure detail.

There are three sizes of crack which are important for the characterization of fatigue crack propagation: the initiation, the detectable and the final acceptable size (dimension just before failure due to material fatigue). An important parameter for fatigue damage is the number of load cycles of the structure causing stress fluctuations [12, 13].

The described stochastic analysis of fatigue damage to a supporting structure detail is carried out in two intersecting directions. The first is a probabilistic solution to the propagation of the emerging fatigue crack. The second level addresses the potential for uncertainty in the determination of histograms used in the numerical modeling. When analyzing the crack propagation speed, it is common to describe the fatigue crack dimension by only one dimension: length a. This is because the LEFM is often used to describe the propagation of the crack, and this way derived by Paris-Erdogan, e.g. [14, 15], will also be used here:

$$\frac{da}{dN} = C \cdot (\Delta K)^m. \tag{1}$$

In (1) constants C, m describe fatigue behavior of the used material, a is the mentioned crack length and quantity N represents the number of load impulses.

The initial design should be carried out considering the effects of extreme loads that result from the ultimate limit state. When this requirement is met, it is then possible to perform an assessment of fatigue resistance. In the case of stochastic reliability assessment, it is customary to first define a safety margin function Z, the analysis of which leads to the determination of the resulting failure probability:

$$Z_{(R,E)} = R - E. \tag{2}$$

The safety margin (2) contains two fundamental random variables of the reliability theory: the resistance of the structure R and the effect of the extreme load E.

Based on the analysis of equation (2), three situations can be expected for a load-bearing element that is affected by an operating load condition (schematically expressed in Fig. 2):

a) **state of safe service life** - the structural reliability is not threatened by fatigue phenomena,
b) **acceptable failure rate state** - the load-bearing structural element is weakened by fatigue phenomena, which leads to a reduction in its resistance (bearing capacity),
c) **acceptable failure rate state** - fatigue phenomena in the supporting element are manifested by a change in stress a subsequent decrease of reliability.

During the numerical modeling of the course of fatigue degradation of the load-bearing element, the stress is determined, while the resistance of the structure R is affected by the constant value of the stress at the yield point f_y when an acceptable fatigue crack occurs. The described situation c) better expresses the actual behavior in the structure, which is also the reason why it is preferred in the described methodology over the situation b), as it describes, among other things, the non-linear growth of stress in the degraded element.

In the following description of the mathematical interpretation, it is desirable to modify relation (1). The stress intensity coefficient ΔK affects the method of stressing the supporting element (bending, tension), the size of the described structural part and, of course, the way of crack propagation (from the edge or surface), which significantly affects the state of tension at the crack tip. If the normal stress and the range of their changes $\Delta\sigma$ are constant in the analyzed flange, the following relationship can be used to express the value of the stress intensity coefficient ΔK:

$$\Delta K = \Delta\sigma \cdot \sqrt{\pi \cdot a} \cdot F_{(a)}. \tag{3}$$

The formula contains a calibration function $F_{(a)}$ that describes the speed and mode of propagation of the investigated fatigue crack and can be obtained experimentally, e.g. [16].

It is also necessary to determine a relationship that expresses the dependence of crack growth from length a_1 to length a_2 on the number of load impulses N_1 at the beginning of the analyzed period and N_2 at its end. This can be done based on Eqs. (1) and (3):

$$\int_{a_1}^{a_2} \frac{\mathrm{d}a}{\left(\sqrt{\pi \cdot a} \cdot F_{(a)}\right)^m} = \int_{N_1}^{N_2} C \cdot (\Delta\sigma)^m \mathrm{d}N. \tag{4}$$

In relation (4), it is on the left the resistance of the structure R. If the lower integration limit is a_0 (initial size of fatigue crack) and upper integration limit is intended as a_{ac}

(acceptable size of fatigue crack), the structural resistance $R_{(a_{ac})}$ can be determined. The resistance of the structural element $R_{(a_d)}$ is calculated alike with the upper integration limit a_d (detectable size of fatigue crack).

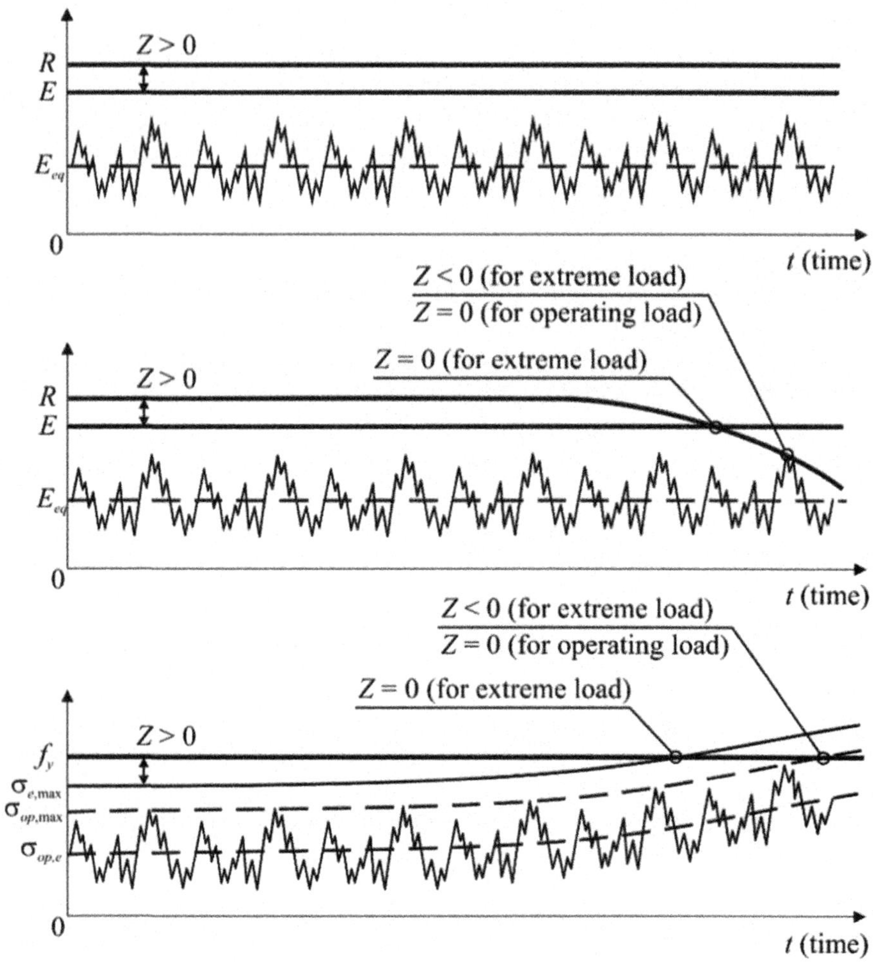

Fig. 2. Safety margin Z related to designed and operating effects of the loading.

Equation (4) can also be used to determine the cumulative load effect E. The part on the right corresponds to the randomly variable extreme effects of the load on the investigated structure:

$$E_{(N)} = \int_{N_0}^{N} C \cdot (\Delta\sigma)^m \mathrm{d}N = C \cdot (\Delta\sigma)^m \cdot (N - N_0). \tag{5}$$

The value of N corresponds to the total number of stress peak oscillations $\Delta\sigma$ that cause the crack length to change from a_0 to a_{ac} (or a_d analogously). The N_0 parameter corresponds to the number of load pulses with the stress range $\Delta\sigma$ at the time when the fatigue crack is initiated (it is usually considered equal to zero).

Stochastic fatigue damage modeling of the analyzed element is based on relation (2), which can be supplemented by quantities derived in (4) and (5):

$$Z_{(\mathbf{X})} = R_{(a_{ac})} - E_{(N)}. \tag{6}$$

The vector of random variables **X** expresses all random factors in the solved task, such as the physical characteristics of the material, the dimensions of the structural detail, the variable load and the dimensions of the fatigue crack itself.

A calculation based on a numerical analysis of the defined safety margin function (6) then allows to obtain an estimate of the probability of failure p_f:

$$p_f = P\left(Z_{(\mathbf{X})} < 0\right) = P\left(R_{(a_{ac})} < E_{(N)}\right). \tag{7}$$

In this way, the failure probability p_f can be obtained for any period of operation of the investigated structure with a specified time step (usually 1 year). The results can then be used, among other things, to determine the remaining service life of the structure being assessed.

The above-mentioned approach is demonstrated in a case of the tension flange of the stringer at the point of connection of the cross member in the steel part of the steel-concrete bridge. The studied bridge is built on the highway near Žilina in Slovakia. In order to compare the effect of formation of two types of fatigue cracks (from the edge and from the surface, see Fig. 1), it is necessary to define appropriate calibration functions $F_{(a)}$ in (4). The useful calibration functions for short edge cracks are derived based on tests in [16].

For the crack propagating from the edge of the analyzed detail of the steel part of the bridge, the calibration function for bending (applies to the calculation of the stress in the weakened section) was used:

$$F_{(a)} = 1.122 - 1.4 \cdot \frac{a}{b} + 7.33 \cdot \left(\frac{a}{b}\right)^2 - 13.08 \cdot \left(\frac{a}{b}\right)^3 + 14.0 \cdot \left(\frac{a}{b}\right)^4. \tag{8}$$

The formula contains 2 parameters: length of the fatigue crack a and flange width b. Using this calibration function leads to more conservative results compared to the usage obtained in the laboratory [16].

The description of crack propagation from the surface is somewhat more complicated. The area of weakening of the analyzed tensile flange is influenced by the shape of the crack propagating from the surface, whose shape is, e.g., according to [17] limited by a curve with the shape of half an ellipse (the axes of the ellipse are described using two parameters: a - the depth of the crack and c - half the width of the crack, see Fig. 1) and depending on the number of load cycles N it can be described as:

$$A_{cr(N)} = \frac{1}{2} \cdot \pi \cdot a_{(N)} \cdot c_{(N)}. \tag{9}$$

When analyzing the initialization and behavior of a fatigue crack incl. Crack area calculation (9) when propagating from the surface, it is usually necessary to determine both dimensions of the crack (as opposed to a crack from the edge). The increasing depth of the crack a, depending on the increasing value of its width c, changes by an increment:

$$\Delta a = \left[\frac{1}{1.1 + 0.35 \cdot \left(\frac{a}{t}\right)^2 \cdot \sqrt{\frac{a}{c}}} \right]^m \cdot \Delta c. \tag{10}$$

As it turned out during the investigation, both dimensions a and c in (10) of the crack from the surface are dependent on each other, so the growth of this type of crack can then be expressed by one formula:

$$c = 0.3227 \cdot \frac{a^2}{t} + 1.0202 \cdot a + 0.00699 \cdot t. \tag{11}$$

When analyzing the fatigue damage of the supporting element, it is not entirely possible to determine whether the fatigue crack has already initiated incl. its actual size. Furthermore, due to the possibility of other inaccuracies and their influence on the calculation of crack propagation, it would be beneficial to conduct a sophisticated control in situ to determine the current state at a given time t of operation of the load-bearing element from the point of view of the formation and detection of the fatigue crack itself. This would help to determine the acceptable crack size, which in turn would influence opportune moment for the required inspection. Of course, fatigue cracks may not be detected during this inspection. Even in such a case, the result of the inspection can serve as an important basis for estimating the probability of the occurrence of a fatigue crack in the future.

Such probabilistic analysis is then based on the conditional probability of the occurrence of a crack at time $T > t$ and on the definition of three random events (12) to (14) during the lifetime of the structure, which are closely related to the propagation of a fatigue crack:

- $U_{(t)}$ **phenomenon (the crack was not detected)**: No fatigue damage was detected at the time t of the structural inspection. It does not have to mean that the fatigue crack does not exist, its size $a_{(t)}$ is just smaller than the technology used is able to identify the crack (a_d – detectable size). It therefore applies:

$$a_{(t)} \le a_d. \tag{12}$$

- $D_{(t)}$ **phenomenon (the crack was detected but in terms of structural reliability is not yet dangerous)**: During the inspection, fatigue damage was already detected, but the size of the crack is still smaller than the calculated acceptable size of the crack a_{ac}. This situation can be expressed as:

$$a_d < a_{(t)} \le a_{ac}. \tag{13}$$

- $F_{(t)}$ **phenomenon (failure)**: At time t, a fatigue crack was detected, the size of which has already exceeded the length of the acceptable crack a_{ac}, and the structure is therefore in a state of failure. Applies to:

$$a_{ac} < a_{(t)}. \tag{14}$$

The probabilities of the three random events described in (12) to (14) can be calculated for any operation time t of the analyzed structure using a probabilistic approach. The authors chose for the calculation the already mentioned DOProC method, which turned out to be very suitable for solving this problem thanks to its high efficiency and great accuracy (very desirable for small values of the probability of failure p_f). The first step of the probabilistic analysis consisted in determining the time t_I of the first inspection with a focus on fatigue damage of the incriminated bearing element. During the calculation with the time step Δt (set for 1 year of operation), the failure probability values $P(F_{(t)})$ were gradually determined. The resulting inspection time is then determined at the moment when the failure probability reaches the proposed failure probability p_d. As already mentioned, the results of this inspection can be used for the design of subsequent inspections in time $T > t_I$. This proposal is based on the use of conditional probability, namely:

$$P\left(F_{(T)}|U_{(t_I)}\right) = \frac{P\left(F_{(T)}\right) - P\left(F_{(t_I)}\right) - P\left(D_{(t_I)}\right) \cdot P\left(F_{(T)}|D_{(t_I)}\right)}{P\left(U_{(t_I)}\right)}. \tag{15}$$

The relationship (15) represents the calculation of the probability of failure $F_{(T)}$ at time T, assuming that a fatigue crack was not detected at the time of the first inspection t_I (phenomenon $U_{(t_I)}$).

Table 1. Used stochastic quantities expressed by the input histograms.

Variable	PDF	Parameters (μ/σ)
Range of stress peaks $\Delta\sigma$ [MPa]	Norm	30/3
Number of stress fluctuations in 1 year of operation N [–]	Norm	$10^6/10^5$
Stress at the yield point f_y [MPa]	LogNorm	280/28
Nominal normal stress in tension flange σ [MPa]	Norm	200/20
Initial fatigue crack size a_0 [mm]	LogNorm	0.2/0.05
Estimation of the minimum detectable crack size a_d [mm]	Norm	10/0.6

The shown probabilistic calculation used the input quantities from Table 1. The required level of reliability is expressed by the usual value of the reliability index $\beta = 2$ for these situations, which can be easily transformed into the corresponding target rate of failure $p_d = 0.02277$. The input quantities expressed as deterministic parameters were as follows: material constants $C = 2.2 \cdot 10^{-13}$ and $m = 3$, flange dimensions $b_f = 400$ mm and $t_f = 25$ mm. Table 2 then summarizes the results of the stochastic analysis in the form of a proposal for a system of inspections that should take place in the indicated years of operation. Interesting information is also the method of numerical integration during stochastic modeling, when the adaptive method was used with good efficiency.

Currently, the authors are trying to apply the described methodology to the use of high-strength steel (HSS) in bridge construction. The use of HSS in bridges (grades

Table 2. Design of the inspection system of the analyzed bridge structure (in years).

Inspection No	Failure crack from the edge	Failure crack from the surface
1	49	110
2	57	123
3	61	131
4	64	137
5	67	142
6	69	147

S460, S690, and S960) has so far only been investigated very sporadically in terms of fatigue resistance and residual life. After extensive parametric studies based on stochastic modeling using experimentally obtained material parameters, differences in the behavior of the analyzed detail of the bridge structure made from these three HSS strength grades were found (steel S960 behaved best in terms of fatigue resistance).

4 Conclusion

In the article, a brief description of the methodology for the analysis of the reliability of the load-bearing element was carried out, focusing on fatigue damage due to the propagation of fatigue cracks from the edge/surface in the load-bearing elements of steel structures and bridges that are exposed to cyclic loading. The calculation model is based on LEFM and contains the definition of three basic fatigue crack sizes (initial, detectable and acceptable) and three random phenomena that can occur in the analyzed detail.

Due to the requirement to predict the propagation of fatigue cracks during the operation of the structure, the use of probabilistic methods is offered, which allow to take into account possible uncertainties in the input parameters in an effective way. The calculation shown in the paper makes use of the Direct Optimization Probability Calculation (DOProC), which seems to be well-suited to this type of probability calculation. One of the reasons for choosing the DOProC method as a calculation tool in the solved task is the fact that the accuracy of the calculation by this method is very high and burdened only by the numerical error and the error resulting from the discretization of the original PDF of the input random variables and, of course, also the output quantities. The advantage of this calculation method is the significant computational complexity in tasks with a large number of random variables, which was not the case here. The very effective stochastic modeling was additionally enhanced by the FCProbCalc software used, with which all probabilistic calculations were performed in a user-friendly and operationally highly efficient environment.

The described methodology and its application could potentially lead to significant improvements in the estimation of maintenance costs for structures and bridges affected by cyclic loading. The first steps were taken to establish cooperation with entities focused on the maintenance of roads and highways, but so far without significant success. Current research works are mainly focused on the aforementioned application of the described methodology to determine the accumulation of fatigue damage in HSS structures. This is a very extensive and very current topic, which the author's team plans to extend to research on the effect of corrosion damage on fatigue resistance.

Acknowledgement. The research was supported by Czech Science Foundation Grant No. 21-14886S (Influence of material properties of high strength steels on durability of engineering structures and bridges).

References

1. Melchers, R.E.: Structural Reliability Analysis and Prediction, 2nd edn. John Wiley & Sons Ltd., Chichester (1999)
2. Maljaars, J., Steenbergen, H.M.G.M., Vrouwenvelder, A.C.W.M.: Probabilistic model for fatigue crack growth and fracture of welded joints in civil engineering structures. Int. J. Fatigue **38**, 108–117 (2012)
3. Ishikawa, H., Tsurui, A., Tanaka, H., Ishikawa, H.: Reliability assessment of structures based upon probabilistic fracture mechanics. Probab. Eng. Mech. **8**(1), 43–56 (1993)
4. Janas, P., Krejsa, M., Sejnoha, J., Krejsa, V.: DOProC-based reliability analysis of structures. Struct. Eng. Mech. **64**(4), 413–426 (2017)
5. Chen, N.Z., Wang, G., Soares, C.G.: Palmgren-Miner's rule and fracture mechanics-based inspection planning. Eng. Fract. Mech. **78**(18), 3166–3182 (2011)
6. Kim, S., Frangopol, D.M.: Inspection and monitoring planning for RC structures based on minimization of expected damage detection delay. Probab. Eng. Mech. **26**(2), 308–320 (2011)
7. Soliman, M., Frangopol, D.M., Kim, S.: Probabilistic optimum inspection planning of steel bridges with multiple fatigue sensitive details. Eng. Struct. **49**, 996–1006 (2013)
8. Liu, M., Frangopol, D.M., Kwon, K.: Fatigue reliability assessment of retrofitted steel bridges integrating monitored data. Struct. Saf. **32**(1), 77–89 (2010)
9. Kwon, K., Frangopol, D.M.: Bridge fatigue assessment and management using reliability-based crack growth and probability of detection models. Probab. Eng. Mech. **26**(1), 471–480 (2011)
10. Law, S.S., Li, J.: Updating the reliability of a concrete bridge structure based on condition assessment with uncertainties. Eng. Struct. **32**(1), 286–296 (2010)
11. Catbas, F.N., Susoy, M., Frangopol, M.D.: Structural health monitoring and reliability estimation: long span truss bridge application with environmental monitoring data. Eng. Struct. **30**(9), 2347–2359 (2008)
12. Anderson, T.L.: Fracture Mechanics: Fundamentals and Applications, 3rd edn. CRC Press, Taylor & Francis Group, Boca Raton, Florida (2005)
13. Ni, Y.Q., Ye, X.W., Ko, J.M.: Monitoring-based fatigue reliability assessment of steel bridges: analytical model and application. J. Struct. Eng. **136**(12), 1563–1573 (2010)
14. Paris, P., Erdogan, F.: A critical analysis of crack propagation laws. J. Fluids Eng. **85**(4), 528–533 (1963)
15. Pugno, N., Ciavarella, M., Cornetti, P., Carpinteri, A.: A generalized Paris' law for fatigue crack growth. J. Mech. Phys. Solids **54**(7), 1333–1349 (2006)
16. Seitl, S., Miarka, P., Malíková, L., Krejsa, M.: Comparison of calibration functions for short edge cracks under selected loads. Key Eng. Mater. **754**, 353–356 (2017)
17. Chryssanthopoulos, M.K., Righiniotis, T.D.: Fatigue reliability of welded steel structures. J. Constr. Steel Res. **62**(11), 1199–1209 (2006)

In-Vitro Assessment on Surface Functionality of CoCrMo Alloy Manufactured by Laser Powder Bed Fusion

Suryank Dwivedi[1](✉), Pratik Kumar Shaw[1], Akash Nag[2], Tanmay Tiwari[3], Alok Kumar Das[1], and Amit Rai Dixit[1]

[1] Department of Mechanical Engineering, Indian Institute of Technology (Indian School of Mines), Dhanbad 826004, India
suryankdwivedi11@gmail.com

[2] Faculty of Mechanical Engineering, VSB-Technical University of Ostrava, Ostrava-Poruba 70800, Czech Republic

[3] Center for Precision Manufacturing, Department of Mechanical Engineering, University of Akron, Ohio 44325, USA

Abstract. Recent advancements in laser powder bed fusion (LPBF) techniques have revolutionized the manufacturing of biomedical devices using metallic biomaterials. For hip-joint prosthetics, CoCrMo has emerged as a preferred material. Therefore, this study examines the surface functionality of the as-printed and mechanically post-processed 3D-printed CoCrMo specimen. Surface functionality includes the study of surface integrity, wettability, and in-vitro biotribology in the physiological environment. The 3D surface topographical analysis was conducted on both samples. The obtained 3D profiles revealed that the as-printed samples exhibited small-scale balling effects, resulting in micro-pores that persisted after polishing. The topographical parameters, such as R_a, R_q, R_z, R_{ku}, and R_{sk}, were measured, and polished samples showed better surface characteristics. Surface wettability tests indicated a transition from hydrophobic to hydrophilic behavior, with lower surface free energy observed in the as-printed parts compared to the post-processed counterpart. Further, in-vitro biotribology tests were conducted under the physiological environment for the hip-joint prosthesis condition. The biotribological results indicated a reduced coefficient of friction and frictional force for the as-printed compared to the polished sample. The 3D wear profiles also showed visible wear marks for the polished samples, which exhibited higher volumetric wear loss than the as-printed counterpart. These findings emphasize the importance of achieving optimal surface roughness to enhance surface functionality in 3D-printed components through surface finishing methods.

Keywords: Laser Powder Bed Fusion · CoCrMo · Surface Functionality · Wettability · Biotribology · Friction and Wear

M. Pagac et al. (Eds.): SCFF 2024, LNNS 1062, pp. 116–126, 2024.
https://doi.org/10.1007/978-3-031-65656-9_12

1 Introduction

The continuously rising demand for orthopedic implants in recent years has prompted ongoing research into the quality of implants, aiming to minimize the need for revision surgeries. In 2010, the global orthopedics market was around $12 billion, capturing 37.5% of the total biomaterials market. In 2021, it reached $16.01 billion and is expected to reach $39.11 billion in the next 10 years with a growth rate of 10.36% [1, 2]. The orthopedic metallic implants manufactured by conventional and advanced manufacturing routes are subject to rigorous safety standards and protocols to ensure their functionality and compatibility within the human body's physiological environment. These measures are critical for minimizing adverse reactions, thereby enhancing the reliability and lifespan of these implants. Typically, orthopedic implants are designed with a predetermined lifespan, typically around 15–20 years, after which revision surgery becomes necessary [3]. Consequently, it is imperative for researchers to comprehend the essential criteria and characteristics that govern the performance of metallic implants. It includes the relationship between surface roughness, wettability, and bio-tribology (study of friction, wear, and lubrication in biological systems). A comprehensive examination of these aspects is crucial for developing metallic implants that can effectively fulfill their intended purpose while ensuring the safety of patients.

In the recent past, researchers have adopted a sustainable manufacturing route, i.e., Additive manufacturing (AM), which provides design freedom and customization with enhanced bulk and surface properties [4–6]. Bulk properties contribute to improved microstructural and mechanical behavior, while surface functionality governs various surface properties that directly influence wetting mechanisms, tribological characteristics, bioactivity, and osseointegration during the initial stages of bone formation [7, 8]. Previous studies dictated that surface functionality plays a pivotal role in determining the success of the implant under a physiological environment. The biological response of the natural bone and artificial implant significantly impacts the cell behavior at this interface in both in-vivo and in-vitro states and is influenced by the surface area provided by the implant surface [9–11]. Previously performed studies have shown that a larger contact area between the implant and tissue enhances mechanical anchorage and cell adhesion compared to a smooth surface. This increase in surface area is often attributed to the surface roughness of the implant, which in turn affects the rate of osseointegration and biomedical fixation. The surface roughness of the implanted part is typically classified into macro, micro, and nano levels. Literature suggests that macro-roughness ($10\ \mu m < Ra < 20\ \mu m$) facilitates a durable connection between the implant and natural bone, ensuring long-term mechanical stability and potentially reducing the risk of periimplantitis and ion release. Conversely, micro-roughness ($1 < Ra < 10\ \mu m$) influences the contact area between the implant and bone, potentially reducing inflammation and ion release risks while affecting bone tissue response. On the other hand, nano-roughness ($Ra < 1\ \mu m$) promotes early bone response, protein adsorption, and osteoblast adhesion [12–14]. Consequently, it can be inferred that AM techniques can effectively control roughness at the macro and micro levels, necessitating the adoption of advanced surface finishing methods to achieve nano-level roughness.

However, in the existing literature, the relationship between surface roughness and the functionality of biomedical implants remains limited and requires further exploration. Therefore, in this study, samples were fabricated using the laser powder bed fusion (LPBF) technique with CoCrMo metal powder. Samples were examined in both as-printed and mechanically polished conditions to investigate the effect of surface characteristics on roughness parameters, surface wettability, surface free energy, and biotribological performance under a physiological environment.

2 Materials and Methods

In the present investigation, a DMP Flex metal 3D printer with a maximum laser source of 500 W was utilized to produce CoCrMo samples via the LPBF technique. Printing was conducted with the original equipment manufacturer (OEM) parameters set as follows: Laser power at 200 W, scanning speed at 2500 mm/s, layer thickness at 30 μm, bi-directional rotating strategy, and hatch spacing at 80 μm. These parameters correspond to a volumetric energy density of 44.44 J/mm^3 and are consistent with those previously adopted by several researchers [15, 16]. The printing chamber was degassed before fabrication, and an inert environment with Ar gas was maintained throughout the process. The samples were designed with dimensions of 10 mm × 10 mm × 1 mm using CAD software, and digital input was provided using 3D XPert slicing software. The samples were printed parallel to the build plate and placed in the middle of the build plate. The printed samples underwent post-processing, including de-powdering, heat treatment, and sandblasting operations. These processes aimed to achieve dimensional stability by eliminating residual stresses and removing any unmelted metal powders from the surface. After post-processing, the samples were cut using a WEDM machine and then cleaned twice in a bath-type ultrasonic sonicator with isopropyl alcohol followed by DI water for 30 min each.

To investigate the effect of surface roughness on surface functionality further, the samples were mechanically grounded using a rotating-shaft apparatus at 200 RPM against different grit sizes of aluminum oxide polishing paper, with each grit size paper applied for 30 min. The 3D surface topography was captured using a non-contact white light 3D profilometer (Make: Zygo, Model: New View 9000), and roughness values such as R_a, R_q, R_z, R_{ku}, and R_{sk} were measured and compared with the as-printed counterpart. Furthermore, the wetting mechanism and surface free energy (SFE) for both as-printed and mechanically polished samples were analyzed using simulated body fluid (SBF). A droplet was dispensed through a needle-syringe arrangement and allowed to settle for 30 s. Droplet images were captured using the drop analyzer (Model: DSA 25, Make: KRUSS) at a room temperature of 25 °C ± 1 °C. Further, SFE was calculated using the pair of SBF and diiodomethane based on Fowke's theory.

Finally, the biotribological performance of both as-printed and polished samples for hip-joint prosthetics was assessed under fetal bovine serum (FBS) at a temperature of 37 ± 1 °C, mimicking physiological conditions. The biotribology test utilized a ball-on-plate reciprocating set-up on a universal tribometer (Make: Rtec, Model: MFT5000), as presented in Fig. 1. The literature dictates that hip joints typically experience a Hertzian contact pressure of 106.7 MPa under normal conditions, corresponding to a 10 N load

[17]. Hence, a Ti6Al4V ball was used as a counterbody, and a 10N load was applied during testing. Tests were conducted at frequencies of 2 Hz, 4 Hz, 6 Hz, 8 Hz, and 10 Hz, translating to sliding speeds of 20 mm/s, 40 mm/s, 60 mm/s, 80 mm/s, and 100 mm/s, respectively, with a stroke length of 5 mm, each lasting 5 min. The coefficient of friction (COF), generated frictional force, and wear depth were measured for each condition. Additionally, to investigate the wear mechanism of both as-printed and polished surfaces, the samples were cleaned in a sonicator using acetone and DI water, dried in a furnace at 80 °C, and subsequently analyzed using 3D profilometry.

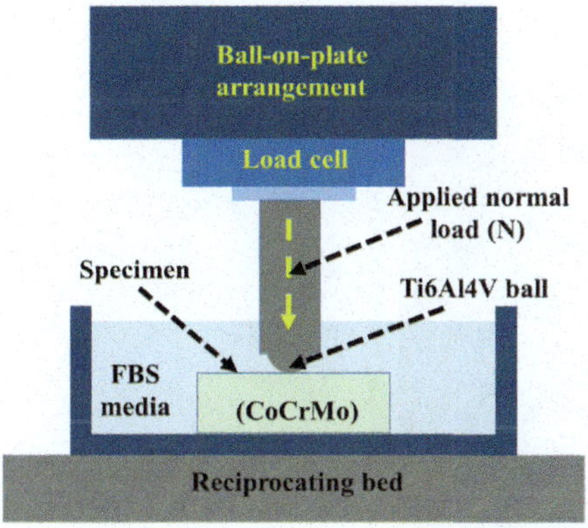

Fig. 1. A schematic arrangement of the used ball-on-plate tribo pair.

3 Results and Discussion

3.1 Surface Characteristics of the 3D-Printed CoCrMo Alloy

The quantitative measurement of both the as-printed and mechanically polished samples was conducted through topographical examination using a white light 3D profilometer. The 3D surface topographical characterization was performed after the heat treatment and sandblasting operations, as illustrated in Fig. 2. The surface topography of the as-printed sample (Fig. 2a) reveals a fabrication process employing bi-directional laser scanning, with layers deposited sequentially. However, Fig. 1a shows partially melted CoCrMo metal particles adhered between the laser melt tracks, indicating a high hatch spacing that could be reduced to achieve full densification in the printed part. Further, the 3D images of the mechanically polished sample were also captured and depicted in Fig. 2b. The image reveals the presence of scattered pores, resulting in localized porosity within the fabricated part.

The 3D profiles were captured at the top surface and various random locations. Surface characteristics, including average roughness (R_a), root mean square roughness (R_q), and average peak-to-valley roughness (R_z) values, were measured for both samples. Results indicate that R_a values for the as-printed and polished samples are 13.69 μm and 1.11 μm, respectively, while R_q values are 17.45 μm and 1.44 μm, respectively.

Similarly, R_z values are consistent with the R_a and R_q values, measured at 22.67 μm and 1.62 μm for the as-printed and polished samples. Based on the obtained average surface roughness values, it is evident that the R_a values for both the as-printed and polished samples fall within the macro (10 μm < R_a < 20 μm) and micro (1 μm < R_a < 10 μm) surface roughness ranges. Previous studies have demonstrated that macro roughness is crucial in establishing mechanical interlocking between natural and artificial implants, thereby enhancing the bonding strength in long-term applications. Whereas, micro-roughness facilitates early-stage interactions between natural bone tissues and implant materials, offering a larger contact area to mitigate peri-implant inflammation and ion-release risk [12, 13].

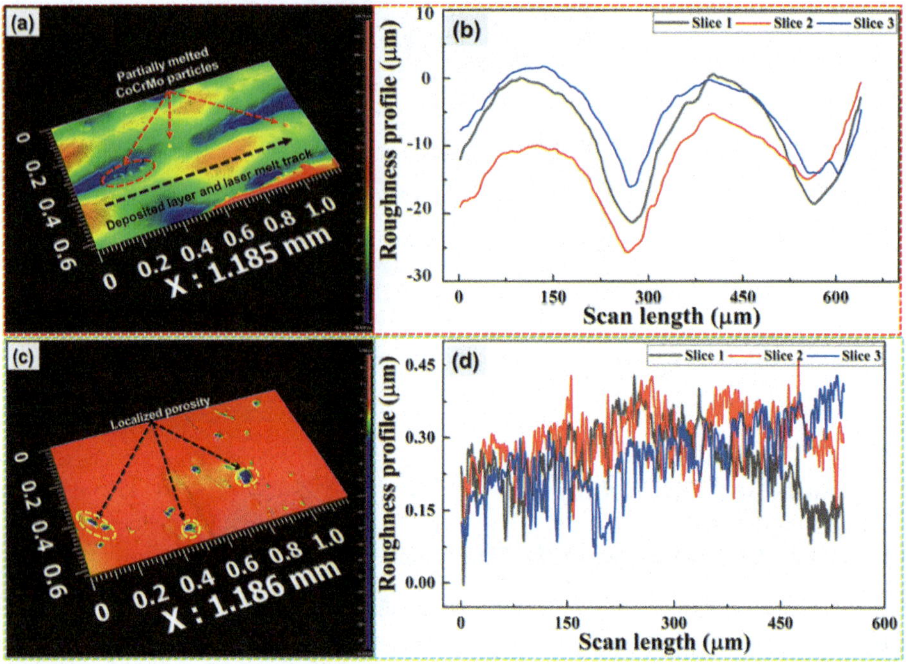

Fig. 2. 3D surface profile and 2D cross-sectional profile of the (a and b) as-printed and (c and d) mechanically polished samples.

To further investigate the effect of topographical behavior on surface texture and load-bearing capacity, Skewness (R_{sk}) and Kurtosis (R_{ku}) roughness factors were extracted from the 2D cross-sectional profiles of both the as-printed and polished samples. R_{sk} and R_{ku} are utilized to characterize the shape and sharpness of surface features. The R_{ku} values obtained from the 2D cross-sectional profiles of the as-printed and polished samples are 3.19 and 1.62, respectively. According to the open literature, an R_{ku} value of approximately 3 suggests the presence of both sharp peaks and intended features with uniform distribution across the manufactured part. A higher R_{ku} value for the as-printed part indicates the presence of numerous peaks and more profound valleys, as confirmed by the 2D cross-sectional profile (Fig. 2b). Similarly, a lower R_{ku} value was observed for the polished sample, presenting a lower peak and uniform surface profile as compared to the as-printed counterpart. Additionally, R_{sk} values were extracted from the cross-sectional profiles of both samples, resulting in values of 1.49 and −1.22 for the as-printed and polished samples, respectively. These results indicate a positive skewness factor for the as-printed sample and a negative value for the polished sample. Previous studies have shown that a positive R_{sk} value corresponds to stiffer peaks with shallow valleys, resulting in higher roughness values. Conversely, a negative Rsk value is associated with deeper valleys and truncated peaks, providing increased contact points and surface area during motion transfer. This plays a crucial role in load transfer between the implant and natural bone.

3.2 Wetting Behavior and Surface-Free Energy

The surface functionality was evaluated through surface wettability and (SFE) analysis, determining the wetting behavior of the manufactured surfaces crucial for implant success in physiological environments. Previous studies have highlighted the importance of hydrophilic surfaces for enhanced osseointegration during initial bone growth stages. Therefore, surface wettability examinations were conducted using SBF via a needle-syringe set-up at various locations. Droplet profiles, as depicted in Fig. 3(a and b), revealed hydrophobic wetting behavior in the as-printed sample, contrasting with the hydrophilic behavior observed in the polished counterpart. The contact angle values obtained for the as-printed and polished samples are 100.3° and 47.0°, respectively, illustrating the influence of surface roughness. The Wenzel and Cassie-Baxter models explained this effect by describing droplet-spreading behavior on rough surfaces. According to the Wenzel model, a droplet creates only one contact line with the solid surface upon impact, while the Cassie-Baxter model proposes that the presence of surface asperities leads to the formation of two contact lines: between the droplet and entrapped air and between the entrapped air and solid surface. This phenomenon impedes droplet spreading, resulting in hydrophobicity [18].

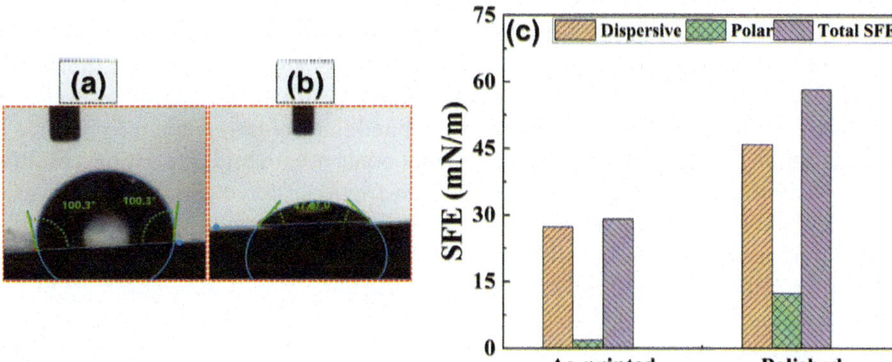

Fig. 3. Droplet profiles of the (a) as-printed and (b) polished samples and (c) surface free energy of both specimens.

The CoCrMo alloy is used as a permanent implant for Hip prosthetics. Therefore, cell attachment, spreading, and proliferation are necessary for adequate osseointegration within the physiological environment, which can be determined using SFE calculation.

SFE calculation is vital in assessing biocompatibility, as it influences the interaction between host tissue and the implant's solid surfaces. SFE relies on factors such as wetting behavior, liquid surface tension, and the polar and dispersive components of the mating surface. Therefore, a set of liquids, such as SBF and diiodomethane, was taken as a standard liquid, and the Fowkes theory was adopted to calculate the SFE. The calculated values are presented in Fig. 3c, and it can be seen that polished samples show higher SFE (Total SFE: 58.124 mN/m) compared to the as-printed (Total SFE: 27.69 mN/m) part. The obtained outcomes are in line with the contact angle result and would be helpful in achieving improved biocompatibility.

3.3 In-Vitro Biotribology for Hip-Joint Condition

To date, hip-joint prosthetics have utilized various combinations, including metal-on-metal, metal-on-polymer, ceramic-on-ceramic, and ceramic-on-polymer. CoCrMo-based alloys, renowned for their potential in metal-on-metal combinations, are commonly employed as permanent implants [19–21]. In this study, 3D-printed CoCrMo samples were subjected to testing under physiological conditions (Lubricant: Bovine serum, Temperature: 37.5 °C, RH: 50%). Tribological characteristics such as COF, frictional force, and wear depth values were obtained and depicted in Fig. 4. The experiments considered hip-joint loading conditions with varying sliding speeds, including 20 mm/s, 40 mm/s, 60 mm/s, 80 mm/s, and 100 mm/s. Figure 4(a and b) illustrates the variation in the COF over time for both as-printed and polished samples.

The results indicate that the polished sample exhibited a higher COF value compared to the as-printed part, particularly at low sliding speeds. However, the COF value decreased for the polished samples with increasing sliding speed, reaching a minimum at 100 mm/s. This trend was also observed for the as-printed counterparts. Specifically, the average COF values for polished samples were 0.163, 0.107, 0.067, 0.041, and 0.002 at 20 mm/s, 40 mm/s, 60 mm/s, 80 mm/s, and 100 mm/s, respectively. Similarly, the average COF values for as-printed samples were 0.132, 0.088, 0.048, 0.029, and 0.002 at the same sliding speeds, respectively. These findings suggest forming a lubricating film at higher sliding speeds, reducing surface contact for both samples. However, at low sliding speeds, the polished specimen exhibited line contact, while the presence of asperities on the as-printed surfaces led to point contact with bridging action, resulting in lower COF values. Furthermore, the average frictional force and wear depth were also quantified and illustrated in Fig. 4(c and d). The figure demonstrates a decrease in frictional force with increasing sliding speed, becoming negligible at high speeds (i.e., 100 mm/s). Furthermore, slightly higher frictional force values were observed for the polished sample, suggesting an influence of surface roughness on the contact mechanism between the mating pairs.

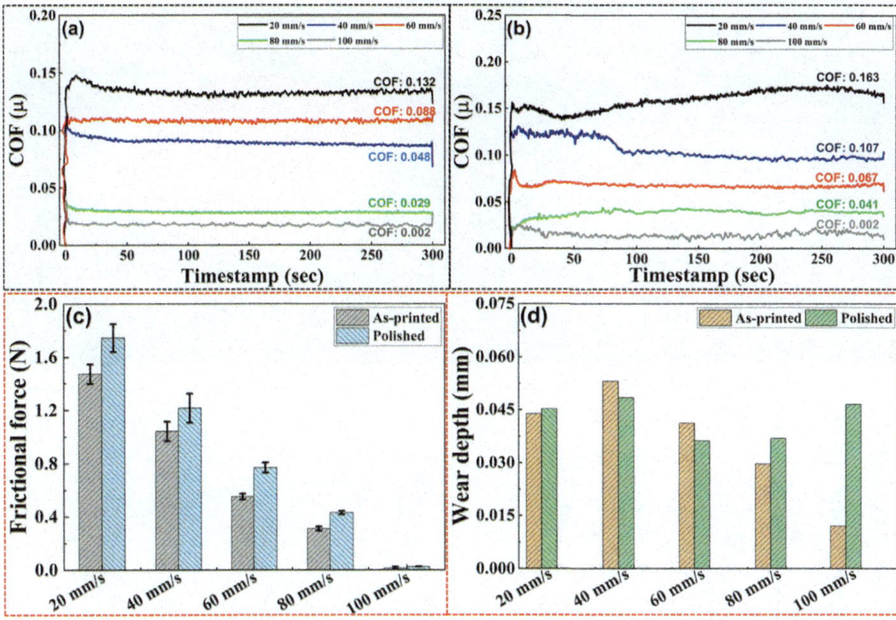

Fig. 4. The variation in (a and b) COF with time, (c) frictional force, and (d) average wear depth for as-printed and polished samples.

3.4 Wear Mechanism of As-Printed and Polished Specimens

To comprehend the correlation between the measured tribological characteristics of the as-printed and polished samples, wear tracks were examined using a 3D profilometer and documented in Fig. 5(a–c). Figure 5a showcases the 3D worn-out profile of the as-printed surface, exposing localized surface wear attributed to an uneven surface profile. Moreover, the high roughness value results in formed peaks acting as bridges, with valleys serving as lubricant reservoirs, ensuring adequate lubrication during sliding motion. However, the 3D worn-out profile and 2D cross-sectional profile of the polished sample illustrated in Fig. 5 (b and c), exhibit visible scratch marks formed during sliding motion, indicating persistent line contact throughout the friction test. This contributes to higher COF and frictional force values for the polished sample than the as-printed part.

4 Conclusions

The CoCrMo alloy-based metallic biomaterial was successfully fabricated for biomedical applications using the LPBF technique. The surface functionality of the printed samples was comprehensively evaluated through surface topographical, wetting, SFE, and biotribological characterization methods. From the present study, the following major outcomes can be drawn:

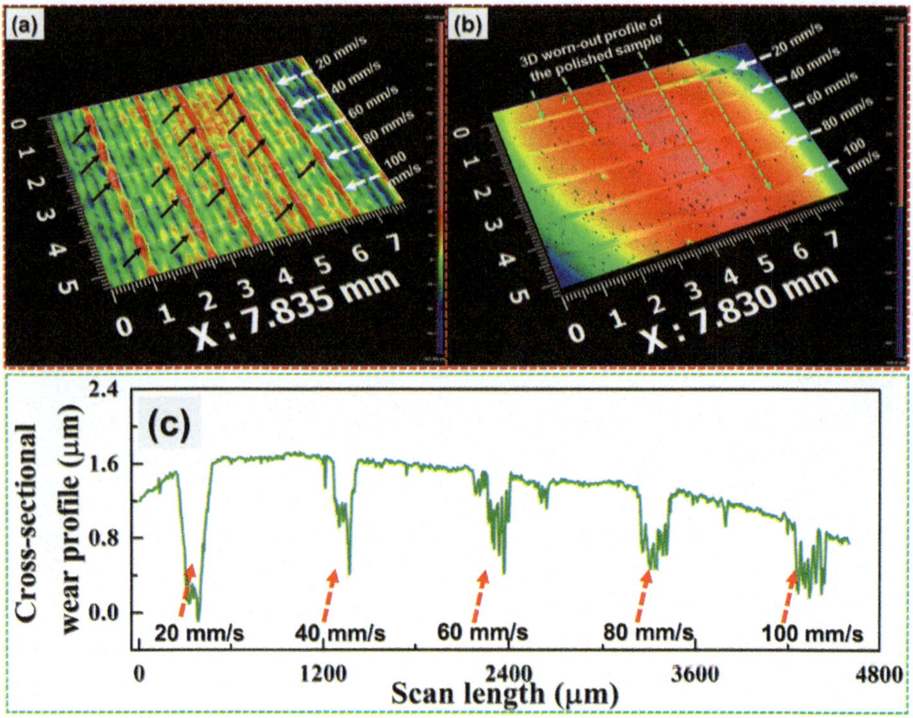

Fig. 5. 3D worn-out profiles for the (a) as-printed and (b) polished samples, and (c) 2D cross-sectional profile of the polished sample.

1. The 3D surface profile of the printed sample displays macro-roughness, with elevated peaks seen in the surface profile due to a positive skewness value. Moreover, polished surfaces show micro-roughness, evidenced by a negative skewness value, suggesting a greater load-bearing capacity.
2. The as-printed surface displays hydrophobic wetting behavior, while the polished sample exhibits hydrophilic wetting behavior, contributing to higher surface free energy (SFE), essential for facilitating adequate osseointegration.
3. The impact of surface roughness, lubrication, and sliding speed was analyzed for both as-printed and polished specimens. The results demonstrate the beneficial influence of surface roughness and sliding speed on the COF, frictional force, and wear depth values for the as-printed sample.
4. The wear morphology of the as-printed sample illustrates a bridging action, providing both a point of contact and acting as a lubricant reservoir during the sliding motion.

Acknowledgments. The authors thank the Central Research Facility of the IIT (ISM), Dhanbad, India for providing the metal 3D printing and characterization facilities. The authors also thank the Department of Science and Technology (DST), India, for the 3D profilometer facility under the FIST project grant SR/FST/ET-II/2018/222(C).

References

1. Germaini, M.M., Belhabib, S., Guessasma, S., Deterre, R., Corre, P., Weiss, P.: Additive manufacturing of biomaterials for bone tissue engineering—a critical review of the state of the art and new concepts. Prog. Mater. Sci. **130**, 100963 (2022)
2. https://www.strategicmarketresearch.com/market-report/orthopedic-biomaterials-market
3. Geetha, M., Singh, A.K., Asokamani, R., Gogia, A.K.: Ti based biomaterials, the ultimate choice for orthopaedic implants—a review. Prog. Mater. Sci. **54**(3), 397–425 (2009)
4. Hajnys, J., Pagáč, M., Měsíček, J., Petru, J., Król, M.: Influence of scanning strategy parameters on residual stress in the SLM process according to the bridge curvature method for AISI 316L stainless steel. Materials **13**(7), 1659 (2020)
5. Phadke, N., Raj, R., Srivastava, A.K., Dwivedi, S., Dixit, A.R.: Modeling and parametric optimization of laser powder bed fusion 3D printing technique using artificial neural network for enhancing dimensional accuracy. Mater. Today Proc. **56**, 873–878 (2022)
6. Kunčická, L., Kocich, R., Németh, G., Dvořák, K., Pagáč, M.: Effect of post process shear straining on structure and mechanical properties of 316 L stainless steel manufactured via powder bed fusion. Addit. Manuf. **59**, 103128 (2022)
7. Dwivedi, S., Dixit, A.R., Das, A.K.: A relationship between manufacturing routes-microstructure-biotribology of selective laser melted stainless steel 316L under dry and simulated body fluid. Int. J. Precis. Eng. Manuf.-Green Technol. 1–15 (2023)
8. Dwivedi, S., Dixit, A.R., Das, A.K.: Wetting behavior of selective laser melted (SLM) biomedical grade stainless steel 316L. Mater. Today Proc. **56**, 46–50 (2022)
9. Lodhi, M.K., Deen, K.M., Greenlee-Wacker, M.C., Haider, W.: Additively manufactured 316L stainless steel with improved corrosion resistance and biological response for biomedical applications. Addit. Manuf. **27**, 8–19 (2019)
10. Karimi, J., Antonov, M., Kollo, L., Prashanth, K.G.: Role of laser remelting and heat treatment in mechanical and tribological properties of selective laser melted Ti6Al4V alloy. J. Alloy. Compd. **897**, 163207 (2022)
11. Dwivedi, S., Dixit, A.R., Das, A.K., Adamczuk, K.: Additive texturing of metallic implant surfaces for improved wetting and biotribological performance. J. Market. Res. **20**, 2650–2667 (2022)
12. Arkusz, K., Pasik, K., Halinski, A., Halinski, A.: Surface analysis of ureteral stent before and after implantation in the bodies of child patients. Urolithiasis **49**(1), 83–92 (2021)
13. Mazurek-Popczyk, J., Palka, L., Arkusz, K., Dalewski, B., Baldy-Chudzik, K.: Personalized, 3D-printed fracture fixation plates versus commonly used orthopedic implant materials-biomaterials characteristics and bacterial biofilm formation. Injury **53**(3), 938–946 (2022)
14. Dwivedi, S., Dixit, A.R., Das, A.K., Nag, A.: A novel additive texturing of stainless steel 316L through binder jetting additive manufacturing. Int. J. Precis. Eng. Manuf.-Green Technol. **10**(6), 1605–1613 (2023)
15. Barucca, G., Santecchia, E., Majni, G., Girardin, E., Bassoli, E., Denti, L., Mengucci, P., et al.: Structural characterization of biomedical Co–Cr–Mo components produced by direct metal laser sintering. Mater. Sci. Eng. C, **48**, 263–269 (2015)
16. Demir, A.G., Previtali, B.: Additive manufacturing of cardiovascular CoCr stents by selective laser melting. Mater. Des. **119**, 338–350 (2017)
17. Afoke, N.Y., Byers, P.D., Hutton, W.: Contact pressures in the human hip joint. J. Bone Joint Surg. **69**(4), 536–541 (British volume) (1987)
18. Dwivedi, S., Dixit, A.R., Das, A.K., Srivastava, A.K.: Time-dependent wetting behavior of the micro-textured stainless steel 316L using the mechanical indentation method. In: Proceedings of the Institution of Mechanical Engineers, Part E: Journal of Process Mechanical Engineering, 09544089231205782 (2023)

19. Ruggiero, A., Gómez, E., Merola, M.: Experimental comparison on tribological pairs UHMWPE/TIAL6V4 alloy, UHMWPE/AISI316L austenitic stainless and UHMWPE/AL2O3 ceramic, under dry and lubricated conditions. Tribol. Int. **96**, 349–360 (2016)
20. Ruggiero, A., De Stefano, M.: Experimental investigation on the bio-tribocorrosive behavior of Ti6Al4V alloy and 316 L stainless steel in two biological solutions. Tribol. Int. **190**, 109033 (2023)
21. Shaw, P.K., Dwivedi, S., Dixit, A.R., Tyagi, R.: Trends of tribology in biomedical application of additively manufactured parts. In: Tribological Aspects of Additive Manufacturing, pp. 156–176. CRC Press (2024)

Metrological Analysis of 3D Scanning Results for Samples with Slight Openings Manufactured by FFF/FDM 3D Printing Technology

Aleksandra Bochenek[1](✉), Dominik Malara[1,2], Tomasz Kozior[1,2], Jerzy Bohnia[2], and Michał Nawotka[1]

[1] Metrological IT Laboratory, Digital Technology Metrology Department, Central Office of Measures, 00-139 Warsaw, Poland
aleksandra.bochenek@gum.gov.pl

[2] Department of Metrology and Unconventional Manufacturing Methods, Faculty of Mechatronics and Mechanical Engineering, Kielce University of Technology, 25-314 Kielce, Poland

Abstract. Main focus of the article is place at challenges in 3D scanning for various small openings in flat plane. As manufacture method was chosen fused filament fabrication (FFF/FDM) method. Samples were scanned both in horizontal and vertical position, with a few different angles of scanning. Conducted research point out several interesting dependences and limitations of 3D scanning such as, dependency of the depth of the scanning and angle of the scanner and differences of shape of the holes based on placement of the sample.

Keywords: Additive technology · 3D scanning · Metrology · FFF/FDM

1 Introduction

As technology develops, users' expectations regarding available applications increase. Both in terms of measurement methods and production. Constant development provides new areas of research and allows us to take on newer challenges. For instance, such field is 3D scanning. For most applications, the main focus comes down to mapping larger objects [1] or complex, finer shapes [2]. It is possible to find many works that focus on data preservation using scanning due to the non-invasiveness of this method [3, 4]. It is also useful when restoring damaged elements and inspecting the adopted solutions [5–7].

An important branch of the field is also the analysis of small structures inside larger surfaces, such as holes of several millimeters. Being on the edge and slightly outside the working range of the caliper, the most popular solution is testing with a coordinate measuring machine. They are characterized by high accuracy but require significant work and time to carry out measurements, and the measurements have a very high level of accuracy, which is not always necessary for the user's needs. Being between these two well-known measurement tools opens the possibility of using a 3D scanner and examining its capabilities.

M. Pagac et al. (Eds.): SCFF 2024, LNNS 1062, pp. 127–137, 2024.
https://doi.org/10.1007/978-3-031-65656-9_13

2 Materials and Methods

2.1 Samples Manufacturing

Sample's project was made with use of CAD 3D software SolidWorks (2022, Dassault Systèmes, Vélizy-Villacoublay, France). Design contains flat square plate (dimensions: 120 × 120 × 8 mm) with different kinds of holes (Fig. 1). There are 15 circular holes in 3 groups of 5 with diameters from 5 to 1 mm and 15 square holes (also in groups) with length of the square side equal from 5 to 1 mm. Each group have different angle of inside wall. First set has holes with walls perpendicular to the surface of the sample. Second and third sets have that angle tilted by 15 degrees, accordingly, directed to the inside center of the hole (inside-tilt) and outside of the center (outside-tilt). All of those openings are through the whole sample thickness. Due to the geometrical nature of inside-tilt holes, there are not through and are closed on the bottom for all sizes accept the biggest one. Moreover, there were designed 2 freeform patterns on the opposite sides of the sample. They are copies of each other with change of width – from 1 to 3 mm. They are 5 mm deep for both, along the entire length.

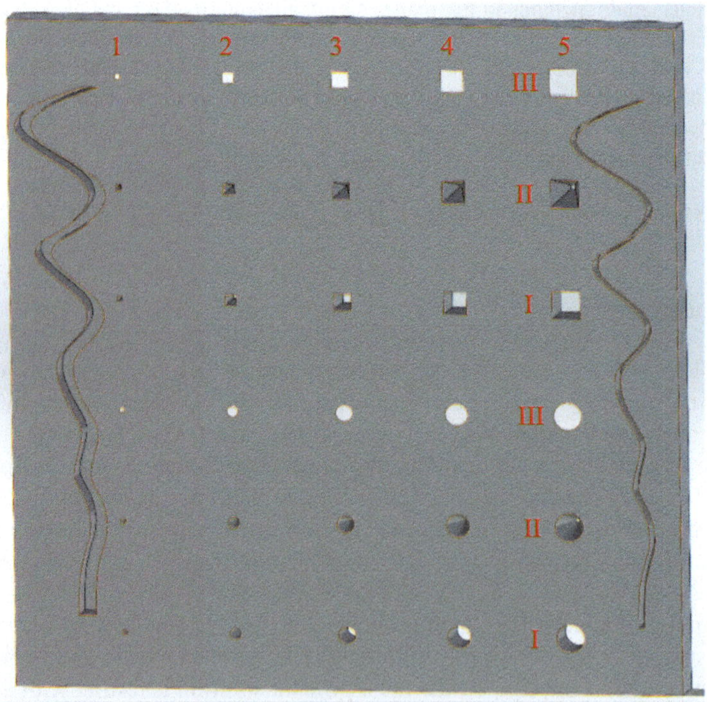

Fig. 1. Sample designed in SolidWorks software. 1–5 indicate values of dimensions (diameters and side of the square); row I marks wall with no tilt, II - inside tilt, III - outside tilt.

The specificity of an element of given sizes and different shapes allows you to examine the limitations of measurement technology using an optical scanner.

Sample's project was saved as STL file in form of mesh of 10208 triangles, with deviation around 0.0094 mm and angle placed at 3.583 degrees. Next there were printed 2 samples by fused filament fabrication (FFF/FDM) method with use of Sketch and Replicator 3D printers, both produced by MakerBot (New York City, USA). PLA was selected as a material. Layer hight was set at 0.2 mm, extruder temperature was set at 220 °C, for Sketch temperature of worktable was set at 70 °C, Replicator does not have heated table and raft was used for both printers. Samples are presented at Fig. 2.

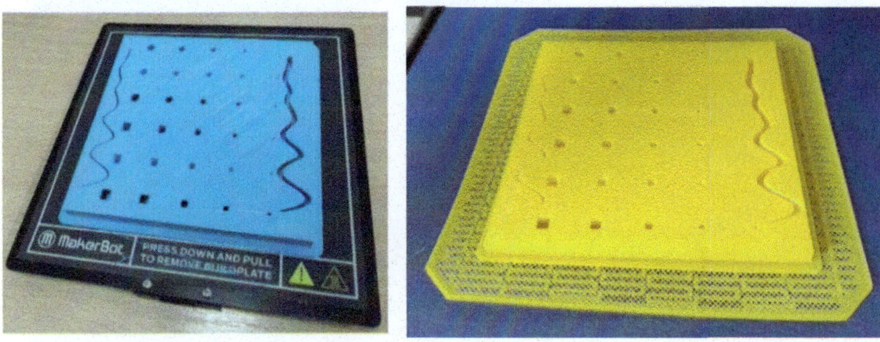

Fig. 2. Printed samples on worktables. Blue - Sketch, Yellow – Replicator.

Openings were designed to examine the accuracy of the 3D scanner for relatively small holes and freeforms. Use of standard caliper would be very limited for measures, on the other hand, use of coordinate measuring machine would give desired results but also would be much more time consuming and even to precise for potential needs.

Among 30 designed openings only 18 of them were analyse due to imperfections of the manufacturing process – holes with 2 and 1-mm diameter/length of the side were printed with high deformations visible at Fig. 3.

2.2 Measurement Methodology

Examination of the samples was conducted by ATOS II 3D (Atos, Bezons, France) scanner and then processed by dedicated software – ZEISS INSPECT Optical 3D (ZEISS, Oberkochen, Baden-Württemberg, Germany). The permissible measurement error described by the manufacturer is in range from ± 0.01 mm to ± 0.03 mm for different scanning ranges, camera resolution is 2 × 5,000,000 pixels and density of measured points is 0.02 ÷ 0.79 mm with their registration up to 1,400,000. Depending on the simplicity of the geometry or its complexity, the number of points in the cloud changes during a specific scan. The quality of the scanned sample is also important, as any imperfections increase the density of the point cloud locally.

First sample (blue) was placed horizontally on rotating table and by doing 10 separate scans (one for each angle of rotation) the image of the sample was stitched from those individual scans. Process was performed for few different angles of placing the scanner. For horizontally scanned sample angle between scanner and worktable was set at 45, 60 and 80 degrees. Second sample (yellow) was placed vertically on the table and scanned

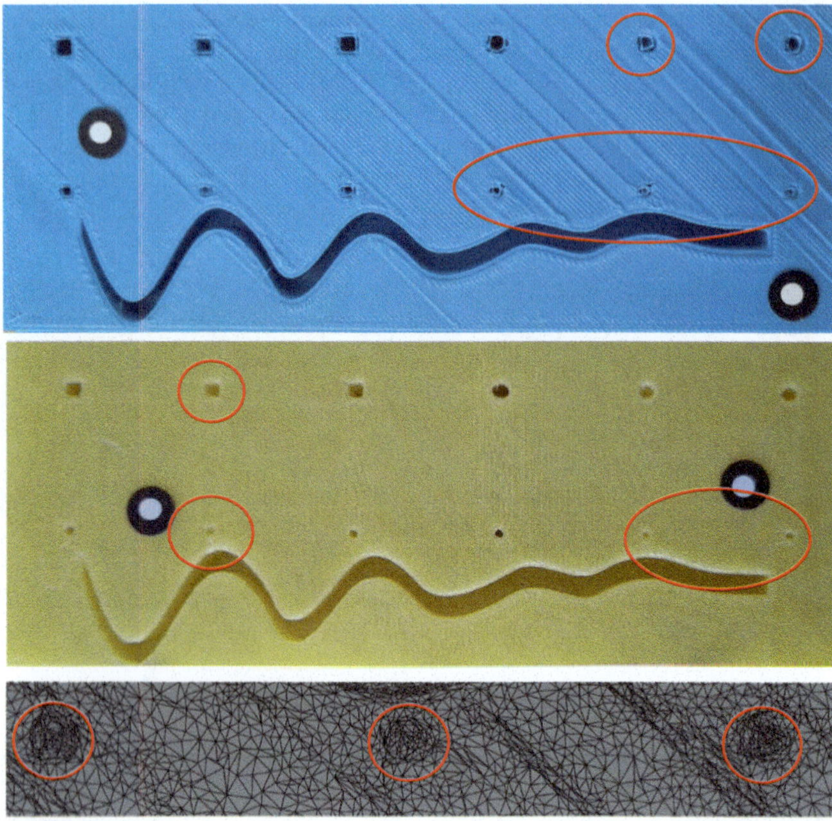

Fig. 3. Imperfections for the smallest size holes on sample and on mesh. Most visible are marked by red marker.

in angles 30 and 45 for the scanner. In the next step scans were inspected in SolidWorks software. All openings were examined in the same points of interest, presented at Figs. 4 and 5.

3 Results

3.1 Analysis of the Openings

Analysis of the triangle mesh objects received from scanning after stitching provides interesting information and allows to observe areas of the sample from various points of view. All horizontal scans share some features like uneven, sharp edges of the end of openings as visible at Figs. 6 and 7. As vertically scanned samples had data collected for both sides of the sample and this effect is absent, because information for the "back side" is provided during scanning.

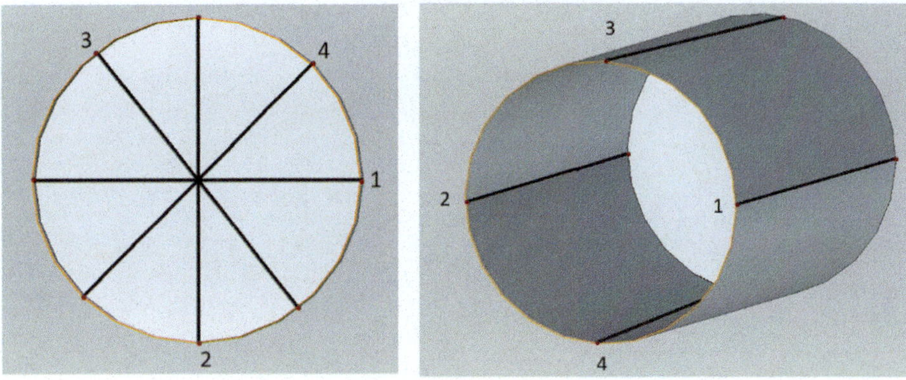

Fig. 4. Sequence of points for circular holes, examination of the shape and depth of the sample.

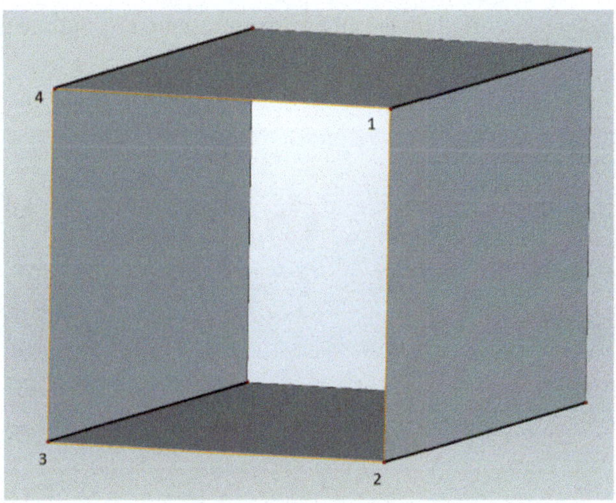

Fig. 5. Sequence of points for square holes, examination of the shape and depth of the sample. Length of the side of a square was measured from 1 to 2, 2 to 3 etc.

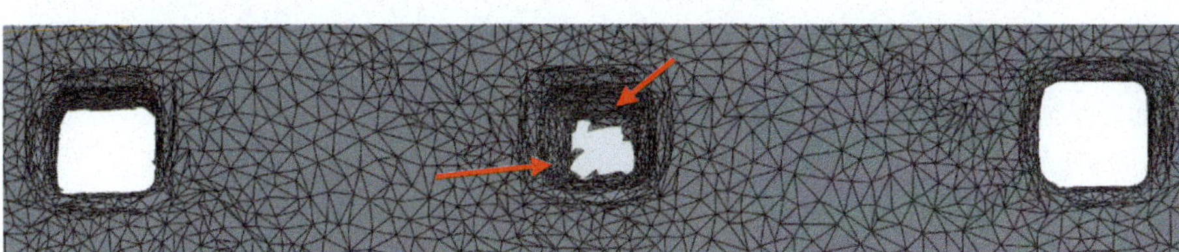

Fig. 6. Front side of square holes with different angles, same size.

Another shared quality for horizontal scans is that they were well mark imperfections of printing process. As visible at Fig. 8 around the openings there are small gaps in

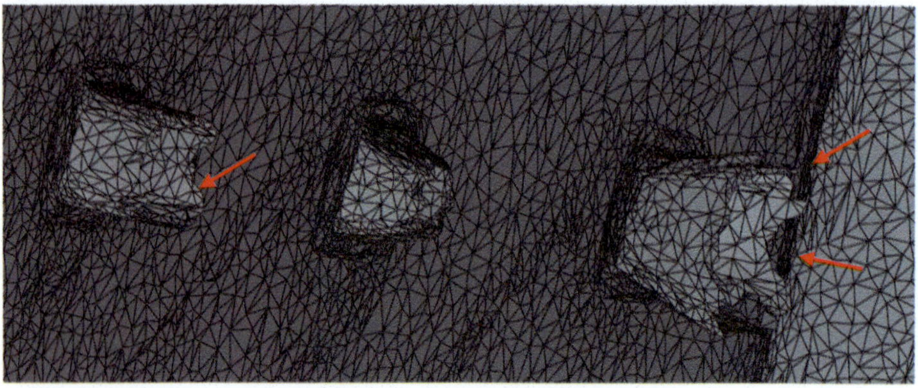

Fig. 7. Back side of the square holes with different angles, same size.

contours of the holes. At virtual images they are represented by outline dents around the holes (Fig. 9).

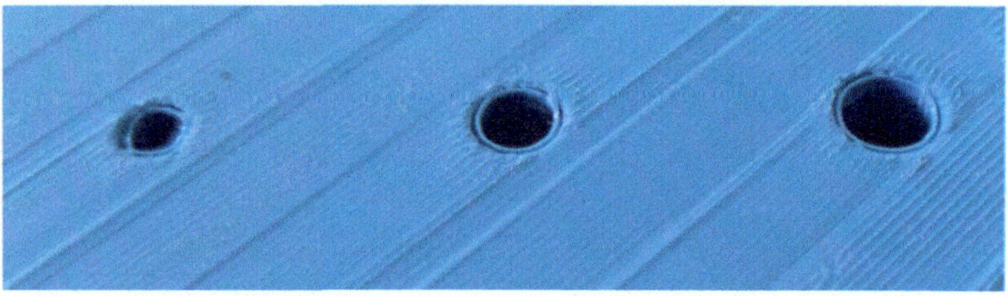

Fig. 8. Imperfections of printing process around the holes.

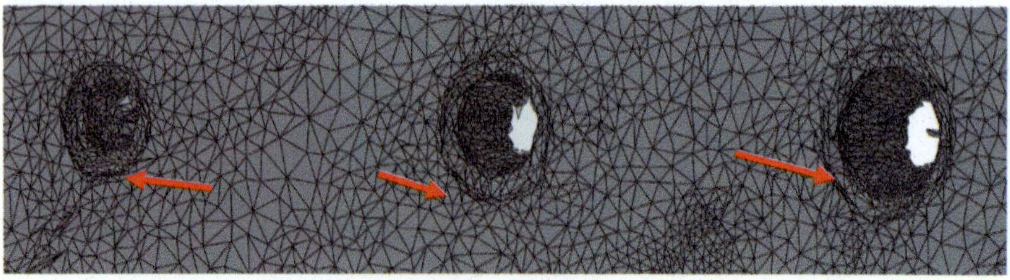

Fig. 9. Representation of the imperfections in printing process around the holes.

The 45 degree angle scan for vertical sample turned out to be most deformed and inaccurate to the real image. Square openings appeared as circular ones and surface had unclear mesh, almost like shredded mesh as presented at Fig. 10. Inspection of individual scans form before stitching the final result also did not bring better results as at this point already shape was distorted. Due to those issues, this particular scan was excluded from comparative analysis of the dimensions.

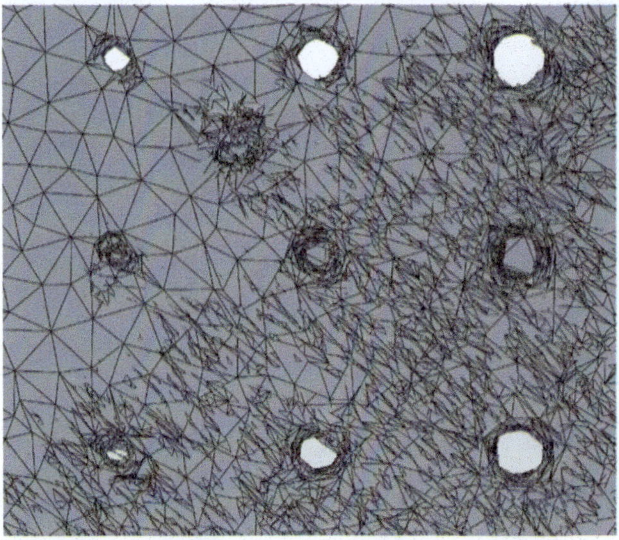

Fig. 10. Deformed square holes for vertical scan at 45-degree angle.

In the matter of depth of the openings, generally the smaller the hole, the lower depth value was achieved. Best results were observed for walls with no tilt accept 45 degree angle horizontal scan and for outside tilt beside 80 degree angle horizontal scan.

From the scans were collected values of dimensions. For each hole there was calculated arithmetic average value of diameter for circular opening or length of the side for the square. Information of those values and how they are related to the original default value are placed at Fig. 11, divided for shapes and angles of the inside wall of the openings.

After further analysis of individual measurements there was noted that most of the holes have tendency to slight elongation in direction parallel to the freeform. However, the deformation is not bigger than the 0.7 mm and mainly present for outside tilted openings. In the same axis for vertical scans depth of the walls have values higher even for about 2 mm.

3.2 Analysis of the Freeform

For all of the scans wider freeform is clearly visible. Horizontal scans achieved higher precision in case of depth for the freeform, at most of its points model have openings at the bottom, but walls are around 5 mm deep as in the original project. However, there are few ghost points – artifacts in the structure where no surface should be placed, example is set at Fig. 12. Those points exist for all of the scanning angles, but least amount is present for 80 degree angle scan. Highest density of ghost points is observed between "peaks" of the freeform.

Thinner freeform is quite irregular in depth, in most cases it is only marked on the surface, with maximum around 1 mm deep instead of 5 mm. In that case there are also missing bottom parts and floor of the freeform is partially opened, least amount of

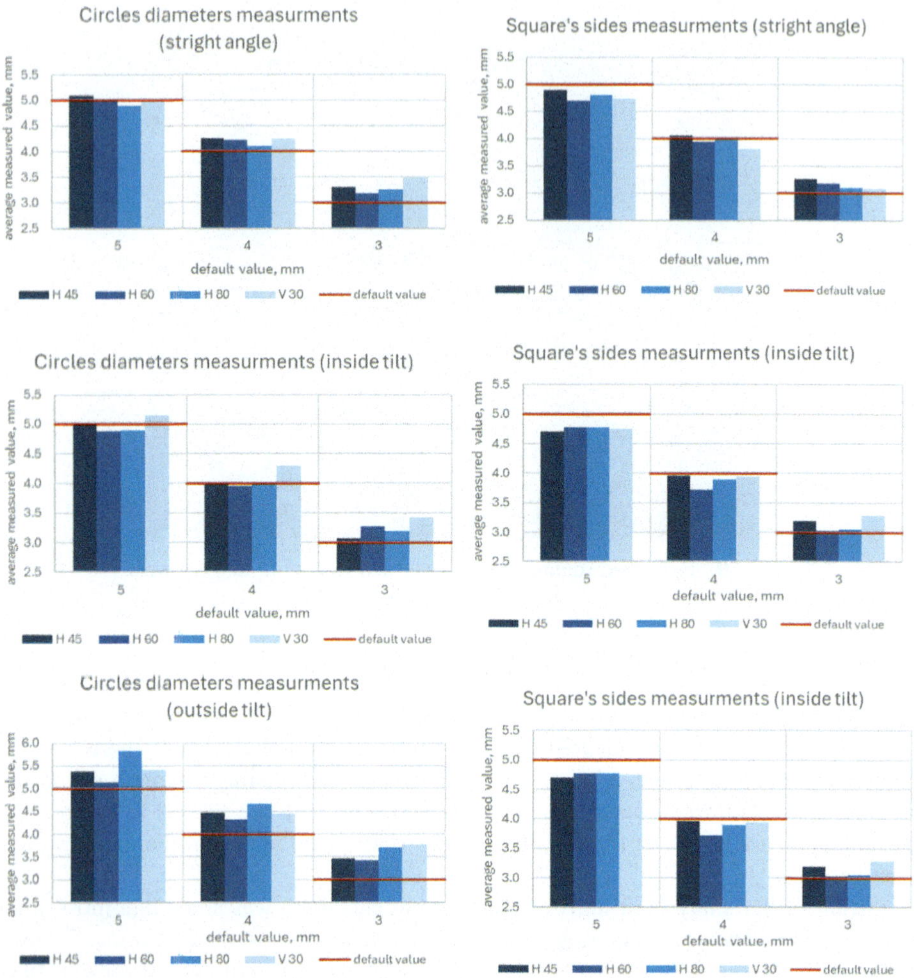

Fig. 11. Characteristics of the dimensions of the openings, where H – horizontal scan, V – vertical scan, numbers in series describe the angle of the scan.

openings is for 45 degree angle scan (Fig. 13). Nevertheless, shape is still preserved for the entire length of the structure.

In case of vertical scans, wider freeform is visible, depths fluctuate between 4 and 5 mm, bottom of the structure is mostly absent. Results are similar to the horizontal scans. It is important to underline the fact that sample was standing on the side closest to the thinner freeform during measurements. For 45 degree angle, thinner freeform is noticeable but less than 1 mm deep. For 30 degree angle depth achieves values around 3 mm but is not consistent, there is clear link between walls of the structure - Fig. 14, roughly around the rotation axis of the sample.

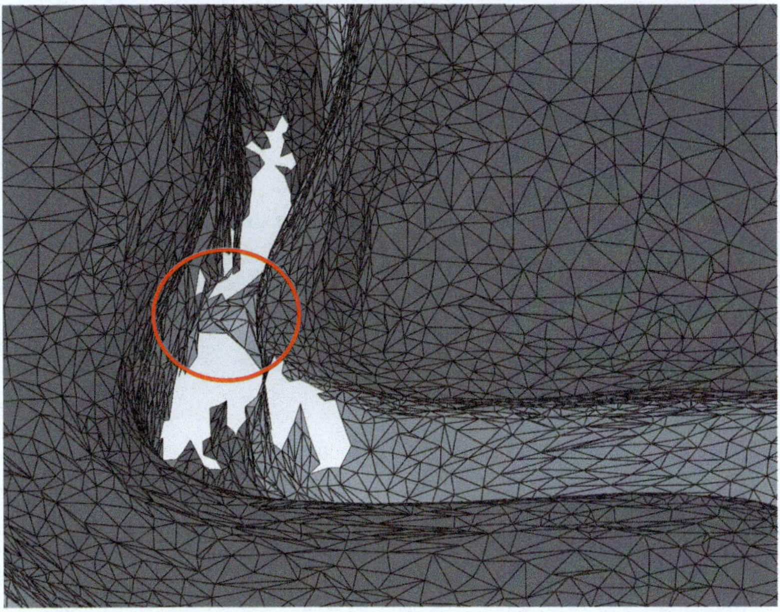

Fig. 12. Example of ghost point in free form for 80 degree scanning angle.

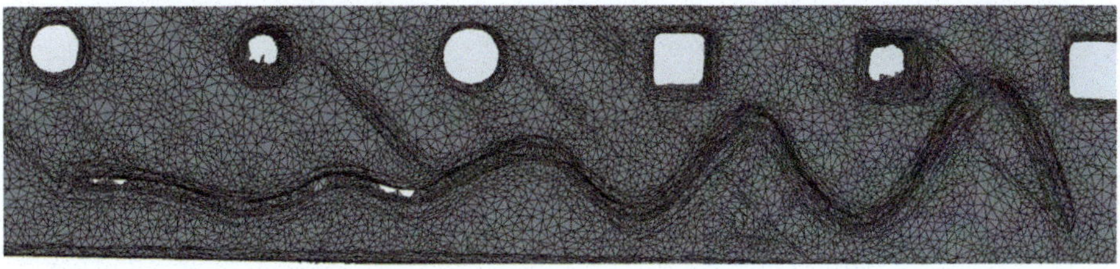

Fig. 13. Thinner freeform for 45 degree angle scan.

4 Discussion

Results presented at Fig. 11 show that dimensions values of small structures can fluctuate in both sides but there are noticeable tendencies for shapes of the openings but also for wall tilts of the holes. In general walls with no tilt presented values closest to original ones and squares beside the smallest one had values lower than default.

Depth of the measurements was the most complex to observe as involves more parameters. It is dependent on angle of scanning and wall tilt but also in same cases squares achieved better values than circles. To be able to further discuss depth it is necessary to proceed more research.

Freeforms had the best results for horizontal scans. Use of freeform gives many possibilities to evaluate out not only geometry but most of all changes in depth, under condition that structures are wide enough for the examination. However, there were

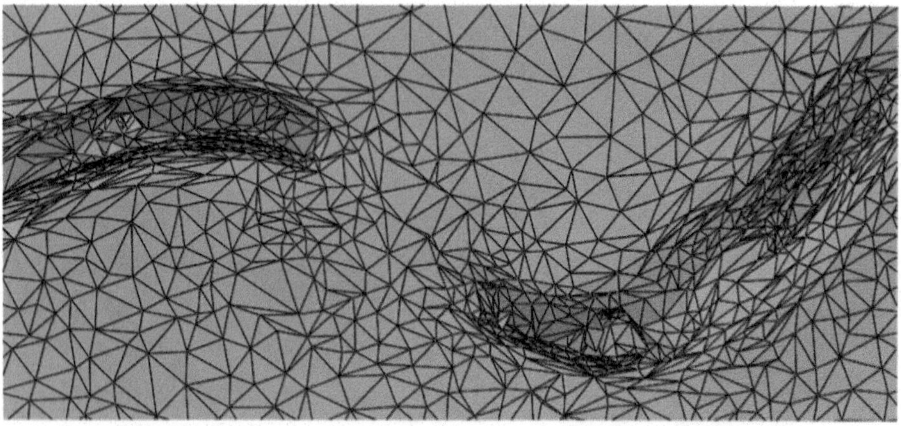

Fig. 14. Closed walls of the thinner structure for 30 degree angle scan.

repeated problem of ghost points, which require additional caution during measurements and analysis.

The selected manufacturing technology allows you to observe the limitations of the scanning processes and detect sample imperfections such as lack of surface flatness, curl, and shape deviations.

5 Conclusions

Conducted research shows limitations of the 3D scanning method, especially how essential may be the selection of right angle for scanning specific kinds of openings. Firstly, for the smallest size of the opening, lower quality mesh was received. When add manufacture imperfections, the information of the holes are distorted to much to properly measure and compare.

For the same shape and angle but different wall tilt of the hole results can shift from the best to the least accurate in the full set. For irregular shapes such as freeform, choice of horizontal scan was proven to be the most beneficial. Nevertheless, to ensure that there are required obligated test in various orientations of the sample and freeform on the plane in relation to the scanner in vertical position. However, even for horizontal position bottom layers of freeforms were absent or create at corrupt depth with presence of ghost points.

The practical application of the research results concerns the use of research for low-time geometric measurements, thus avoiding time-consuming measurements using a coordinate measuring machine. This may enable the selection of the correct measurement method knowing the limitations and capabilities of this technology, especially for measuring geometric features that are difficult to measure. The CMM has limitations related to the dimensions of the probe tips of the measuring machine and the grip part.

One of the possible paths to research development is to measure limit values of the tilt and scanning angles, in order to describe further limitations of measuring method.

References

1. Sargentis, G.-F., et al.: 3D scanning/printing: a technological stride in sculpture. Technologies **10**, 9 (2022). https://doi.org/10.3390/technologies10010009
2. Piperi, E., Galantucci, L., Kaçani, J., Bodi, I., Spahiu, T.: From 3D Scanning to 3D Printing: Application in Fashion Industry (2016)
3. Farhan, M., Wang, J.Z., Bray, P., Burns, J., Cheng, T.L.: Comparison of 3D scanning versus traditional methods of capturing foot and ankle morphology for the fabrication of orthoses: a systematic review. J. Foot Ankle Res. **14** (2021). https://doi.org/10.1186/s13047-020-00442-8
4. Paneva, M., Panev, P., Stoimenov, N., Gyoshev, S.: Methodology for 3D scanning of objects. WSEAS Trans. Appl. Theor. Mech. (2023). https://doi.org/10.37394/232011.2023.18.20
5. Nowacki, J., Sieczkiewicz, N., Nocoń, M.: The use of 3D scanning technology in measurements of welding distortions. Biuletyn Instytutu Spawalnictwa 27–36 (2019). https://doi.org/10.17729/ebis.2019.1/3
6. Bastir, M., et al.: Workflows in a virtual morphology lab: 3D scanning, measuring, and printing. J. Anthropol. Sci. **31**(96), 107–134 (2019). https://doi.org/10.4436/JASS.97003. Epub 2019 Aug 30 PMID: 31472012
7. Li, Z., Chang, B., Chang, S., Zhang, H., Liang, Z., Liao, Z., Wang, L., Du, D.: In-situ monitoring of interlayer surface with electron beam luminescence in wire-based additive manufacturing, Measurement **229** (2024). https://doi.org/10.1016/j.measurement.2024.114418

Metrological Research of the Dimensional Accuracy of Fused Deposition Modelling Technology in the Context of Prototyping Small-Size Casting Patterns

Paweł Zmarzły(✉)

Kielce University of Technology, al. Tysiąclecia Państwa Polskiego 7, 25-314 Kielce, Poland
pzmarzly@tu.kielce.pl

Abstract. A paper presents research results on the possibility of applying selected additive technologies to produce casting patterns. A special research model was designed, the geometry of which reflects the features of typical casting patterns, i.e. rounding, casting tilts, and cylindrical shapes. During the production of castings, these types of geometric features can cause significant problems. The Fused Deposition Modelling Technology has been used to produce the research model, considering two variables of printing parameters: printing direction (Pd) and thicknesses of a single material layer (Lt). The next step was a comprehensive assessment of the dimensional accuracy of the research models made using the CMM. Next, the casting patterns have been used to produce casting molds using bituminous molding sand. Each model was used for molding 100 times. After forming cycles to conduct comparative research, the parameters describing the dimensional accuracy of the examined casting patterns were assessed again. Additionally, a visual assessment of the defects and damage to the samples resulting from the forming process was performed. The study concluded that FDM technology is suitable for building small-sized casting patterns. Research has indicated that the formation process may cause insignificant damage to the surface of the casting models, which should not impact the final casting. It should be noted that the forming process itself resulted in reductions in the values of linear dimensions of casting patterns, which is a certain guideline for designers of additive manufacturing processes.

Keywords: Foundry · Casting Patterns · Fused Deposition Modelling Technology · dimensional accuracy · CMM

1 Introduction

Foundry is a commonly used technology for manufacturing various machine parts. This process involves making metal components by filling casting molds with molten metal. There are several casting processes, such as sand casting, spin casting, casting die, and pressure casting. Lost-wax casting can also be used to make precision parts [1, 2].

The sand molding method is the oldest and most popular method due to its simplicity and low cost of producing castings. In the case of this method, it is necessary to create a

M. Pagac et al. (Eds.): SCFF 2024, LNNS 1062, pp. 138–146, 2024.
https://doi.org/10.1007/978-3-031-65656-9_14

casting pattern, i.e. an element that is to replicate the shape of the casting in the casting mold. Casting patterns are made of wood, metal alloys, or plastics [3]. The casting pattern is placed in a molding box and then covered with molding sand, which is compacted manually or by machine. Due to that casting patterns are used many times to make casting molds, they must be characterized by proper strength and precision.

Producing casting patterns using traditional machining methods is a time-consuming process that requires experienced modelers. Moreover, in the case of constructing the first (prototype) castings, in many cases, it is necessary to change the geometry of the casting patterns multiple times, which, in the case of using traditional manufacturing methods, lengthens the production process of the final casting. These disadvantages can be minimized by using additive technologies to create casting patterns. Currently, efforts are being made to obtain final castings with high dimensional and shape accuracy to minimize the requirement for machining operations, which generates additional costs [4–6]. This is specifically important in mass production, e.g. of components for electric and hybrid vehicles [7]. Due to their versatile character, 3D printing technologies are used in many practical applications, e.g. for building hip prosthesis [8]. They are also used in foundry, in particular in the production of foundry molds. The article [9] presents research aimed at the influence of selected printing parameters of the selective laser sintering technology (i.e., laser power, scan speed, and scan spacing (hatch)) on the quality of production of sand casting molds. Upadhyay et al. [10] presented a wide overview of additive methods that can be used to produce sand-casting molds. In particular, the issue of defining the relationship between the 3D printing process parameters, the materials used, and the properties of the printed form was discussed here. Hodder and Chalaturnyk [11] presented the possibility of applying the M-Flex Sand Printer 3D printer to directly produce sand molds. Four different types of sand masses were tested as the building material for the casting mold. They showed that traditional sand mass can be used for additive manufacturing of casting molds, with certain limitations. However, such casting molds are characterized by lower dimensional accuracy and higher values of parameters that describe surface roughness. The paper [12] presents concepts for the integration of modern additive technologies with traditional methods of producing molds and casting patterns. Research has shown that the application of 3D printing allows the production of casting molds with acceptable dimensional accuracy.

Besides creating casting molds, 3D printing technologies can be used to produce casting patterns, which will then be used to make sand casting molds. Badanova et al. [13] presented extensive research on the assessment of the dimensional and shape accuracy of casting patterns and molds made using SLA technology. It has been shown that the shape accuracy described by the roundness deviation is much lower than the dimensional accuracy. The work [14] presents research work aimed at assessing the dimensional accuracy and surface roughness of foundry models made using FDM technology. Two types of materials were tested: PLA and ABS. Research has shown that despite the application of a high degree of filling, casting patterns produced using PLA material have better dimensional accuracy and lower surface roughness. The work [4] presents the influence of layer thickness on the surface waviness of casting patterns produced using three additive technologies, namely FDM, PJM, and SLS technologies. Research has shown that casting patterns made using PJM technology are characterized by the smallest values of surface waviness parameters.

Literature analysis has shown that most research works are focused on the use of 3D printing technology, in particular FDM technology, to create foundry molds. There is limited research work comprehensively assessing the possibility of using FDM technology to produce practical casting patterns and assessing how the forming process affects the change of parameters describing dimensional and shape accuracy. There are a lot of technological parameters in additive manufacturing technologies [15]. Many works have shown that the printing direction (Pd) and the thickness of a single layer (Lt) of building material are crucial [16–19]. Therefore, the paper presents research results considering how the selection of printing direction and the thickness of a single layer of building material affect the dimensional accuracy of real casting patterns, and the impact of the forming process on changing these dimensions is assessed. It should be mentioned that the results presented in the article are part of extensive research aimed at assessing the possibility of adapting selected additive technologies for the rapid building of casting patterns [20].

2 Materials and Method

2.1 Casting Pattern

To fulfill their functions, casting patterns must have certain characteristic casting features, i.e. appropriate casting tilts and surface rounding. This facilitates the subsequent removal of the model from the casting mold. Therefore, a research model was designed that took into account these features. In addition, a functional shape that is often used in castings, i.e. valve-tee, was taken into account. Figure 1 shows drawings of the research model with the analyzed geometric features (X1–X5).

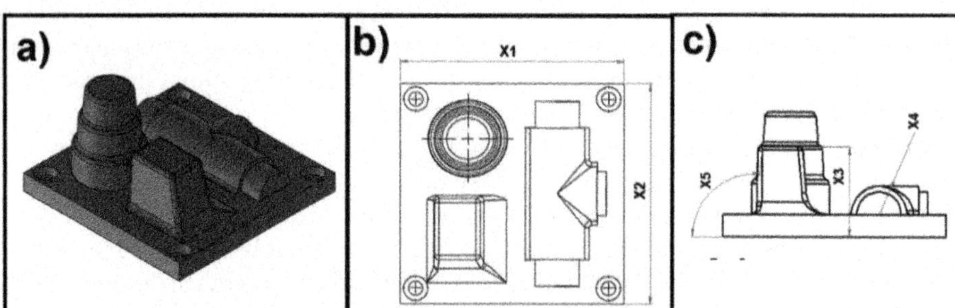

Fig. 1. Research model: a) isometric view in 3D, b) top view, c) side view.

2.2 Fused Deposition Modelling Technology

Because FDM technology is very popular and has the potential to be implemented in the foundry industry, it was decided to build a designed research model (casting pattern) using this technology. The A Dimension 1200 ES printer offered by Stratasys was used to make the casting patterns. Two variable technological parameters were analyzed, namely the single thickness of the material layer (Lt = 0.254 mm, Lt = 0.33 mm) and the printing direction (Pd = 0°, Pd = 90°). ABS P430 was used as the building material, and the support material was P400R Soluble Support, which was removed after printing by bathing in a hydrocarbon solution in a pressure washer. Figure 2 shows a casting pattern produced using two printing variants Pd = 0° and Pd = 90°.

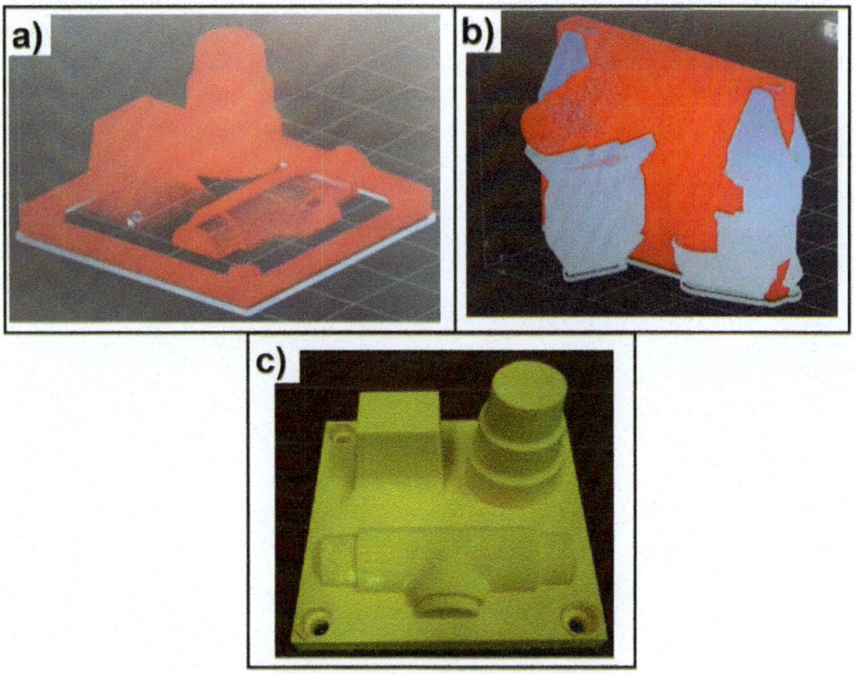

Fig. 2. Casting pattern printed in FDM technology, where the support material is shown in blue color and the building material in red color: a) casting pattern printed using Pd = 0°, b) casting pattern printed using Pd = 90°, c) cleaned casting pattern.

To assess the possibility of practical application of the printed casting patterns, they were used to create foundry molds using bituminous mass in molding boxes. The molding sand was compacted manually using a pneumatic compactor. Before forming, each model was fully metrological tested and then subjected to 100 forming cycles. Next, the damage caused by molding was visually assessed and the geometric features were re-examined.

2.3 Metrological Measurements

To ensure heightened measurement precision and repeatability, a Prismo Navigator coordinate measuring machine from Zeiss equipped with a VAST Gold S-ACC scanning probe was used to evaluate the dimensional accuracy of casting patterns. The portal measuring machine allows for measurements in the range of X = 900 mm; Y = 1200 mm; Z = 700mm. The Maximum Permissible Error of linear dimensions measurements is MPE_E = 0.9 + L/350 µm, therefore this measuring system can be successfully used to measure the geometric features of elements manufactured using FDM technology. To keep the same measurement conditions for geometric features before and after the forming process, a measurement plan was developed and the measurements were performed automatically. The article presents an analysis of the most important geometric features of the casting pattern in terms of correct forming, i.e. overall dimensions (X1–X3), valve radius X4, and casting tilt X5. Figure 3 shows the process of measuring the geometric features of the casting model using CMM.

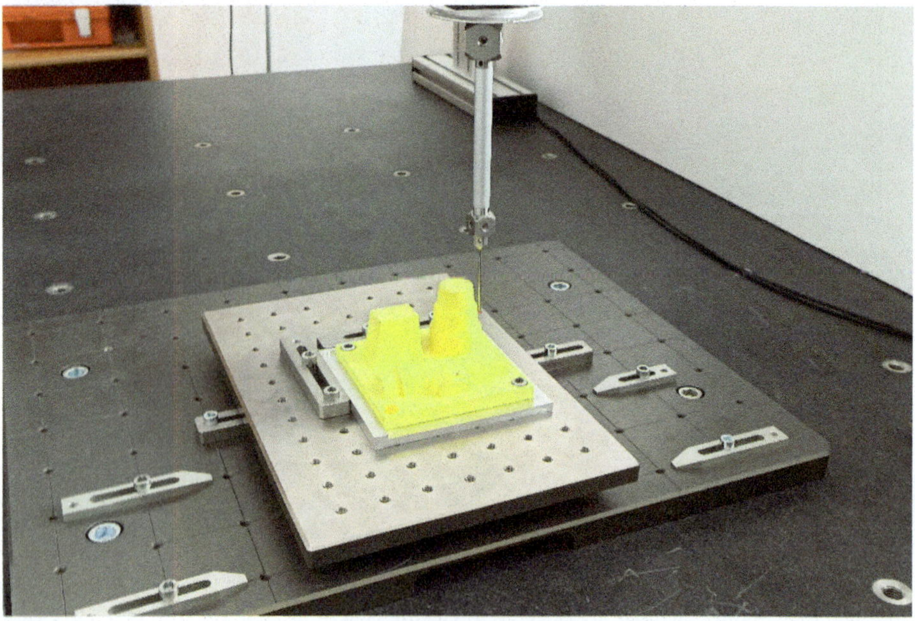

Fig. 3. Measurement of the geometric features of the casting pattern using a coordinate measuring machine.

3 Results and Discussion

The metrological analysis of the manufactured casting patterns was divided into three main parts. First, the dimensional accuracy of the primary models was assessed. Then, the models were used for the forming process described in Sect. 2. The second stage consisted of a visual examination of any wear and defects of the models after the forming process. The third stage involved re-measuring the features of the casting patterns after the molding process. Table 1 shows the average values from measurements of selected geometric features X1-X5 in Fig. 1. The nominal values for the analyzed features are as follows: X1 = 100 mm; X2 = 100 mm; X3 = 40 mm; X4 = 12.5 mm; X5 = 95°.

Table 1. Research results.

Pd, °	0		0		90		90	
Lt, mm	0.254		0.33		0.254		0.33	
	Before molding	After molding	Before molding	After molding	Before molding	After molding	Before molding	After molding
X1,mm	99.97	99.92	99.98	99.92	100.35	100.29	100.24	100.19
X2,mm	99.96	99.90	99.94	99.90	99.96	99.92	100.05	100.03
X3,mm	40.13	40.06	40.37	40.09	40.12	40.09	4.22	40.25
X4,mm	12.49	12.48	12.46	12.45	15.53	12.55	12.54	12.56
X5	94°58'	93°51'	95°14'	94°46'	94°43'	94°14'	94°35'	94°41'

Analyzing the test results presented in Table 1, it can be concluded that the application of a printing direction of Pd = 0° allows to obtain the values of the base dimensions of the X1 and X2 models that are closest to the nominal values, i.e. 100 mm. Similar accuracies were obtained for both single layer thickness Lt = 0.254 mm and Lt = 0.33 mm. For these technological parameters, the measured values were lower than the nominal values, which may be due to the shrinkage of the material due to the cooling process. However, considering the Pd=90° printing variant, it is possible to notice higher values than the nominal values, which is a direct result of the location of the research model on the building platform (see Fig. 2b). Assessing the impact of the forming process on the change in the dimensions of the base of the casting pattern (X1 and X2), it is possible to notice a reduction in dimensions in all analyzed printing parameters compared to the values before modeling. This may be due to the influence of the molding sand on the pattern surfaces. As a result of forming, the surfaces of the model may be abraded with molding sand, which may result in a reduction in overall dimensions. Nevertheless, by analyzing the X3 dimension determining the height of the model, it can be concluded that for all examined technological parameters, values exceeding the nominal values were obtained. Here, we can also see higher accuracy achieved for models printed using the smallest single layer thicknesses of the building material, i.e. Lt = 0.254 mm for both Pd = 0° and Pd = 90°. As in the case of the dimensions determining the base dimensions of the X1 and X2 models, for the X3 the forming process reduces the analyzed dimension for most of the tested printing technological parameters.

Investigating the value of the X4 radius, which is nominally equal to X4 = 12.5 mm, it can be concluded that the application printing direction Pd = 0° and thickness layer Lt = 0.254 mm allowed to obtain the highest accuracy of the assessed geometric feature R = 12.49 mm. However, for the remaining technological parameters, the dimensional deviation from the nominal value was similar and amounted to a maximum of 0.05 mm. If the casting patterns were located on the working platform at an angle Pd = 0°, the forming process resulted in a reduction of the radius of the tested valve. However, for the printing orientation Pd = 90°, the opposite phenomenon was observed.

Considering the influence of FDM printing parameters on the casting tilt described by the X5 feature, it is not possible to draw as clear conclusions as for the other analyzed features. It can be stated that the dimension closest to the nominal value of $X5 = 95°$ was obtained with a single layer thickness $Lt = 0.254$ mm and a printing direction $Pd = 0°$. In the case of this geometric feature, the value of the Lt parameter did not have a major impact on the accuracy of the casting tilt described by feature X5. Similarly to the external dimensions X1, X2, the height of the model X3, and the radius X4, the forming process resulted in a reduction in the value of the casting tilt X5.

Assessing the surface state of the casting patterns after 100 molding cycles, only small scratches were noticed on the external surfaces of the model, which were most exposed to the molding sand as a result of the molding process (see Fig. 4). No cracks or other permanent damage were noticed. It can be concluded that the model can be used for a further series of moldings. Furthermore, in most of the analyzed cases, the forming process resulted in an increase in the roughness of the tested surfaces.

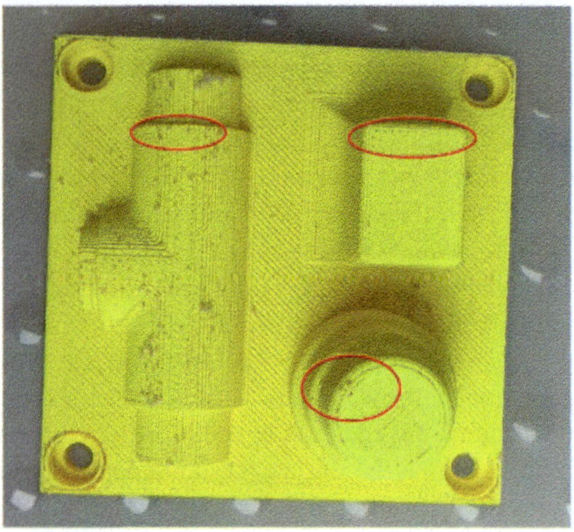

Fig. 4. Casting pattern after molding process.

4 Conclusions and Outlook

The intensive growth of the production industry requires the foundry to be more flexible. In the case of traditional casting in sand molds, the key factor is the efficient production of casting patterns. Therefore, modern methods of producing casting patterns should be desired, because the traditional approach is time-consuming, expensive, and requires qualified personnel. The use of traditional manufacturing methods, e.g. machining, has a negative impact on the environment due to its high energy consumption. Most of these inconveniences can be eliminated by using additive technologies, also known as 3D printing, to make casting patterns.

The main goal of the research presented in the paper was to assess the impact of FDM printing parameters, i.e. printing direction and single layer thickness, on the dimensional accuracy of small-size casting patterns. In addition, it was assessed how the forming process influences the change in the measured geometric features.

The research results showed that FDM technology can be successfully used to build casting patterns. A greater impact on the accuracy of the obtained features was noted for modifying the location on the build platform than for the thickness of a single layer of the applied material. In most of the analyzed geometric features, the dimensions obtained were closest to the nominal value for the location Pd = 0° and Lt = 0.254 mm. However, for this location, the application of higher values of a single layer of applied material does not significantly reduce the accuracy of the tested geometric features. Due to the reduction in printing time, the parameters Pd = 0° and Lt = 0.33 mm are recommended for producing prototype casting patterns. In most cases, the forming process resulted in reductions in the values of the analyzed dimensions, which may result from the wear process (abrasion of the external surfaces) of the model due to contact with the bituminous mass.

Further research will include assessing the possibility of using FDM technology to produce large-sized casting models, e.g. foldable models. In addition, additional technological parameters will be tested.

Funding. The research was financed by the National Centre for Research and Development as part of the Lider XI project, number LIDER/44/0146/L-11/19/NCBR/2020, under the following title: "An analysis of application possibilities of the additive technologies to rapid fabrication of casting patterns".

References

1. Beeley, P.: Foundry Technology, 2nd edn. Butterworth Heinemann (2001)
2. Nwaogu, U.C., Tiedje, N.S.: Foundry coating technology: a review. Mater. Sci. Appl. **02** (2011). https://doi.org/10.4236/msa.2011.28155
3. Wang, S., Miranda, A.G., Shih, C.: A study of investment casting with plastic patterns. Mater. Manuf. Process. **25** (2010). https://doi.org/10.1080/10426914.2010.529585
4. Zmarzły, P., Gogolewski, D., Kozior, T.: Assessment of surface waviness of casting patterns made using 3D printing technologies. Bull. Polish Acad. Sci. Tech. Sci. (2023). https://doi.org/10.24425/bpasts.2023.144585
5. Gogolewski, D., Makiela, W., Nowakowski, L.: An assessment of applicability of the two-dimensional wavelet transform to assess the minimum chip thickness determination accuracy. Metrol. Meas. Syst. **27**, 659–672 (2020). https://doi.org/10.24425/mms.2020.134845
6. Grochała, D., Grzejda, R., Parus, A., Berczyński, S.: The wavelet transform for feature extraction and surface roughness evaluation after micromachining. Coatings **14** (2024). https://doi.org/10.3390/coatings14020210
7. Skuza, A., Jurecki, R., Szumska, E.: Analysis of the operating parameters of electric, hybrid, and conventional vehicles on different types of roads. Open Eng. **13** (2023). https://doi.org/10.1515/eng-2022-0443
8. Gogolewski, D., Zmarzły, P., Kozior, T.: Multiscale analysis of functional surfaces produced by L-PBF additive technology and titanium powder Ti6Al4V. Materials (Basel) **16** (2023). https://doi.org/10.3390/ma16083167
9. Casalino, G., De Filippis, L.A.C., Ludovico, A.D., Tricarico, L.: An investigation of rapid prototyping of sand casting molds by selective laser sintering. J. Laser Appl. **14** (2002). https://doi.org/10.2351/1.1471561

10. Upadhyay, M., Sivarupan, T., El Mansori, M.: 3D Printing for Rapid Sand Casting—A Review (2017). https://doi.org/10.1016/j.jmapro.2017.07.017
11. Hodder, K.J., Chalaturnyk, R.J.: Bridging additive manufacturing and sand casting: utilizing foundry sand. Addit. Manuf. **28** (2019). https://doi.org/10.1016/j.addma.2019.06.008
12. Shah, M., Patel, R.D., Pande, S.: Additive manufacturing integrated casting—a review. Mater. Today Proc. **62** (2022). https://doi.org/10.1016/j.matpr.2022.03.413
13. Badanova, N., Perveen, A., Talamona, D.: Study of SLA printing parameters affecting the dimensional accuracy of the pattern and casting in rapid investment casting. J. Manuf. Mater. Process. **6** (2022). https://doi.org/10.3390/jmmp6050109
14. Hafsa, M.N., Ibrahim, M., Wahab, M.S., Zahid, M.S.: Evaluation of FDM pattern with ABS and PLA material. In: Applied Mechanics and Materials (2014). https://doi.org/10.4028/www.scientific.net/AMM.465-466.55
15. Pagac, M., Hajnys, J., Ma, Q., Jancar, L., Jansa, J., Stefek, P., Mesicek, J.: A review of Vat photopolymerization technology: materials. Polymers (Basel) **13**, 598 (2021)
16. Kozior, T., Kundera, C.: Viscoelastic properties of cell structures manufactured using a photo-curable additive technology—pjm. Polymers (Basel) **13** (2021). https://doi.org/10.3390/polym13111895
17. Kozior, T., Kundera, C.: Rheological properties of cellular structures manufactured by additive PJM technology. Teh. Vjesn. **28** (2021). https://doi.org/10.17559/TV-20191007145545
18. Gogolewski, D.: Multiscale assessment of additively manufactured free-form surfaces. Metrol. Meas. Syst. **30** (2023). https://doi.org/10.24425/mms.2023.144393
19. Gogolewski, D., Zmarzły, P., Kozior, T., Mathia, T.G.: Possibilities of a hybrid method for a time-scale-frequency analysis in the aspect of identifying surface topography irregularities. Materials (Basel) **16** (2023). https://doi.org/10.3390/ma16031228
20. Zmarzły, P., Gogolewski, D., Kozior, T.: An analysis of application possibilities of the additive technologies to rapid fabrication of casting patterns. https://lider-odlewnictwo-druk3d.tu.kielce.pl/

New Hybrid Algorithm in Surface Topography Evaluation

Damian Gogolewski(✉)

Faculty of Mechatronics and Mechanical Engineering, Kielce University of Technology, al. Tysiąclecia Państwa Polskiego 7, 25-314 Kielce, Poland
dgogolewski@tu.kielce.pl

Abstract. The paper evaluates the possibility of using Fused Deposition Modeling additive technology to produce casting patterns. The study evaluated the effect of the forming process on the variation of surface irregularity distributions. A hybrid model based on power spectral density and one-dimensional discrete wavelet transformation was used as a tool. The study was carried out considering many parameters of the wavelet transform evaluating their influence on the filtering results. The use of both, two algorithms for the evaluation of surface topography together allows in-depth analysis and diagnostics of the surface in terms of the impact of variation on the subsequent properties of products. The analysis showed that there is a relatively small influence of a certain number of cycles of the forming process; however, changes in the magnitude of individual surface irregularities before and after the forming process were noted in the spectra. The research fills a research gap in the field of comprehensive diagnosis and evaluation of profiles and the influence of the forming process on the variation of surface irregularity distributions of additively produced casting patterns.

Keywords: Wavelet transformation · Power spectral density · Surface topography · Additive technology · Foundry industry

1 Introduction

The surface topography of the machine components plays a key role in terms of the future operation of final products. Individual irregularities affect reliability, durability or functional properties [1, 2]. The development of surface metrology determined by the evolution of science and technology, ever-increasing miniaturization or the pursuit of ever-increasing efficiency of machinery and equipment results in the need to develop alternative methods to evaluate the distribution of surface irregularities. Adequate diagnostics of surface topography, identification of additional features, irregularities or defects allow the introduction of additional treatments to improve the quality of machine part components [3, 4]. In recent years, methods have been developed [5, 6] that allow the evaluation of surface irregularity distributions. Particularly important from this point of view are multiscale methods [7, 8] that allow the analysis of individual surface features with respect to scale. By defining the scale of observation of these features, it is possible to find relationships between functional properties, surface quality, and the process and its parameters.

M. Pagac et al. (Eds.): SCFF 2024, LNNS 1062, pp. 147–155, 2024.
https://doi.org/10.1007/978-3-031-65656-9_15

One of the multiscale methods applied in the assessment of surface irregularity distributions is the wavelet transformation [9–11]. Used in both discrete and continuous approaches, it allows the evaluation of signals containing random irregularities and the identification of key features in them that can affect the subsequent performance of finished products. A wide spectrum of the mother wavelet defined by specific properties offers potentially large possibilities in the aspect of diagnosing surface with a random, non-periodic distribution of irregularities. Based on the analysis of the current state-of-the-art, it can also be concluded that the hybrid combination of several algorithms allows in-depth, complete, comprehensive evaluation of signals/profiles [12, 13]. Therefore, it is crucial to search for hybrid approaches to identify profile features that are not captured during analysis using only one algorithm. In recent years, it can also be noted that an increasingly used method in the assessment of surface topography is power spectral density [14–16]. The method allows analysis of surface irregularities taking into account frequency components and allows quantitative comparison of surface quality by harmonics. Power spectral density represents a square value of amplitudes of surface roughness profiles as a function of wavelength [17].

The development of modern methods for evaluating surface topography is of particular importance in terms of identifying defects in components manufactured using additive technologies. Additive technologies are increasingly applied in various industrial fields [18, 19]. Therefore, the evaluation of both tribological and mechanical properties [20, 21] as well as surface topography [22, 23] and dimensional and shape accuracy [24, 25] is crucial in terms of the practical applicability of unconventionally manufactured parts.

The paper evaluates the possibility of applying a new hybrid algorithm based on power spectral density carried out for signals resulting from the use of a one-dimensional discrete wavelet transform. The novelty of the paper is the use of a complex approach to the evaluation of surface using a modern, multiscale hybrid method, taking into account filtration parameters. The research fills a research gap in the field of comprehensive diagnosis and evaluation of profiles and the influence of the forming process on the variability of surface irregularity distributions of casting models produced additively.

2 Materials and Methods

The potential to use additive technology to build casting patterns and assessing the impact of the forming process on surface topography was evaluated for one of the most popular additive technologies. Fused Deposition Modeling (FDM) technology was selected for the study. The study was carried out taking into account the parameters of the printing process. The FDM technology used a Dimension 1200ES printer and ABS material P430, the technological parameters were: layer thickness Lt = 0.254 mm or Lt = 0.33 mm and the printing direction 0° or 90°. In order to more precisely determine the possibility of applying additive technologies to the production casting patterns, the shape of the samples was defined by the typical features of a casting model, including the specific inclination, rounding and typical features of models that are produced in the foundry industry. An isometric view of the sample is shown in Fig. 1.

Surface profile measurements were carried out using a Talysurf PGI1200, Taylor Hobson stylus profilometer. Due to the size of the irregularity of the surface profiles, as

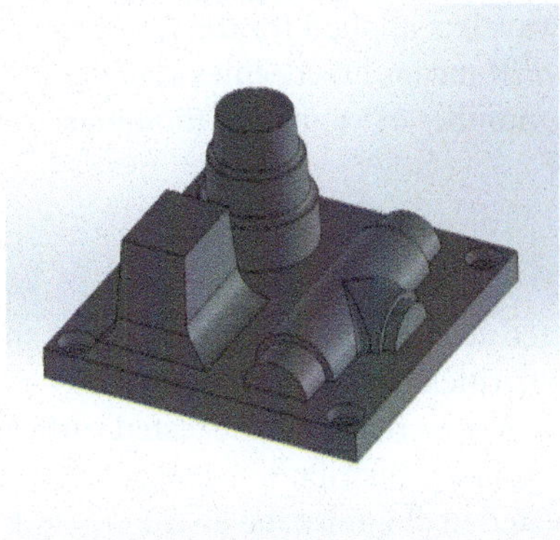

Fig. 1. An isometric view of the sample – CAD model

well as the dimensional limitations of the samples, it was reasonable to set the evaluation length as large as possible. In accordance with DIN EN ISO 4288, the sampling length for a potential filtration should be set at 8 mm. The measurement was carried out at a speed 0.5 mm/s and a sampling density 1 μm. This instrument is equipped with a diamond blade with an angle of 90° and a rounding radius of 2 μm. An example of a profile before and after the forming process is shown in Fig. 2.

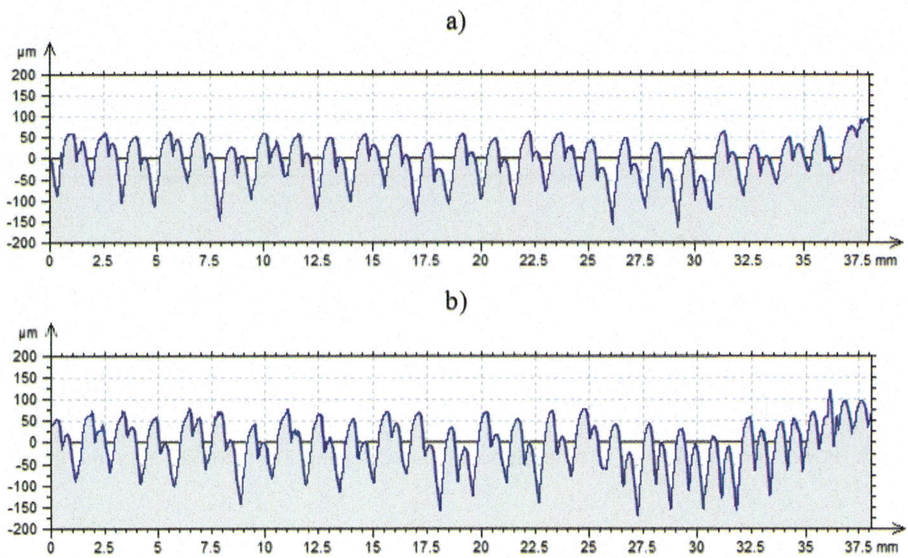

Fig. 2. Example of surface profiles a) before b) after, the forming process

The study using a multiscale method, wavelet transformation, was carried out using selected mother wavelets. The differentiation of the various forms of the wavelets, their specific properties made it possible to assess the impact of different types of wavelets on the filtration results. The wavelets applied for the research were: db1, db6, db10, coif2, coif5, sym2, sym8, where *dbN* means Daubechies wavelet families *coifN* – coiflet, *symN* - symlet, and *N* means the number of vanishing moments.

3 Results

The research was conducted to evaluate the influence of the forming process on the variability of surface irregularities distributions. The study evaluated the character, size and quantity of individual features created on the tested surface. The first aspect analyzed was the evaluation of irregularity distribution using power spectral density. Random surface profiles were evaluated to diagnose the distribution of irregularities. An example spectrum is shown in Fig. 3.

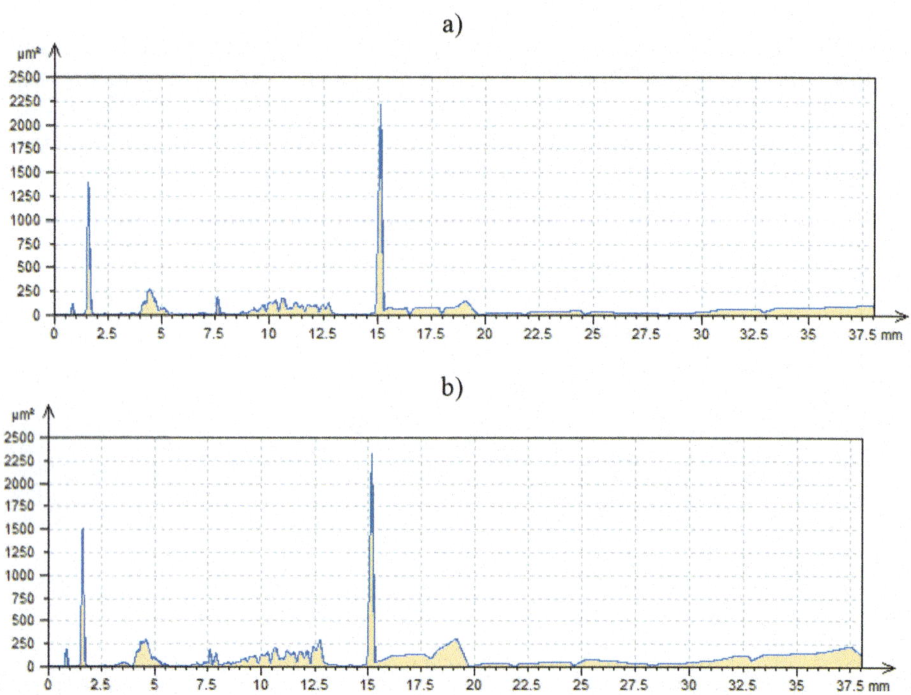

Fig. 3. Power spectral density of an example of surface profiles a) before b) after, the forming process

Analysis of power spectral density for particular samples showed that there are differences for samples before and after the forming process. However, it can be noted that the variations are relatively small. In the example comparison shown in Fig. 2, it can be concluded that the forming process resulted in a more irregular surface after the process. An increase in the height of individual irregularities can be noted, but the characteristic peaks in the figure before the forming process were reflected in the distribution after the process. For both profiles, the dominant irregularities of a periodic nature correspond to a value of about 1.7 mm, but the occurrence of a larger distortion corresponding to about 15 mm was also noted. The selected measurement strategy made it possible to detect a potential error of the manufacturing process, the deformation of the finished products.

The random, irregularity in the distribution of irregularities of samples produced additively determines the need for a more complex and comprehensive assessment. Therefore, it was considered advisable to develop an algorithm that would allow the evaluation of topography irregularities at different scales. Potentially, the analysis concerning the scale will allow diagnostic evaluation of the manufacturing process, including, in particular, in terms of assessing errors in the process of building, solidification of successive layers of the model. The methodology of the hybrid approach is based on the performance of the power spectral density of signals generated at successive levels of wavelet decomposition. The evaluation of the signal considering the irregularity scale allow the identification of specific information that is often invisible in the PSD analysis of the input signal. Figure 4 shows the results of power spectral density for the detail signals generated by the wavelet decomposition process.

PSD analysis carried out for detail signals created after wavelet decomposition showed that the selection of wavelet parameters is important in terms of the results obtained. Mother wavelets with a shorter support are more sensitive to individual irregularities and at subsequent levels of analysis filter them out in the form of sharper characteristics, local relatively narrow peaks. Nevertheless, a in-depth analysis of the surface profiles of the models before and after the forming process showed that there are no relatively large differences in the resulting results. Depending on the level of decomposition, there were only slight changes in the values obtained in the results of analysis using power spectral density.

The width of the support and the properties of the individual wavelets affect the smoothing of the signal during wavelet decomposition. Low-pass and high-pass filtering using wavelets with longer support resulted in the appearance of additional components for the entire length in the form of a non-regular band of relatively small amplitude. An in-depth analysis of such showed that the resulting distortion was caused by the forming process, the influence of the molding compound on the irregularities of the surface topography. This information was not highlighted for a wavelet with a shorter support. This trend was observed for most of the analyzed profiles after the forming process.

The comparative study conducted allows to conclude that as the width of the support increases, the individual information obtained from the analysis of detail signals using power spectral density is highlighted with different intensities. For wavelets with short support, the individual information was highlighted in the form of sharp peaks, while no significant differences were noted in the aspect of profiles before and after the forming process. The different situation is in the case of mother wavelets in a larger support width, which allow more smooth filtration. For these wavelets, the dominant components were analogously filtered out, however, also visible in the images created using power spectral density is the information potentially created by the forming process. For individual wavelengths, relatively small values of amplitudes were recorded for the profile after

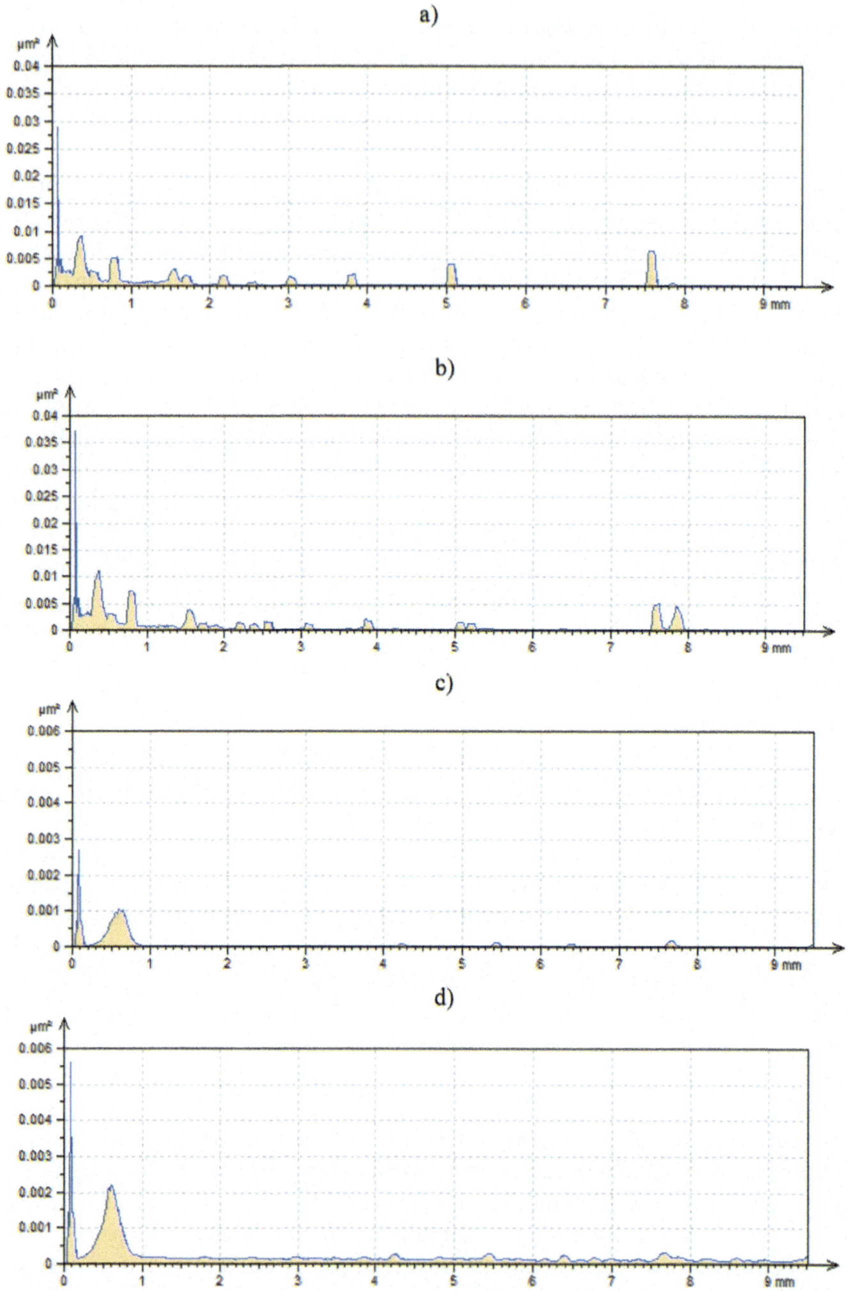

Fig. 4. Power spectral density of an example of detail signal on the fifth level of decomposition a) before the forming process, mother wavelet db1, b) after the forming process mother wavelet db1, c) before the forming process mother wavelet db10, d) after the forming process mother wavelet db10

the forming process, which were not recorded for the profiles measured before forming. Thus, it can be concluded the forming process affected the distribution of surface irregularities, and the proposed algorithm provides information that can potentially affect the exploitation, functional properties of final products.

4 Conclusions

The paper evaluates the possibility of using modern filtering methods for signal diagnosis. The study was carried out for the surface profiles of samples produced by FDM additive technology, taking into account the influence of the forming process on the variation, distribution of surface topography irregularities.

1. The use of an algorithm that is a hybrid of power spectral density and wavelet transformation made it possible to indicate differences in individual distributions with respect to scale. The proposed hybrid algorithm allows the identification the components of profiles that can potentially affect the subsequent exploitation properties of final parts. The combination of both methods makes it possible to provide additional information about the evaluated profiles and the influence of the forming process on their variability.
2. Based on the study, it was noted that the use of wavelets with a smaller support width obtained a sharper spectral density, while the appearance of additional components for the entire length in the form of an irregular band of relatively small amplitude was noted as the length of the support increased.
3. A complex surface topography assessment of additively manufactured parts with simultaneous analysis of multiscale filtering parameters is novel approach and fills a research gap related to in-depth surface diagnostics. The research makes a positive impact to the development of advanced diagnostic systems for fast and complex evaluation of 2D surface irregularities.

Future research will focus on optimizing the selection of wavelet decomposition parameters to identify the components of surface irregularities as well as assessing the possibility of identifying additional features to non-periodic profiles.

Acknowledgements. The research was financed by the National Centre for Research and Developments part of the Lider XI project, number LIDER/44/0146/L-11/19/NCBR/2020, under the title "An analysis of application possibilities of the additive technologies to rapid fabrication of casting patterns".

References

1. Zmarzły, P.: Influence of the internal clearance of ball bearings on the vibration level. Eng. Mech. **2018**, 961–964 (2018). https://doi.org/10.21495/91-8-961
2. Zmarzły, P.: Influence of bearing raceway surface topography on the level of generated vibration as an example of operational heredity. Indian J. Eng. Mater. Sci. **27**, 356–364 (2020)

3. Maleki, E., Bagherifard, S., Bandini, M., Guagliano, M.: Surface post-treatments for metal additive manufacturing: progress, challenges, and opportunities. Addit. Manuf. **37**, 101619 (2021). https://doi.org/10.1016/j.addma.2020.101619
4. Zmarzły, P.: Analysis of technological heredity in the production of rolling bearing rings made of AISI 52100 steel based on waviness measurements. Materials (Basel) **15**, 3959 (2022). https://doi.org/10.3390/ma15113959
5. Bigerelle, M., Galliere, A., Santana, Y.Y., Morvan, H., Dubar, M., Trelcat, J.F., Boilet, L., Paris, E.: A multiscale topographical surface analysis of high entropy alloys coatings by laser melting. Materials (Basel) **16** (2023). https://doi.org/10.3390/ma16020629
6. Adamczak, S., Zmarzły, P., Stępień, K.: Identification and analysis of optimal method parameters of the V-block waviness measurements. Bull. Pol. Acad. Sci. Technol. Sci. **64**, 325–332 (2016). https://doi.org/10.1515/bpasts-2016-0037
7. Edjeou, W., Cerezo, V., Zahouani, H., Do, M.T.: Contribution of multiscale analysis to the understanding of friction evolution of aggregates surfaces. Surf. Topogr. Metrol. Prop. **11** (2023). https://doi.org/10.1088/2051-672X/acb95d
8. Maleki, I., Wolski, M., Woloszynski, T., Podsiadlo, P., Stachowiak, G.: A comparison of multiscale surface curvature characterization methods for tribological surfaces. Tribol. Online **14**, 8–17 (2019). https://doi.org/10.2474/trol.14.8
9. Li, L., Zhang, W., Wang, J., Ruan, X., Li, L., Xie, M.: Multiscale characterization and contact performance analysis of machining surfaces. Sci. Rep. **13**, 1–15 (2023). https://doi.org/10.1038/s41598-023-36907-6
10. Gogolewski, D., Zmarzły, P., Kozior, T.: Multiscale analysis of functional surfaces produced by L-PBF additive technology and titanium powder Ti6Al4V. Materials (Basel) **16**, 3167 (2023). https://doi.org/10.3390/ma16083167
11. Pahuja, R., Ramulu, M.: Study of surface topography in abrasive water jet machining of carbon foam and morphological characterization using discrete wavelet transform. J. Mater. Process. Technol. **273**, 116249 (2019). https://doi.org/10.1016/j.jmatprotec.2019.05.030
12. Gogolewski, D., Zmarzły, P., Kozior, T., Mathia, T.G.: Possibilities of a hybrid method for a time-scale-frequency analysis in the aspect of identifying surface topography irregularities. Materials (Basel) **16**, 1228 (2023). https://doi.org/10.3390/ma16031228
13. Dong-wei, W., Wei-dong, W., Jia-jun, H., Wei-guo, Z., Lai, L.: Experimental study of cavitation noise characteristics in a centrifugal pump based on power spectral density and wavelet transform. Flow Meas. Instrum. **94**, 102481 (2023). https://doi.org/10.1016/j.flowmeasinst.2023.102481
14. Krolczyk, G.M., Maruda, R.W., Nieslony, P., Wieczorowski, M.: Surface morphology analysis of duplex stainless steel (DSS) in clean production using the power spectral density. Meas. J. Int. Meas. Confed. **94**, 464–470 (2016). https://doi.org/10.1016/j.measurement.2016.08.023
15. Li, Q., Deng, Y., Li, J., Shi, W.: Roughness characterization and formation mechanism of abrasive air jet micromachining surface studied by power spectral density. J. Manuf. Process. **57**, 737–747 (2020). https://doi.org/10.1016/j.jmapro.2020.07.039
16. Podulka, P.: Selection of methods of surface texture characterisation for reduction of the frequency-based errors in the measurement and data analysis processes. Sensors **22**, 791 (2022). https://doi.org/10.3390/s22030791
17. Kubiak, K.J., Bigerelle, M., Mathia, T.G., Dubois, A., Dubar, L.: Dynamic evolution of interface roughness during friction and wear processes. Scanning **36**, 30–38 (2014). https://doi.org/10.1002/sca.21082
18. Pagac, M., et al.: A review of vat photopolymerization technology: materials, applications, challenges, and future trends of 3D printing. Polymers (Basel) **13**, 1–20 (2021). https://doi.org/10.3390/polym13040598
19. Upadhyay, M., Sivarupan, T., El Mansori, M.: 3D printing for rapid sand casting—a review. J. Manuf. Process. (2017). https://doi.org/10.1016/j.jmapro.2017.07.017

20. Kozior, T.: Rheological properties of polyamide PA 2200 in SLS technology. Teh. Vjesn. (2020). https://doi.org/10.17559/TV-20190225122204
21. Kundera, C., Martsynkowskyy, V., Gudkov, S., Kozior, T.: Effect of rheological parameters of elastomeric ring materials on dynamic of face seals. Procedia Eng. 307–313 (2017). https://doi.org/10.1016/j.proeng.2017.02.230
22. Gogolewski, D., Kozior, T., Zmarzły, P., Mathia, T.G.: Morphology of models manufactured by SLM technology and the Ti6Al4V titanium alloy designed for medical applications. Materials (Basel) **14**, 6249 (2021). https://doi.org/10.3390/ma14216249
23. Tato, W., Blunt, L., Llavori, I., Aginagalde, A., Townsend, A., Zabala, A.: Surface integrity of additive manufacturing parts: a comparison between optical topography measuring techniques. Procedia CIRP **87**, 403–408 (2020). https://doi.org/10.1016/j.procir.2020.02.082
24. Beniak, J., Križan, P., Šooš, Ľ., Matuš, M.: Research on shape and dimensional accuracy of FDM produced parts. IOP Conf. Ser. Mater. Sci. Eng. **501**, 012030 (2019). https://doi.org/10.1088/1757-899X/501/1/012030
25. Tiwari, K., Kumar, S.: Analysis of the factors affecting the dimensional accuracy of 3D printed products. Mater. Today Proc. **5**, 18674–18680 (2018). https://doi.org/10.1016/j.matpr.2018.06.213

Novel Topology Optimization Strategy Utilizing Multi-stage Acceleration Method

Martin Sotola(✉), Pavel Marsalek, and David Rybansky

VSB—Technical University of Ostrava, 17. Listopadu 2172/15, Ostrava, Czechia
martin.sotola@vsb.cz

Abstract. Topology optimization is a widely used technique for obtaining optimal material distribution. The material distribution usually must have a minimum (maximum) objective function and fulfil constraint functions. Optimization requires tuning of parameters to get optimal shape. The process is a time-consuming task due to the need for multiple runs. Neglecting the process might lead to "false-optimal" results. This paper aims to present a novel strategy using a multi-stage acceleration method. Novelty is presented in refining the initial guess of the following stage using results from the previous one. The multi-stage method is presented in three ways; Transferring between coarser mesh to fine one, restarting using a new filter, or restarting continuation. The usability of the strategy is presented using practical optimization of spring-loaded camming devices, especially its cams. Topology optimization is performed using the software MATLAB.

Keywords: Topology Optimization · Camming Device · Multi-Stage Method · Time Reduction

1 Introduction

Finding effective material distribution is a crucial task for any designer trying to pursue mass reduction. Finding a balance between fulfilling all criteria (such as stiffness, mass, stress conditions etc.) is a time-consuming but important process. One of the tools is using some form of structural optimization. Ground research can be found at the beginning of the 20th century [1]. Special optimization is topology optimisation (TO). TO is trying to find optimal material distribution according to the minimum objective function and meeting constraint functions. Usability of TO is conditional on using numerical methods such as finite element method, finite volume method, meshfree methods etc.

A typical problem in structural optimization is finding minimum compliance with volume constraint. It means finding the optimal distribution of material to have maximal stiffness using reduced volume. It should be noted if the volume is reduced compared to the current solution, it usually leads to more compliant designs even though we maximize stiffness. A problem using the finite element method can be defined as [2]

$$\min_{x} F = \sum_{i}^{N} \int_{\Omega_i} \{\varepsilon_e(u(x))\}^T \{\sigma_e(u(x))\} dV, \tag{1}$$

M. Pagac et al. (Eds.): SCFF 2024, LNNS 1062, pp. 156–165, 2024.
https://doi.org/10.1007/978-3-031-65656-9_16

$$G_0(x) = \sum_i v_i x_i - v_{frac} \sum_i v_i \leq 0, \tag{2}$$

$$0 \leq x_i \leq 1, \quad i = 1, \ldots, N, \tag{3}$$

where F is the objective function, $\{\varepsilon_e\}$ is the strain vector, $\{\sigma_e\}$ is the stress vector, $u(x)$ is the displacement, x is the design variable, G_0 is the volume constraint, v_i is the volume of element, v_{frac} is the volume fraction and N is the number of elements. Equations (1) and (2) are formulating problems. The design variables are influencing overall stiffness and therefore resulting displacement. Equation (3) defines the possible range of design variables for all elements.

Topology optimization has applications in various fields, from prosthetics [3, 4], aerospace [5] to sports applications. In sports, TO helped optimize bicycle parts [6], and TO also helped reduce the weight of motorcycle parts [7]. Using TO leads to lighter design, however, comparing reduction to the whole assembly is neglectable. Out of all possible fields, optimizing mountaineering equipment has large potential due to the enormous effect of weight reduction. Optimization of mountaineering protection equipment was performed in the paper by authors Rybansky et al. [8], using TO.

Temporary rock-climbing protection devices are divided into two groups: passive and active. Active ones usually utilize some form of mechanism, on the other hand, passive ones are "simple" wedges. Common active protection is a spring-loaded camming device (SLCD), see Fig. 1a. SLCD is usually constructed using two pairs of cams. Cams have prescribed shapes defined by logarithmic spirals. An example of a cam can be seen in Fig. 1b.

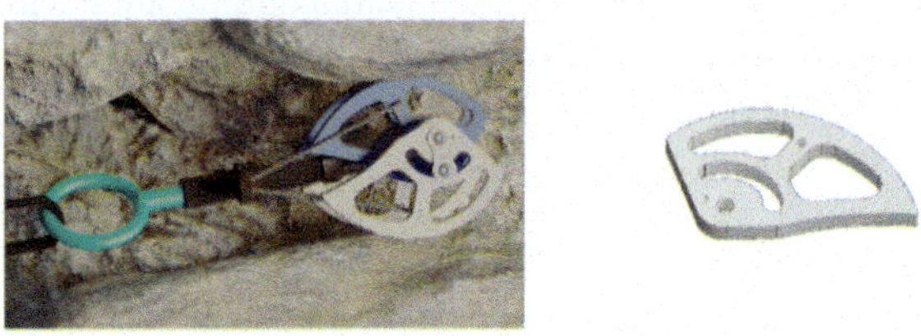

(a) (b)

Fig. 1. Render of spring-loaded camming device and its cam

However, performed TO had limitations to optimization and final design. Optimization is performed using plane stress which limits possible results to have a uniform thickness. Another limitation is in prepared boundary conditions which do not yield fast solutions. Tuning optimization parameters or finding new suitable optimization formulations was restricted mainly by solution time. But users of TO must always try to find better solutions and better shape.

Topology optimization is an overall demanding task to solve. A large portion of solving time is dedicated to solving equilibrium equations. Many authors are trying to reduce this complexity.

Authors Zheng et al. [9] presented a possible reduction of degrees of freedom of inactive elements. Inactive elements usually have small non-zero stiffness, but their displacements can be skipped in solution. Criteria determining inactivity can be using design variables (especially nodal values) or using displacement saturation. Authors Amir et al. [10] proposed the usage of an iterative solver for solving equilibrium equations and using suitable preconditioners. These authors focused on lowering complexity of equilibrium equations. The principles can be applied on any mesh.

2 Material and Methods

Every optimization has its specifics and therefore shape is highly dependent on optimization formulation as mentioned in previous research by authors [11]. Performing multiple runs of optimization can be time-consuming. Overall optimization complexity can be reduced even more. This part focuses on introducing novel multi-stage acceleration method and prepared benchmark.

2.1 Multi-stage Acceleration Method

The novel method is aimed at changing the initial guess of design variables. Currently, the initial guess is uniform using volume fraction throughout all design domains. Alternatively, it is possible to prepare random distributions of design variables or deliberately prepare distributions. The latter alternative leads to possible acceleration and getting higher resolution results.

Before mentioning the process, it is suitable to remember the conservative approach to optimization. The approach works with sequential optimization with slightly lowering volume up to the desired value. It usually leads to robust design. However, it has two main problems. The first problem is that the evolution of results might not lead to the global minimum. The second crucial problem is preparing new geometry and its finite element model which might not be a simple task. Therefore, this approach is not economical and is time-consuming.

However, one could use design variables from the first stage instead of creating a new geometry. Obtained design variables can be used as initial guesses for the following stage. A novel approach could be viewed as a modified conservative one based on mapping results between stages. To maximize potential stages should have two meshes (coarser and fine). The novel approach has also origins in the work of authors Lin et al. [12]. The authors used two-phase optimization; the first phase had a coarser structured mesh. Elements were split into 4 new ones creating a new mesh. Afterwards results from the first mesh were mapped onto a new mesh. Structural meshes however cannot be used on various applications. This paper prepares a methodology for mapping values between two unstructured meshes using nodal design variables. The principle of mapping is like the filtering method and its weight matrix. In this case, mapping is performed using the following formula.

$$x_{f_j} = \frac{\sum_{i \in Ne_j} w_{i_c j_f} x_{c_i}}{\sum_{i \in Ne_j} w_{i_c j_f}}, \tag{4}$$

where x_{f_j} is the design variable of j th node of fine mesh, x_{c_i} is the design variable of i th node of coarser mesh, $W_{i_cj_f}$ is the term of rectangular weight matrix for transferring between stages $[W]$ and Ne_j is defining the neighbourhood of node. The construction of matrix $[W]$ follows a similar process as the filter matrix for design variables. Therefore, terms are calculated as

$$W_{i_cj_f} = \begin{cases} R_W - dist(i_c, j_f), & \text{if } dist(i_c, j_f) \leq R_W, \\ 0, & \text{if } dist(i_c, j_f) > R_W, \end{cases} \tag{5}$$

where R_W is the weight radius, the operator *dist* is calculating distance between nodes. The weight radius is chosen like the filter radius; in this paper, they are identical.

One downside is that it is not possible to use symmetry (as in the case of filtering matrix). However, one can still use parallel computing using sub-regions. After mapping, the transfer is finished by converting nodal values to elemental design variables. The whole process is shown in diagram Fig. 2 Figure also displays neighbourhood.

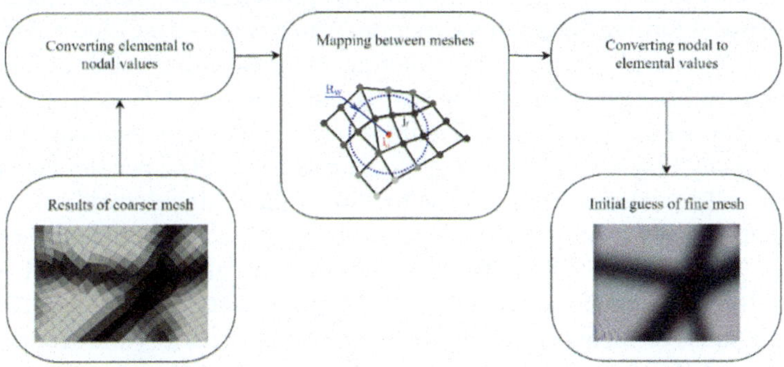

Fig. 2. Diagram of transferring for multi-stage method

This process can be repeated multiple times. To utilize the potential of the method, one should limit the number of optimization iterations. An ideal number of iterations is in the range from 20 to 50. Usually, the first few iterations provide crucial changes to distribution. In this paper, number of iterations is set to 30.

The novel method does not aim to reduce the complexity of equilibrium equations directly but solving it on a coarser mesh with follow-up transformations leads to reduced solution times.

2.2 Optimization Benchmark

For a presentation of a novel acceleration method, optimization of the cam of SLCD is chosen. As mentioned in the introduction, this task has a limitation, mainly solution time is problematic even for plane analysis. The analysis uses multiple load cases which simulate boundary conditions of contact wall-cam with various widths of crack (example can be seen in Fig. 3a).

The benchmark geometry (see Fig. 3b) is obtained from previous research, but it is divided into three regions, one design domain (blue region) and two fixed domains (red

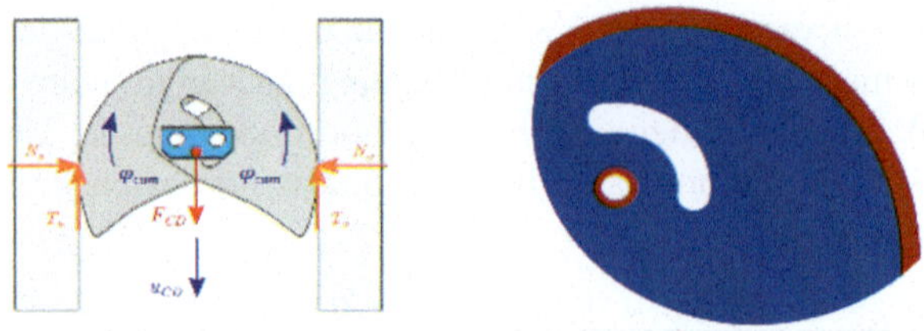

(a) (b)

Fig. 3. Optimization benchmark

region). It is necessary to retain a 5 mm region of contact curve. Also, the region near the pin must have some stiffness. The thickness of the cam is set as 6 mm. This thickness is based on real life cams.

Optimization is set as problem minimum compliance with volume constraint. This allows us to use algorithm Optimality Criterion (which is one of the oldest algorithms for TO). Mesh independency is secured by filtering methods with a radius of 2 mm. In this work, a combination of density [13] and greyscale filter [14] is used. The combination reduces the number of optimization iterations (around 40–60 is enough), however, the results have many grey regions with intermediate densities. The penalty factor of SIMP is chosen as 3. The volume fraction is set to 30% of the design domain.

Two finite element meshes are prepared using ANSYS Workbench Mechanical (Ansys, Inc. Canonsburg, USA). The first mesh has an element size of 1 mm (4610 elements, 4762 nodes) and the second one 0.24 mm (72,865 elements, 73,460 nodes). Linear plane stress elements are used. There are two reasons why the second mesh has large amounts of elements. The first reason is to obtain a result with the best resolution. The second reason is to simulate space with many degrees of freedom. A method suitable for fine mesh should be applicable for volume analysis. Mesh visualization is on the Fig. 4. Additionally, bar elements are used in pinhole for simulating stiff connections. Material data are in Table 1. The material model in this analysis is a linear model defined by Young's modulus and Poisson ratio for aluminium alloy AW2011.

Table 1. Material properties of aluminum alloy AW2011 [8]

Description	Symbol	Value	Unit
Young modulus	E	72 500	MPa
Poisson ratio	μ	0.33	–
Yield stress	σ_Y	260	MPa
Ultimate stress	σ_U	370	MPa
Density	ρ	2 820	Kg m^{-3}

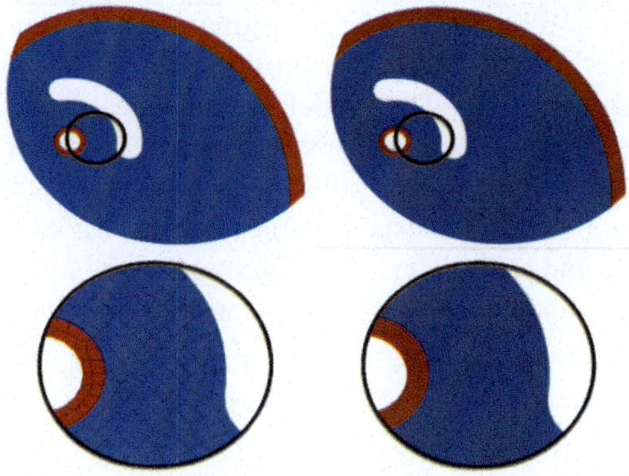

Fig. 4. Finite element meshes, left coarser, right fine

If one would focus on preparing a realistic simulation, the analysis should be modelled with three types of nonlinearities: elastoplastic material model, large deformation, and contact. The main nonlinearity is contact wall-cam. Nonlinearities generally make simulation difficult to solve, therefore it is necessary to try linearizing problems before optimization.

In the previous paper [8], the linearization of contact was by two supports (see Fig. 5a). The first support is in the pin pilot node and represents the symmetry of SLCD. The restricted degree of freedom has determined direction by cam angle and angle between the pin and contact point (see Fig. 5b). The acting force ($F = 1.25$ kN) is perpendicular to this direction. The second support represents a contact point and removes both degrees of freedom. This linearization has one flaw. Changing the contact point for various load cases (various widths of the crack) changes the angle of force and first support. This rotation is performed by transform matrix or using Lagrange multiplier. Both methods however lead to a non-constant stiffness matrix which changes for each load case. For solving the primary variable (displacement), it is necessary to invert the stiffness matrix for each load case. This linearization is labelled as original optimization.

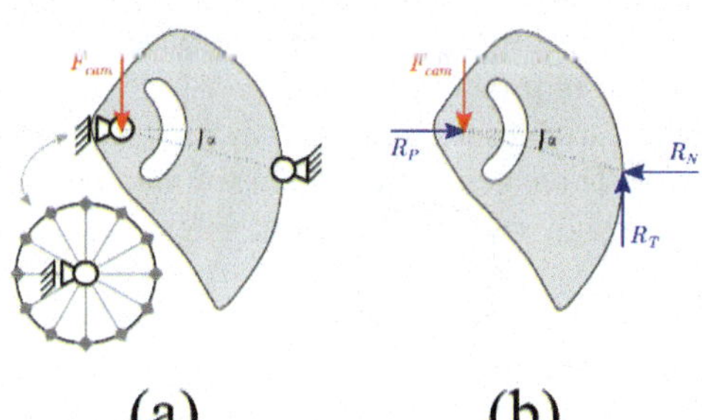

Fig. 5. Original linearization

In this paper, new linearization is proposed. The first support is in the pin and restricts both degrees of freedom (see Fig. 6a). Acting force is prepared before simulation and has a value representing reaction forces of original linearization. These forces are in equilibrium (especially in moment equilibrium), but the finite element method requires to have fully constraint analysis. The model still requires one boundary condition or influencing or adding an artificial spring for stabilization. The latter one is chosen for this paper. Spring is added to the last contact point (Fig. 6b) with a fixed value $k = 10^6$ N/mm. The proposed linearization has a constant stiffness matrix for all load cases (during one optimization iteration) and can utilize preparing one inverse matrix. Direct solutions of linear equations are recommended. This linearization is labelled as optimization using spring.

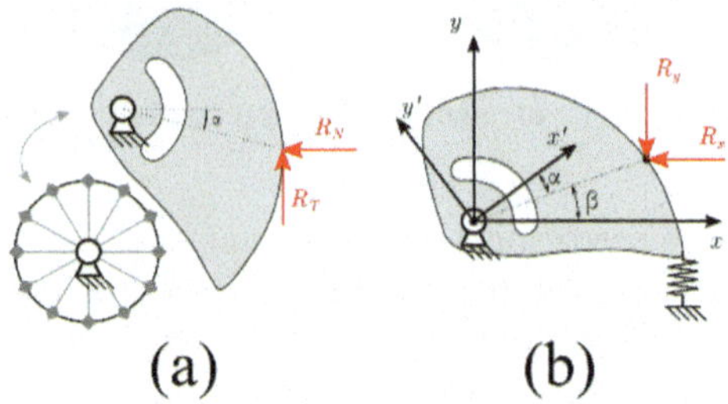

Fig. 6. Linearization using spring

3 Results

For evaluation, it is crucial to monitor the following quantities: number of iterations, normalized objective function and solution time. Volume is not important in the case of plane analysis since the design variable can be used as a multiplier of uniform thickness. These quantities are written in Table 2. For original linearization, linearization using spring and multi-stage optimization. Two things are immediately apparent. Solution time is shortened by new linearization and mainly by multiple stages. The objective function has similar or lower values which is desired. Lowering its value leads to stiffer designs. Therefor multi-stage optimization yields best results. Objective function change in time is displayed in Fig. 7. Almost vertical line is caused by coarser mesh. Value gets almost to the minimal value of previous results.

Figure 8 shows design variables for each optimization (for multi-stage optimizations, both stages are presented). Three geometries seem similar, but second-stage results have clearer details. Getting design variables without intermediate values leads to usable designs.

Alternatively, multiple stages can be used as a restart for analysis. Restart can be performed after a certain amount of iteration. Restart is an ordinary process of optimisation. One can change the formulation and use the initial guess of the previous stage or just continue with the current formulation but with more iterations. Results of each alternative are also written in Table 2 and shown in Figs. 9 and 10. These alternatives improve distribution and minimize intermediate areas. However, they have longer solution time than new linearization results due to the extra seven iterations and process of mapping.

Table 2. Results of optimization

Optimization	Iterations	Objective function (normed)	Time (s)
Original linearization	53	3.88	39,197
New linearization	53	3.88	1099
Multi-stage, transfer	60	3.60	606
Multi-stage, filter	60	3.78	1232
Multi-stage, transfer	60	3.63	1202

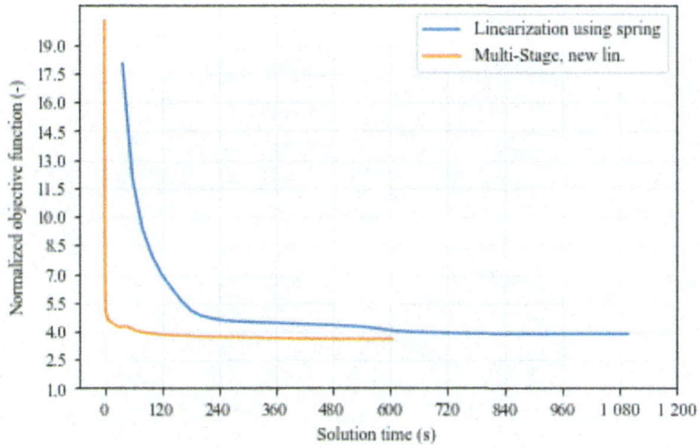

Fig. 7. Normalized objective function versus solution time

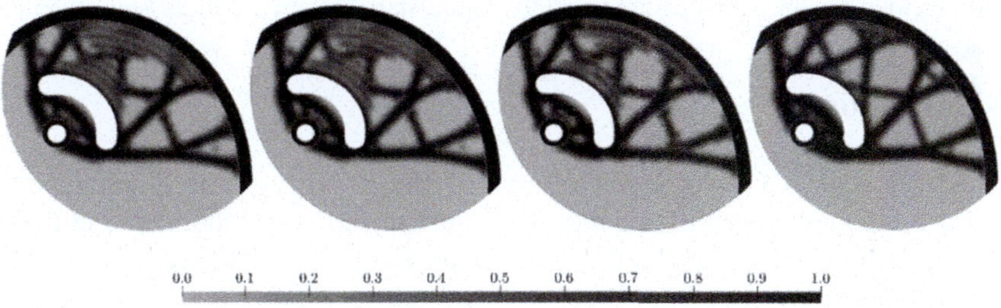

Fig. 8. Field of design variable, from left original linearization, new linearization, multi-stage with transform first stage and second stage

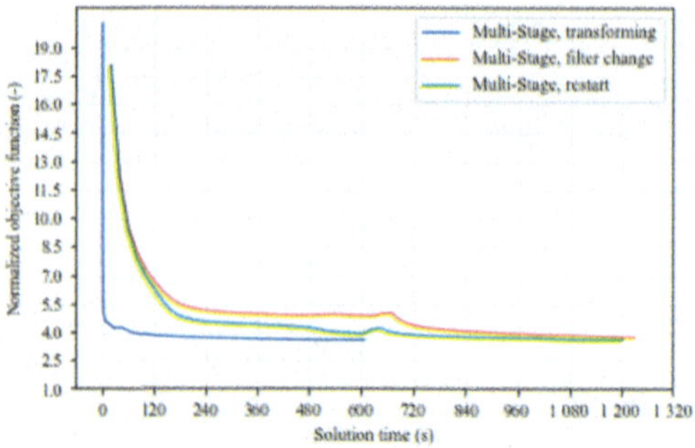

Fig. 9. Normalized objective function versus solution time

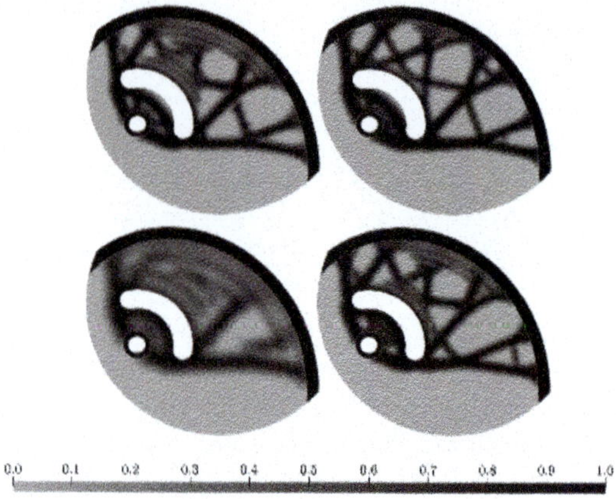

Fig. 10. Field of design variable, top row multi-stage restart, bottom row multi-stage filter change

4 Conclusion

This paper describes a novel topology optimization strategy for accelerating solution time. The main method is utilizing multi-stage optimization by refining the initial guess of each stage. The initial guess is based on the previous result. The method has strength when multiple meshes are used. These meshes should gradually get finer to obtain better detail.

The optimization benchmark of cam optimization is prepared for evaluation of strategy. Cam optimization has main limitations of solution time known from previous research. Therefore, it is a suitable candidate for testing methods. For this paper, new linearization is proposed resulting in shortening time. Optimization is performed using plane analysis using two meshes (coarser and fine).

The authors recommend using a multi-stage method using multiple meshes to achieve detailed shapes in a shorter time than using just fine mesh. Multi-stage optimization leads to a design with fewer intermediate design variables and has clearer shape. This is manifested by smaller objective function (stiffer design). Solution time utilizing two meshes is shortened. This variant is the effective way of optimization. Multi-stage alternatively can be used as a restart tool. It leads to higher solution time but also a smaller objective function. It should be noted that the multi-stage method is also applicable on volume meshes.

Acknowledgements. This work was supported by Specific Research "Experimental and Numerical Modeling for Solving Problems in Mechanics and Biomechanics" (SP2024/037).
This work is an output of the project Computational modelling of ductile fracture of identical wrought and printed metallic materials under ultra-low-cycle fatigue created with financial support from the Czech Science Foundation under the registration no. 23-04724S.

References

1. Michell, A.G.M.: LVIII. The limits of economy of material in frame-structures. Lond. Edinburgh Dublin Philos. Mag. J. Sci. **8**(47), 589–597 (2010)
2. Bendsøe, M.P., Sigmund, O.: Topology Optimization: Theory, Methods and Applications, Second Edition, Corrected Printing. Springer, Berlin (2004)
3. Sotola, M., Stareczek, D., Rybansky, D., Prokop, J., Marsalek, P.: New Design Procedure of Transtibial ProsthesisBed Stump Using Topological Optimization Method," Symmetry, vol. 12, no. 11. MDPI AG, p. 1837, Nov. 06, 2020
4. Fraldi, M., Esposito, L., Perrella, G., Cutolo, A., Cowin, S.C.: Topological optimization in hip prosthesis design. Biomech. Model. Mechanobiol. **9**(4), 389–402 (2010)
5. Shi, G., Guan, C., Quan, D., Wu, D., Tang, L., Gao, T.: An aerospace bracket designed by thermo-elastic topology optimization and manufactured by additive manufacturing. Chin. J. Aeronaut. **33**(4), 1252–1259 (2020)
6. Zeleny, P., Cadek, M.: Topology optimization of a bicycle part. MM Sci. J. **2015**(03), 696–700 (2015)
7. Kudrna, L., Ma, Q.P., Hajnys, J., Mesicek, J., Halama, R., Fojtik, F., Hornacek, L.: Restoration and possible upgrade of a historical motorcycle part using powder bed fusion. Materials **15**(4) (2022)
8. Rybansky, D., Sotola, M., Marsalek, P., Poruba, Z., Fusek, M.: Study of optimal cam design of dual-axle spring-loaded camming device. Materials 18 (2021)
9. Zheng, W., Wang, Y., Zheng, Y., Da, D.: Efficient topology optimization based on DOF reduction and convergence acceleration methods. Adv. Eng. Software **149** (2020)
10. Amir, O., Stolpe, M., Sigmund, O.: Efficient use of iterative solvers in nested topology optimization. Struct. Multidiscip. Optim. **42**(1), 55–72 (2010)
11. Sotola, M., Marsalek, P., Rybansky, D., Fusek, M., Gabriel, D.: Sensitivity analysis of key formulations of topology optimization on an example of cantilever bending beam. Symmetry **13**(4), 712 (2021)
12. Lin, C.-Y., Chou, J.-N., Sigmund, O.: A two-stage approach for structural topology optimization. Adv. Eng. Softw. **30**(4), 261–271 (1999)
13. Bourdin, B., Chou, J.-N., Sigmund, O.: Filters in topology optimization. Int. J. Numer. Meth. Eng. **50**(9), 2143–2158 (2001)
14. Groenwold, A.A., Etman, L.F.P.: A simple heuristic for gray-scale suppression in optimality criterion-based topology optimization. Struct. Multidiscip. Optim. **39**(2), 217–225 (2009)

Preliminary Design of a Single-Phase Linear Generator for Free-Piston Engine Application

Nguyen Huynh Thi[1], Nguyen Van Trang[1(✉)], Huynh Thanh Cong[2], and Dao Huu Huy[1]

[1] Ho Chi Minh City University of Technology and Education, Ho Chi Minh City, Vietnam
trangnv@hcmute.edu.vn
[2] Vietnam National University-Ho Chi Minh City, Ho Chi Minh City, Vietnam

Abstract. This article proposes a single-phase permanent magnet linear generator (PMLG) applied to a free piston engine (FPE) to produce electricity. It has an outer stator and an inner rotor (mover) that is driven by a motor with a linear frequency ranging from 10 to 50 Hz. The generator model has been simulated in 3-D using the Finite Element Method (FEM) through Ansoft Maxwell V16 (AM-V16) software to determine the magnetostatic properties. The MATLAB software is then utilized to obtain basic output parameters such as current, voltage, and power, using these parameters. The simulation and experimental results show that the voltage, current, and power differ insignificantly from 10 to 30 Hz. However, the difference is significant in the speed range from 40 to 50 Hz due to magnetic saturation. The peak power and efficiency are achieved at approximately 140 W and 81% with a load $R_L = 10\ \Omega$ at a linear frequency of 50 Hz. The results from this study show that single-phase linear generators are completely applicable to power generation on FPE.

Keywords: Permanent magnet linear generator—PMLG · Electromagnetics analysis · Finite Element Method—FEM

1 Introduction

Nowadays, the Permanent Magnet Linear Generator (PMLG) is considered a potential candidate for energy conversion devices that have linear frequency. Different from the traditional generator, PMLG employs variable linear motion rather than rotary motion, resulting in a simpler structure and minimal energy loss during the conversion process by eliminating redundant transmission devices [1–3]. With the rapid advancement of high energy density magnetic materials, PMLG is garnering increasing attention for numerous applications due to its enhanced power and reliability [4–6] such as regenerating energy from ocean waves [7–9]. Furthermore, the PMLG serves as a device that converts mechanical energy into electrical energy for use in the Free Piston Linear Engine (FPLE). This technology offers the benefits of decreasing harmful gas emissions [10] and achieving high efficiency by minimizing friction when compared to traditional heat engines. Its variable compression ratio makes it adaptable to a wide range of fuels [11–15]. Presently, the FPLE represents a focal point in research endeavors aimed at enhancing vehicle performance and extending range [16–19].

M. Pagac et al. (Eds.): SCFF 2024, LNNS 1062, pp. 166–175, 2024.
https://doi.org/10.1007/978-3-031-65656-9_17

This article uses a single-phase PMLG with a simple structure for various applications such as creating energy from ocean waves and free piston engines. The generator is designed and examined across a linear frequency range of 10–50 Hz, with a maximum linear displacement of – 20 to 20 mm. The 3D model of generator is simulated using the Finite Element Method (FEM) with AM-V16 to analyze magnetostatics. The simulation results obtained from AM-V16 are then inputted into MATLAB software to determine output parameters such as current, voltage, and power. Finally, the experimental results are compared to evaluate the reliability of the simulation results.

2 Construction of a Single-Phase Linear Generator Model

The article introduces a flat single-phase generator comprising two generating surfaces, with its three-dimensional structure depicted in Fig. 1. The initial dimensions of both the rotor (mover) and stator assemblies are detailed in Table 1 and illustrated in Fig. 2.

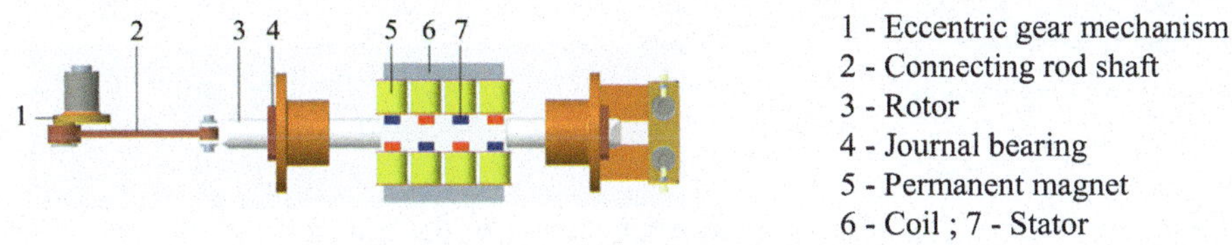

Fig. 1. PMLG applied to FPE system

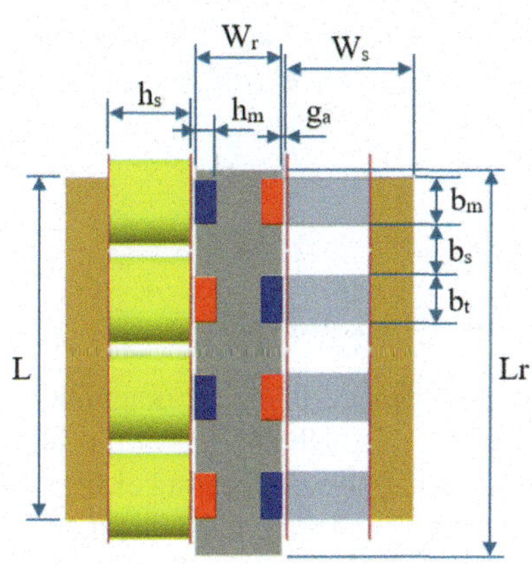

Fig. 2. Cross-sectional structure of PMLG

The PMLG comprises two main components: the rotor and the stator (mover). The rotor features two sides equipped with permanent magnets, with the polarization direction being opposite on the two opposing faces and alternating on the same face. Constructed from aluminum, the rotor is designed to minimize weight. Linear frequency motion is achieved through an eccentric gear mechanism, serving as a substitute for a

Table 1. Specification of PMLG

Variables	Symbol	Value
Stator width	W_s	40 mm
Length of the stator core	L	160 mm
Tooth width	bt	20 mm
Slot width	b_s	20 mm
Height of magnets	h_m	10 mm
Magnets width	b_m	20 mm
Tooth pitch	τ_t	40 mm
Tooth high	h_s	35 mm
Air gap	g_a	1 mm
Length of the rotor	L_r	180 mm
Rotor width	Wr	40 mm
Number of armatures	M_s	2
Number of turns per coil	N_{ph}	135 × 8 coil

free-piston engine. The permanent magnet section of the rotor is crafted from NdFeB rare earth magnets. Horizontal motion of the rotor is facilitated by a motor with a maximum power output of 1.1 kW. The outer stator consists of two multilayer ferrite cores aligned in parallel with the permanent magnets' surfaces. A 1mm clearance is maintained between the magnet and the stator. Each ferrite core is segmented into four teeth, encircled by wound coils. The copper wire's diameter is set at 1.5mm, with 135 turns of wire per tooth.

3 Simulation Study

3.1 Basic Parameters of the Generator

The functioning of a PMLG is grounded in electromagnetic induction and linear motion motor principles. When a magnet is attached to a rotor, it induces a consistent alteration in the magnetic field across the coil, thereby creating electromotive force and resulting in alternating current as the output. The electromechanical characteristics of the generator are significantly affected by the core sizes and winding parameters. To solve this problem, the AM-V16 is used to address all concerns related electrostatics, which are stated in Maxwell's differential equations.

The basic principles governing permanent magnet linear generators are derived from Maxwell's equations, especially those relating to magnetic fields.

From Ampere's law:

$$\nabla \times H = J \tag{1}$$

where J is the current density, H is the field intensity.

According to Gauss law:

$$\nabla \times \mathrm{B} = 0 \quad (2)$$

where B is the magnetic flux density.

Magnetic flux density B can be related to H as:

$$\mathrm{B} = \mu \times \mathrm{H} \quad (3)$$

where μ is the material permeability.

If the material is nonlinear, the permeability is a function of B:

$$\mu = \frac{B}{H} \quad (4)$$

The magnetic flux density B can be expressed in terms of the potential energy vector A as follows:

$$\mathrm{B} = \nabla \times \mathrm{A} \quad (5)$$

where A is the magnetic vector potential.

From Eqs. (4) and (5), it is obtained:

$$\nabla \times \left(\frac{1}{\mu(B)} \times A \right) = 0 \quad (6)$$

The FEM is a numerical technique used to solve Eq. (6), especially for static problems with nonlinear B-H relationships. Figure 3 illustrates the simulated model generated using AM-V16.

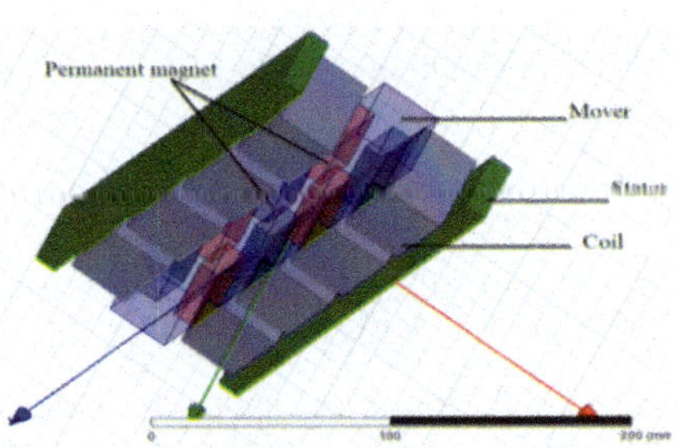

Fig. 3. Simulation model on AM-V16

The simulation results depicted in Fig. 4 indicate that the maximum flux passing through the coil is Br = 1.3T. The magnetic force lines form a closed loop around the coil, thereby generating an induced electromotive force.

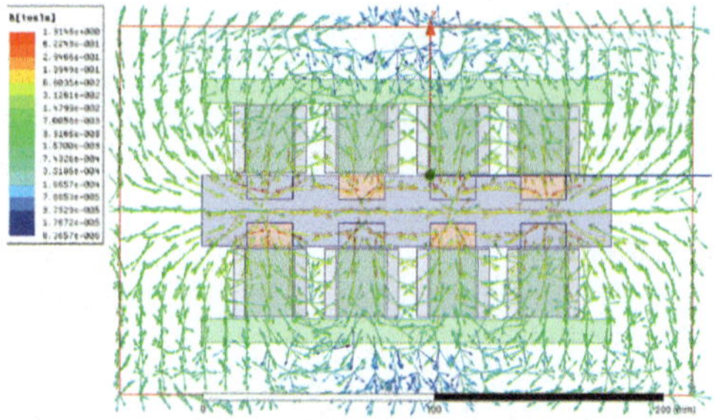

Fig. 4. Direction vector B

3.2 Output Parameters Calculation of Generator

The simulation parameters obtained from FEM, along with the geometric parameters determined in Table 1 will be imported into MATLAB software to model the output parameters of generator. The purpose of the simulation is to determine the current, voltage and power characteristics under a resistive load of RL = 10 Ω. The armature resistance is determined to be Ra = 0.25 Ω and the synchronous inductance of the coil is expressed as Ls = 0.105 H, with all parameters taken from experimental data. The performance of a single-phase generator operating at constant speed is then tested using the equivalent circuit depicted in Fig. 5.

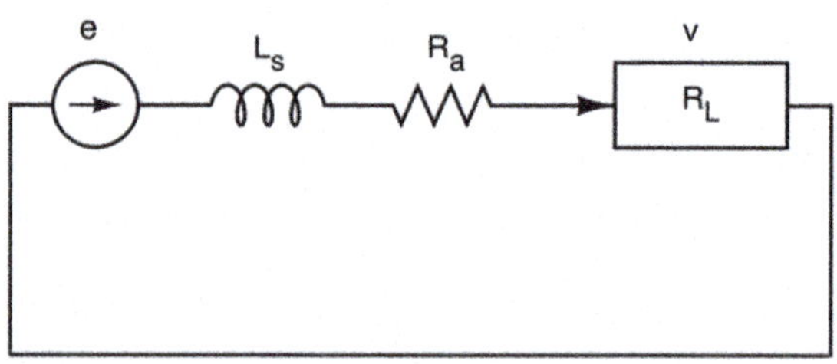

Fig. 5. Equivalent circuit of a single-phase generator winding

The Carter coefficient calculates the equivalent clearance [20]:

$$K_c = \frac{\tau_t(5g_a + b_s)}{\tau_t(5g_a + b_s) - b_s^2} \quad (7)$$

where g_a is the air gap between the rotor and stator, b_s is the slot width; τ_t is the tooth pitch, it may be expressed as the sum of the slot width b_s and tooth width b_t.

The equivalent air gap is calculated:

$$g_{eq} = K_C \times g_a \quad (8)$$

The air gap flux density, taking into account the increase in reluctance, can be expressed as follows:

$$\emptyset = \frac{B_r h_m H_c \mu_0}{h_m H_c \mu_0 - g_{eq} B_r} \quad (9)$$

where Br represents the allowable magnetic flux density in the translator core t, Hc = 90,5000 A/m represents the forced magnetic field intensity, and Hm represents the height of the magnet. The permeability of free space, denoted as μ_0: $\mu_0 = 4\pi \times 10 - 7$ T.m/A.

In Fig. 5, the values of phase induced voltages is:

$$e = R_a \times i + L_s \frac{di}{dt} + V \tag{10}$$

where R_a is the armature resistance, L_s is the synchronous inductance, V is the voltage (V), and i is the current (A).

$$V = i \times R_L \tag{11}$$

where R_L is the resistive load.

The induced voltages in the windings for single-phase is:

$$e_{ph} = K_E \times \cos(\frac{\pi}{\tau} z) \times v(t) \tag{12}$$

where z is the uniformly variable displacement of the mover:

$$z = A \times \sin(2\pi ft - \frac{\pi}{2}) \tag{13}$$

where A represents the displacement of the mover, and f denotes the frequency of the displacement movement.

Speed of the mover v(t) is:

$$v(t) = z' = 2\pi \times A \times f \times \cos(2\pi ft - \frac{\pi}{2}) \tag{14}$$

$$K_E = M_s \times W_s \times N_{ph} \times \Phi \times v_{av} \tag{15}$$

where M_s is the number of armatures, W_s is the stator width, N_{ph} is the number of turns per coil, v_{av} is the average speed of the mover.

Differential equation for phase current

$$\frac{di}{dt} = \frac{e - v - i \times R_a}{L_s} \tag{16}$$

The power . $P_{out} = e_{ph} \times i$ (17)

The displacement with maximum speed from 1 to 6 m/s corresponding to from 10 to 50 Hz.

4 Simulation and Experimental Results

Using the design data and simulation outcomes, an experimental prototype of the PMLG is developed. The linear motion of the mover is driven by a 1.1 kW motor through an eccentric gear mechanism. To adjust the linear frequency and monitor the voltage and current parameters of the motor, the VEICHI AC10 inverter is employed. The experimental arrangement is illustrated in Fig. 6. Experimental data is collected using the dedicated pulse meter X431TOP, equipped with 4 channels, a maximum frequency of 20 MHz, and capable of measuring voltages up to ± 150 V.

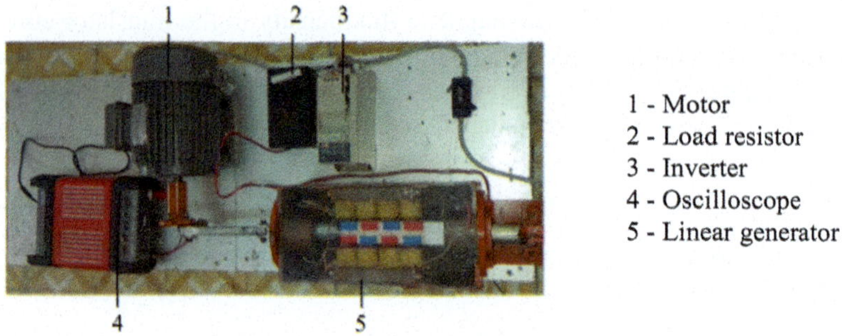

Fig. 6. Experimental model of PMLG

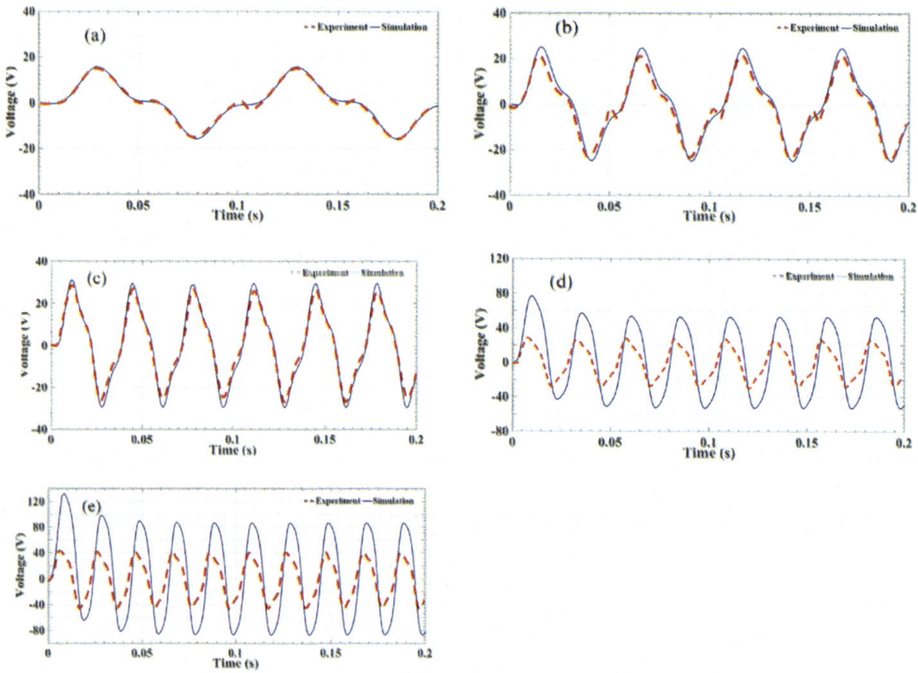

Fig. 7. Voltage waveform of the simulation and experimental results at linear frequency of 10 Hz (a), 20 Hz (b), 30 Hz (c), 40 Hz (d), and 50 Hz (e)

The simulation and experiment have been performed to compare and analyze voltage, current, and power parameters based on load $R_L = 10\ \Omega$ and different frequencies, the frequency is set as 10 Hz, 20 Hz, 30 Hz, 40 Hz, and 50 Hz respectively.

Displacement of PMLG is restricted from − 20 to 20 mm. The waveforms of the simulation and experiment are almost the same, there are still small deviations because the speed mover is difficult to control precisely due to the inertia force. The limit of the operation is 50 Hz because of the highest speed of the motor and the mechanical strength. The simulation results are compared with those experimental results under the same conditions and are shown in Fig. 7.

When the piston frequency increases from 10 to 30 Hz corresponding to the speed of motor movement increasing from 1 to 3 m/s, the voltage increases rapidly from 15 to 30 V according to experimental data and from 17 to 38 V according to the data simulation. In this period, the experimental and simulation do not have much difference, the same for current, and power. The linear frequency from 40 to 50 Hz corresponds to the maximum speed of the motor from 3 to 6 m/s, simulated voltage increased from 60 to 90 V, current increased from 3.5 to 8 A, and peak power at 50 Hz = 720 W. However, the experimental results show that the parameters of the generator insignificantly increased, the voltage increased from 35 to 40 V, and the current increased from 2.8 to 3.5 A with peak power of 140 W at 50 Hz. Current and power are shown in Fig. 8, the magnetic saturation of the generator has reached the limit.

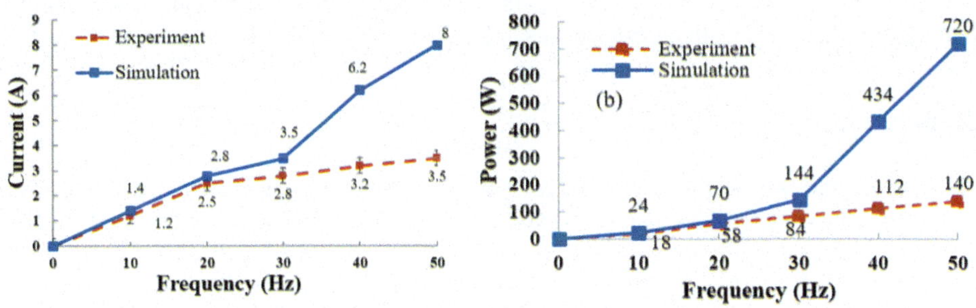

Fig. 8. Current (a) and power (b) at different frequencies

In this model, the generator efficiency is calculated:

$$h = \frac{W_g}{W_m} \times 100\% \tag{18}$$

where W_g is the efficiency of the generator, W_m is the power of the motor used to drive the mover and W_m is collected directly on the inverter $W_m = 171$ W at a linear frequency 50 Hz. Thus, efficiency is h ≈ 81% at 50 Hz power loss due to mechanical friction and magnetic through the air gap.

The displacement speed of the mover can change the determination of voltage, current, and power. In addition, reducing the armature resistance R_a and the synchronous inductance of the coil L_s can increase the power of the PMLG, but this depends on the quality of the winding material. On the one hand, because the generator uses a permanent magnet, magnetic saturation occurs, so to increase the power of the generator, it is recommended to use an electromagnet instead of a permanent magnet or use a linear generator 3-phase.

5 Conclusion

This paper introduces a single-phase PMLG structure specifically designed for Free Piston Linear Engines (FPLEs). It presents 3D schematics, experimental models, and dimensional parameters. Using the Finite Element Method (FEM), simulation parameters including flux distribution density and inductance were obtained using both Ansoft Maxwell V16 and MATLAB software. These simulations are instrumental in determining the power and current characteristics of the PMLG. The simulation and experimental results align closely over the frequency range of 10–30 Hz, showcasing linear increases in voltage, current, and power parameters. However, discrepancies emerge at higher frequencies, specifically from 40 to 50 Hz, attributable to magnetic saturation effects. Peak power and efficiency are attained at approximately 140 W and 81%, respectively, under a load of $R_L = 10\ \Omega$ at a linear frequency of 50 Hz. The experiments therefore demonstrate the suitability of this structure for low-frequency linear oscillation applications, such as ocean wave energy regeneration, free piston engine energy conversion and energy recovery from car vibrations.

References

1. Zou, J., Zhao, M., Wang, Q., Zou, J., Wu, G.: Development and analysis of tubular transverse flux machine with permanent magnet excitation. IEEE Trans. Ind. Electron. **59**(5), 2198–2207 (2012)
2. Jing, H., Maki, N., Ida, T., Izumi, M.: Electromechanical design of an MW class wave energy converter with an HTS tubular linear generator. IEEE Trans. Appl. Supercond. **28**(4), Art. no. 4902504 (2018)
3. Huang, L., Hu, B., Hu, M., Liu, C., Zhu, H.: Research on primary excitation fully superconducting linear generators for wave energy conversion. IEEE Trans. Appl. Supercond. **29**(5), Art. no. 5203405 (2019)
4. Gandhi, A., Parsa, L.: Thrust optimization of a five-phase fault-tolerant flux-switching linear synchronous motor. In: Proceedings of IEEE Industrial Electronics Society Conference, pp. 2067–2073 (2012)
5. Chen, H., Zhan, Y., Wang, H., Nie, R.: A tubular permanent magnet linear generator with novel structure. IEEE Trans. Plasma Sci. **47**(6), 2995–3001 (2019)
6. Chen, H., Zhan, Y., Wang, H., Nie, R.: A novel single-phase tubular permanent magnet linear generator. IEEE Trans. Appl. Superconduct. **30**(4), Art. no. 5202705 (2020)
7. Hodgins, N., Keysan, O., McDonald, A.S., Mueller, M.A.: Design and testing of a linear generator for wave-energy applications. IEEE Trans. Ind. Electron. **59**, 2094–2103 (2012)
8. Kimoulakis, N.M., Kladas, A.G., Tegopoulos, J.A.: Power generation optimization from sea waves by using a permanent magnet linear generator drive. IEEE Trans. Magn. **44**(6), 1530–1533 (2008)
9. Hanipah, M.R., Mikalsen, R., Roskilly, A.P.: Recent commercial free-piston engine developments for automotive applications. Appl. Therm. Eng. **75**, 493–503 (2015)
10. Feng, H., Guo, C., Jia, B., Zuo, Z., Guo, Y., Roskilly, T.: Research on the intermediate process of a free-piston linear generator from cold start-up to stable operation: numerical model and experimental results. Energy Convers. Manage. **122**, 153–164 (2015)
11. Hansson, J., Leksell, M: Performance of a series hybrid electric vehicle with a free-piston energy converter. IEEE Xplore (2007)

12. Mikalsen, R., Roskilly, A.P.: A review of free-piston engine history and applications. Appl. Therm. Eng. **27**, 2339–2352 (2007)
13. Jia, B., Mikalsen, R., Smallbone, A., Paul Roskilly, A.: A study and comparison of frictional losses in free-piston engine and crankshaft engines. Appl. Therm. Eng. (2018)
14. Mikalsen, R., Roskilly, A.P.: The control of a free-piston engine generator, part 1: fundamental analyses. Appl. Energy **87**, 1273–1280 (2010)
15. Yuan, C., Jing, Y., Liu, C., He, Y.: Effect of variable stroke on fuel combustion and heat release of a free piston linear hydrogen engine. Int. J. Hydrogen Energy **44**, 20416–20425 (2019)
16. Vysoký, O.: Linear combustion engine as main energy unit for hybrid vehicles. In: Proceedings of Transtec Prague. Prague: Czech Technical University, pp. 236–244 (2007)
17. Lim, O., Hung, N.B., Oh, S., Kim, G., Song, H., Iida, N.: A study of operating parameters on the linear spark ignition engine. Appl. Energy **160**, 746–760 (2015)
18. Li, L., Luan, Y., Wang, Z., Deng, J., Wu, Z.: Simulations of key design parameters and performance optimization for a free-piston engine. In: SAE, p. 1105 (2010)
19. Sun, P., et al.: Hybrid system modeling and full cycle operation analysis of a two-stroke free-piston linear generator. Energies **10**, 213 (2017)
20. F. W. Carter.: Air-gap induction. Electr World Eng. **38**(22), 884–888 (1901)

Quality Characteristics of Borosilicate Glass Machined Through Abrasive Water Jet: An Optimization and Experimental Approach

Piyush Kumar[1,3], Suryank Dwivedi[1], Akash Nag[2], Ranjib Biswas[3], and Amit Rai Dixit[1](✉)

[1] Department of Mechanical Engineering, Indian Institute of Technology (Indian School of Mines), Dhanbad 826004, India
amitraidixit@iitism.ac.in

[2] Faculty of Mechanical Engineering, VSB—Technical University of Ostrava, Ostrava 70800, Czech Republic

[3] Department of Mechanical Engineering, Jadavpur University, Kolkata 700032, India

Abstract. Borosilicate glass is classified as a crucial engineering material due to its exceptional qualities, such as a low thermal expansion coefficient, reduced density, chemical inertness, electrical non-conductivity, and optical clarity. However, its brittleness and hardness pose significant obstacles to conventional machining methods. Abrasive water jet machining (AWJM) offers a viable alternative, as it effectively addresses these challenges by minimizing both thermal distortion and induced stress on the material being machined.This study seeks to analyze the effect of various machining parameters on the material removal rate (MRR) during AWJM of borosilicate glass. Employing a central composite design (CCD) methodology based on response surface methodology (RSM), experiments were performed using four factors, and five levels were considered for each. The input factors include abrasive mass flow rate (AFR), water pressure (WP), stand-off distance (SOD), and traverse speed (TS) for conducting the experiment. Analysis of variance (ANOVA) revealed that the AFR significantly influenced the MRR, followed by WP and SOD. Single-objective optimization identified an MRR of 13.664 g/min, with specific parameter settings: a WP of 1409.381 bar, an AFR of 52.853 g/min, a TS of 384.153 mm/min, and a SOD of 35.656 mm. The desirability function approach (DFA) was also employed to optimize the input variables and finally validated the MRR through a confirmation test. Therefore, this study offers crucial insights for optimizing machining parameters to enhance cutting efficiency and productivity in industrial applications involving AWJM of borosilicate glass.

Keywords: Borosilicate glass · Unconventional Machining · Abrasive Water Jet Machining · Optimization · MRR

1 Introduction

Borosilicate glass is well-known for its exceptional properties and finds diverse applications owing to its composition, including silica sand, soda ash, and alumina, resulting in a material with a higher melting temperature than ordinary glasses [1]. This

M. Pagac et al. (Eds.): SCFF 2024, LNNS 1062, pp. 176–186, 2024.
https://doi.org/10.1007/978-3-031-65656-9_18

type of glass offers resilience against thermal shock and chemical corrosion, making it suitable for various applications where durability and clarity are the utmost important factors [2]. Its versatility and durability make it suitable for various industrial applications, including lighting equipment, electronics, coatings, laboratory glassware, pharmaceutical packaging, and low-temperature co-fired ceramics (LTCC) applications [3, 4].

Machining borosilicate glass to create various features poses challenges due to its brittle, non-conducting, and chemically inert nature. Significant dust emissions occur during glass manufacturing, with particles averaging about 300 mm in size [5]. Abrasive water jet machining (AWJM) emerges as an optimal and environmentally friendly solution. This method effectively carries away dust generated during machining with the water jet. There is minimal thermal and mechanical distortion on the cut surface, too. Conventional machining methods typically involve significant forces exerted on the material due to tool-workpiece contact, leading to vibration and chattering. Moreover, the high brittleness of borosilicate glass makes it prone to fracture, resulting in material damage and potentially inferior quality of the surface. Furthermore, its chemical inertness and electrical non-conductivity present challenges for machining using other unconventional processes. AWJM resolves these issues by eliminating direct tool-workpiece contact, thereby reducing machining forces and minimizing induced stresses. The cold-cutting process employed in AWJM prevents thermal distortion during machining. Additionally, the flexibility of the AWJM process enables the cutting of complex shape contours, a task often challenging with conventional machining methods [6].

AWJM functions on the principle of harnessing high-velocity water streams, typically reaching speeds of around 200 m/s, to achieve material removal predominantly through shearing and tearing mechanisms [7]. This process entails directing a focused water jet and abrasive particles towards the workpiece while maintaining a certain distance between the work surface and the nozzle. This arrangement facilitates material erosion due to the kinetic energy imparted by the abrasive-laden water jet. The necessary pressure for the water jet is generated by pumps capable of elevating the pressure to levels as high as 400 MPa, drawing water from a reservoir equipped with an accumulator to stabilize pressure fluctuations during operation. As the pressurized water traverses through a micro-sized nozzles, potential energy is converted into kinetic energy, leading to a water velocities of approximately 1000 m/s. Abrasive particles are introduced into the water jet stream at high velocities within a mixing chamber, allowing for precise material cutting upon impact with the workpiece. For effective cutting operation, various input parameters need to be controlled and optimized.

Researchers have extensively analyzed the impact of various input parameters, such as abrasive mass flow rate (AFR), water pressure (WP), traverse speed (TS), and stand-off distance (SOD), on the quality characteristics of machined surfaces. Srivastava et al. [8], Nag et al. [9], and Hollch et al. [10] demonstrated that the AFR and SOD significantly influence MRR. The authors found that MRR increases with higher AFR and decreases with increasing SOD. Additionally, Gunamgari et al. conducted optimization studies on input variables and MRR, identifying TS as the most critical parameter, followed by AFR and SOD, particularly in machining high-strength aluminum 7068 alloys [11]. Conversely, when machining stainless steel 304, the author concluded that WP and TS are crucial in influencing MRR [12]. Mardi et al. studied the machining of Mg-based nanocomposite, revealing that higher WP resulted in a better surface finish [13]. In Bagchi et al.'s study, it was discovered that for hard material, i.e., Nimonic C263, high jet WP, low TS, and reduced SOD result in enhanced MRR [14]. The impact of AFR and TS on MRR and SR was investigated, and the author inferred that elevating AFR led to an augmentation in MRR and a reduction in surface roughness, whereas a decrease in TS resulted in increased MRR and a coarser surface texture [15]. Furthermore, several studies highlighted that utilizing heavy machinery, high-pressure generators, limited nozzle lifespan, the hazardous nature of abrasives, and non-recyclability contribute to the high running cost, emphasizing the need for optimization [16–20].

Previous studies have utilized statistical modeling techniques to analyze and optimize the process parameters for cutting industrial metal parts, composites, and ceramics [21–26]. Gowthama et al. conducted a study on optimizing the process parameters of AWJM applied to Al 7071, utilizing the design of experiments. The authors found that TS and AFR significantly impacted both surface roughness and MRR. Additionally, it was observed that decreasing the SOD resulted in improved surface smoothness [21]. In a separate investigation, Kumar et al. explored specimens of 2, 4, 6, 8, and 10 wt.% tungsten carbide reinforced composite, manufactured via stir casting technique [22]. They employed the analysis of variance (ANOVA) method to assess the efficacy of the model developed through response surface methodology (RSM). The developed model exhibited satisfactory adequacy at a 95% confidence level. MRR was predominantly influenced by TS, followed by the percentage of reinforcement and SOD, respectively. Conversely, surface roughness was primarily affected by the percentage of tungsten carbide, followed by traverse speed and stand-off distance, respectively. Literature shows that most of the work is limited to optimizing the process parameters and machineability of metal and metal matrix composites. Recently, amorphous-based borosilicate ceramic has emerged as a promising material for LTCC parts due to its high density compared to other glass materials.

To date, extensive research has been conducted on crystalline, brittle, and ductile materials. However, the cutting behavior of amorphous materials, such as borosilicate, remains less understood due to their high brittleness, which poses challenges for conventional machining methods. To address this gap, AWJM was performed on borosilicate glass. This study aims to optimize process variables using RSM based on central composite design (CCD) to assess the impact of WP, AFR, TS, and SOD on the MRR. ANOVA was used to determine the significance of these parameters on MRR, with further optimization conducted through single-response optimization analysis using MINITAB. The optimization process employed a desirability-based response method to enhance objective characteristics effectively.

2 Materials and Method

2.1 Experiment Setup and Material

The study utilized borosilicate glass as the workpiece material, employing an AWJM setup (Model: 1313B, Make: Flow Mach 3) with a maximum pressure capability of 6500 bar (94,000 psi), as shown in Fig. 1. The machine demonstrates linear straightness accuracy of 0.0381 mm per 1m, cut speed of 7.6m/min, traverse speed of 12.7m/min, and repeatability of 0.050 mm, featuring an orifice diameter of 0.254 mm and a focusing tube diameter of 0.762 mm while operating with an AFR of 0.36296 kg/min. The setup includes an abrasive hopper, an abrasive feeder system equipped with a pneumatically controlled valve, and a workpiece table measuring 1219.2 mm × 1219.2 mm. Almandine garnet, with a mesh size of 80, was consistently chosen as the abrasive due to its wide industrial applicability. In the experimental procedure, various parameter configurations were employed based on the experimental model tailored for machining the borosilicate substrate. Each experiment involved machining a commercially available borosilicate piece measuring 20 mm × 13 mm × 33 mm. The experiments were conducted in triplicate, and the mean values of the response measurements from each parameter set were considered as the output data.

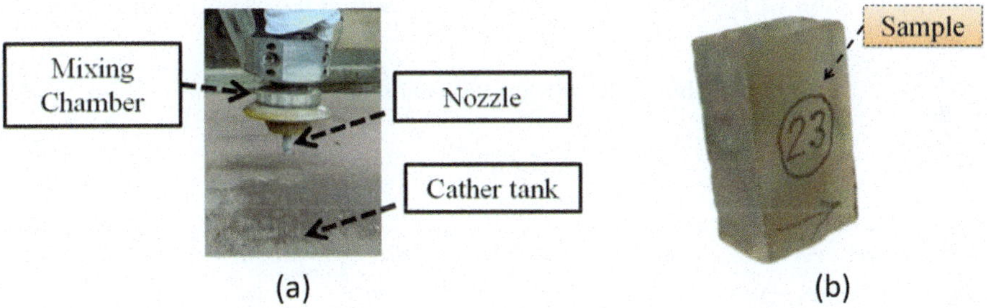

Fig. 1. Abrasive water jet machine with borosilicate glass sample.

2.2 Calculation of MRR

The MRR is calculated by dividing the difference in the weight of the workpiece before and after machining by the machining time, resulting in the MRR expressed in grams per minute, as depicted in Eq. (1). Weight loss was measured using a precision balance (SARTORIUS AG GERMANY, CPA 10035, Model: 26109103) with an accuracy of 0.001g. The duration of the machining process was monitored using a stopwatch. MINITAB software was employed to formulate a mathematical model and optimize parameters to achieve the highest material removal rate during the machining of borosilicate glass.

$$MRR = \frac{WB - WA}{t} \quad (1)$$

WA and WB represent the weight of the workpiece after and before machining, respectively, measured in grams, while t denotes the duration of the machining process in minutes.

2.3 Experimental Design Based on RSM

The research utilized a central composite rotatable second-order design methodology within the framework of response surface methodology. This involved employing response surface modeling to exhibit a mathematical correlation among the response variable, represented as y_u, and different machining parameters. The main objective was to determine the most effective operational settings for the system. A general second-order polynomial response surface model was employed to assess the impact of parameters on the response criteria, which is expressed as:

$$y_u = \beta_0 + \sum_{i=1}^{k}(\beta_i x_{iu}) + \sum_{i=1}^{k}\left(\beta_{ii} x_{iu}^2\right) + \sum_{i<j}\left(\beta_{ij} x_{iu} x_{ju}\right) + e_u \tag{2}$$

Here, y_u represents the response, x_{iu} denotes the coded values of the ith machining parameters for the uth experiment, k is the number of machining parameters, β_i, β_{ii}, β_{ij} are the second-order regression coefficients, and e_u is the residual indicating the experimental error of the uth observation. The experiments involved defining specific ranges for process variables such as WP, AFR, TS, and SOD, as outlined in Table 1.

Table 1. Parameters with units, and levels

Parameters	Units	Levels				
		– 2	– 1	0	1	2
Water pressure	bar	1000	1125	1250	1375	1500
Abrasive mass flow rate	g/min	18	27	36	45	54
Traverse speed	mm/min	200	250	300	350	400
Stand-off distance	mm	30	37.5	45	52.5	60

2.4 Mathematical Model for Predicting MRR

Table 2 displays the ANOVA results for MRR with a confidence level of 95%. The findings reveal that among the examined factors, AFR has the most substantial influence on MRR, followed by SOD and WP, which show similar effects. Conversely, TS has the least impact on MRR. Due to the curvature observed on the response surface, the quadratic equation is considered appropriate for modeling MRR. The mathematical expression of MRR with uncoded factors is provided below.

$$\begin{aligned} \text{MRR} = & -44.8 + 0.0495\ \text{WP} + 0.253\ \text{AFR} + 0.0351\ \text{TS} \\ & + 0.422\ \text{SOD} - 0.000018\ \text{WP} * \text{WP} - 0.001579\ \text{AFR} * \text{AFR} \\ & - 0.000038\ \text{TS} * \text{TS} - 0.005961\ \text{SOD} * \text{SOD} + 0.000002\ \text{WP} * \text{AFR} \\ & - 0.000000\ \text{WP} * \text{TS} + 0.000036\ \text{WP} * \text{SOD} - 0.000011\ \text{AFR} * \text{TS} \\ & - 0.00023\ \text{AFR} * \text{SOD} - 0.000097\ \text{TS} * \text{SOD} \end{aligned} \tag{3}$$

The metrics evaluating the model's goodness-of-fit, encompassing R^2, adjusted R^2, and predicted R^2, stand at 0.9814, 0.9652, and 0.9159, respectively, all converging close to unity. This alignment among the R^2 values strongly indicates the model's adequacy. Moreover, the lack-of-fit F-value does not exhibit statistical significance in comparison to the pure error, indicating a positive outcome.

Table 2. Analysis of variance for MRR

Source	DF	Adj. SS	Adj. MS	F-Value	*P*-Value
Model	14	70.99	5.07	60.43	0
Linear	4	65.84	16.46	196.18	0
Square	4	5.1	1.27	15.19	0
2-Way Interaction	6	0.04	0.01	0.09	1
Error	16	1.34	0.08		
Lack-of-Fit	10	0.97	0.1	1.54	0.31
Pure Error	6	0.38	0.06		
Total	30	72.33			
Model Summary		S	R-sq	R-sq (adj.)	R-sq (pred)
		0.29	98.14%	96.52%	91.59%

3 Results and Discussion

Thirty-one experiments were performed in accordance with the design of experiment provided by the RSM technique, as outlined in Table 1. The machined sample is shown in Fig. 2.

Fig. 2. Sample after experiment

3.1 Parametric Analysis of Characteristics

The interaction between AFR and TS is depicted in Fig. 3a, with WP and SOD maintained at 1250 bar and 45 mm, respectively. The study revealed that the MRR increases with higher TS across all AFR levels. This trend is attributed to the abrasive water jet's broader coverage of the workpiece surface, which removes more material due to the increased distance traversed. The research also indicated that MRR increases with increased AFR at all traverse speed levels. The elevation in AFR introduces a greater quantity of abrasive particles into the process, thereby enhancing the cutting action on the workpiece. The heightened presence of abrasive particles leads to a larger number of cutting edges actively involved in material removal, thereby augmenting the overall MRR. This effect directly stems from the improved cutting efficiency facilitated by a higher quantity of abrasive particles interacting with the workpiece.

The combined effect of WP and SOD is illustrated in Fig. 3b, with the abrasive mass flow rate and traverse speed maintained at 36 g/min and 300 mm/min, respectively. Initially, the MRR increases with rising water pressure but begins to decline beyond 1375 bar, regardless of the AFR. Elevated WP enhances jet velocity, increasing the kinetic energy of abrasive particles striking the work material, resulting in higher removal of material. However, excessive pressure leads to increased suction of abrasive particles, causing collisions among them, diminishing kinetic energy, and decreasing the MRR.

Moreover, the MRR increases as the SOD decreases across all water pressure levels. Increasing the SOD causes jet expansion and external drag, which reduces the kinetic energy of abrasives and lessens their impact on the work material, thereby decreasing MRR. At higher SOD, the jet may disintegrate into droplets, generating free abrasive particles that rebound upon impact, further reducing MRR. Beyond a stand-off distance of 45 mm, MRR slightly decreases due to inadequate distance for proper material impact, leading to inefficient cutting. However, as the SOD increases further, MRR improves to achieve higher values. Understanding and optimizing the interplay between WP and SOD are essential for maximizing MRR and ensuring efficient AWJ-cutting processes.

3.2 Single-Objective Optimization Analysis

Figure 4 presents the optimization outcomes aimed at maximizing the MRR using the developed mathematical model represented by Eq. (3). The approach employed to enhance the response involves assigning equal importance to the linear desirability function's lower and upper bounds. A weight value of 1 is assigned to the linear desirability function (D). The optimization process was conducted using Minitab v17 software to determine the optimal responses for abrasive water jet machining on borosilicate glass. The results of optimizing the maximum MRR through RSM are depicted in Fig. 4. The linear desirability function (D) maintains a value of 1, indicating that all parameters are operating within their specified limits. By setting the WP, AFR, TS, and SOD to their optimal values of 1409.3807 bar, 52.8525 g/min, 384.1530 mm/min, and 35.6557 mm, respectively, a maximum MRR of 13.6641 g/min was achieved.

3.3 Confirmation Test

Further, to validate the model, three experiments were carried out for MRR using the parameter settings derived from the single-objective optimization process for MRR. The comparison was made between these experimental results and the predicted optimal values. The analysis revealed a close alignment between the predicted and actual

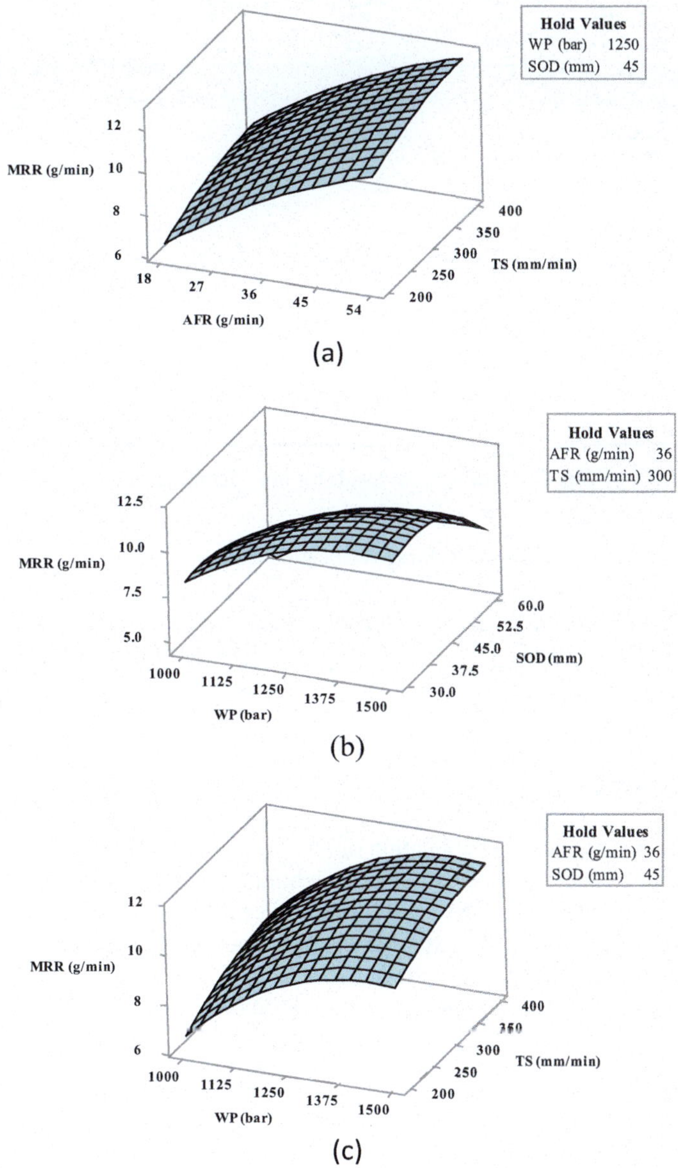

Fig. 3. Material removal rate plot with water pressure, abrasive mass flow rate, stand-off distance, and traverse speed.

experimental outcomes. A detailed comparison of the experimental results, predicted optimal values, and the associated prediction errors can be found in Table 3. The obtained outcomes provide a comprehensive overview of the model's performance in predicting MRR under the given parameter settings.

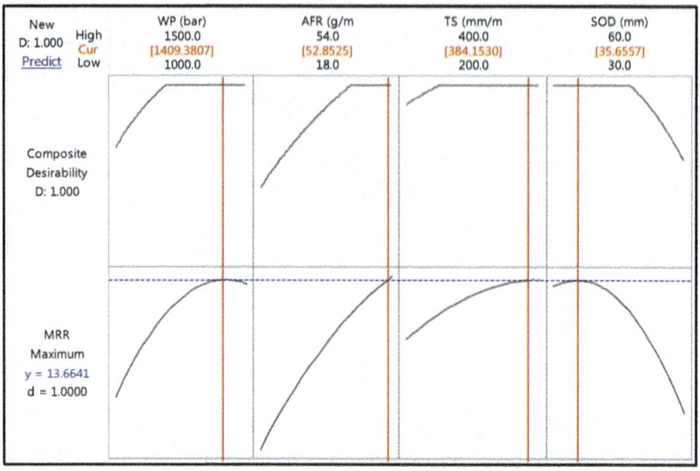

Fig. 4. Optimization result for material removal rate.

Table 3. Final verification experiments

Machining responses	Predicted value at optimal parametric settings	Experimental value at optimal parametric settings	% of error
MRR (g/min)	13.66	12.63	7.54

4 Conclusions

This research investigates the application of AWJM on borosilicate glass. This approach efficiently removes dust generated during machining using the water jet. Furthermore, it results in minimal thermal distortion and reduced stress induction on the cut surface. The study systematically explores the influence of various process parameters on response variables, particularly the MRR. The experiments were conducted using a single-objective optimization approach, focusing on optimized input parameters such as AFR, WP, TS, and SOD. The findings are summarized as follows:

1. The study demonstrates that machining borosilicate glass using AWJM can be performed without notable brittle fracture.
2. ANOVA of the output responses reveals that all selected input parameters WP, TS, AFR, and SOD significantly impact the output response. Among them, AFR emerges as the most influential parameter on MRR, followed by WP and SOD, while TS exhibits the least significance.
3. Increasing WP, AFR, and TS enhances MRR, whereas increasing SOD decreases MRR. However, excessively low SOD may also diminish MRR due to insufficient material impact, resulting in inefficient cutting.
4. Single-objective optimization leads to an MRR of 13.6641 g/min, achieved with specific input parameter values: WP of 1409.3807 bar, AFR of 52.8525 g/min, TS of 384.1530 mm/min, and SOD of 35.6557 mm.

Acknowledgement. The authors extend their regards to the Department of Science Technology (DST) for providing financial support under the project grant DST (SERB) (CRG/2022/000643). The authors also thank the Saha Institute Of Nuclear Physics, Kolkata, India, for providing abrasive water jet machining facilities.

References

1. Shelby, J.E.: Introduction to glass science and technology, 3rd edn. Royal Soc Chem, Cambridge (2023)
2. Kumar, S., Singh, K., Kumar, D.: SiO2/B2O3 glass formers effect on transparency and mechanical properties of soda-lime borosilicate glasses for automobile applications. J. Non Cryst. Solids **618**(122530), 122530 (2023)
3. Nasa, P.: A review on pharmaceutical packaging material. World J. Pharm. Res. **3**(5), 344–368 (2014)
4. Luo, X., et al.: Eco-friendly tape casting of borosilicate glass/Al_2O_3 sheets for LTCC applications. Ceram. Int. **48**(18), 25975–25983 (2022)
5. Vallero, D.A.: Fundamentals of Air Pollution, 5th edn. Academic, San Diego (2014)
6. Tiwari, T., et al.: Parametric investigation on abrasive waterjet machining of alumina ceramic using response surface methodology. IOP Conf. Ser. Mater. Sci. Eng. **377**, 012005 (2018)
7. Uhlmann, E., Sammler, F., Richarz, S., Heitmüller, F., Bilz, M.: Machining of carbon fibre reinforced plastics. Procedia CIRP **24**, 19–24 (2014)
8. Srivastava, A.K., et al.: Surface integrity in tangential turning of hybrid MMC A359/B4C/Al2O3 by abrasive waterjet. J. Manuf. Process. **28**, 11–20 (2017)
9. Nag, A., et al.: Hybrid aluminium matrix composite AWJ turning using olivine and Barton garnet. Int. J. Adv. Manuf. Technol. **94**, 2293–2300 (2018)
10. Hloch, S., Valicek, J., Simkulet, V.: Estimation of the smooth zone maximal depth at surfaces created by Abrasive Waterjet. Int. J. Surf. Sci. Eng. **3**(4), 347–359 (2009)
11. Rana, M., Akhai, S.: Multi-objective optimization of Abrasive water jet Machining parameters for Inconel 625 alloy using TGRA. Mater. Today **65**, 3205–3210 (2022)
12. Akhai, S., Rana, M.: Taguchi-based grey relational analysis of abrasive water jet machining of Al-6061. Mater. Today **65**, 3165–3169 (2022)
13. Mardi, B.K., Dixit, A.R., Srivastava, A.K., Mallick, A., Scucka, J., Hlaváček, P., Hloch, S., Zeleňák, M.: Effect of water pressure during abrasive waterjet machining of Mg-based nanocomposite. In: Applications of Fluid Dynamics: Proceedings of ICAFD 2016, pp. 605–612. Springer Singapore (2018)
14. Gunamgari, B.R., Kharub, M.: Experimental investigation on abrasive water jet cutting of high strength aluminium 7068 alloy. Mater. Today **69**, 488–493 (2022)
15. Srivastava, A.K., Nag, A., Dixit, A.R., Tiwari, S., Srivastava, V.S.: Parametric study during abrasive water jet turning of hybrid metal matrix composite. In: Advances in Manufacturing Engineering and Materials, pp. 72–84. Springer International Publishing, Cham (2019)
16. Chithirai Pon Selvan, M., Midhunchakkaravarthy, D., Senanayake, R., Ramaswamy Pillai, S., Reddy Madara, S.: A mathematical modelling of Abrasive Waterjet machining on Ti-6Al-4V using artificial neural network. Mater. Today **28**, 538–544 (2020)
17. Joel, C., Jeyapoovan, T.: Optimization of machinability parameters in abrasive water jet machining of AA7075 using Grey-Taguchi method. Mater. Today **37**, 737–741 (2021)
18. Adam Khan, M., Gupta, K.: Machining Ni-Cr-Fe based superalloy using abrasive water jet cutting process and its surface studies. Mater. Today **19**, 2139–2143 (2019)

19. Srinivas, D.S., Babu, N.R.: Role of garnet and silicon carbide abrasives in abrasive waterjet cutting of aluminum-silicon carbide particulate metal matrix composites. Int. J. Appl. Res. Mech. Eng. 109–122 (2011)
20. Selvam, R., Arunkumar, N., Karunamoorthy, L.: An investigation on machining characteristics in abrasive water jet machining of hybrid laminated composites with SiC nano particles. Mater. Today **39**, 1701–1709 (2021)
21. Gowthama, K., Somashekar, H.M., Suresha, B., Rajole, S., Ravindran, N.: Optimization of abrasive water jet machining process parameters of Al 7071 using design of experiments. Mater. Today **52**, 2102–2108 (2022)
22. Ravi Kumar, K., Sreebalaji, V.S., Pridhar, T.: Characterization and optimization of Abrasive Water Jet Machining parameters of aluminium/tungsten carbide composites. Measurement (Lond.) **117**, 57–66 (2018)
23. Nag, A., Dixit, A.R., Petrů, J., Váňová, P., Konečná, K., Hloch, S.: Maximization of wear rates through effective configuration of standoff distance and hydraulic parameters in ultrasonic pulsating waterjet. In: Facta Universitatis, Series: Mechanical Engineering (2022)
24. Srivastava, A.K., Nag, A., Dixit, A.R., Tiwari, S., Srivastava, V.S.: Parametric study during abrasive water jet turning of hybrid metal matrix composite. In: Advances in Manufacturing Engineering and Materials: Proceedings of the International Conference on Manufacturing Engineering and Materials (ICMEM 2018), pp. 72–84, Springer International Publishing, Nový Smokovec, Slovakia (2019)
25. Derringer, G., Suich, R.: Simultaneous optimization of several response variables. J. Qual. Technol. **12**(4), 214–219 (1980)
26. Kumar, K.R., Sreebalaji, V.S.: Desirability based multi objective optimization of abrasive wear and frictional behaviour of aluminium (Al/3.25Cu/8.5Si)/fly ash composites. Tribol. Mater. Surf Interfaces **9**(3) (2015)

Reverse Engineering of a Historic Motorcycle Cylinder Head Through 3D Scanning and Printing

Lukáš Kudrna[1(✉)], Štěpán Pravda[1], Tomáš Navrátil[1], Ladislav Hrabec[1], and František Fojtík[2]

[1] Faculty of Mechanical Engineering, Department of Machining and Industrial Design, VSB – Technical University of Ostrava, 17 Listopadu 2172/15, Ostrava 708 00, Czech Republic
lukas.kudrna@vsb.cz

[2] Faculty of Mechanical Engineering, Department of Applied Mechanics, VSB – Technical University of Ostrava, 17 Listopadu 2172/15, Ostrava 708 00, Czech Republic

Abstract. This paper explores the reverse engineering process applied to the cylinder head of the historic Praga 500 BD motorcycle, employing contemporary industrial techniques, including 3D scanning and 3D printing. Initially, an optical 3D scanning device was used to create a model of the original part. However, distortions in proportions were identified, likely resulting from aging and insufficient modifications and repairs. Subsequently, adjustments to the obtained 3D model were made using Autodesk Fusion 360 software to rectify these distortions, aligning the part with the dimensions and shapes specified in the manufacturer's original technical drawings. The validated 3D model was then utilized to produce a 3D printed part using a PRUSA MK4 3D printer. This printed part served to verify physical dimensions and test its fit onto the engine cylinder to check dimensions and clearances. After these checks, the same part printed on the 3D printer will serve as a casting mould to produce the final metal part. The metal part is then machined.

Keywords: Reverse Engineering · 3D Scanning · 3D Printing · Cylinder Head · Historic Motorcycle

1 Introduction

Reverse engineering represents a specific type of design that does not adhere to classical design practices. It involves extracting the original design of a part or system and building, or synthesizing representations that are less implementation dependent. There are two main subareas of reverse engineering—design recovery and redocumentation [1]. The process of design recovery is experiencing growing support from contemporary industrial measurement methods, notably through the integration of advanced technologies like 3D scanning and Coordinate-Measuring Machine (CMM). These sophisticated tools play a pivotal role in accurately capturing intricate details, dimensions, and geometries of existing components, facilitating a more precise and comprehensive reconstruction of

M. Pagac et al. (Eds.): SCFF 2024, LNNS 1062, pp. 187–196, 2024.
https://doi.org/10.1007/978-3-031-65656-9_19

the original design [2]. In contemporary reverse engineering practices, redocumentation primarily emphasizes Computer-Aided Design (CAD) and 3D modeling, particularly as many parts are subsequently manufactured using additive technologies [2, 3], sheet metal drawing [4, 5], or sheet metal bending [6]. While the current importance of recreating technical drawings may seem reduced, it remains a critical outcome in the realm of reverse engineering. This is especially evident for components demanding precise machining, even after utilizing additive manufacturing techniques. In the production of limited series of reverse-engineered products, traditional machining methods continue to be widely employed.

Some parts need to be machined to the required size, shape, and surface finish for coating purposes [7]. In this case, it will only be machined to the required size and shape.

1.1 Searches (Praga 500 BD Motorcycle)

The most technically innovative motorcycle in 1920s Czechia was the BD 500 (Fig. 1), manufactured by Praga (formerly Breitfeld-Danek) and designed by engineer Frantisek Koch [8, 9].

Mr. Jaroslav Frantisek Koch was not only the chief engineer of the BD 500 project but also a skilled racer and traveler. He actively participated in testing prototypes of the motorcycle by extensively taking them to various racing competitions and embarking on long-distance journeys across Europe [8, 9].

Between 1926 and 1935, approximately 2500 motorcycles were manufactured, fulfilling roles in both racing and touring. These robust multipurpose machines were widely distributed and sold throughout former Czechoslovakia. They found applications in the military, police force, and among individuals seeking a comfortable, reliable, and affordable means of transportation [8, 9].

The most unique think about the BD 500 motorcycle was its engine. Particularly when talking about its valvetrain. In the 1920s, when the BD 500 was designed, the most prevalent valvetrain type was the side-valve system (SV). This configuration was not only popular in motorcycles but also widely employed in automobile and truck engines, owing to its ease of manufacturing using the prevalent methods of that time. If additional power gain was necessary, some manufacturers produced the later, more popular, and very versatile OHV (Overhead Valve) engines. Yet, they often had to contend with issues related to high vibration caused by the reciprocating masses of the valvetrain components in push-rod systems [8, 9].

On the other hand, the BD 500 engine employed the DOHC (Double Overhead Camshaft) system, a notably advanced technology, especially for its era. Unlike modern engines with this type of valvetrain, the BD 500 system drove the camshaft using a bevel gear. In addition to this unique valve train system, the BD 500 engine essentially adhered to the standard characteristics of an air-cooled four-stroke single-cylinder engine. The engine had a displacement of 499 cm^3 with a compression ratio of 4.7:1, delivering a maximum power of 15 PS at 4400 RPM [8, 9].

The aluminum piston, featuring a concave bottom, was equipped with three piston rings. The connecting rod, made from chromium-nickel steel, was fitted with a roller bearing. In the cylinder assembly, two camshafts operated camshaft lobes that directly opened valves within a conical combustion chamber, eliminating the need for lifters. A bevel shaft drove the main gear with a 1:1 ratio, which then transmitted power to both camshaft gears at a 1:2 ratio. Another notable aspect lies in the utilization of Micarta for constructing the main gear [8]. As a thermoset composite, Micarta plays a crucial role in diminishing valvetrain noise while maintaining mechanical strength and ensuring thermal stability [8].

Fig. 1. Praga 500 BD motorcycle as displayed in a vintage brochure [9].

1.2 Reconstruction of the Historical Motorcycle Parts

The Turin Charter stands as the sole authoritative documentation outlining the guidelines for the reconstruction of historic motorcycle parts. The charter's objective is to preserve and protect the history of vehicles, covering aspects such as technical design, appearance, functions, documented history, and their relationship to society and the social environment. Achieving a profound understanding of the issue, fostering greater appreciation, ensuring the preservation of vehicles, and enabling their operation on public roads necessitate the use of diverse research methods. This includes leveraging existing scientific, technical, and historical knowledge, as well as engaging with a variety of institutions active in this field [10].

The Turin Charter was created under FIVA. FIVA is an advisory body to UNESCO [10].

FIVA (the Fédération International des Vehicules Anciens) – International Federation of Historic Vehicles [10].

This document describes a recommended marking system for reconstructed parts:

- **NB: "newly built"**—an accurate as possible copy in terms of form, materials, and make, reproduced directly from a documented original [10].
- **FR: "free reconstruction"**—reconstruction without using any historic model in terms of form, material, or work technique. The part, however, fulfills the technical function of an historic component utilized earlier [10].
- **CS: "conservational stabilization"**—a later structural reinforcement added to stabilize the historic substance [10].

1.3 Flowchart

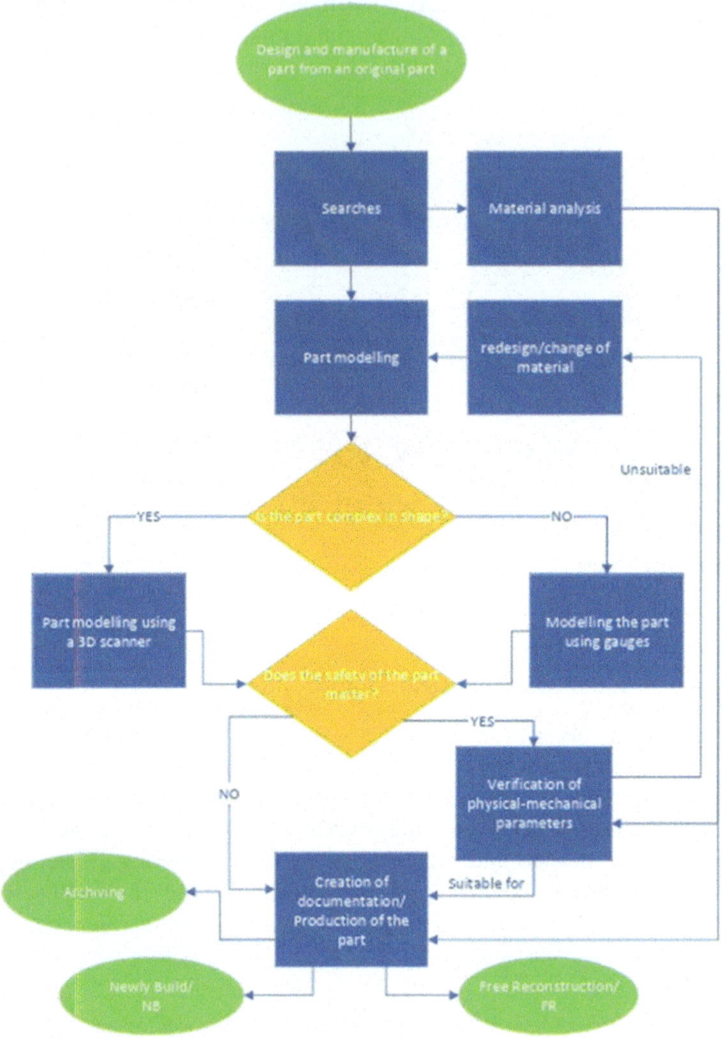

In this case it will be: Searches ⟶ Material analysis ⟶ Part modelling using a 3D scanner ⟶ Creation of documentation/Production of the part ⟶ Newly Build/NB

2 Materials and Methods

In this section, we outline the procedure for reconstructing the cylinder head sourced from the historic motorcycle Praga BD 500. This process encompasses reverse engineering, incorporating 3D scanning, CAD design, and 3D printing. Furthermore, it involves material analysis of the original part and suggests a manufacturing approach aligned with the directives stipulated in the previously mentioned Turin Charter, specifically tailored for a newly built part.

2.1 Original Cylinder Head Condition

The original cylinder head extracted from the motorcycle in its pre-restoration condition underwent improper repairs. Notably, some cooling ribs were broken and subsequently rewelded as can be seen in Fig. 2. The inappropriate welding technique applied to the cast iron part resulted in cracking. Nonetheless, it retains crucial information about the dimensions of the original part, making a substantial contribution to the design recovery process.

Fig. 2. Original cylinder head, 1—unprofessionally welded crack 2—broken cooling ribs [12]

2.2 Material Analysis

Material analysis is crucial for determining the chemical composition of the material for new part manufacturing. In this operation, a rib fragment from the original cylinder head measuring approximately 15 × 20 mm was utilized. The method employed for measuring the chemical composition was optical emission spectrometry with glow discharge (GDOES), utilizing a GDA 750 device from Spektrum.

From the comprehensive chemical analysis of the material (Table 1), the content of carbon (3.24 wt%), silicon (2.06 wt%), and phosphorus (0.484 wt%) can be discerned, indicating that it is gray cast iron. Gray cast iron exhibits favorable properties in terms of machinability, strength, and vibration damping [12].

2.3 Part Modelling Using a 3D Scanner

The 3D scanning process in RE involves three main steps: scanning, point processing and developing a geometric model for a specific application. In these steps, the physical

Table 1. Material composition of the material [12]

C	Mn	Si	P	S	Cr	Ni	Mo	Cu
3.24	0.78	2.06	0.484	0.081	0.025	0.046	0.004	0.154
Ti	Co	B	Pb	V	W	Zr	Al	Nb
0.077	0.021	0.002	0.003	0.07	0.01	0.01	0.004	0.005

object is transformed into a point cloud and then converted into a CAD model. The point processing phase involves importing the raw point cloud data and noise reduction, as well as reducing the number of points using filtering algorithms. Point cloud data can be merged if a part is captured in multiple scans. Since most CAD software is not designed to handle large amounts of point cloud data, separate software is required for the complex process of transforming point clouds into surfaces that can be used in CAD software [11]. A Creaform Handyscan 700 laser scanner was used to capture all the dimensions of the original cylinder head. Due to the imperfect representation of all surfaces and the large number of points, it was necessary to adjust the mesh to a complex shape. This process involved filling the holes and reducing the number of points using Creaform VX Elements software. To account for the unscanned portion between the ribs, holes were patched and the number of point clouds was adjusted from 750,000 to 250,000 points [12]. After scanning, the point cloud was transferred to Autodesk Fusion 360 software for further adjustments and refinement, ensuring alignment with the original dimensions of the piece as documented in historical sources. The finished final 3D model can be seen in Fig. 3. A comparison of the raw point cloud model and the final 3D model obtained from the Fusion 360 software can be seen in Fig. 4 [12].

Fig. 3. Raw 3D scan of the original part [12]

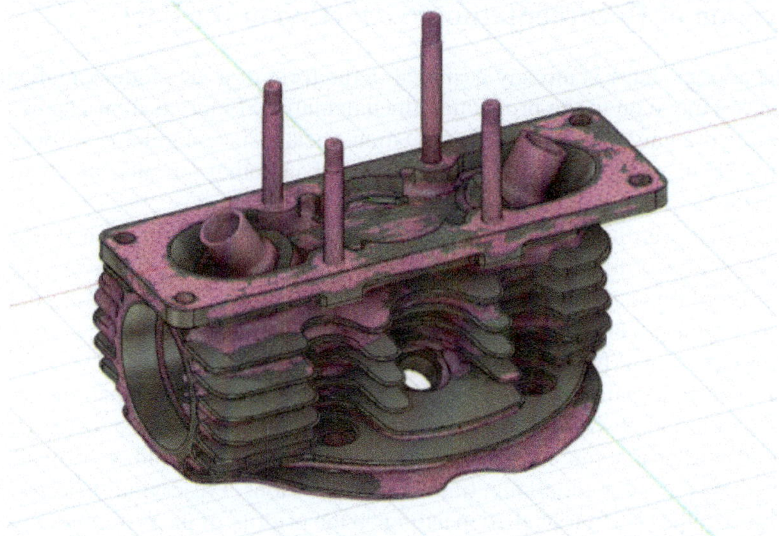

Fig. 4. Comparison of the raw point cloud model (pink color) and the final 3D model (grey color) [12]

2.4 3D Printing

Once the 3D model was completed, the additive manufacturing process of 3D printing the plastic "test" part began. This part was used to verify the fit and all other clearances after assembly with the remaining parts of the original motorcycle engine. The filament used to produce this part was the standard Prusament PLA Marble Grey filament. Figure 5 is a comparison of the original cylinder head and the 3D printed part. The fit and all dimensions of the 3D printed part were verified using a 3D scan. We then proceeded to plan the manufacturing process using a similar 3D printed part with casting allowances.

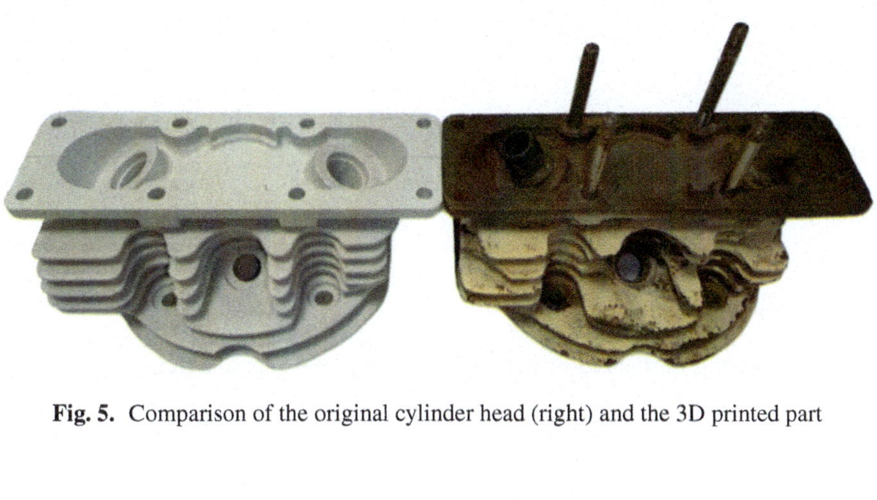

Fig. 5. Comparison of the original cylinder head (right) and the 3D printed part

3 Creation of Documentation/Production of the Part

Since the correct shape is already designed in the form of a 3D model and the material is known, we move on to the production documentation. Figure 6 shows an example of a production drawing of the Praga 500 BD motor head. Current graphic software allows converting 3D models to 2D, directly into dwg, dxf and other formats, which are further worked with [13].

The last step is the actual production of the part. The production of the part is commissioned according to the technology. In the case of a casting, the production must first be commissioned to the foundry and then machined. Next comes the surface treatment, which can be electroplating (chrome plating, nickel plating, etc.) or painting, etc. In our case, it will be a heat-resistant black paint [13].

Almost every part has a different production technology. The production technology will result from the previous chapters and from the experience of the restorer, technologist or designer. In some cases, the original production technology or materials are abandoned, whether for economic, time or safety reasons. Preserving the original technology is often the costliest of all, depending on the priority of the restorer or investor. In most cases, this will emerge from the Restoration Plan, which is drawn up before work begins [13].

In the first step, the models are printed on a 3D printer from PLA (Polylactic acid) material using the FDM (Fused Deposition Modeling) method, and then the models will be cemented and sanded. The models prepared in this way can then be moulded into a moulding compound, resulting in a disposable sand mould. The models are removed from the mould before casting and can be used to prepare the next mould. This method was used in the production of the brake pedal of a historic motorcycle [13, 14].

4 Newly Build/NB

In our case, it is newly built. This is because the part will have the same shape, material and manufacturing technology. A part reverse engineered in this way is designed according to the strictest criteria of the Turin Charter [10].

5 Results

- The results as documentation and the research conducted provide a good basis for further extension research.
- It can serve as a "muster" for the implementation of some restoration works.
- The results will serve as project plans and contributions for the creation of new projects aimed at the restoration of technical monuments.

6 Conclusion

In our case, it is a derivative process of designing and manufacturing apart from an original part. These are usually already damaged parts or parts borrowed from various institutions or individuals. After modelling and thorough inspection, the next steps follow the flow chart (Sect. 1.3).

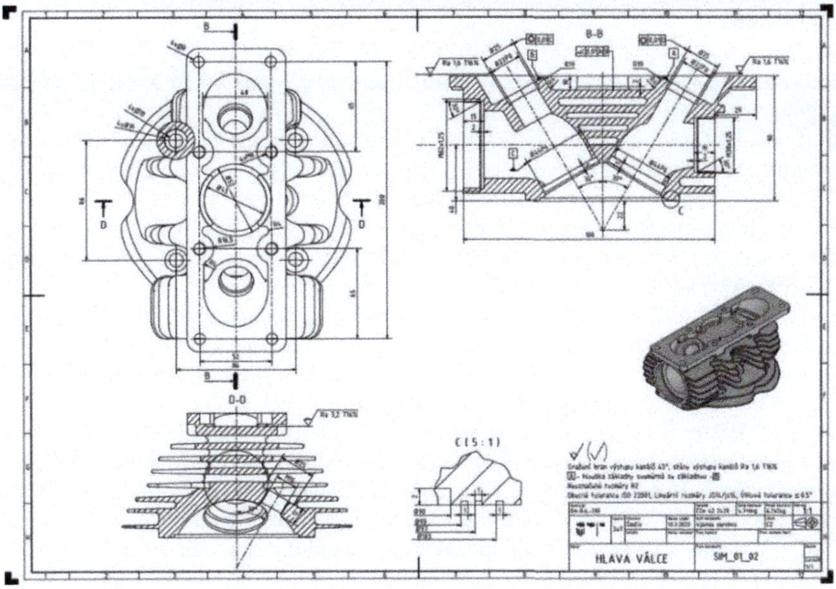

Fig. 6. Picture of the drawing for machining the casting [12]

This procedure has been used for many parts, but for this purpose the engine head from a Praga 500 BD motorcycle (1929–1931) was chosen as the most suitable for illustration. The methodology was developed based on practical experience. From the above it can be concluded that the above methodology of designing and manufacturing apart from an original part can be successfully applied.

If the work is carried out according to the presented procedure, it will increase the knowledge of the manufacturer, the technology and allow a sensitively assembled whole to be properly documented and archived for future generations, ensuring the correct functionality of the part.

In our case, the last step to be taken is the actual production based on the data obtained.

The work is in accordance with the Turin Charter, which is an advisory body to UNESCO.

Acknowledgement. This research was funded by "Research and innovation of modern processes and technologies in industrial practice", grant number SP2024/001 and" Experimental and Numerical Modeling for Solving Problems in Mechanics and Biomechanics ", grant number SP2024/037 by MŠMT ČR (Ministry of education youth and sports).

References

1. Rekoff, M.G.: On reverse engineering. IEEE Trans. Syst. Man Cybern. **SMC-15**(2), 244–252 (1985)
2. Carbone, V., Carocci, M., Savio, E., et al.: Combination of a vision system and a coordinate measuring machine for the reverse engineering of freeform surfaces [online]. AMT **17**, 263–271 (2001). https://doi.org/10.1007/s001700170179
3. Várady, T., Martin, R.R., Cox, J.: Reverse engineering of geometric models—an introduction. Comput. Aided Des. **29**(4), 255–268 (1997). ISSN 0010-4485. https://doi.org/10.1016/s0010-4485(96)00054-1
4. Čada, R., Pektor, T.: Development of a low-cost drawing tool for a press brake. In: Proceedings of the 30th Anniversary International Conference on Metallurgy and Materials METAL 2021, Brno, 26.-28.5, pp 241–246. Tanger, Ltd., Brno (2021). ISBN 978-80-87294-99-4, ISSN 2694-9296. https://doi.org/10.37904/metal.2021.4117
5. Čada, R., Pektor, T.: Design of stamping drawing technology from thin sheet. In: Proceedings of the 31st Anniversary International Conference on Metallurgy and Materials METAL 2022, Orea Congress Hotel Brno, 18.-19.5, pp 262–267. Tanger, Ltd., Brno (2022). ISBN 978-80-88365-06-8, ISSN 2694-9296. https://doi.org/10.37904/metal.2022.4474
6. Čada, R., Pektor, T.: Design of the production technology of a bent component. Appl. Sci. **13**(24), 13033. Online ISSN (eISSN) 2076-3417. https://doi.org/10.3390/app132413033
7. Sternadelová, K., Krupová, H., Matýsek, D., Mohyla, P.: Quality assessment of the vitreous enamel coating applied to the weld joint. MM Sci. J. [online] 6333–6338 (2023). ISSN 18031269. https://doi.org/10.17973/MMSJ.2023_03_2022105
8. Reissner, M.: Století motocyklu. Průhonice, 335p. Professional Publishing (2019). ISBN 978-80-88260-34-9
9. Českomoravská Kolben-Daněk a.s., Praha. Motocykl "BD" 500 ccm OHC. Brochure P.R.O č.06–1929.-2818-Auto, Českomoravská Kolben-Daněk a.s. (1929)
10. Charter of Turin—English Version: The International Federation for Historic Vehicles—FIVA, Torino, Italy (2022) [online]. Available from: https://1.fiva.org/wp-content/uploads/2022/02/Charter-of-Turin-English-Version.pdf [cit. 25 Mar 2024]
11. Helle, R.H., Lemu, H.G.: A case study on use of 3D scanning for reverse engineering and quality control. Mater. Today Proc. **45**(Part 6), 5255–5262. https://doi.org/10.1016/j.matpr.2021.01.828
12. Šimčík, M.: Motorcycle cylinder head reconstruction. Bachelor thesis, VSB—Technical University of Ostrava, Ostrava (2023). Available from: http://hdl.handle.net/10084/150889 [Cit. 11 April 2024]
13. Kudrna, L., Ma, Q. P., Hajnyš, J., Měsíček, J. Halama, R., Fojtík, F., Horňaček, L.: Restoration and possible upgrade of a historical motorcycle part using powder bed fusion. Materials [online] **15**(4), 24 (2022). ISSN 1996-1944. https://doi.org/10.3390/ma15041460
14. Fojtík, F., Potrok, R., Hajnyš, J., Ma, Q.P., Kudrna, L., Měsíček, J.: Quantification and analysis of residual stresses in braking pedal produced via laser-powder bed fusion additive manufacturing technology. Materials [online], **16**(17), 18 (2023). ISSN: 1996-1944. https://doi.org/10.3390/ma16175766

Review of Directed Energy Deposition (DED) in Component Repair

Roman Blaha(✉), Jiri Hajnys, Jakub Mesicek, Abdesselam Mechali, Dominik Krisak, and Marek Pagac

Department of Machining Assembly and Engineering Metrology, Faculty of Mechanical Engineering, VSB—Technical University of Ostrava, Ostrava, Czech Republic
roman.blaha@vsb.cz

Abstract. Directed Energy Deposition (DED) is a groundbreaking technology in the aerospace industry, particularly in repairing components. DED offers precise restoration capabilities with minimal distortion, ensuring that components maintain their original specifications while fostering metallurgical bonding. This paper conducts a comprehensive review of academic and industrial studies to examine the feasibility and effectiveness of DED. DED provides significant advantages over conventional repair methods, showcasing remarkable accuracy in restoring erosion-damaged components such as AM355 steel T700 blisks. The endorsement of industry leaders, such as Rolls-Royce Deutschland's use of DED for repairing worn flanges and damping wire grooves, underscores its potential to deliver substantial returns on investment. Despite these advancements, challenges remain in optimizing process parameters and deposition strategies to further enhance part quality. However, ongoing research initiatives are actively addressing these challenges, positioning DED to become a critical technology in aerospace maintenance and repair operations. This paper offers valuable insights into the current state of DED technology, highlighting its benefits, challenges, and future prospects in aerospace component repair. Ultimately, the widespread adoption of DED promises to revolutionize aerospace maintenance practices, providing cost-effective solutions while minimizing environmental impact.

Keywords: Directed energy deposition · laser power · additive manufacturing · repairing

1 Introduction

Directed Energy Deposition (DED) refers to an advanced additive manufacturing technology that has gained significant recognition in the field of additive manufacturing. Classified as an additive technology by ISO/ASTM 52900 [1], DED utilizes concentrated thermal energy from sources such as laser beams or electron beams to melt material directly onto a work surface, enabling the creation of intricate three-dimensional structures layer by layer. In DED, material can be supplied in either powder or wire

M. Pagac et al. (Eds.): SCFF 2024, LNNS 1062, pp. 197–205, 2024.
https://doi.org/10.1007/978-3-031-65656-9_20

form, each offering distinct advantages and limitations that influence the characteristics of the final product. This process involves depositing successive layers of material, which are fused and bonded together upon solidification and cooling, resulting in the formation of a solid and homogeneous structure [2].

The versatility and precision of DED have made it a preferred choice across a range of industries, including aerospace, automotive, medical, and toolmaking, due to its ability to produce complex geometric shapes with high accuracy [3]. With ongoing advancements in materials and technology, the potential for further enhancement and expansion of DED's capabilities is significant.

This paper focuses on a specific variant of DED technology, particularly its application in aerospace component repair. DED has emerged as a transformative solution, offering unparalleled accuracy and minimal distortion in the restoration of damaged aerospace components [4].

Through a comprehensive review of academic and industrial studies, we examine the feasibility, efficacy, challenges, and opportunities associated with DED in repair applications. We also highlight ongoing research initiatives aimed at optimizing DED processes to enhance part quality, thereby facilitating its widespread adoption in maintenance and repair operations. The pie chart in Fig. 1 demonstrates the market distribution of metal additive manufacturing systems as a percentage of the total number of businesses in the industry. This illustration provides insight into the prevalence and adoption rates of metal additive manufacturing technologies across different businesses.

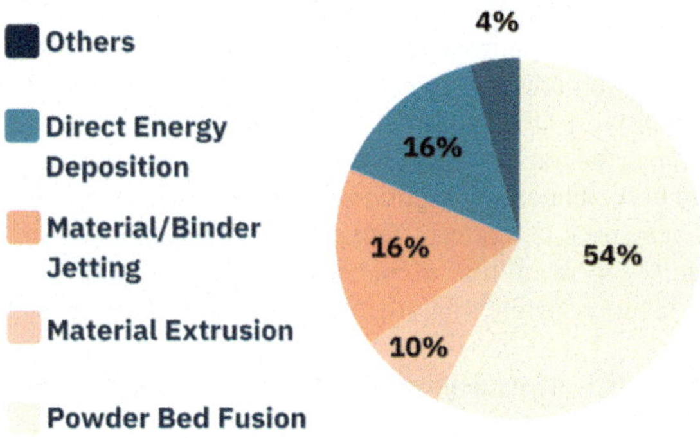

Fig. 1. The market distribution of metal additive manufacturing systems in 2020 [5].

2 Direct Energy Deposition

DED additive manufacturing technology is a noteworthy approach in the additive manufacturing industry, Its operational principle entails the efficient utilization of concentrated thermal energy, which can be produced by various sources, such as laser beams, electron beams, or arc welding. This thermal energy serves as the foundation for melting the material, which is directly applied to the work surface [6].

The material can be supplied to the work area in two primary forms: as a powder, which is precisely measured and evenly distributed, or as a wire, which is gradually fed into the melting process. Both options have specific advantages and limitations that must be carefully considered in light of the desired characteristics of the printed object. The printing process involves the successive layers of material being deposited on top of each other, with each layer being subsequently fused and bonded to the previous layers. Once the material solidifies and cools, a solid and homogeneous structure is formed, which serves as the foundation for creating the desired print.

This technology makes it possible to produce complex geometric shapes with high precision and detailed characteristics, making it an attractive choice for a range of industrial applications, including aerospace, automotive, medical, and toolmaking. As advances in materials and technology continue, the capabilities of this innovative manufacturing method are expected to improve and expand further [7].

2.1 Systems of Laser Additive Manufacturing (LAM)

Laser Additive Manufacturing (LAM) systems are complex manufacturing processes that involve the creation of three-dimensional metal objects by sequentially melting or sintering metal powders based on a digital model. The process, known as Powder Bed Fusion (PBF), entails the application of a thin layer of powder followed by fusion according to specifications from a Computer-Aided Design (CAD) file [8]. The primary component of this technology is a heat source, typically a laser or electron beam, which is controlled based on the geometry and design of the model.

This technology is used for example in topology optimization [9]. The main techniques used in a PBF-based laser additive manufacturing system include Direct Metal Laser Sintering (DMLS), Selective Laser Melting (SLM), Selective Laser Sintering (SLS), Laser CUSING (Concept Laser), and Laser Metal Fusion (LMF). These techniques permit the creation of intricate metal parts with high precision and detail, making them widely employed in various industries, including automotive and medical applications [10].

The devices are equipped with a single laser, but some solutions offer the possibility of using 3–9 lasers. These machines utilize either infrared or blue light as their light source. The print head or worktable is capable of rotating around 3–5 axes. One of the significant benefits of these devices is their ability to perform multi-material printing. The most commonly used materials for printing include steel, titanium, aluminum, and nickel [11–14].

2.2 Feed Systems

Powder feed systems, a metallic powder is fused onto a base material using heat. This is accomplished by depositing fine particles through a nozzle onto the construction surface, following the contour of the component's cross-section. A variety of techniques, such as Laser Engineered Net Shaping (LENS), Direct Metal Deposition (DMD) and Laser Metal Deposition (LMD), are employed for powder feed processes. The deposited layers typically measure over 0.1 mm in thickness.

These systems offer greater build volumes and can manufacture larger components compared to (PBF) systems. The nozzle may remain stationary while the workpiece moves, or a fixed nozzle may be used while the part undergoes movement to achieve the

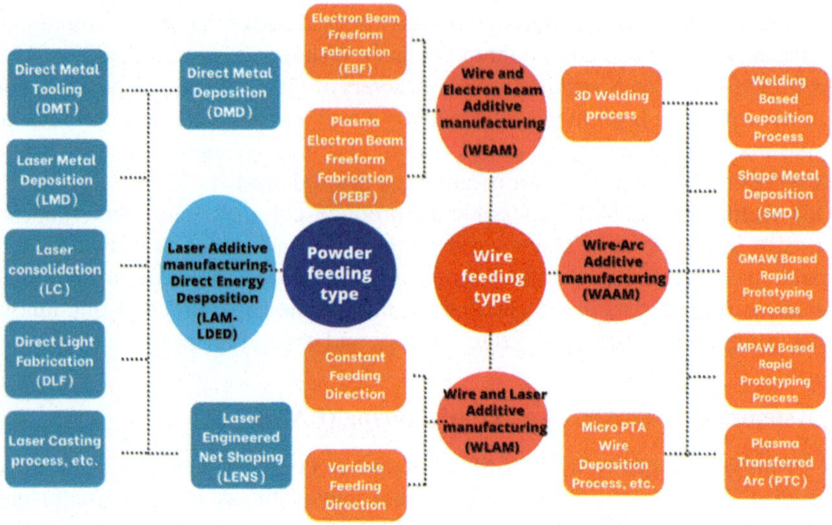

Fig. 2. Classification of DED process [15]

desired shape. Notably, 3D metallic parts generated through DED exhibit finer grain textures than those produced by conventional manufacturing methods due to the accelerated cooling rates following the melting process [16].

An alternative approach to DED systems utilizes metal wire as the feedstock, which is fused layer by layer using heat from a source, typically a laser or arc. This process, as illustrated in Fig. 2 involves repeatedly fusing the material to construct the desired three-dimensional structure. Wire feed systems are well-suited for deposition tasks due to the rapid mass transfer enabled by solid wire, making them ideal for applications requiring high deposition rates. Additionally, they offer the capability to fabricate large construction envelopes.

However, wire-based feedstock presents certain challenges for these systems, including reduced component precision and subpar surface finish. The increased heat input associated with wire feed systems exacerbates concerns about distortion and residual stress. Melting the feeder wire necessitates greater heat, potentially resulting in a larger molten pool volume, which could further exacerbate residual stress and thermal deformation in the design [17] (Fig. 3).

2.3 Applications of DED

A cost-effective strategy for aerospace components involves employing repair applications instead of replacements. This approach yields considerable financial savings by mending damaged parts. DED process is a transformative solution due to its precise nature and minimal distortion. DED ensures unparalleled accuracy in restoring components, keeping them close to their original specifications. It also facilitates metallurgical bonding, making it the preferred repair process in the aerospace industry. Moreover, DED reduces material waste during repair operations, which benefits the environment.

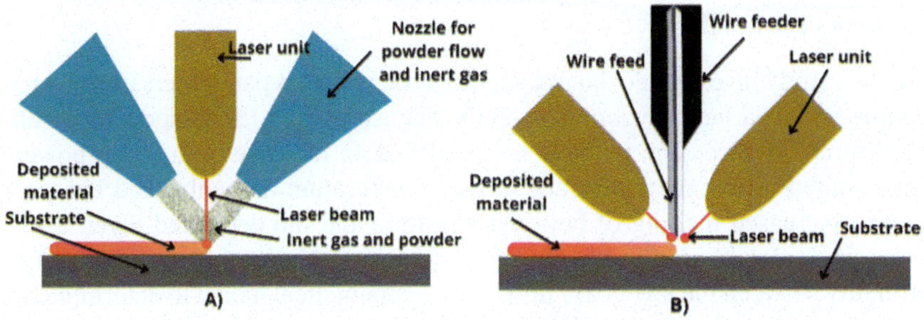

Fig. 3. Feed system process: A) powder feed system, B) wire feed system [15].

Numerous studies, including both academic and industrial investigations, have examined the feasibility and efficacy of the DED process for aerospace repair applications. These studies have demonstrated its ability to restore components with remarkable accuracy. For example, Optomec used DED to repair erosion-damaged AM355 steel T700 blisks, showcasing its potential for substantial returns on investment [18]. Similarly, Rolls-Royce Deutschland has endorsed the DED process for a range of repair operations, including the restoration of worn flanges and damping wire grooves in aerospace components [19] and others [20–24].

Although the DED process presents numerous advantages, its widespread adoption in the aerospace industry is not without challenges. Optimizing process parameters and deposition strategies to further enhance part quality remains a crucial challenge. However, ongoing research initiatives continue to address these challenges, offering a glimpse into a future where DED serves as a cornerstone of aerospace maintenance and repair operations [25]. In a study conducted by Bennett et al. [26], it was revealed that dies repaired using the LAM-DED (Laser Additive Manufacturing—Directed Energy Deposition) process exhibit a significantly longer service life compared to those repaired through traditional methods. The enhancement range was reported to be between 3.4 and 8.0 times, emphasizing the effectiveness and durability of the LAM-DED technique in contrast to conventional repair processes.

Additionally, Saboori et al. [27] conducted a study in which the LAM-DED process was employed for the repair of austenitic stainless steel vessels. This usage of advanced additive manufacturing methods in mending critical components further demonstrates the versatility and applicability of such techniques across various industrial sectors.

Furthermore, Wahab et al. [28] utilized the Laser Metal Deposition (LMD) process to restore the sprocket of a scraper conveyor. By leveraging this leading-edge technology, they were able to effectively refurbish a crucial component, ensuring its continued operation and longevity within the conveyor system. This application showcases the potential of additive manufacturing processes in addressing maintenance and repair challenges in mechanical systems, offering cost-effective and efficient solutions for extending the operational lifespan of equipment.

2.4 Process Parameters

The present study investigated the impact of various process parameters on the geometry of a single track, particularly focusing on the deposition of AISI M4 powder on an AISI 1045 substrate surface. The findings revealed that an increase in the high-power laser energy resulted in a higher substrate surface temperature and enhanced heat transfer, leading to increased melting of both the substrate and the deposited layer, ultimately resulting in greater dilution depth. However, excessive dilution can negatively affect the bond quality between the layers. By utilizing a mathematical model to determine optimal process parameters, the accuracy and quality of the deposited part can be improved. Furthermore, the study derived a formula for predicting the shape of the deposited bead, which can serve as a practical tool in the design of multi-track and multi-layer deposition processes, as well as layer thickness setting for multi-layer deposition [29].

This study, which compared three clads, found slight differences while maintaining a laser feed rate of 500 mm/min and a constant powder flow rate throughout the printing process. However, in the cases of 10-750-1.5, 12-750-1.5 and 14-750-1.5, balling appeared, indicating the absence of a wafer, and the metal balls did not bond and fell off the substrate. In contrast, cases 10-750-1, 12-750-1 and 14-750-1 showed some balling but were bonded to the substrate [30].

And another study by Pei-Yi Lin which dealt with porosity. Porosity initially decreased slightly to 0.329% before increasing as the scanning speed reached 11 mm/s, resulting in a dilution of 0.2%. The discrepancy between the predicted values and the average measured values under identical conditions was 18.28%. Three one-factor-at-a-time experiments demonstrated that laser power, laser scanning speed, and powder feed rate influenced the porosity and dilution of DED-constructed components. The study concluded that these factors could be combined as two novel parameters: specific energy (SE) and powder density (PD). These parameters, along with the experimental results, were utilized to establish a mathematical model for forecasting product quality, which serves as a valuable tool for adjusting DED processing parameters and may constitute the initial step in the development of intelligent machinery for the DED fabrication process [31].

3 Conclusion

DED is a cutting-edge technology in the aerospace industry that enables the precise repair of components. DED ensures that the original specifications of components are maintained while promoting metallurgical bonding. This paper provides a comprehensive review of both academic and industrial studies to evaluate the feasibility and effectiveness of DED. The technology offers significant advantages over conventional repair methods, particularly in the repair of erosion-damaged components such as AM355 steel T700 blisks. Industry leaders, such as Rolls-Royce Deutschland, have endorsed DED for repairing worn flanges and damping wire grooves, further emphasizing its potential to deliver substantial returns on investment.

Despite these advancements, there are ongoing challenges in optimizing process parameters and deposition strategies to enhance part quality. However, research initiatives are actively working to address these challenges, positioning DED as a critical technology in aerospace maintenance and repair operations. The paper offers valuable insights into the current state of DED technology, including its benefits, challenges, and future prospects in aerospace component repair. The research emphasized the substantial impact of process parameters on the configuration, bond strength, and porosity of DED components. They discovered optimal combinations of these parameters to increase the precision and excellence of the parts, while tackling problems such as excessive mixing and clumping. Mathematical models were formulated to forecast the shape of the beads and the quality of the final product, providing essential tools for adjusting the parameters of DED processing and enhancing the intelligence of machinery used in manufacturing processes. Overall, the widespread adoption of DED promises to revolutionize aerospace maintenance practices by providing cost-effective solutions while minimizing environmental impact.

Funding. This study was conducted in association with the project Innovative and Additive Manufacturing Technology—New Technological Solutions for 3D Printing of Metals and Composite Materials (reg. no. CZ.02.1.01/0.0/0.0/17_049/0008407) financed by Structural Funds of the European Union. *Article has been done in connection with project Students Grant Competition SP2024/087 "Specific Research of Sustainable Manufacturing Technologies" financed by the Ministry of Education, Youth and Sports and Faculty of Mechanical Engineering VŠB-TUO.*

References

1. ISO/ASTM, 52900:2021–11: Additive manufacturing—general principles—fundamentals and vocabulary. https://doi.org/10.31030/3290011, (n.d.)
2. Baykasoğlu, C., Akyildiz, O., Tunay, M., To, A.C.: A process-microstructure finite element simulation framework for predicting phase transformations and microhardness for directed energy deposition of Ti6Al4V. Addit. Manuf. **35**, 101252 (2020). https://doi.org/10.1016/J.ADDMA.2020.101252
3. Li, Z., et al.: Microhardness and wear resistance of Al_2O_3-TiB2-TiC ceramic coatings on carbon steel fabricated by laser cladding. Ceram. Int. **45**, 115–121 (2019). https://doi.org/10.1016/J.CERAMINT.2018.09.140
4. Ahn, D.-G.: Directed energy deposition (DED) process: state of the art. Int. J. Precis. Eng. Manuf. Green Technol. **8**, 703–742 (2021). https://doi.org/10.1007/s40684-020-00302-7
5. Vafadar, A., Guzzomi, F., Rassau, A., Hayward, K.: Advances in metal additive manufacturing: a review of common processes, industrial applications, and current challenges. Appl. Sci. **11** (2021). https://doi.org/10.3390/app11031213
6. Casalino, G., Karamimoghadam, M., Contuzzi, N.: Metal wire additive manufacturing: a comparison between arc laser and laser/arc heat sources. Inventions **8** (2023). https://doi.org/10.3390/inventions8020052
7. Braga, D.F.O., Azevedo, L., Cipriano, G., Moreira, P.M.G.P.: Hybrid Selective Laser Melting, direct energy deposition process strength-based optimization for hot-forming tool. Procedia Struct. Integr. **54**, 568–574 (2024). https://doi.org/10.1016/J.PROSTR.2024.01.120
8. Frazier, W.E.: Metal additive manufacturing: a review. J. Mater. Eng. Perform. **23**, 1917–1928 (2014). https://doi.org/10.1007/s11665-014-0958-z
9. Mesicek, J., Jancar, L., Ma, Q.-P., Hajnys, J., Tanski, T., Krpec, P., Pagac, M.: Comprehensive view of topological optimization scooter frame design and manufacturing. Symmetry (Basel) **13** (2021). https://doi.org/10.3390/sym13071201

10. Zhao, N., et al.: Direct additive manufacturing of metal parts for automotive applications. J. Manuf. Syst. **68**, 368–375 (2023). https://doi.org/10.1016/J.JMSY.2023.04.008
11. OPTOMEC Machines: https://optomec.com/lens-cs-600-system/. Last accessed 25 Mar 2024
12. Meltio Machines: https://meltio3d.com/metal-3d-printers/meltio-m600/. Last accessed 25 Mar 2024
13. DMG Mori Machines: https://cz.dmgmori.com/novinky-media?tax=14712&redirect=i. Last accessed 25 Mar 2024
14. Trumpf machines: https://www.trumpf.com/cs_CZ/produkty/stroje-systemy/laserova-svarovaci-zarizeni-a-bunka-pro-oblouk-svarovani/trulaser-cell-3000/. Last accessed 25 Mar 2024
15. Ahn, D.-G.: Directed energy deposition (DED) process: state of the art. Int. J. Precis. Eng. Manuf. Green Technol. **8**, 703–742 (2021). https://doi.org/10.1007/s40684-020-00302-7
16. Singh, A., Kapil, S., Das, M.: A comprehensive review of the methods and mechanisms for powder feedstock handling in directed energy deposition. Addit. Manuf. **35**, 101388 (2020). https://doi.org/10.1016/J.ADDMA.2020.101388
17. Fu, J., Gong, L., Zhang, Y., Wu, Q., Shi, X., Chang, J., Lu, J.: Microstructure and mechanical properties of Ti-6Al-4V fabricated by vertical wire feeding with axisymmetric multi-laser source. Appl. Sci. **7** (2017). https://doi.org/10.3390/app7030227
18. Optomec: Blisk repair solution. Accessed 3 Dec 2021 (2014). https://www.optomec.com/wp-content/uploads/2014/04/Optomec_LENS_Blisk_Repair_Datasheet.pdf
19. Gasser, A., Backes, G., Kelbassa, I., Weisheit, A., Wissenbach, K.: Laser additive manufacturing: laser metal deposition (LMD) and selective laser melting (SLM) in turbo-engine applications. Laser Tech. J. **7**(2), 58–63 (2012)
20. Jones, J.B., McNutt, P., Tosi, R., Perry, C., Wimpenny, D.I.: Remanufacture of turbine blades by laser cladding, machining and in-process scanning in a single machine (2012)
21. Qi, H., Azer, M., Singh, P.: Adaptive toolpath deposition method for laser net shape manufacturing and repair of turbine compressor airfoils. Int. J. Adv. Manuf. Technol. **48**(1–4), 121–131 (2010). https://doi.org/10.1007/s00170-009-2265-7
22. Kistler, N.A., Corbin, D.J., Nassar, A.R., Reutzel, E.W., Beese, A.M.: Effect of processing conditions on the microstructure, porosity, and mechanical properties of Ti-6Al-4V repair fabricated by directed energy deposition. J. Mater. Process. Technol. **264**, 172–181 (2019). https://doi.org/10.1016/j.jmatprotec.2018.08.041
23. Zhang, X.C., Li, W., Chen, X.Y., Cui, W.Y., Liou, F.: Evaluation of component repair using direct metal deposition from scanned data. Int. J. Adv. Manuf. Technol. **95**(9–12), 3335–3348 (2018). https://doi.org/10.1007/s00170-017-1455-y
24. Liu, Q., Janardhana, M., Hinton, B., Brandt, M., Sharp, K.: Laser cladding as a potential repair technology for damaged aircraft components. Int. J. Struct. Integr. (2011)
25. Gradl, P.R., Cervone, A., Gill, E.: Surface texture characterization for thin-wall NASA HR-1 Fe–Ni–Cr alloy using laser powder directed energy deposition (LP-DED). Adv. Indus. Manuf. Eng. **4**, 100084 (2022). https://doi.org/10.1016/J.AIME.2022.100084
26. Bennett, J., Garcia, D., Kendrick, M., Hartman, T., Hyatt, G., Ehmann, K., You, F., Cao, J.: Repairing automotive dies with directed energy deposition: industrial application and life cycle analysis. J. Manuf. Sci. Eng. **141** (2018). https://doi.org/10.1115/1.4042078
27. Saboori, A., Aversa, A., Marchese, G., Biamino, S., Lombardi, M., Fino, P.: Application of directed energy deposition-based additive manufacturing in repair. Appl. Sci. **9** (2019). https://doi.org/10.3390/app9163316
28. Rahito, D.A.W., Azman, A.H.: Additive manufacturing for repair and restoration in remanufacturing: an overview from object design and systems perspectives. Processes **7** (2019). https://doi.org/10.3390/pr7110802
29. Lee, E.M., Shin, G.Y., Yoon, H.S., Shim, D.S.: Study of the effects of process parameters on deposited single track of M4 powder based direct energy deposition. J. Mech. Sci. Technol. **31**, 3411–3418 (2017). https://doi.org/10.1007/s12206-017-0239-5

30. Lim, C.W.J., Zhang, Y., Huang, S., Chan, W.L.: A dimensionless analysis to select directed energy deposition process parameters for proper clad formation. Int. J. Adv. Manuf. Technol. **125**, 947–963 (2023). https://doi.org/10.1007/s00170-022-10393-y
31. Lin, P.-Y., Shen, F.-C., Wu, K.-T., Hwang, S.-J., Lee, H.-H.: Process optimization for directed energy deposition of SS316L components. Int. J. Adv. Manuf. Technol. **111**, 1387–1400 (2020). https://doi.org/10.1007/s00170-020-06113-z

The Influence of Orientation by SLM Additive Manufacturing Process on Surface Integrity

Vladimír Bechný[1(✉)], Miroslav Matuš[1], Richard Joch[1], Mário Drbúl[1], Andrej Czán[1], Miroslav Cedzo[1], and Dominik Krišák[1,2]

[1] Faculty of Mechanical Engineering, University of Zilina, Univerzitná 8215/1, 01026 Zilina, Slovakia
vladimir.bechny@fstroj.uniza.sk

[2] Center of 3D Printing Protolab, Department of Machining, Assembly and Engineering Technology, Faculty of Mechanical Engineering, VSB-TU Ostrava, 17. Listopadu 2172/15, 70800 Ostrava, Czech Republic

Abstract. The research involves a detailed analysis of the surface integrity of printed samples generated and produced by additive manufacturing based on Selective Laser Melting technology. The technological principle involves melting a fine metal powder with a heat source that generates a focused laser beam. The fine metal powder is stored in a filling hopper, where it is selectively applied layer by layer into the main hopper. After one layer has been scanned, the laser stops working and the main tray drops by the height of one layer. The new powder is then added to this height, which is evenly applied and compacted to the required density. The entire process is reiterated until the components reach the desired height. This technology represents a favourable alternative to conventional technologies, as the components produced have similar properties to those of traditional technologies. It is an eco-friendly manufacturing method where unused metal powder is recyclable through a predefined process for the next printing cycle, making it nearly zero-waste. This research aims to analyze and contrast the impact of various orientations of printed parts within the workspace on the surface quality of produced samples. Five orientations were employed in the Selective Laser Melting fabrication process for experimental purposes. From the conducted measurements, it is evident that the orientation of parts within the workspace substantially affects the surface integrity of components manufactured through additive processes.

Keywords: Additive manufacturing production · Selective laser melting · Roughness parameters

1 Introduction

Additive manufacturing represents an innovative approach to fabricating plastic, composite, and metal components across diverse industries. This manufacturing technique operates based on material layering, typically utilizing powder or filament. It begins with designing the part in CAD software, followed by exporting it in STL format. Subsequently, the model undergoes preparation in slicing software, dividing it into layers that the device can process [1–3].

M. Pagac et al. (Eds.): SCFF 2024, LNNS 1062, pp. 206–215, 2024.
https://doi.org/10.1007/978-3-031-65656-9_21

The model can be generated through structural design within a CAD system or by scanning a physical object and subsequently creating the model from the scan data. Proper setup of the entire process is crucial to ensure the desired product quality [4–6]. Important parameters in configuring the device include the part's orientation within the printing space, nozzle temperature, layer thickness and inner filler shape, among others. These parameters have a substantial impact on both the mechanical properties and the overall quality of the produced parts [7–9].

Selective Laser Melting (SLM) stands as one of the more recently established additive manufacturing techniques, involving the creation of three-dimensional objects layer by layer utilizing a high-power laser. Uniform deposition of metal powder occurs on the surface of the printing substrate [10–13]. The laser is then aimed at the areas where the powder needs to be heated to melting temperature, thus creating a small metallic liquid melt. Once a single layer has been melted, the working platform is gradually lowered allowing further layers of powder to be gradually applied and the 3D object itself to be built up. Once the melting of the individual layers is complete, the material is rapidly cooled, allowing for a solid bond with the previous layer. All those steps are under flow of inert gas preventing oxidation and some other negative effects. SLM technology is used in the manufacture of complex metal parts such as aerospace and automotive components, medical implants, and various tools where high strength and precision of the material is required [14, 15]. This technology enables the creation of detailed and functional metal parts with almost zero geometric constraints [16, 17].

In the Powder Bed Fusion processes, particularly the SLM process, the processing parameters are the most crucial values which determine the outcome of the process. Beyond the processing parameters mentioned above, factors such as scanning strategy, scanning speed, laser power, and the flow of inert gas significantly impact the outcome, as stated in articles [18–21]. It is noteworthy that these variables vary considerably depending on the material being processed. Their careful calibration is irreplaceable for achieving and maintaining the desired product quality.

The roughness of the surface of the AM metal components produced through Selective Laser Melting (SLM) is influenced by several factors, such as powder properties, laser parameters, and post-processing methods. The typical range for SLM parts is Ra = 2–25 μm, which is mostly influenced by the layer thickness, laser power, and scan speed. It is generally observed that a higher laser power and slower scan speed lead to smoother surfaces. Aside from SLM process parameters, post-processing methods like tumbling, sandblasting, machining, polishing, or shot peening also exert considerable influence in diminishing surface roughness. Achieving an optimal surface finish holds paramount importance for SLM parts, necessitating precise mechanical properties and an aesthetically pleasing appearance. For further insights into each post-processing technique, detailed information is available in the referenced articles [22, 23].

The research aims to contribute novel insights to the additive manufacturing domain, specifically focusing on the impact of part orientation within the workspace on the resultant surface integrity of samples fabricated via SLM technology.

2 Materials and Methods

The experiments aimed to explore how the orientation of parts produced via SLM technology influences overall surface integrity. Building upon prior research, a cylindrical sample was identified and designed. Figure 1 illustrates the orientations of the sample, which were manufactured in five variations in axes X, Y, Z, XY, and XYZ.

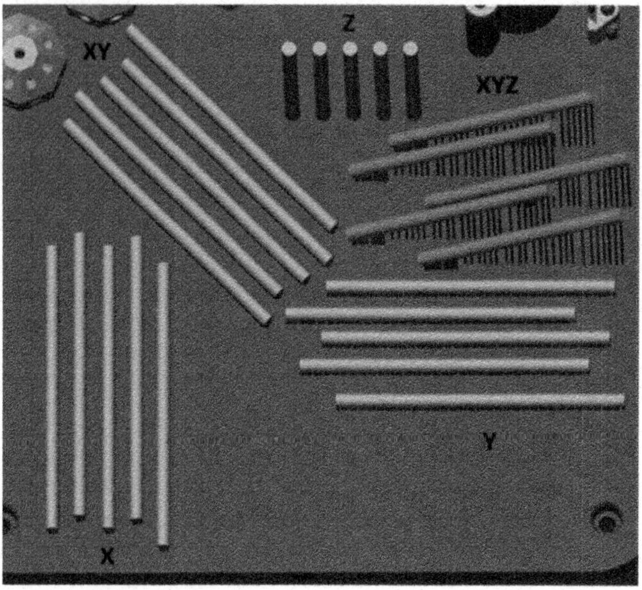

Fig. 1. Orientation of parts in the work chamber

2.1 Design of Experiment (DoE)

For the experiment, samples were fabricated in the form of cylindrical rods (Fig. 2) with a circular cross-section, possessing a diameter of 5 mm and a length of 100 mm. To maintain an unbiased comparison, the thickness of all test samples was standardized to 50 μm. Each orientation was produced with three samples.

2.2 Applied Technology and Used Device

Selective Laser Melting (SLM) is an advanced additive manufacturing technique that utilizes a high-powered laser to selectively melt and fuse metallic powders, layer by layer, to create intricate three-dimensional objects. It is a key method in the realm of additive manufacturing, particularly for producing complex metal parts with high precision.

In SLM, a CAD model guides the laser as it precisely melts and fuses thin layers of metal powder. As each layer is completed, the build platform descends, and a new layer of powder is spread over the previous one. This process repeats until the entire object is formed (Fig. 3).

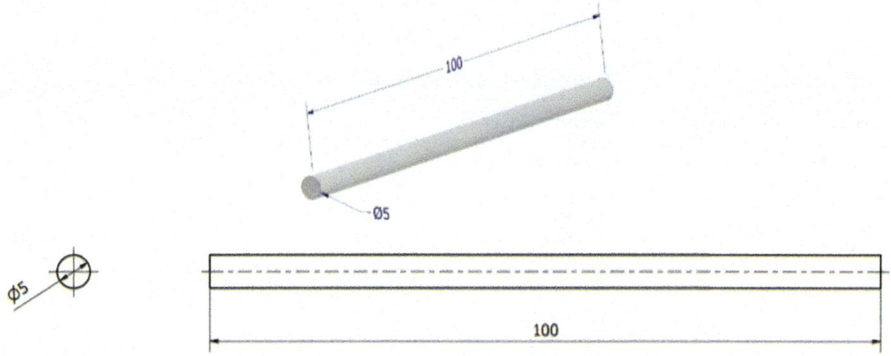

Fig. 2. Shape of the test sample

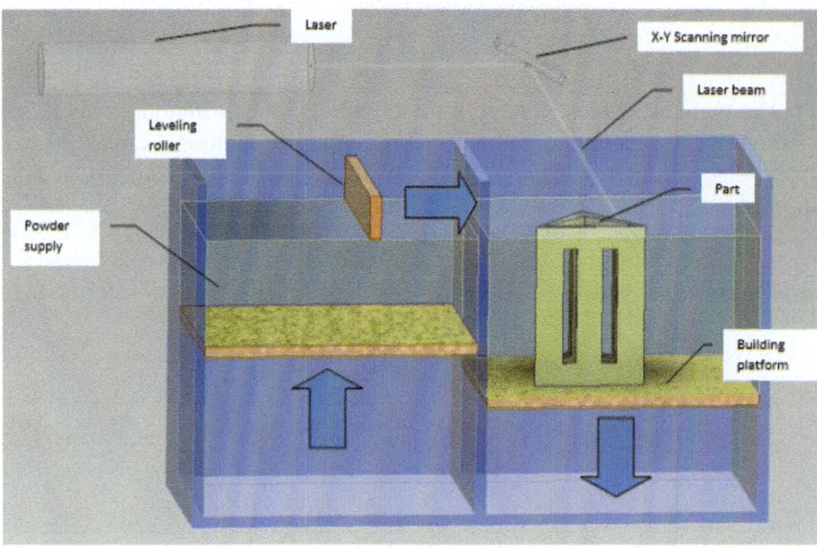

Fig. 3. Manufacturing process of metal parts produced by SLM technology

The Renishaw AM400 was selected for sample production, which is an advanced additive manufacturing machine for a variety of industrial applications. Renowned for its accuracy and reliability, the AM400 uses advanced laser melting technology to produce high quality metal parts with complex geometries. With a large build volume and efficient powder handling system, it offers flexibility and scalability for a variety of production needs. The equipment is the preferred choice for the aerospace, automotive and medical industries seeking to optimize manufacturing processes and material utilization (Table 1).

2.3 Used Material

For the experimental samples, 316L stainless steel material was chosen. It is an austenitic corrosion-resistant steel, which is resistant to grain sensitization due to its low carbon

Table 1. Technical parameters of Renishaw AM400

Machine parameters	Parameter description/value
Producing technology	Selective laser melting
Direction of production	Universal (X, Y, Z)
Laser power Scanning velocity Hatching distance Layer thickness Energy input	200 W 650 $mm \cdot s^{-1}$ 110 μm 50 μm 55,94 $J \cdot mm^{-1}$
Powder bed dimensions Inert gas	250 × 250 × 400 mm Argon
Materials	Stainless steel 316l, tool steels, nickel alloys, chromium alloy, AlSi10Mg, Ti6Al4V ELI

content. The material is also suitable for additive manufacturing due to its toughness, corrosion resistance, hardness, and tensile strength. The material is used for machine components and production tools, durable prototypes, medical instruments, and the food industry (Table 2).

Table 2. Chemical composition of 316L metal powder as per the attestation certificate (wt. %) [24]

Cr	Fe	Mn	Mo	N	Ni	O	S	Si	P
16–18	Bal	≤ 2	2–3	≤ 0.1	10–14	≤ 0.1	≤ 0.03	≤ 0.03	≤ 0.045

2.4 Measurement of Roughness Parameters

Surface integrity was measured by Alicona Infinite Focus G5 confocal microscope. With this 3D measuring instrument, components can be observed with high resolution and accuracy. Infinite Focus is one of the fast and accurate optical 3D measuring systems. With one sensor, it is possible to confirm surface roughness and the dimensional accuracy of components. Components are tracked with high vertical resolution and opto-visibility. The device enables form and roughness measurements even for large and heavy components. The system is additionally employed also for fully automated measurements in the production.

In the measurement and subsequent evaluation of surface roughness, the analysis focused on the surface roughness and the analysis of important parameters Sa, Sz and Spk, S10z. The most important parameter was the roughness parameter S10z which represents 10 points surface height, the average height value of five peaks with the highest global peak height, added to the average depth value of five valleys with the greatest global depth. The produced samples were measured at 20× magnification in accordance with ISO 21920 standard, whereas the measured surface was evaluated on a measured section of 2.5 × 2.5 mm. After surface scanning, it was necessary to filter the sample shape and flatten the scanned surface to ensure a proper evaluation of the sample roughness.

3 Results

During the execution of the experiments, the integrity of the surface of the manufactured samples was measured and evaluated. Surface roughness parameters were measured and evaluated using the Alicona InfiniteFocus G5 optical measuring microscope. Data recorded by this microscope were subsequently processed into a table (Table 3). The values presented in the table illustrate how orientation affects the quality of the surfaces of the manufactured samples.

Table 3. Evaluation of selected surface roughness parameters

Sample	Sa [μm]	S10z [μm]	Spk [μm]	Ssk
XY1	12.99	109.30	20.47	0.74
XY2	15.24	176.96	22.69	0.07
XY3	16.06	164.29	20.07	−0.42
XYZ1	10.54	115.02	19.14	0.56
XYZ2	10.36	118.06	19.73	0.77
XYZ3	9.52	91.70	15.70	0.26
Y1	15.57	144.99	26.53	0.30
Y2	18.64	146.53	17.48	−1.11
Y3	18.10	176.22	36.25	0.04
X1	13.13	119.38	22.43	0.13
X2	15.11	143.15	18.07	−0.28
X3	15.47	145.24	18.94	−0.10
Z1	12.22	89.82	13.68	−0.30
Z2	11.82	89.39	13.46	0.16
Z3	12.03	103.13	16.39	0.45

The selected values of roughness parameters Sa and S10z presented in the table (Fig. 4) illustrate the best how orientation affects the quality of the manufactured samples. From the graphical comparison of roughness parameters Sa and S10z, we can observe different values depending on the orientation of the parts. Increased values are observed especially in the XY orientations. Conversely, the best surface quality is achieved by samples produced in the Z orientation.

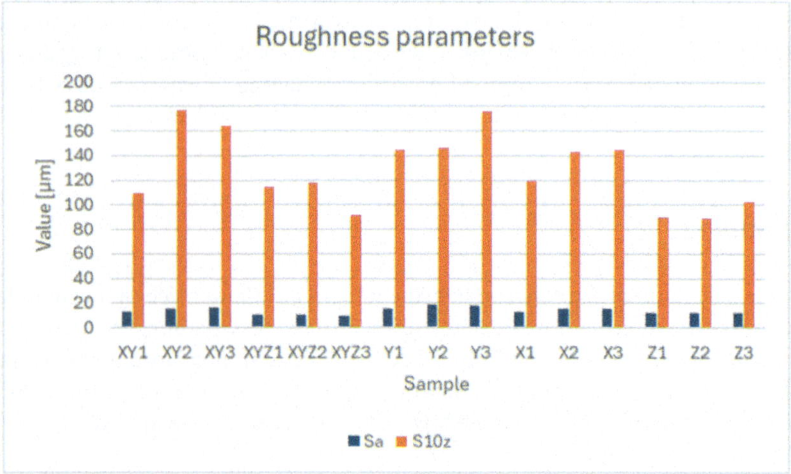

Fig. 4. Graphical comparison of measured roughness parameters Sa and S10z

The measured results were processed into graphical evaluations with respect to the parameter S10z. This parameter played the most important role in the evaluation itself. In Fig. 5, we present a graphical representation illustrating the selected surface roughness of samples based on the orientation of the parts within the working chamber, the left sample showing the worst surface and the right sample the best surface. We focused our analysis on parts with the highest-rated roughness compared to those with the lowest. It can be asserted that samples produced in the Z orientation outperformed those in other orientations. Sample Z2 yielded the most favourable outcome with the value of S10z roughness parameter 89.39µm. Conversely, the sample with the worst achieved results was in the XY orientation (sample XY2), where the measured value of the roughness parameter S10z was 176.69 µm.

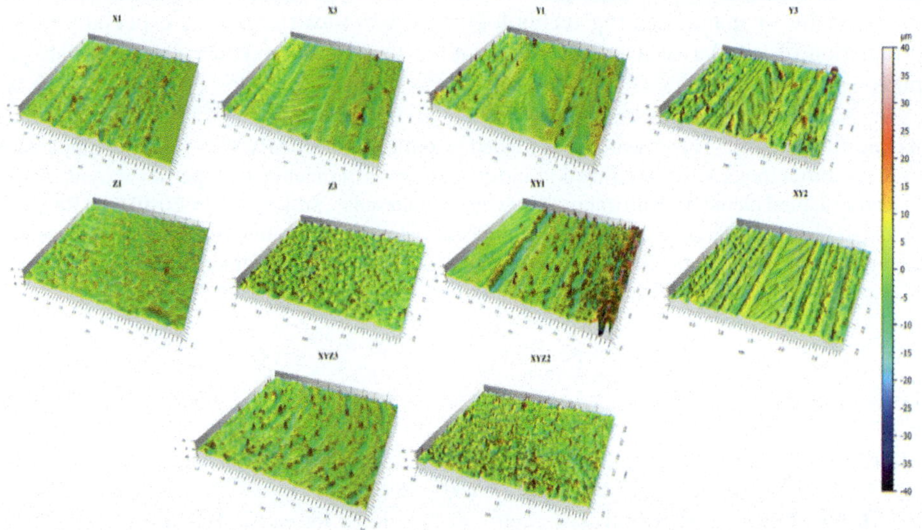

Fig. 5. Colour map of the best and worst samples in terms of surface roughness

4 Conclusion

Through experimental measurements conducted using the Alicona InfiniteFocus G5, the surface roughness parameters of the samples produced by the SLM technology were measured and assessed. We can observe that the values listed in the table demonstrate the influence of orientation on the resulting quality of the surfaces of the manufactured samples. Thus, in the conducted experiments, samples were manufactured with consideration to their orientation in the working chamber, encompassing various directions.

Based on experiments we can conclude and recommend the following:

- The surface integrity of samples manufactured by additive manufacturing process varies significantly depending on the orientation of the parts within the working chamber
- The sample at orientation Z (sample Z2) achieved the best values for surface roughness parameters, with a measured value of 89.39 μm for the roughness parameter S10z. Conversely, the sample in the XY orientation (sample XY2) yielded the worst results, with a measured value of 176.69 μm for the roughness parameter S10z
- For optimal surface integrity, particularly concerning the overall surface quality of parts produced by additive manufacturing processes, we recommend metal part manufacturers orient these parts along the Z-axis.
- The surface integrity of components produced by additive processes is influenced not only by the orientation of the parts within the working chamber but also by the properties of the powder, laser parameters, and post-processing methods.

The SLM technology represents an optimal and suitable selection of additive manufacturing technology, also a great alternative to traditional manufacturing technologies such as CNC machining or casting. SLM technology can significantly help complement the portfolio of manufacturing technologies in the industrial sector, especially in the production of smaller series of products and in the manufacturing of complex-shaped parts.

Acknowledgement. This research was funded by project project APVV 15-0405: "Complex use of X-ray diffractometry for identification and quantification of functional properties of dynamically loaded structural elements from important technical materials", project APVV 20-0561: "Research on the implementation of new measurement methods for the calibration of measurement systems for industrial metrology practice", Kega project 033ŽU-4/2022: "Implementation of the language of geometric product specification in the field of coordinate 3D metrology".

References

1. Gibson, I., Rosen, D. W., Stucker, B., Khorasani, M., Rosen, D., Stucker, B., Khorasani, M.: Additive manufacturing technologies **17**, 160–186 (2021)
2. Hajnys, J., Pagac, M., Mesicek, J., Petru, J., Spalek, F.: Research of 316L metallic powder for use in SLM 3D printing. Adv. Mater. Sci. **20**(1), 5–15 (2020)
3. Wu, Y., Fang, J., Wu, C., Li, C., Sun, G., Li, Q.: Additively manufactured materials and structures: a state-of-the-art review on their mechanical characteristics and energy absorption. Int. J. Mech. Sci. 246 (2023)
4. Ngo, T.D., Kashani, A., Imbalzano, G., Nquyen, K.T., Hui, D.: Additive manufacturing (3D printing): a review of materials, methods, applications and challenges. Compos. B Eng. **143**, 172–196 (2018)
5. Bourell, D., et al.: Materials for additive manufacturing. CIRP Ann. **66**(2), 659–681 (2017)
6. Joch, R., Šajgalík, M., Drbúl, M., Holubják, J., Czán, A., Bechný, V., Matúš, M.: The application of additive composites technologies for clamping and manipulation devices in the production process. Materials **16**(10) (2023)
7. Kafle, A., Luis, E., Silwal, R., Pan, H.M., Shrestha, P.L., Bastola, A.K.: 3D/4D printing of polymers: fused deposition modelling (FDM), selective laser sintering (SLS), and stereolithography (SLA). Polymers **13**(18) (2023)
8. Gardner, L.: Metal additive manufacturing in structural engineering–review, advances, opportunities and outlook. Structures **47**, 2178–2193 (2023)
9. Sadaf, M., Bragaglia, M., Slemenik Perše, L., Nanni, F.: Advancements in metal additive manufacturing: a comprehensive review of material extrusion with highly filled polymers. J. Manufact. Mater. Process. **8**(1), 14 (2024)
10. Paraschiv, A., Matache, G., Condruz, M.R., Frigioescu, T.F., Ionică, I.: The influence of laser defocusing in selective laser melted in 625. Materials **14**(13), 3447 (2021)
11. Nancharaiah, T.: A review paper on metal 3D printing technology. Adv. Mater. Process. Manuf. Appl.: Proc. iCADMA **2020**, 251–259 (2021)
12. Gao, B., Zhao, H., Peng, L., Sun, Z.: A review of research progress in selective laser melting (SLM). Micromachines **14**(1), 57 (2022)
13. Kong, D., et al.: Bio-functional and anti-corrosive 3D printing 316L stainless steel fabricated by selective laser melting. Mater. Des. **152**, 88–101 (2018)
14. Song, X., Zhai, W., Huang, R., Fu, J., Fu, M. W., Li, F.: Metal-based 3D-printed micro parts & structures. Encycl. Mater.: Metals Alloy. 448–461 (2021)
15. Zou, Z., Liu, Z., Yang, L., Tang, Y., Qiao, Y., Lu, D.: Corrosion behavior of different building planes of selective laser melting 316L stainless steel in 0.1 M HCl solution. J. Market. Res. **28**, 4738–4753 (2024)

16. Das, P., Halama, R., Natarajan, A.V., Khutia, N., Dey, P.P., Kunčická, L., Sari, F.: Constitutive modelling and damage prediction of AlSi10Mg alloy manufactured by SLM technology with emphasis on ratcheting in LCF regime. Int. J. Fatigue 181 (2024)
17. Klimek, L., Bułhak, B., Śmielak, B.: A comparison of the structure and selected mechanical properties of Cr/Co alloys obtained by casting and selective laser melting. J. Funct. Biomater. **15**(3), 61 (2024)
18. Hajnys, J., Pagac, M., Kotera, O., Petru, J., Scholz, S.: Influence of basic process parameters on mechanical and internal properties of 316L steel in SLM process for Renishaw AM400. MM Sci. J. 2790–2794 (2019)
19. Hajnys, J., Pagáč, M., Měsíček, J., Petru, J., Król, M.: Influence of scanning strategy parameters on residual stress in the SLM process according to the bridge curvature method for AISI 316L stainless steel. Materials **13**(7), 1659 (2020)
20. Bláha, R., Krisak, D., Hajnys, J., Mesicek, J., Pagac, M., Jancar, L., Gazda, M.: Experimental verification of process parameters influencing the mechanical properties of M300 in the SLM process. MM Sci. J. (2023)
21. Limbasiya, N., Jain, A., Soni, H., Wankhede, V., Krolczyk, G., Sahlot, P.: A comprehensive review on the effect of process parameters and post-process treatments on microstructure and mechanical properties of selective laser melting of AlSi10Mg. J. Market. Res. **21**, 1141–1176 (2021)
22. Mesicek, J., et al.: Abrasive surface finishing on SLM 316L parts fabricated with recycled powder. Appl. Sci. **11**, 2869 (2021)
23. Mechali, A., Mesicek, J., Ma, Q.-P., Hajnys, J., Gautam, P., Blaha, R., Krisak, D., Petru, J.: Abrasive surface treatment of ALSI10 mg parts made by L-PBF. MM Sci. J. (2024)
24. Renishaw.: https://www.renishaw.com/resourcecentre/en/details/data-sheet-ss-316l-0407-powder-for-additive-manufacturing--90802. Accessed 02 March 2024

Topology Optimization of the Actuating Arm

Eliska Posmykova[1](✉), Jakub Mesicek[1], Daniel Kotzur[1], Marek Pagac[1], Jana Petru[1], and Tomasz Kozior[2]

[1] Department of Machining Assembly and Engineering Metrology, Faculty of Mechanical Engineering, VSB—Technical University of Ostrava, 70800 Ostrava, Czech Republic
eliska.posmykova.st@vsb.cz

[2] Department of Metrology and Unconventional Manufacturing Methods, Faculty of Mechatronics and Mechanical Engineering, Kielce University of Technology, 25–314 Kielce, Poland

Abstract. This paper explores the use of topology optimization (TO) to improve the actuating arm's performance in drum brakes while minimizing material usage. The Fusion 360 software has been utilized to compare the outcomes of using three different materials: AlSi10Mg, AISI316L, and Ti6Al4V. The TO process optimizes the component's shape while preserving critical functional areas, with careful consideration of load cases, constraints, and mesh granularity to ensure computational efficiency and accuracy. The objectives of the research were to achieve a 20% mass reduction while maintaining or enhancing durability. Our results demonstrate that the desired mass reduction, while maximizing stiffness, was successfully attained. After evaluating various material variants, the 316L material was evaluated as the optimal TO variant.

Keywords: Topology optimization · Mechanical engineering design · Additive Manufacturing Production · Selective Laser Melting · Roughness Parameters · Powder bed fusion

1 Introduction

Topology optimization (TO) is a crucial aspect of modern engineering practices. It provides an effective way to improve performance while minimizing using materials. This study focuses on the application of TO of a critical component in drum brake systems—the actuating arm [1].

The TO techniques in Fusion 360 software have been used to refine the actuating arm component. By balancing mass reduction with durability through simulations and material evaluations, the aim was to optimize the actuating arm's design for real-world applications. This investigation has implications for automotive engineering, using additive manufacturing practices, highlighting the effectiveness of TO in mechanical design [2]. This article [3] presents a method to optimize infill structure placement in FDM 3D printing. Multi-infill geometry outperforms traditional rectilinear infill in loading capacity and stiffness. The optimization method changes the shape of the part only where necessary, while maintaining the functional requirement of other areas [4]. See Fig. 1 for a visual representation.

M. Pagac et al. (Eds.): SCFF 2024, LNNS 1062, pp. 216–225, 2024.
https://doi.org/10.1007/978-3-031-65656-9_22

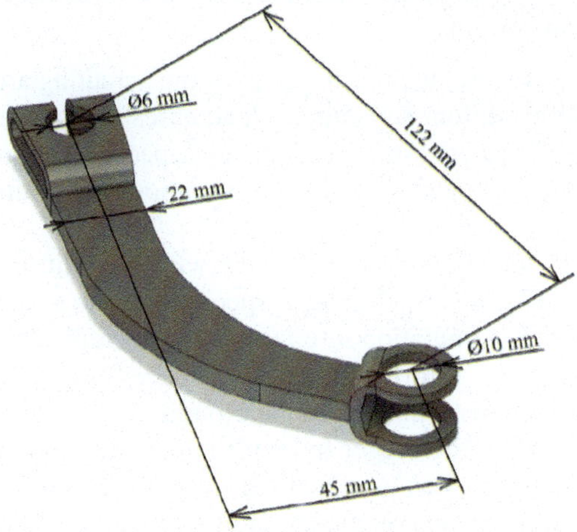

Fig. 1. The initial design of part before the TO

2 Materials and Methods

This paper performs a software-based Technical Optimization of a component and compares the results for three materials: AlSi10Mg, AISI316L, and Ti6Al4V. Based on the comparison, the appropriate material is recommended [5]. Each material's unique mechanical properties influence the optimization process and final component performance. The values in Table 1 were used to compare the materials, and the most suitable material was selected based on the optimization results [6].

Table 1. Mechanical properties of possible material choices [5].

	AlSi10Mg	AISI316L	Ti6Al4V
Average density [$kg \cdot m^{-3}$]	2600	8000	4420
Layer thickness	25	50	30/60
Yield strength $Rp_{0.2}$ [MPa]			
Horizontal	400 ± 13	662 ± 2	1077 ± 2
Vertical	366 ± 30	574 ± 10	1060 ± 19
Tensile strength R_m [MPa]			
Horizontal	265	520	980
Vertical	240	494	890
Young modulus [GPa]			
Horizontal	68	190	114
Vertical	60	60	415

2.1 Simulation Preparation

Utilizing Fusion 360 software, the CAD model of the actuating arm served as the basis for simulation and optimization. By defining load cases and constraints, the conditions necessary for effective TO were established. A key aspect of this preparation was identifying areas within the component requiring preservation due to functional requirements [7].

The 3D model was transferred into a new tab where conditions for shape optimization were entered. Load cases, including a specified force acting on the component, were incorporated into the simulation setup to accurately replicate real-world operating conditions. These load cases informed the design space for optimization, ensuring that critical areas of the component remained intact to maintain functionality [8].

A load case that acts on the bigger hole of the part was added. The Bearing Force function under the Loads tab was used and the force value was set as specified to F = 450N and this force is distributed on the larger cylindrical part [9].

The constraints and load cases were established, and then the information about the parts of the model that were left intact (the areas where constraints and load cases are acting) was supplied to the program. These areas must be preserved due to functionality. In those preserved parts, the mass distribution on the model cannot be modified during the TO process. Both preserved areas are depicted in Fig. 2 [10].

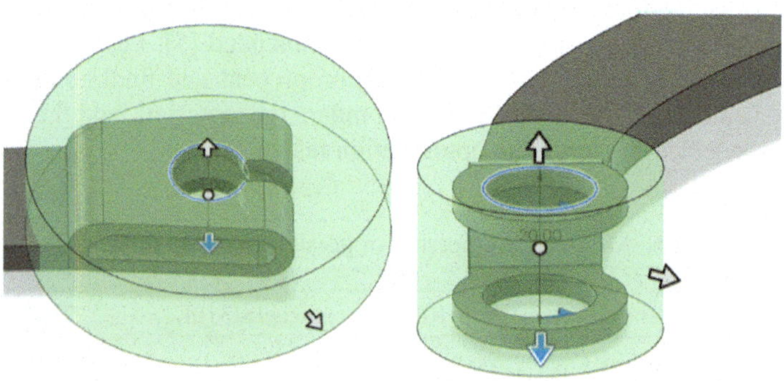

Fig. 2. The definition of design space and preserved areas

The granularity of the mesh design of the part was carefully considered to balance computational efficiency with model accuracy. Fine-tuning the mesh size allowed for precise optimization while managing computational resources effectively [11]. When designing a local mesh of the part, there must be a decision made about, how big the cell size of the mesh will be. The finer the mesh is chosen, the longer it will likely take to compute and design. However, the created optimized model will be more accurate [12].

3 Results and Discussions

The TO simulation has been done three separate times, each simulation for one material. After the TO had been performed, the manufacturability of all three variants was evaluated. the shape of each variant has been enhanced [1].

3.1 Topology Optimization

As mentioned earlier, the design space (the area where the simulation and shape modification in the form of the TO was performed) and the non-design space (preserved areas, and functional parts of the actuating arm) were specified [13].

The primary objectives of the TO process were to achieve a minimum 20% reduction in mass compared to the original component without compromising durability the 20% limit was chosen because greater weight savings are not possible in most TO cases. Additionally, the factor of safety was to be maintained at or above the original level to ensure structural integrity under operational loads [14]. In Fig. 3, green areas signify regions where material is unnecessary, while red areas signify regions where material removal is not advisable.

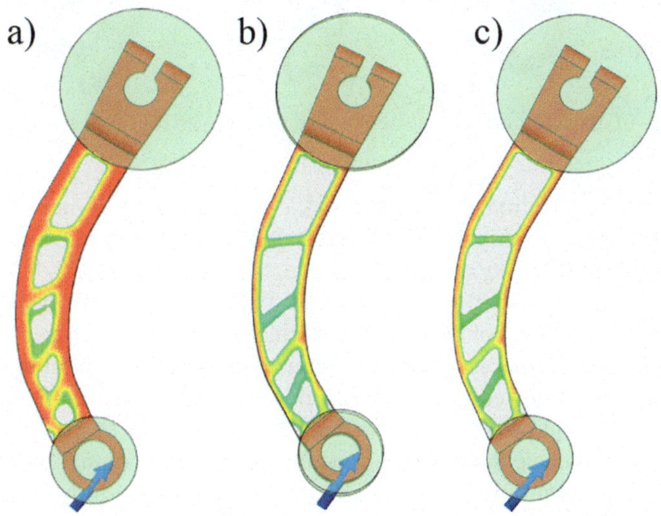

Fig. 3. Variants of optimization using different materials analyzed with FEM, **a** AlSi10Mg, **b** AISI316L, **c** Ti6Al4V

The shapes of the three variants generated by the TO are not suitable for manufacturing. Therefore, the optimization results were used to minimize the amount of material being used and create a model with a better ability to be manufactured, at the same time [15].

In Fig. 4, there are shown the different actuating arms results, which have been reworked after TO in the Fusion 360 software. The AlSi10Mg variant has the largest percentage of mass compared to the original component. On the other hand, the Ti6Al4V variant can use a lesser amount of material, due to the mechanical properties of the material being used [16].

Based on the comparison table (Table 2) and graph (Fig. 5), all three optimization options utilizing selected materials meet the requirement of reducing the original component's weight by at least 20%. The weight of the AlSi10Mg optimization variant is reduced by 27.18%. It is important to note that removing more than 20% of the mass

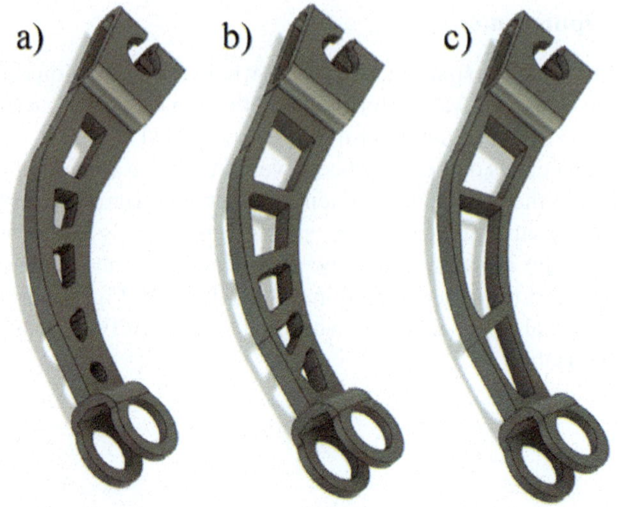

Fig. 4. Variants of optimization using different materials, **a** AlSi10Mg, **b**) AISI316L, **c** Ti6Al4V

in this particular application may pose a risk of eliminating too much material, thus making it inadvisable for this particular application. Prior to optimization, the original component weighed 0.103 kg.

For the AISI316L variant, it is reduced to 65.05%. Tor the Ti6Al4V variant, the weight reduction is 33.98% compared to the original part. In Fig. 6. There is a comparison of mass and the safety factor for all three optimization variants. The minimum values of the safety factor are indicated in Table 2 [2]. Based on a qualified estimate, the price of printing one rocker arm prototype for individual materials was determined. The values of the Mass Ratio indicate the percentage of the mass of the original part that remains. Due to the mechanical properties of the material, the aluminum alloy variant of TO has the most mass of all three of them. In contrast, the Ti6Al4V has the lowest mass ratio [17].

Table 2. The comparison of achieved results of TO from different materials.

	AlSi10Mg	AISI316L	Ti6Al4V
Mass [kg]	0.028	0.067	0.035
The factor of safety [–]	1.914	1.275	3.814
Mass ratio [%]	27.18	65.05	33.98
Price for manufactured part [EUR] (as of 2024)	35.45	40	275

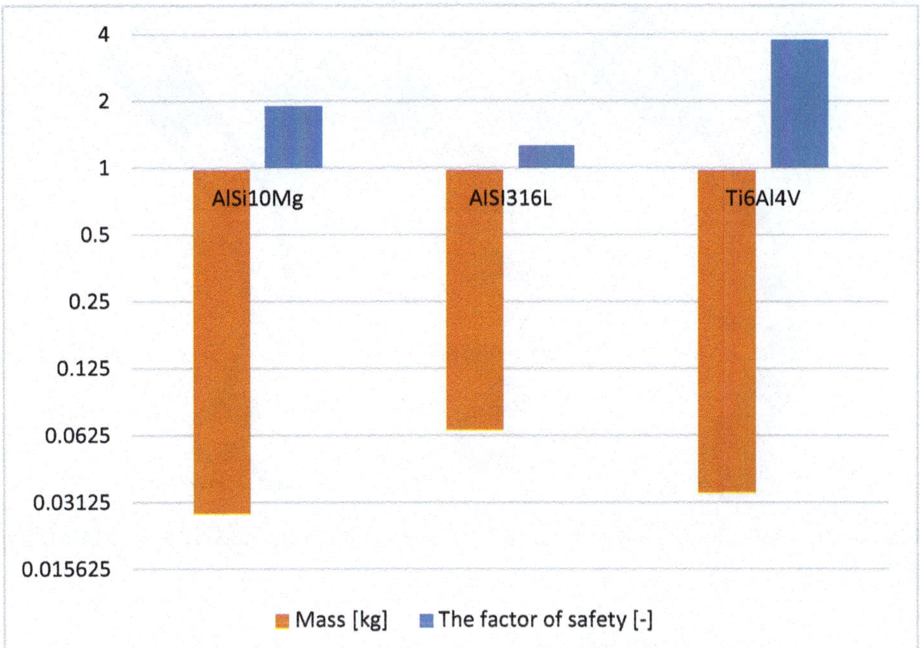

Fig. 5. The comparison of achieved results of TO from different materials

3.2 Verification of Reshaped Models

To improve manufacturability, the topology optimization results for all three versions were adjusted. The CAD models were validated through FEM analysis, and Fig. 6 shows the modified models. The color map displays the safety factor distribution, with minimum values indicated in Table 2.

According to the results of TO, the AISI316L variant was chosen as the most suitable one for the application of the part. The final CAD model used for the actual assembly is shown in Fig. 7 [1].

In Fig. 8, there is the finished actuating arm that resulted from the TO with consequential modifications made for improved manufacturability. The component has a modified shape but remains lightweight, with the desired mechanical properties guaranteed (previously tested using the FEM analysis). In the visualization, there is shown the utilization of the modeled component in the assembly of drum brakes.

It can be observed in Fig. 9 a close-up of the component, highlighting the intricate mesh network of elements. This image specifically showcases the most heavily strained section of the model, which has been carefully preserved for analysis. The Von Mises stress in this region reached a value of 149.061 MPa [18].

3.3 Selection of the Final Optimization Variant

Based on the obtained results, it was evaluated that the most suitable optimization variant for the selected application of the actuating arm for drum brakes is the variant made of

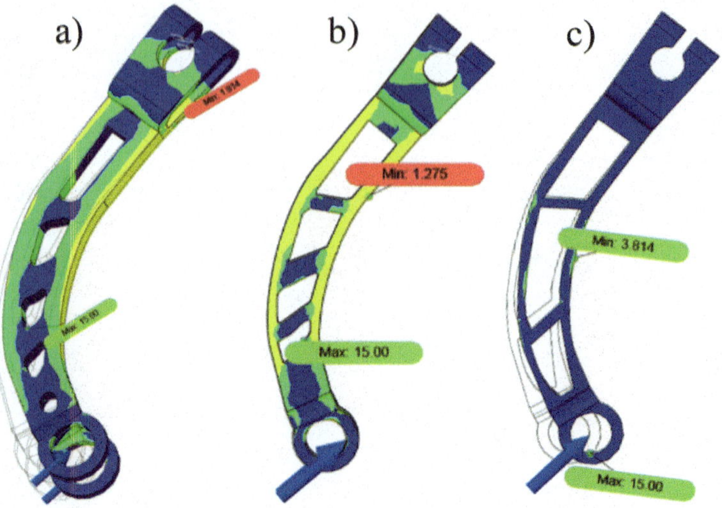

Fig. 6. The results of the FEM analysis of variants of optimization, **a** AlSi10Mg, **b** AISI316L, and **c** Ti6Al4V

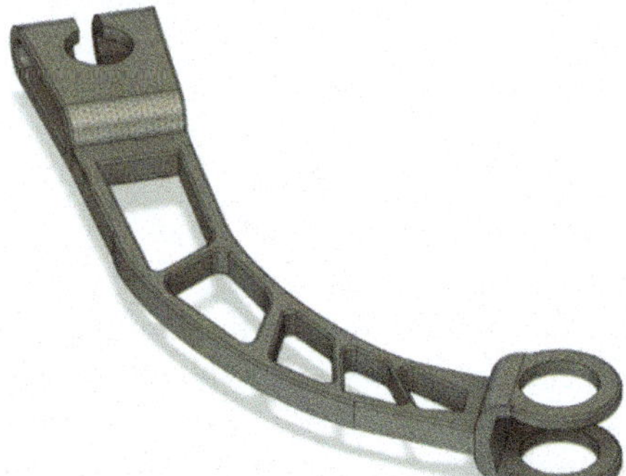

Fig. 7. The final CAD model of the actuating arm

corrosion-resistant steel AISI316L. Below in the Fig. 10, there is a comparison of material AISI316L (a), and Ti6Al4V (b) after performing the final validation FEM analysis [19]. The Ti6Al4V variant was found to be over-dimensioned for the given application in terms of its parameters. It is more economical to choose corrosion-resistant steel as the final material [7].

4 Conclusions

Topology optimization (TO) is a powerful tool in modern engineering that improves performance and reduces material consumption. In this study, TO was used to optimize the actuating arm in drum brake systems, to enhance structural efficiency without compromising durability. The actuating arm's performance was evaluated using Fusion 360

Fig. 8. Visualization of the optimized part in the assembly

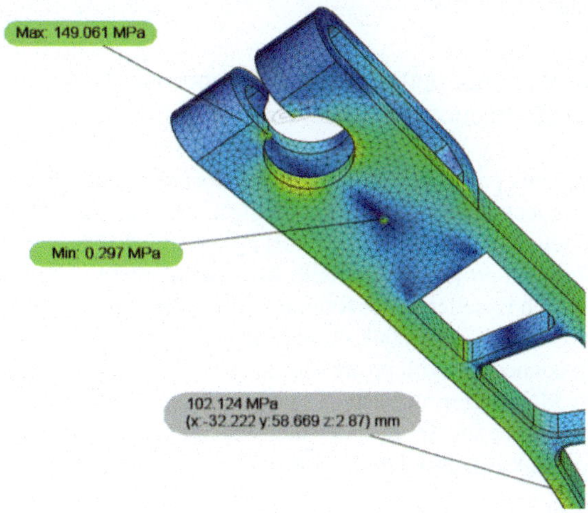

Fig. 9. The critical area of the finished optimized model

software across three materials—AlSi10Mg, AISI316L, and Ti6Al4V—with different mechanical properties. Material selection played a crucial role in the optimization process and final component performance. The topologically optimized part was created for the needs of the assembly for which it is intended to be used.

Subsequent iterations involved refining the optimized shapes for manufacturability, ensuring that the final designs were not only structurally efficient but also feasible for production. Among the materials tested, AISI316L emerged as the most suitable choice for the actuating arm application, balancing reduced mass with adequate strength and durability. Overall, this study underscores the significance of TO in advancing mechanical engineering design, particularly in optimizing complex components for performance and efficiency. The insights gained from this research provide valuable guidance for

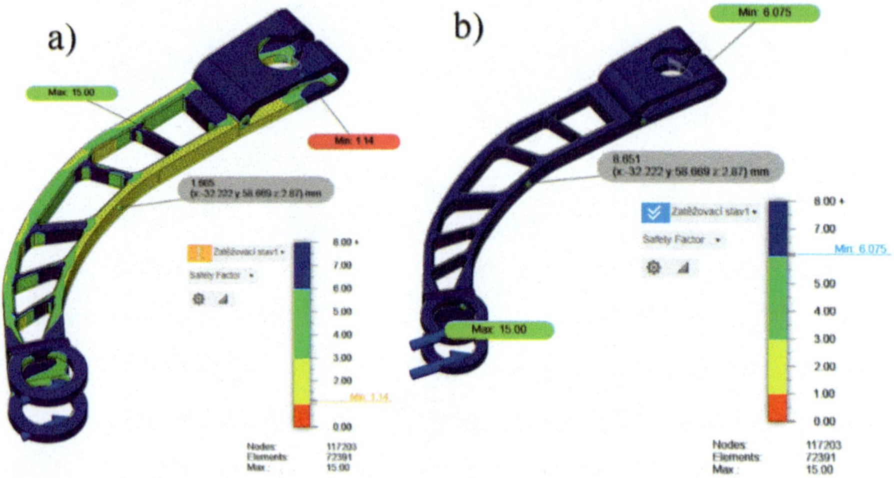

Fig. 10. The FEM analysis results of the compared variants of optimization, **a** AISI316L variant, and **b** Ti6Al4V variant

future endeavors in topology optimization and contribute to the ongoing evolution of additive manufacturing practices in the automotive industry.

Acknowledgements. This paper was supported in association with the project Innovative and additive manufacturing technology—new technological solutions for 3D printing of metals and composite materials, reg. No. CZ.02.1.01/0.0/0.0/17_049/0008407 financed by Structural Funds of the European Union and project. This article was also funded by the European Union under the REFRESH—Research Excellence For Region Sustainability and High-tech Industries project number CZ.10.03.01/00/22_003/0000048 via the Operational Programe Just Transition. Article has been also done in connection with project Students Grant Competition SP2023/088 „Specific Research of Modern Manufacturing Technologies for Sustainable Economy" financed by the Ministry of Education, Youth and Sports and Faculty of Mechanical Engineering VŠB-TUO. The authors would like to thank Advanced Engineering s.r.o. for support in the topology optimization study.

References

1. Cedzo, M., Joch, R., Timko, P., Holubják, J., Czánová, T., Šajgalík, M.: Topology optimization of gripping jaws of industrial robot. Manuf. Technol. **23**(1), 25–31 (2023). https://doi.org/10.21062/mft.2023.009
2. Cervinek, O., Werner, B., Koutný, D., Vaverka, O., Pantělejev, L., Paloušek, D.: Computational approaches of quasi-static compression loading of SS316L lattice structures made by selective laser melting. Materials **14**(9), 2462 (2021). https://doi.org/10.3390/ma14092462
3. Lesiak, P., et al.: Three-dimensional-printed mechanical transmission element with a fiber Bragg grating sensor embedded in a replaceable measuring head. Sensors **22**, 3381 (2022). https://doi.org/10.3390/s22093381
4. Sotola, M., Marsalek, P., Rybansky, D., Fusek, M., Gabriel, D.: Sensitivity analysis of key formulations of topology optimization on an example of cantilever bending beam. Symmetry (Basel) **13**, 712 (2021). https://doi.org/10.3390/sym13040712

5. Gibson, I., Rosen, D., Stucker, B., Khorasani, M.: Additive manufacturing technologies. Springer International Publishing (2021). https://doi.org/10.1007/978-3-030-56127-7
6. Mesicek, J., et al.: Comprehensive view of topological optimization scooter frame design and manufacturing. Symmetry **13**(7), 1201 (2021). https://doi.org/10.3390/sym13071201
7. Laghi, V., Babovic, N., Benvenuti, E., Kloft, H.: Blended structural optimization of steel joints for Wire-and-Arc additive manufacturing. Eng. Struct. **300**, 117141 (2024). https://doi.org/10.1016/j.engstruct.2023.117141
8. Nocoń, W.M., Gruber, K.: Topology optimization and FEA analysis of a brake pedal for additive manufacturing, pp. 1–15 (2024). https://doi.org/10.1007/978-3-031-44282-7_1
9. Jiang, B., Wang, Y., Niu, H., Cheng, X., Zhao, P., Bao, J.: GNNs for mechanical properties prediction of strut-based lattice structures. Int. J. Mech. Sci. **269**, 109082 (2024). https://doi.org/10.1016/j.ijmecsci.2024.109082
10. Sun, S., et al.: Topology optimization, additive manufacturing and thermohydraulic testing of heat sinks. Int. J. Heat Mass Transf. **224**, 125281 (2024). https://doi.org/10.1016/j.ijheatmasstransfer.2024.125281
11. Xia, Y., Langelaar, M., Hendriks, M.A.N.: A critical evaluation of topology optimization results for strut-and-tie modeling of reinforced concrete. Comput.-Aided Civil Infrastruct. Eng. **35**(8), 850–869 (2020). https://doi.org/10.1111/mice.12537
12. Pirinu, A., Primo, T., Del Prete, A., Panella, F.W., De Pascalis, F.: Mechanical behaviour of AlSi10Mg lattice structures manufactured by the Selective Laser Melting (SLM). Int. J. Adv. Manuf. Technol. **124**(5–6), 1651–1680 (2023). https://doi.org/10.1007/s00170-022-10390-1
13. Srivastava, J., Kawakami, H.: Systematic review of difference between topology optimization and generative design. IFAC-PapersOnLine **56**(2), 6561–6568 (2023). https://doi.org/10.1016/j.ifacol.2023.10.307
14. Gogolewski, D., Kozior, T., Zmarzły, P., Mathia, T.G.: Morphology of models manufactured by SLM technology and the Ti6Al4V titanium alloy designed for medical applications. Materials **14**(21), 6249 (2021). https://doi.org/10.3390/ma14216249
15. Raičević, N., Grbović, A., Kastratović, G., Vidanović, N., Sedmak, A.: Fatigue life prediction of topologically optimized torque link adjusted for additive manufacturing. Int. J. Fatigue **176**, 107907 (2023). https://doi.org/10.1016/j.ijfatigue.2023.107907
16. Pagac, M., et al.: Prediction of model distortion by FEM in 3D printing via the selective laser melting of stainless steel AISI 316L. Appl. Sci. **11**(4), 1656 (2021). https://doi.org/10.3390/app11041656
17. Gu, X., Yu, Q., Dong, Y., He, S., Qu, J.: Structural topology optimization for additive manufacturing with free choice of self-supporting and infill-supporting structures. Comput. Methods Appl. Mech. Eng. **421**, 116788 (2024). https://doi.org/10.1016/j.cma.2024.116788
18. Naik, A., Shanmugam, S., Desai, S.: Light weighting of a body jig using computational and topology optimization methods (2022). https://doi.org/10.4271/2022-28-0062
19. Praneashram, D., Balakrishna, T.S., Vinodh, S.: Topology optimization as an enabler of additive manufacturing: a case study, pp. 287–296 (2023). https://doi.org/10.1007/978-981-19-7612-4_24

Wettability Analysis of 3D Printed Bio-Inspired Structures Utilizing Vatphotopolymerization Additive Technique

Ratnesh Raj[1(✉)], Annada Prasad Moharana[1], Subham Kumar Pandey[1], Aman Prakash[1], Akash Nag[2], and Amit Rai Dixit[1]

[1] Indian Institute of Technology (Indian School of Mines), Dhanbad 824006, India
rtnsh07@gmail.com, amitraidixit@iitism.ac.in

[2] Faculty of Mechanical Engineering, VSB- Technical University of Ostrava, Poruba, Ostrava 70800, Czech Republic

Abstract. Bio-inspired structures mimic biological forms and functions found in nature, applied in engineering and architecture. Vat photopolymerization (VPP)-based additive manufacturing replicates these designs by curing the photopolymer layer by layer, allowing precise fabrication of intricate bio-inspired structures with enhanced functionality and performance. In this study, the bio-inspired structures were fabricated using VPP, and their wetting characteristics were investigated. Utilizing VPP technique, various intricate designs (spherical, conical, cylindrical, and cuboidal shapes) were printed and subsequently analyzed for their wettability using a Drop Shape Analyzer setup. The impact of dimensional parameters on the surfaces' functional characteristics regarding wetting behavior was analyzed. Most of the fabricated designs exhibited a degree of hydrophobicity, typically around 90 ± 2 degrees. Adjusting the height and spacing between structures resulted in higher levels of hydrophobicity. Conversely, hydrophilic nature was observed in cases with smaller height and spacing. Furthermore, it explores transitioning from hydrophilic to hydrophobic properties via parameter adjustments.

1 Introduction

Wettability testing of bio-inspired 3D printed structures is essential for sustainable advancements, optimizing material usage, and enhancing efficiency [1]. Understanding surface wettability aids in creating energy-efficient solutions, such as water filtration and heat exchange, reducing energy consumption [2]. Tailoring wettability properties can enable self-cleaning mechanisms, conserving water and minimizing the need for harsh chemicals. In the medical field, surface wettability influences biocompatibility and interactions with biological fluids, improving patient outcomes and extending structure durability, thereby reducing replacements and conserving resources [3].

Nature has provided inspiration for functional surfaces that have evolved over millions of years, showcasing unique properties like the lotus leaf repelling water and the water strider walking on water. Mimicking nature's designs has garnered significant interest, leading to the development of multifunctional wettable materials with hydrophobic, hydrophilic, and slippery properties for various practical applications.

M. Pagac et al. (Eds.): SCFF 2024, LNNS 1062, pp. 226–234, 2024.
https://doi.org/10.1007/978-3-031-65656-9_23

Characterizing wettability typically involves measuring the contact angle (CA) of liquids on a surface, distinguishing between hydrophobicity (CA > 90°) and hydrophilicity (CA < 90°), or a combination thereof [4]. Various methods, including 3D printing, are employed to achieve artificial wettability, offering rapid fabrication without expensive tooling. 3D printing is a laying process of materials to fabricate intricate shapes[5, 6]. For instance, recent research has demonstrated anisotropic wetting properties in 3D printed structures using polylactic acid and vat-photopolymerization (VPP) technique that offered tunable wettability for biomedical and thermofluidic applications [7, 8].

This study explores wettability through CA measurements of 3D printed structures (hemispherical, conical, cylindrical, and cuboidal) fabricated via VPP, investigating various designs and dimensions to expand the understanding of wettability in engineered structures.

2 Materials and Methodology

2.1 Computer Aided Design and 3D Printing

Initially, Computer aided design (CAD) software, Solidworks was utilized to design hemispherical (HS), conical, cylindrical (cyl), and cuboidal (cub) structures with a minimum gap of 100 microns between shapes and a height of 100 microns for each shape as shown in Fig. 1. However, these dimensions proved challenging for fabrication using VPP-based 3D printing technology. To address this, slight increments of 50 microns were made to both dimensions to enhance feasibility. The height (H) was initially optimized through trial experimentation, followed by the optimization of the gap (S) between shapes. This iterative process involved rigorous adjustments to the CAD design and printing trials. Once feasible printable heights and gaps were achieved, the heights of cylindrical, conical, and cuboidal shapes were increased while maintaining a constant gap, with increments of 100 microns. Similarly, the radius of the hemispherical shape (R) was increased in the same manner, while keeping the gaps between shapes constant. ProtoGRY photopolymer was selected for fabrication, employing its standard optimized printing parameters provided by the Original Equipment Manufacturer (OEM).

2.2 3D Profilometry

The fabricated samples underwent examination using a non-contact type 3D profilometer to identify any dimensional deviations from the intended specifications. The CSI (Coherence scanning Interferometry) mode with extended view was employed to capture images of the fabricated surfaces and subsequently the dimensons at different sections were extracted. The sections were highlighted in the Fig. 2f–l. The analysis was carried out at 5.5× magnification at 1.5×1.5 mm^2 scan area. The geometrical heights were averaged and mentioned in the result Table 1.

2.3 Wettability Test

The assessment of sample wettability was conducted employing the sessile drop method. Deionized (DI) water was utilized as the testing medium. CA calculations were carried

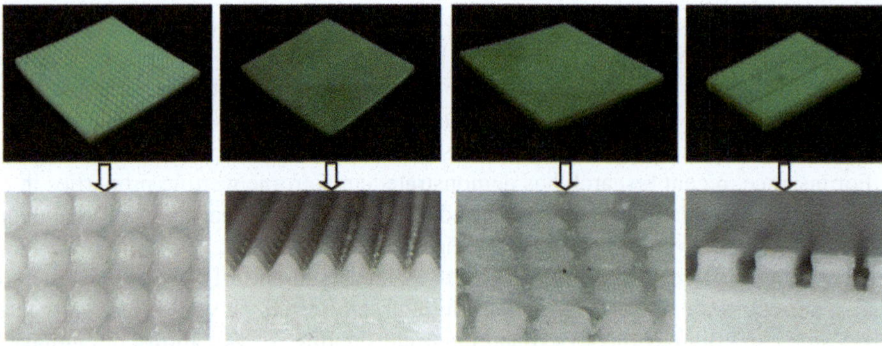

Fig. 1. 3D printed bio-inspired structures

out by capturing sequential drop images using a high-speed camera (IL5, Fastec Inc. USA) equipped with a telecentric lens. CA measurements were conducted at a minimum of three different locations on the samples[9].

3 Results and Discussion

3.1 3D Printing and Profilometry

Samples were fabricated for spherical shapes measuring 200 microns, maintaining a consistent gap between shapes and a constant radius of 600 microns. The height of the spheres ranged from 500 to 700 microns, transitioning from hemispherical to more than hemispherical shapes as shown in Fig. 2(a–c). Conical shapes were created with a gap between shapes and radius set at 130 and 390 microns, respectively. The minimum feasible height for these conical shapes was determined with 130 micron spacing, and heights printed at 600 and 700 microns as shown in Fig. 2(d–e). Here, the % error between the dimesions were observed to be less (refer Table 1). Cylindrical shapes required a minimum gap between shapes of 120 microns and a radius of 400 microns. Two variations in shape gaps were utilized for cylindrical textures, with gaps set at 120 and 300 microns, and heights ranging from 600 to 800 microns. The effective dimensional accuracies were noticed in case of 300 micron spacing depicted in Fig. 2(f–h). Additionally, cuboidal shapes were generated, featuring a minimum gap of 300 microns and variations in gaps set at 300 and 400 microns. The height variation was 600 to 800 microns. Good printability and shape fidelity were visulaized when the 300 micron spacing was used for 800 micron cuboidal height and 400 micron spacing for 600, 700 and 800 micron cuboidal height as shown in Fig. 2(i–l) respectively, as mentioned in Table 1.

3.2 Observation of Wettability Behavior

The majority of fabricated designs exhibit a degree of hydrophobicity typically falling within the range of $90^{\circ} \pm 2^{\circ}$, which represents the lower limit of hydrophobic behavior.

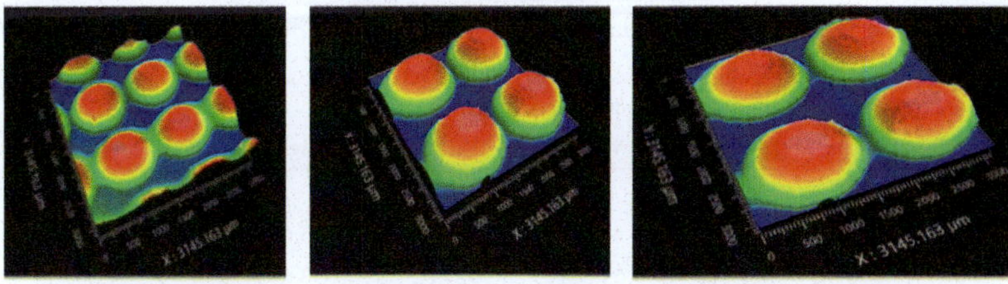

(a) HS: R=500, S=200 (b) HS: R=600, S=200 (c) HS: R=700, S=200

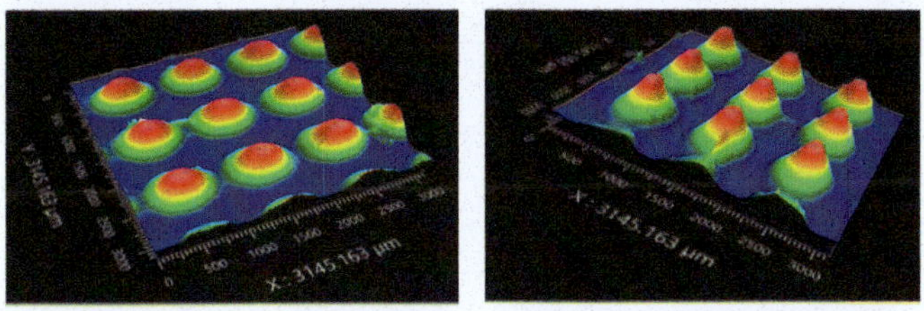

(d) conical: R=390, H=600, S=130 (e) conical: R=390, H=700, S=130

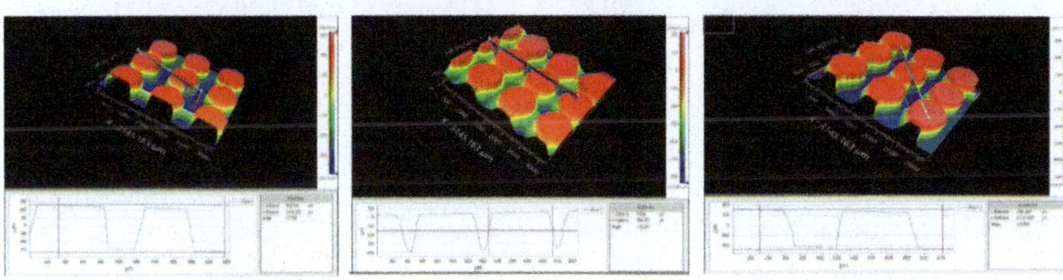

(f) cyl: H=600, S=300 (g) cyl: H=700, S=300 (h) cyl: H=800, S=300

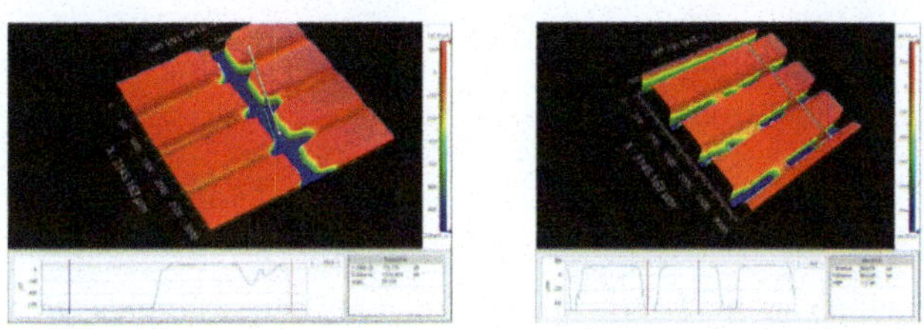

(i) cuboidal: H=800, S=300 (j) cuboidal: H=600, S=400

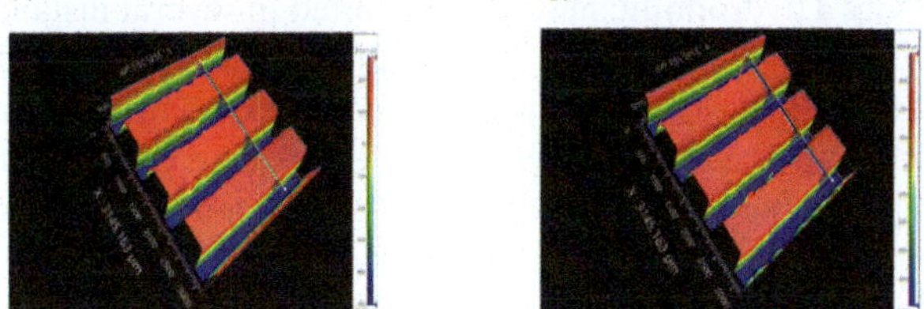

(k) cuboidal: H=700, S=400 (l) cuboidal: H=800, S=400

Fig. 2. 3D profilometry images of different structures

Table 1. The statistical data of dimensions during profilometry test

Design data (units in micro)			Profilometry data (units in micro)				%Error (in printing)		
Spacing	Height	Radius	X-distance	Y-distance	Radius	Height	Radius	Height	
Spherical shape									
200	500	600	642.67	445.63	642.67	445.63	7.11	10.87	
200	600	600	587.82	593.64	587.82	593.64	2.02	1.058	
200	700	600	669.00	624.38	669	624.38	11.5	10.8	
Conical shape									
130	600	390	774.72	475.81	387.36	475.81	0.67	20.69	
130	700	390	797.61	567.43	398.81	567.43	2.25	18.93	
130	800	390	866.63	623.91	433.11	623.91	11.05	22.01	
Cylindrical shape									
120	600	400	883.03	569.07	441.51	569.07	10.37	5.15	
120	700	400	865.43	586.41	432.82	586.41	8.20	16.22	
120	800	400	868.65	659.58	434.32	659.58	8.58	17.55	
300	600	400	886.63	591.87	443.31	591.87	10.82	1.35	
300	700	400	855.35	695.07	442.67	695.07	10.66	0.70	
300	800	400	845.66	787.22	422.83	787.22	5.99	1.59	
Cuboidal shape									
Design data (units in micro)				Profilometry data (units in micro)				%Error (in printing)	
Spacing	Height	Length	Width	X-distance	Y-distance	Height	Width	Height	Width
300	600	4100	500	519	576.3	576.34	519	3.94	3.8
300	700	4100	500	533	677.8	677.87	533	3.16	6.6
300	800	4100	500	515	776.22	776.22	515	2.97	3
400	600	4100	500	519	731.55	624.11	519	4.01	3.8
400	700	4100	500	567	624.11	731.88	567	4.55	13.4
400	800	4100	500	519	772.3	772.38	519	3.45	3.8

However, by increasing the height and spacing between structures, greater hydrophobicity was achieved, with contact angles reaching up to 117.76 degrees. Angles less than 90 degrees, indicating a hydrophilic nature, were noted in several instances and depicted in Table 2. Firstly, in the case of a cone shape with a height of 600, a radius of 390, and a spacing of 130 (Fig. 3b); secondly, in cylindrical shapes with a spacing of 120 and 300 (Fig. 3c–e); and finally, in cuboidal shapes with a spacing of 300 and height 600 to 800 microns (Fig. 3g–i). This behavior is attributed to the minimal spacing between adjacent structures in each of these cases. The proximity of these structures facilitates greater interaction with the liquid, resulting in a lower contact angle indicative of hydrophilicity [10, 11].

Similarly, the hydrophobicity of printed samples was periodically assessed using conical samples with a height of 700–800 microns and spacing of 130 microns (Fig. 3j and k). Increased spacing and height in cylindrical and cuboidal shapes exhibited a similar trend, showing a hydrophobic nature, as depicted in the (Fig. 3l–m). Additionally, when the spacing was increased from 130 microns to 400 microns, water droplets fell between the slots of the geometries, creating a boundary effect and increasing the contact angle.

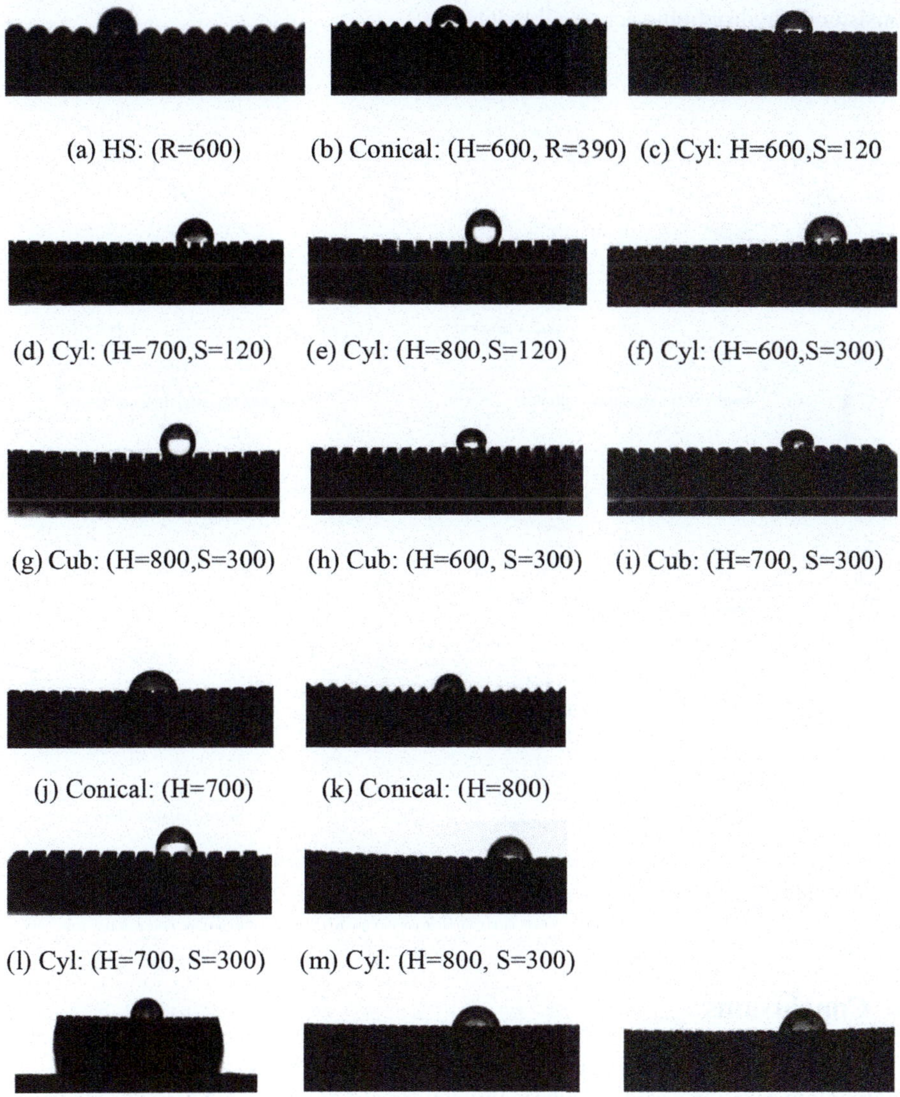

(a) HS: (R=600) (b) Conical: (H=600, R=390) (c) Cyl: H=600,S=120

(d) Cyl: (H=700,S=120) (e) Cyl: (H=800,S=120) (f) Cyl: (H=600,S=300)

(g) Cub: (H=800,S=300) (h) Cub: (H=600, S=300) (i) Cub: (H=700, S=300)

(j) Conical: (H=700) (k) Conical: (H=800)

(l) Cyl: (H=700, S=300) (m) Cyl: (H=800, S=300)

(n) Cub: (H=600, S=400) (o) Cub: (H=700, S=400) (p) Cub: (H=800, S=400)

Fig. 3. Hydrophilic and hydrofobic nature of structures by sessile drop method

The graphical representation of the contact angle vs height of the various structures are indiacted in Fig. 4. For the hemispherically structured sample, there is a consistent linear increase in contact angle. Conical structures maintain angles below 90° up to

700 microns before increasing. Cylindrical structures with 120 micron spacing exhibit minimal variance in contact angles, remaining hydrophilic. However, at 300 micron spacing, they become hydrophobic at 700 and 800 microns. Similar trends are observed in cuboidal structures at 300 micron spacing, while at 400 micron spacing, they are consistently hydrophobic across all heights.

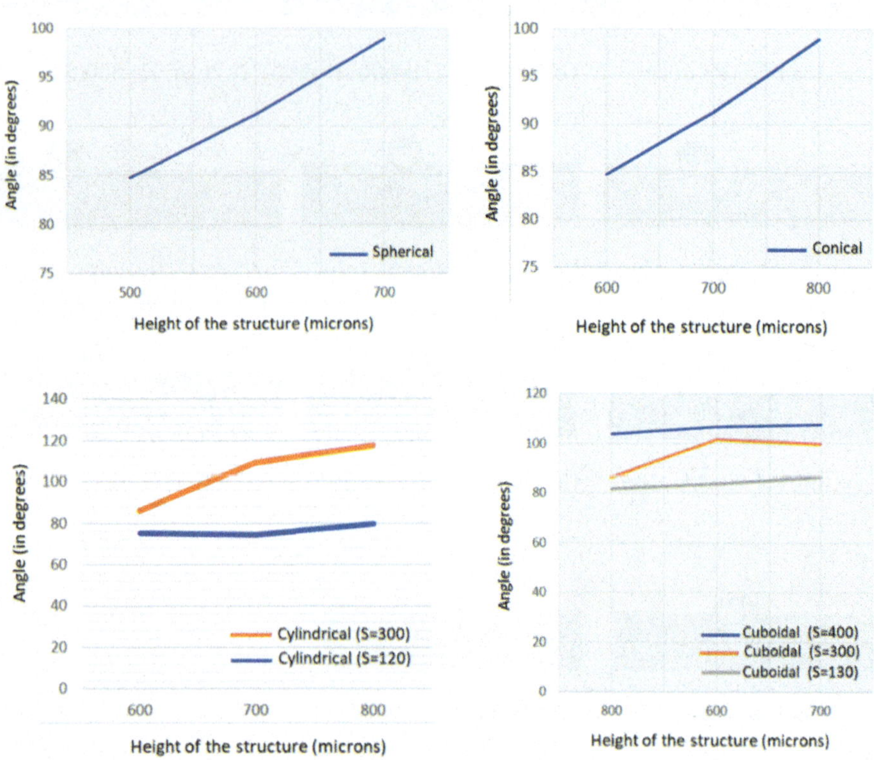

Fig. 4. Graphical illustration of the relationship between structure height and contact angle, highlighting the hydrophilic and hydrophobic characteristics of printed structures

4 Conclusions

The various bio-inspired structures can be easily manufactured through a vat-photopolymerization based additive manufacturing (AM) method. Samples with varying measurements and geometries were prepared for testing. Following the printing process, the printed samples were profiled, and only structures with minimal errors were retained. Additionally, a wettability test was conducted on different specimens to examine their wetting characteristics, revealing the impact of dimensional changes on contact angle and specimen nature. Most of the fabricated designs exhibited a certain degree of wettability, typically around 90 ± 2 degrees. The cylindrical shape with varied dimensional parameters have shown a good agreement with hydrophillic nature and cuboidal shape surfaces

Table 2. The statistical data of dimensions during sessile drop test method

Spacing	Height	Radius		Angle	Nature
Hemispherical shape					
200	500	600		84.71	Hydrophilic
200	600	600		91.23	Hydrophobic
200	700	600		98.89	Hydrophobic
Conical shape					
130	600	390		84.71	Hydrophilic
130	700	390		91.23	Hydrophobic
130	800	390		98.89	Hydrophobic
Cylindrical shape					
120	600	400		75.38	Hydrophilic
120	700	400		74.59	Hydrophilic
120	800	400		79.54	Hydrophilic
300	600	400		86.29	Hydrophilic
300	700	400		109.19	Hydrophobic
300	800	400		117.76	Hydrophobic
Cuboidal shape					
300	600	4100	500	75.66	Hydrophilic
300	700	4100	500	78.93	Hydrophilic
300	800	4100	500	88.92	Hydrophilic
400	600	4100	500	106.87	Hydrophobic
400	700	4100	500	107.67	Hydrophobic
400	800	4100	500	116.97	Hydrophobic

exhibited the good hydrophobic nature. By adjusting the height and spacing between two structures, higher levels of hydrophillicity and hydrophobicity were achieved, with angles reaching up to 86.29 and 117.76 degrees respectively. To support to the wettability test, the successful microfabrication was obtained in cylindrical and conical surfaces using DLP 3D printer.

Acknowledgement. This study was conducted in association with the project Innovative and Additive Manufacturing Technology—New Technological Solutions for 3D Printing of Metals and Composite Materials (reg. no. CZ.02.1.01/0.0/0.0/17_049/0008407) financed by Structural Funds of the European Union. Article has been done in connection with project Students Grant Competition SP2024/08. Specific Research of "Sustainable Manufacturing Technologies" financed by the Ministry of Education, Youth and Sports and Faculty of Mechanical Engineering VŠB-TUO. The authors would also like to express their sincere gratitude to IIT (ISM), Dhanbad for providing testing and fabrication facilities.

References

1. Srivastava, A.K., Kumar, A., Kumar, P. et al.: Research progress in metal additive manufacturing: challenges and opportunities. Int. J. Interact. Des. Manuf. (2023). https://doi.org/10.1007/s12008-023-01661-6
2. Wang, Z., Elimelech, M., Lin, S.: Environmental applications of interfacial materials with special wettability. Environ. Sci. Technol. **50**, 2132–2150 (2016). https://doi.org/10.1021/acs.est.5b04351
3. Gittens, R.A., Scheideler, L., Rupp, F., Hyzy, S.L., Geis-Gerstorfer, J., Schwartz, Z., Boyan, B.D.: A review on the wettability of dental implant surfaces II: biological and clinical aspects. Acta Biomater. **10**, 2907–2918 (2014). https://doi.org/10.1016/j.actbio.2014.03.032
4. Kung, C.H., Sow, P.K., Zahiri, B., Mérida, W.: Assessment and ınterpretation of surface wettability based on sessile droplet contact angle measurement: challenges and opportunities. Adv. Mater. Interfaces. **6**, 1900839 (2019). https://doi.org/10.1002/admi.201900839
5. Raj, R., Dixit, A.R.: Direct ink writing of carbon-doped polymeric composite ink: a review on its requirements and applications. 3D Printing Addit. Manufact. **10**, 828–854 (2023)
6. Raj, R., Moharana, A.P., Dixit, A.R.: Design and fabrication of flexible woodpile structured nanocomposite for microwave absorption using material extrusion additive technique. Addit. Manuf. **79**, 103935 (2024)
7. Chartrain, N.A., Williams, C.B., Whittington, A.R.: A review on fabricating tissue scaffolds using vat photopolymerization. Acta Biomater. **74**, 90–111 (2018). https://doi.org/10.1016/j.actbio.2018.05.010
8. Danielak, A., Islam, A., Cappelletto, N., Garza Agudelo, D.M., Pedersen, D.B.: Design aspects in vat photopolymerization additive manufacturing of hydrophobic surfaces. 3D Printing Addit. Manuf. (2023). https://doi.org/10.1089/3dp.2023.0076
9. Dwivedi, S., Dixit, A.R., Das, A.K., Adamczuk, K.: Additive texturing of metallic implant surfaces for improved wetting and biotribological performance. J. Market. Res. **20**, 2650–2667 (2022). https://doi.org/10.1016/J.JMRT.2022.08.029
10. Yoshimitsu, Z., Nakajima, A., Watanabe, T., Hashimoto, K.: Effects of surface structure on the hydrophobicity and sliding behavior of water droplets. Langmuir **18**, 5818–5822 (2002). https://doi.org/10.1021/la020088p
11. Ko, Y.C., Ratner, B.D., Hoffman, A.S.: Characterization of hydrophilic—hydrophobic polymeric surfaces by contact angle measurements. J. Colloid Interface Sci. **82**, 25–37 (1981). https://doi.org/10.1016/0021-9797(81)90120-X

Artificial Intelligence, Machine Learning, Cloud Computing, Cyber Security, Network Analysis, Big Data, Data Visualization, Blockchain, Metaverse

A Practical Approach for Virtual Desktop Infrastructure Implementation in Educational Institutions

Hung Ho-Dac(✉), Phuong Thanh Nguyen, and Hoan Duc Tran

Thu Dau Mot University, Binh Duong, Vietnam
hunghd@tdmu.edu.vn

Abstract. Laboratories play an important role in training and research at educational institutions. The demand regarding computers for laboratories is increasing in both quantity and performance. Investing in desktop computers with high configuration will encounter some problems in terms of funding, upgradeability, expansion and fixed layout. In the current context, when studying and working tend to be remote, using thin clients to access high-performance computers for work is one of the prominent trends. With virtualization technology, large server systems can be used to create many virtual machines with flexible configurations, helping to utilize infrastructure resources more effectively than traditional approaches. In addition, today's rapidly developing network technologies also contribute to making the use of thin clients to work remotely much more convenient. Therefore, we propose a framework to implement Virtual Desktop Infrastructure for laboratories to optimize infrastructure resources, save investment costs, and create flexibility for end users. The Framework was implemented at Thu Dau Mot University (TDMU) as a proof-of-concept and has achieved positive results. The code bases and scripts are open and available at https://bitbucket.org/tdmudc/vdim/src/main/.

Keywords: Virtual desktop infrastructure · High performance computing · Hyper-V

1 Introduction

In the past ten years, a lot of innovative businesses and academic institutions have embraced the trend of working and studying remotely [1, 2]. The coronavirus epidemic has made remote work necessary and demonstrated the effectiveness of virtual desktop infrastructure (VDI) [3], as a dependable tool for students or remote workers.

VDI is a state-of-the-art technology that provides and manages virtual desktops using virtual machines (VM). A virtual desktop is a preset image of an operating system and its apps that can be accessed from any suitable device, as opposed to being limited to a particular separated desktop. Desktop environments using VDI can be hosted on a central server and made available to users as needed.

The IT department of the institution can effortlessly manage users and generate distinct profiles for various majors. In the traditional desktop interface approach, users

M. Pagac et al. (Eds.): SCFF 2024, LNNS 1062, pp. 237–245, 2024.
https://doi.org/10.1007/978-3-031-65656-9_24

interact with the desktop through the use of applications, browsers, or anything else that can be downloaded into a computer. The operating system is installed on a dedicated device (hardware). However, utilizing VDI is a whole new approach. Same idea, but with significant increases in efficiency and versatility.

In essence, there are two types of VDI: pool based VDI and VDI using designated virtual machines. The VDI deployment based on VMs is a collection of virtual desktops that are established in the data center (DC). In this scenario, users can utilize their thin client to access the virtual desktop with high performance [4]. Basically, thin client is a device with minimal configuration that relies primarily on the support of another computer or its server to fulfill the tasks of a desktop. Using thin clients can save costs significantly, especially in VDI (Fig. 1).

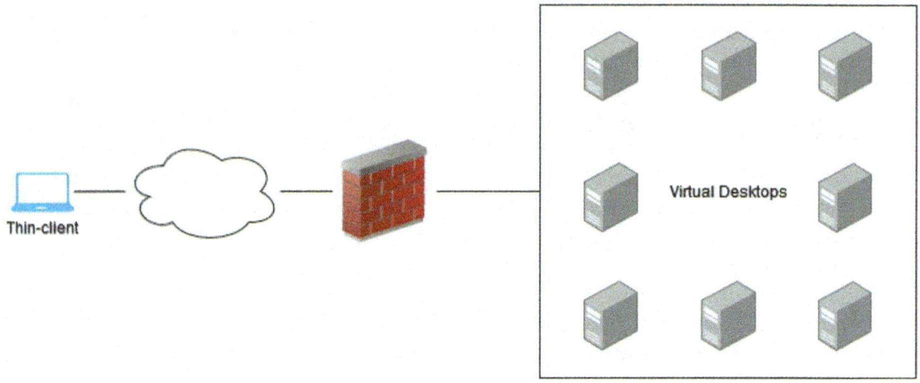

Fig. 1. The concept of virtual desktop infrastructure based on virtual machines

Some students in laboratories typically exploit low resource consumption at a high-performance desktop computer for the laboratories while others exploit heavy loads corresponding to the high-performance configuration. Workstation consolidation could result in significant savings in terms of infrastructure, networking, power consumption, and maintenance expenses, as suggested by the low utilization data. Without sacrificing performance, we would also spend less effort on deployment, security, and fault isolation.

In comparison to high performance desktop computers, desktop virtualization uses thin and zero clients that are based on energy-efficient hardware, have no moving parts, and use a lot less power. Thin client computing gear is typically made to be incredibly dependable and energy efficient. Because of the advancements in components and functionality, the average PC lasts around three years, compared to the five to seven years that thin clients give [5]. This extended lifespan can be explained by the fact that these devices have a lesser likelihood of having hardware problems because they primarily rely on the DC's technology rather than on local components.

DC has been recognized as the main source of carbon emissions and is expected to overtake all other sources of greenhouse gas pollution by 2020 [6]. In terms of energy usage, data centers currently rank among the industries with the highest levels of greed worldwide. The need for these facilities is rising due to the ever-increasing utilization of cloud services. Data centers use as much as 2% [7], of the world's energy, according to the most recent studies. While some of them have embraced green energy generating and consumption practices, non-green energy generation results from the traditional power grid providing the majority of their energy needs. Around 60% of energy is produced using coal and natural gas, according to [8], hence CO_2 emissions into the environment are still very high. Utilizing virtualization at the hypervisor level in the DC significantly lowers energy usage and the carbon footprint.

With its available advantages, VDI can be considered a good solution for deploying and replacing high configuration separated desktops. VDI will help reduce investment and operating costs as well as optimize centralized resources. In this study, we designed and implemented the architecture of VDI deployment for both public and private network.

2 Related Work

Many VDI application studies have been conducted, some of which focus on using VDI to support sustainability. Dasilva et al. [9], conducted a small test on virtual desktop environment using VMware and Wyse technology to portray the clear improvements that hypervisor and desktop virtual can bring to IT infrastructures drawing particular attention to power consumption and the green incentives. In another study, Agrawal et al. [10], presented the capabilities of VDI with the help of thin client machines which not only reduces the cost of infrastructure but also introduces green computing by limiting energy consumption and e-waste.

Besides sustainability factors, some research also focuses on evaluating and improving the performance of approaches using VDI. A modest VDI system consisting of 55 VMs was examined by Shamma et al. [11], using real VDI traces. Despite providing an analysis of workload and traffic patterns, they just examined particular monitoring components, such as a disk-level driver on virtual machines (VMs). Dong et al. [12], introduced new metrics to quantify user-received application performance. The results suggest that certain commercial solutions with a commodity datacenter server show a strong potential for being adapted to residential virtual-desktop computing.

In a study, Chrobak [13] presented a VDI approach in the form of private network. The approach was implemented at the University of Economy in Wroclaw and the author also evaluated the performance of the system.

Studies has largely focused on the application of VDI in private network. In the current context, the demand to use VDI needs to be expanded to access via the Internet. In this study, we present an approach to deploy VDI at Thu Dau Mot University for both private and public networks. Access control and information security components are also included in the actual deployment model.

3 Proposed Approach

In the beginning, VDI faced obstacles such as restricted network capacity, low user adoption, and high deployment costs [14]. But as new technologies emerged, such as server virtualization, better network architecture, and developments in VDI; VDI started to make sense for businesses of all kinds. These days, VDI is a crucial component of many companies, facilitating remote work, increasing efficiency, and simplifying IT administration. VDI has developed further with the advent of cloud computing, providing even greater scalability and flexibility. VDI have some key components including (Fig. 2):

Virtualization-supported cluster: this is the main component that hosts the virtual desktops and gives users access to their desktop environments by providing the necessary processing power and resources. High-performance hardware, including numerous CPUs and lots of RAM, is installed on the host server to guarantee rapid and fluid virtual desktop performance.

Hypervisor: On the virtualization-supported cluster, the hypervisor is in charge of setting up, maintaining, and operating several virtual machines. It makes it possible for several operating systems to operate concurrently on a single physical server by abstracting the underlying hardware. In addition to isolating virtual machines from one another and ensuring effective resource allocation, the hypervisor offers crucial functions like high availability and live migration.

Virtual machines: these are the distinct desktop environments that users can access through remote access. Every virtual desktop has its own virtual machine and may be tailored to the needs of the user. Just like on a real desktop, users may install programs, adjust settings, and save data to customize their virtual desktops. Because virtual desktops may be deployed on-demand, IT managers can easily scale up or down in response to user demands.

Connection broker: the connection broker serves as a go-between for the virtual desktops and the end users. It oversees load balancing, handles user identification, and ensures resources are allocated appropriately. In addition, the connection broker manages session persistence, enabling users to pick up where they left off in virtual desktop sessions—even when they switch devices. It also gives consumers a single point of entry, streamlining the access procedure and guaranteeing a dependable user experience.

Thin clients: users can access their virtual desktops using these devices. Thin clients need a web browser or remote desktop client software in order to access the virtual desktops. By establishing a secure connection with the virtual desktop infrastructure, the client software enables users to engage with their virtual desktops in the same way as they would with a physical computer. Thin clients can easily integrate with the virtual desktop environment by utilizing a variety of peripheral devices, including USB drives, printers, and scanners.

In addition to the key components of VDI, it is necessary to equip additional components for switching, routing, security, and load balancing to support inside/outside connections. Depending on the demand of user groups, VM in VDI can be divided into corresponding groups with appropriate configurations (Fig. 3).

We propose a deployment framework as shown in Fig. 6. Virtual machines are planned into clusters corresponding to user groups with different configuration demand. Virtualized systems and clusters are deployed with HA to ensure availability and load balancing during service. In addition, a number of support components such as AD server, DNS server, router, firewall, and load balancer are also included to ensure connection and security.

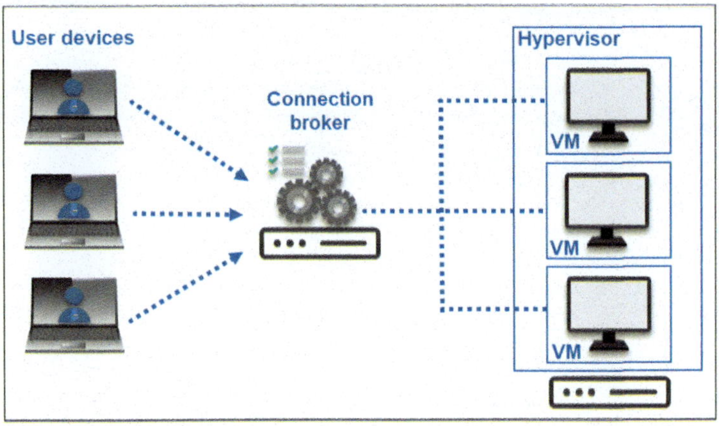

Fig. 2. Components of virtual desktop infrastructure based on virtual machines

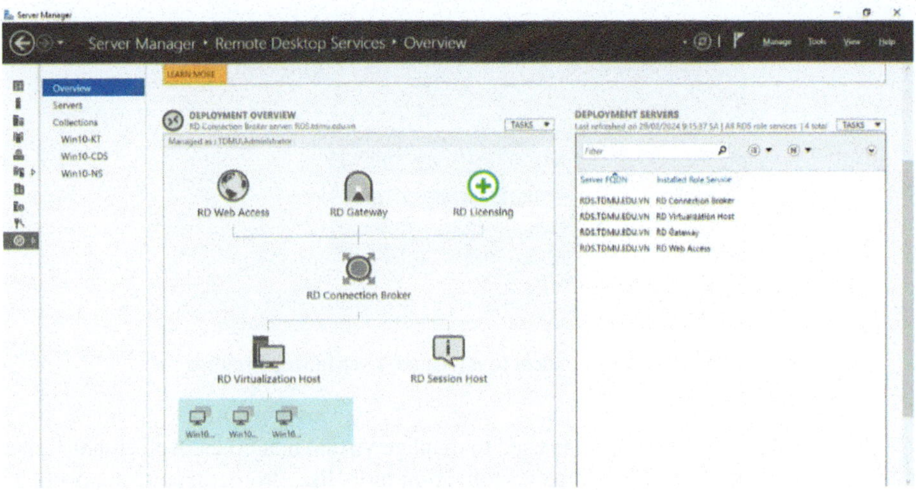

Fig. 3. Remote desktop services management

Users can use the Remote Desktop tool (Fig. 4) or web browser (Fig. 5) to connect to the corresponding allocated VM. This provides convenience and the ability to reduce workload on individual devices (Fig. 6).

4 Experiment

In the PoC implemented at TDMU, we inherit a typo to serve 2 VM groups for the back-office and the labs. Two virtualization systems run in parallel on two HPE server clusters (Table 1) providing virtualization capabilities to serve 120 clients simultaneously.

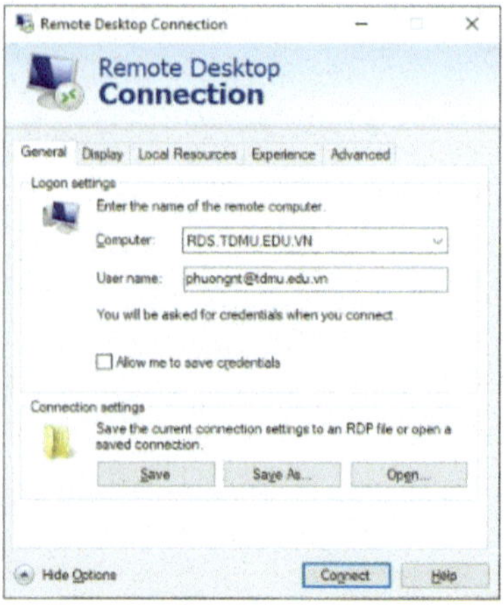

Fig. 4. Users connect to allocated VM via remote desktop tool

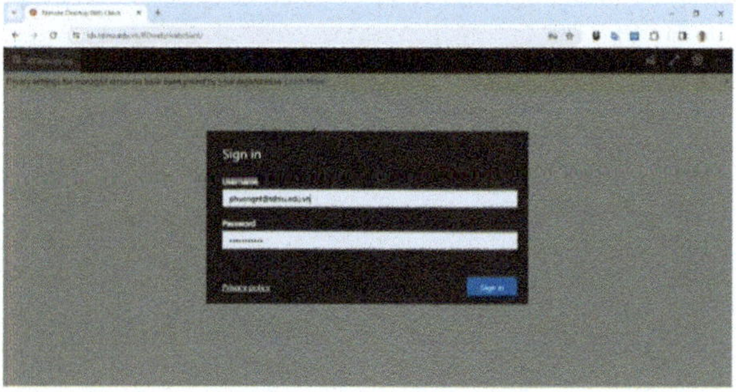

Fig. 5. Users connect to allocated VM via web browser

The Hyper-V supervisor tool is used to manage virtual machines. Additionally, metrics can be extracted and pushed to visualization tools like Prometheus to support centralized management. We also combine the use of AD server, DNS server and RDS server to support authentication and authorization (Fig. 7).

This approach clearly shows efficiency in investment costs (Table 2) as well as supports centralized administration, limiting configuration changes at the terminal side. In order to improve storage efficiency, we incorporate SANS to optimize costs. During operation, PC devices in labs reduce usage time, and students use VMs to perform tasks remotely. In addition, the virtualization mechanism also allows optimizing cluster resources to minimize energy costs through automatic migration and scaling operations.

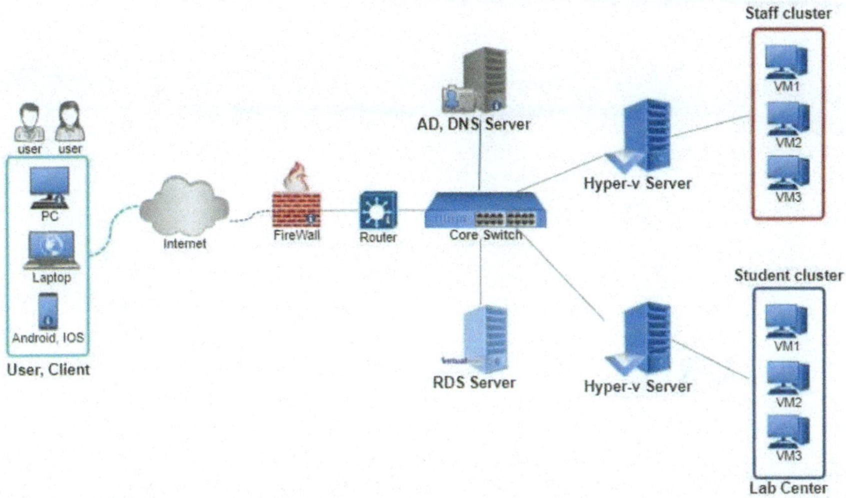

Fig. 6. Proposing deployment framework for VDI in educational institution

Table 1. Cluster specification

Cluster	Specification
HPE SimpliVity 325 Gen10 (3 nodes)	AMD® EPYC® 7000 processor family, 32 × 2 core
	512 GB ram
	6 × 1.92TB SFF SSD
	2 × 300GB SAS 12G
HPE ProLiant XL170r Gen10	8 CPU-160 core
	256 GB ram
	6TB HDD

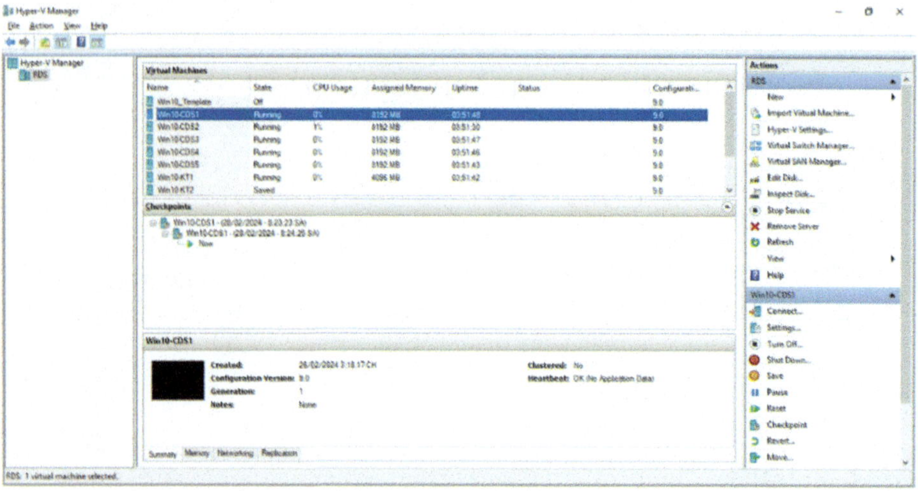

Fig. 7. Hyper-V manager

Table 2. Investment cost benchmark at TDMU

	VDI	Desktop
CPU (Core)	$ 73.6	$69.44
RAM (GB)	$ 27.25	$34.72
Storage (GB)	$0.52	$0.54

5 Conclusion

In this study, we present an approach to deploying VDI for educational institutions. Through the PoC process, the approach has shown viable results and scalability. However, some elements of personalization are not taken into account. In future study, we will apply artificial intelligence techniques to identify personalization features to optimize user experience when using VMs.

References

1. Khlaif, Z.N., Salha, S., Kouraichi, B.: Emergency remote learning during COVID-19 crisis: students engagement. Educ. Inf. Technol. **26**(6), 7033–7055 (2021)
2. Delfino, G.F., Van Der Kolk, B.: Remote working, management control changes and employee responses during the COVID-19 crisis. Account., Auditing Accountability J. **34**(6), 1376–1387 (2021)
3. Ibrahim, A.A.Z.A, et al.: Virtual desktop infrastructures: architecture, survey and green aspects proof of concept. 2016 Seventh International Green and Sustainable Computing Conference (IGSC). IEEE, (2016)
4. Grover, S.: Designing Hyper-V Solutions. Packt Publishing Ltd, (2015)

5. Microsoft.: An Alternative to Thin Clients for the Enterprise. http://www.microsoft.com/en-us/download/details.aspx?id=26677. (27 June, 2011)
6. Lohr, S.: Data Centers Are Becoming Big Polluters, Study Find. http://bits.blogs.nytimes.com/2008/05/01/data-centers-are-becoming-big-polluters-study-finds/. (1 May, 2009)
7. Shehabi, A., Smith, S.J., Sartor, D.A., Brown, R.E., Herrlin, M., Koomey, J.G., et al.: United States Data Center Energy Usage Report. (June 2016)
8. Saundry, P.D.: Review of the united states energy system in transition. Energy Sustain. Soc. **9**, 4 (2019)
9. Dasilva, D.-A., et al.: Enabling green it through building a virtual desktop infrastructure. 2012 Eighth International Conference on Semantics, Knowledge and Grids. IEEE, (2012)
10. Agrawal, S., Biswas, R., Nath, A.: Virtual desktop infrastructure in higher education institution: energy efficiency as an application of green computing. 2014 Fourth International Conference on Communication Systems and Network Technologies. IEEE, (2014)
11. Shamma, M., Meyer, D.T., Wires, J., Ivanova, M., Hutchinson, N.C., Warfield, A.: Capo: recapitulating storage for virtual desktops. In: Proceedings of the 9th USENIX Conference on File and Stroage Technologies, FAST'11, pp. 3–3. Berkeley, CA, USA. USENIX Association, (2011)
12. Dong, H., et al.: Towards enabling residential virtual-desktop computing. IEEE Trans Cloud Comput 11(1), 745–762 (2021)
13. Chrobak, P.: Implementation of virtual desktop infrastructure in academic laboratories. 2014 Federated Conference on Computer Science and Information Systems. IEEE, (2014)
14. Makoviy, K., et al.: Server hardware resources optimization for virtual desktop infrastructure implementation. CEUR Workshop Proceedings, vol. 1904 (2017)

A New Type of State Observer for Automated Generation Control in Multi-Area Power Plants Combined Solar Energy

Dao Huy Tuan, Van Van Huynh(✉), Vo Hoang Duy, and Nguyen Huu Khanh Nhan

Faculty of Electrical and Electronics Engineering, Ton Duc Thang University, Ho Chi Minh City, Vietnam
{daohuytuan,huynhvanvan,vohoangduy, nguyenhuukhanhnhan}@tdtu.edu.vn

Abstract. Disturbance rejection is a fundamental issue with power system load frequency control. In this paper, a frequency regulation strategy based on disturbance observers is proposed for Multi area power plants (MAPP) combined Solar energy. The recommended control strategy consists of the design of Automatic Generation Control (AGC) and a disturbance observer. Building the disturbance observer requires obtaining the nominal model. First, for the given interconnected system, the newly extended states observer (ESO) is meant to estimate both the load uncertainties and the unmeasurable state variables. As an alternative, the frequency response model parameter estimation technique is used in the design of AGC. The suggested method has the advantage of increasing modeling accuracy and frequency regulation effectiveness by considering the dynamics of the inverse response process during the disturbance observer design process. To illustrate the usefulness of the suggested strategy, simulation examples are provided along with a comparison with other traditional approaches. The outcomes of the MATLAB/SIMULINK simulations not only demonstrated the effectiveness of the recently proposed control approach but also significantly altered the parameter fluctuations of the networking power plants (PP).

Keywords: Automatic generation control · Multi area power plants · Extended states observer

1 Introduction

Sustainable and environmentally friendly renewable energy technologies have been the focus of much attention and rapid development in recent years due to the depletion of conventional fossil fuels. Of all the energy sources, solar energy is the most promising. Solar farms have produced more electricity over the last ten years, making a larger overall contribution to the power system. Because of this, the solar farm also significantly affects the local power grid's frequency. Automatic Generation Control (AGC) is a secondary control that manages active power balancing if load variations cause changes in either

M. Pagac et al. (Eds.): SCFF 2024, LNNS 1062, pp. 246–255, 2024.
https://doi.org/10.1007/978-3-031-65656-9_25

the generation or the demand. The stability and security of the electricity system are put at risk when this imbalance results in the frequency deviating from its nominal value [1]. Ensuring power generation maintains pace with power demand during changes in load and source is essential for maintaining a reliable and sufficient power supply for power systems [2–4]. Thus, AGC is crucial to reduce undesired tie-line power flows and frequency fluctuations between nearby interconnected areas [5, 6]. The development and use of large-scale renewable energy sources has sparked a recent research interest in the stability of power system operation. The intermittent and stochastic nature of renewable energy sources has a significant impact on power grid stability, especially when it comes to frequency fluctuation. Therefore, in order to achieve the large-scale grid connection of renewable energy sources, it is crucial to develop efficient control strategies for power system load frequency [7]. Power systems make extensive use of adaptive control [8]. To address the AGC issue in the power system, an optimal controller [9–12] with filtered derivative action and a fractional order integrator is suggested. The design of a PID controller [8–16] for AGC in power systems is presented. The transfer function models for various configurations of power systems. Load fluctuations and uncertainty in renewable energy sources' generation capability considerably accentuate the AGC problem of integrated power systems utilizing renewable energy. This issue might have an answer in control design. Sliding mode control (SMC) is a useful approach because it is invariant to disturbances and uncertainty in the system [17–19]. It can resolve the AGC issue. Numerous SMC-based methods, such as the full-order SMC method [21–23] and the adaptive SMC method [20], have been applied in real-world situations. The consequences of renewable energy have also drawn a lot of attention lately. This study develops Extended States Observer (ESO) techniques to accurately estimate the states of a dynamical model of a full hybrid power system consisting of conventional solar energy and inverter-based power sources and loads in a finite amount of time. This paper focuses on a model where loads and heterogeneous power sources coexist and are connected through power flow exchanges, in contrast to the observer schemes described in the literature [24], which either focus on traditional power grid models or separately consider renewable power sources [25, 26]. As a result, under this scenario, an observer would have to deal with a lot more unknown but constrained fluctuations and uncertainties than in the traditional.

This article aims to provide a method to greatly enhance the performance of frequency tuning. The following are the ways in which the suggested method contributes:

- Upheaval ESO control is suggested for Multi Area Power Plants (MAPP) that integrate solar energy and communication delay, in contrast to current approaches that do not require the disturbance observer system.
- The efficacy of the suggested approach is confirmed using data on solar radiation from real-world ESO.
- To prevent damage to the turbine blades, which could occur if the turbine operates at a frequency different from its typical operation.
- The ESO controller gives us the freedom to define the performance goals of an automation system directly, and it produces the best control for a specified set of goals

- According to the simulation results, the suggested ESO method enhances the dynamic response of the system, giving it the ability to adjust and meet the AGC requirement.

2 The Mathematical Model of Multi-Area Power Plants (MAPP) Model

Area ith dynamic equation:

$$\dot{x}_i(t) = A_i x_i(t) + B_i u_i(t) + \sum_{j=1;j\neq i}^{N} H_{ij} x_j(t) + F_i \Delta P_{di}(t) \quad (1)$$

with $x_i = \begin{bmatrix} \Delta f_i & \Delta P_{m_i} & \Delta P_{V_i} & \int ACE_i & \Delta\delta_i & \Delta P_{S_i} \end{bmatrix}^T$

where $i = 1, \ldots$, N, N denotes the quantity of areas.

$x_i(t) \in R^n$ is the state vector, $u_i(t) \in R^m$ is the control vector, $\Delta P_{di}(t) \in R^k$ variation in the control area's local load, and A_i, B_i, F_i is constant matrix equivalent ith of each area ($A_i \in R^{n\times n}$, $B_i \in R^{n\times m}$, $F_i \in R^{n\times k}$), H_{ij} is the matrix connecting neighboring areas and $x_j(t) \in R^n$ is the status vector of the neighborhood, Δf_i change the power system frequency, ΔP_{mi} is power incremental change $(p.u.MW)$, ΔP_{Vi} is governor-valve position incremental change $(p.u.MW)$, $\int ACE_i$ error integral between line power fluctuation and frequency deviation. $\Delta\delta_i$ is phase angle incremental change (rad), ΔP_{S_i} is the change in solar cell capacity.

3 Proposed Extended States Observer for Automatic Generation Control

The new alternative is a control approach that assumes that all variables can be used for feedback in the control design of the system. In this case, the system's efficiency is not optimal, so estimating impossible variant states is required to achieve the objective. The estimation of state parameters that are not routinely measured is called ESO. A controller that detects and logs different states is called an observer, or simply a state observer. As illustrated below, full state-based observation describes parameters that cover all system states.

The subsystem that is capable of reconstructing the plant's state vector is called an observer. The discrepancy between the measured and estimated output is known as the estimation error or observer error.

Considering a system whose state space equation is as follows:

$$\dot{x} = Ax + Bu \quad (2)$$

$$y = Cx \quad (3)$$

State equation for the observer:

$$\dot{\tilde{x}} = A\tilde{x} + Bu + K_e(y - \tilde{y}) \quad (4)$$

$$\tilde{y} = C\tilde{x} \tag{5}$$

where $\dot{\tilde{x}}$: vector estimated state, $\tilde{y}$: estimated output, y: actual measured output, u: input control, K_e: observer gain matrix.

Substitute Eq. (5) into Eq. (4), we get Eq. (6)

$$\begin{aligned}\dot{\tilde{x}} &= A\tilde{x} + Bu + K_e(y - C\tilde{x}) \\ &= (A - CK_e)\tilde{x} + Bu + K_e y\end{aligned} \tag{6}$$

Taking Eq. (2)–Eq. (6) we get the observation error equation:

$$\dot{x} - \dot{\tilde{x}} = (A - CK_e)(x - \tilde{x}) \tag{7}$$

Let e be the difference between x and $\tilde{x}$

$$e = x - \tilde{x} \tag{8}$$

Substitute Eq. (8) into Eq. (7)

$$\dot{e} = (A - CK_e)e \tag{9}$$

Let $\mu_1, \mu_2, \mu_3 \ldots, \mu_n$ be the desired poles:

$$\left|sI - \left(A^* - C^*K\right)\right| = (s - \mu_1)(s - \mu_2)\ldots(s - \mu_n) \tag{10}$$

We have $A^* - C^*K$ and $A - CK^*$ nearly equal:

$$\left|sI - (A^* - C^*K)\right| = \left|sI - (A - CK^*)\right| \tag{11}$$

Thence inferred:

$$K_e = K^* \tag{12}$$

Extended states observer (ESO) control for multi-area power plant (MAPP) can be written as:

$$\dot{x}(t) = \tilde{A}x(t) + \tilde{B}u(t) + \tilde{F}\Delta P(t) \tag{13}$$

4 Simulation Results and Discussion

To demonstrate the effectiveness of the proposed extended states observer (ESO) control scheme, two simulation scenarios are analyzed for multi-area power plants (MAPP) with non-reheat, reheat, and hydro turbine under different load perturbations and parameter uncertainties. The nominal values of MAPP are given as [27].

Area 1: Non-reheat Turbine $T_{G1} = 0.08$ s, $T_{T1} = 0.05$ s, $T_{P1} = 20$ s, $K_{P1} = 120$ s.

Area 2: Reheat Turbine $T_{G2} = 0.08$ s, $T_{T2} = 0.3$ s, $K_{R2} = 0.5$, $T_{R2} = 10$ s, $T_{P2} = 20$ s, $K_{P2} = 120$ s,

Area 3: Hydro Turbine $T_{G3} = 0.3$ s, $T_{R3} = 5$ s, $T_{RH3} = 28.75$ s, $T_{W3} = 0.3$ s, $T_{P3} = 20$ s, $K_{P3} = 120$ s.

4.1 Scenarios 1

The suggested ESO for AGC strategy applied to Multi area power plants (MAPP) none combined Solar energy relatively large step load perturbations, with nominal power system parameters equal to [27]. A sudden 1% p.u. step load increase occurs in both Areas 1, 2 and 3. For three areas Areas 1, 2 and 3 the frequency deviation is getting close to zero; their respective frequency stabilization times are 6.0, 6.5 and 5.7 s. The frequency deviations (FDs) of the MAPP in Hertz are displayed in Figs. 1, 2 and 3. The tie-line power (TLP) per TATPP unit is displayed in Fig. 4. The frequency overshoot deviations (FOD) of area 1,2 and area 3 are 0.007, 0.009 and 0.01 in (Hz), respectively. You can see the waveforms below. Table 1 displays the results.

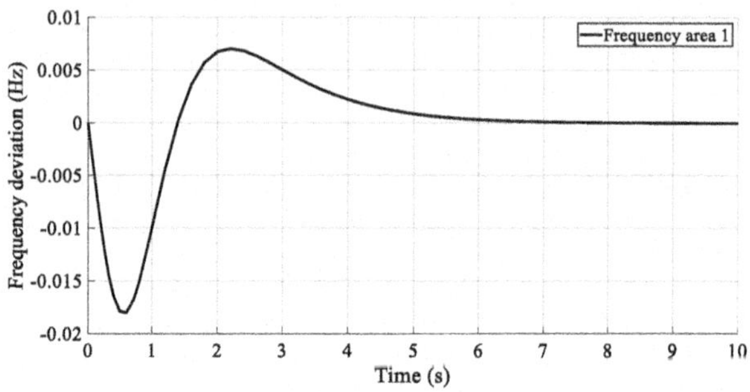

Fig. 1. The variation of frequency in area 1 [Hz] at a 1% step load none solar energy.

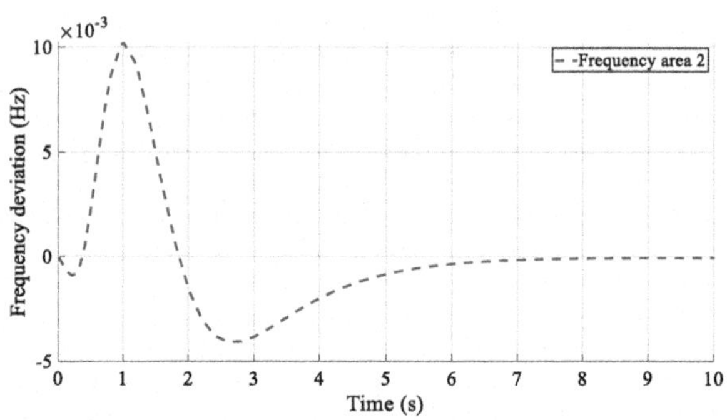

Fig. 2. The variation of frequency in area 2 [Hz] at a 1% step load none solar energy.

4.2 Scenarios 2

The suggested ESO for AGC strategy applied to Multi area power plants (MAPP) combined Solar energy. In order to assess the practicality of the suggested approach, the system's performance is examined in every regulated region with a lumped disturbance. This disturbance comprises a 1% step load change in regions 1, 2, and 3, as well as uncertainties in the system's parameters. The grid is in a steady state for $0 \leq t \leq 9.5$ s, based

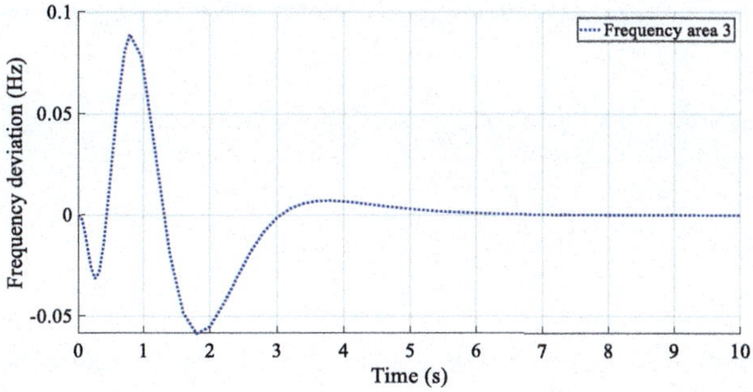

Fig. 3. The variation of frequency in area 3 [Hz] at a 1% step load none solar energy.

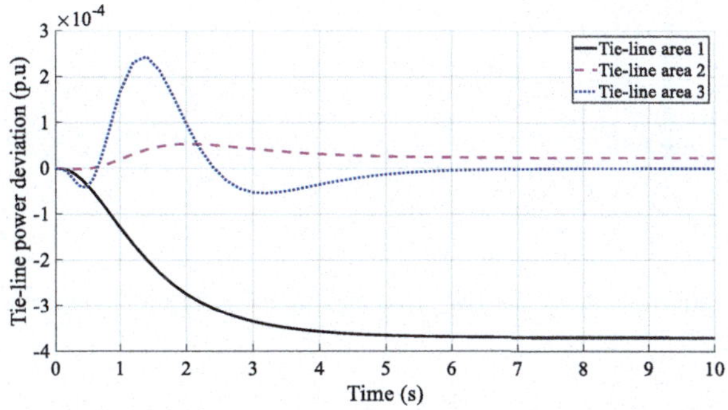

Fig. 4. Variation in tie line power (p.u. MW) none solar energy.

Table 1. Establishing the maximum overshoot MO [Hz] and time T_S[s] comparison.

	The suggested approach		The new approach in [27]	
Measurement	$T_s[s]$	MO [Hz]	$T_s[s]$	MO [Hz]
Δf_1	6.0	0.007	78	0.170
Δf_2	6.5	0.009	75	0.173
Δf_3	5.7	0.010	75	0.137

on the nominal MAPP parameters and the same parameter as in [27]. The frequency overshoot deviations (FOD) of Areas 1, 2, and 3 in Figs. 5, 6 and 7 show the frequency deviations (FDs) of the MAPP in Hertz are 0.009, 0.010 and 0.013 in (Hz), respectively. Areas 1, 2 and 3 have frequency stabilization times of 8.2, 9.0 and 9.5 s, respectively. Figure 8 shows the tie-line power (TLP) per unit of the MAPP as well as the observable control signal. The waveforms mentioned earlier make it evident that both the ACE and frequency error quickly approach zero. These results are in line with the power system's frequency requirements. Table 2 displays the comparative results.

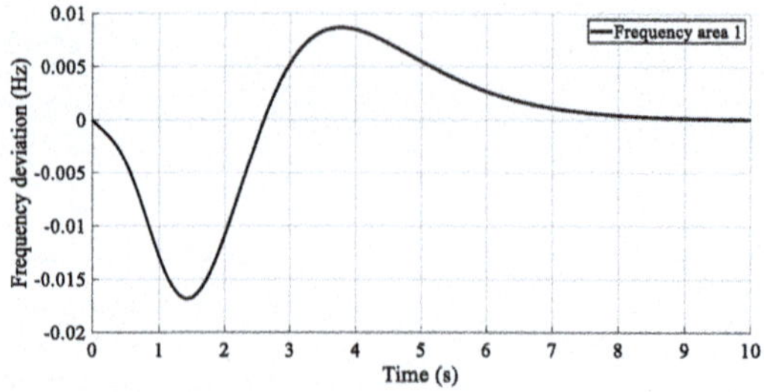

Fig. 5. The variation of frequency [Hz] in area 1 at a 1% step load combined solar energy.

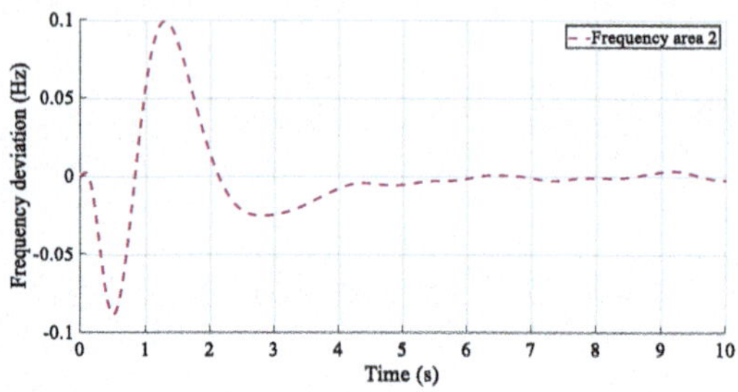

Fig. 6. The variation of frequency [Hz] in area 2 at a 1% step load combined solar energy.

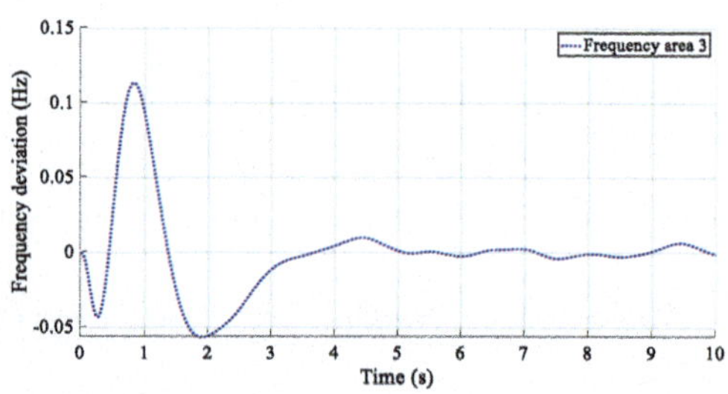

Fig. 7. The variation of frequency [Hz] in area 3 at a 1% step load combined solar energy.

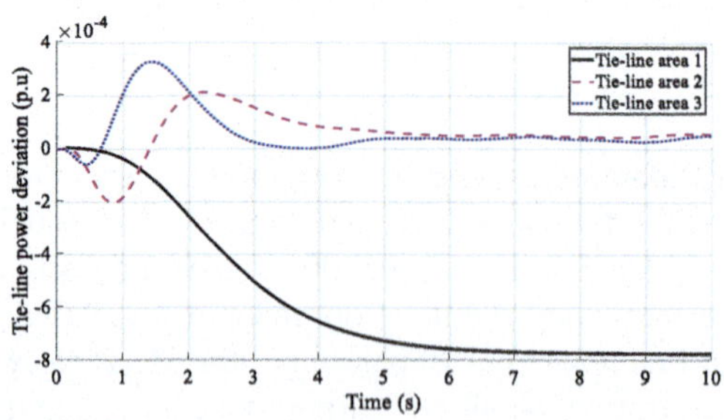

Fig. 8. Variation in tie line power (p.u. MW) combined solar energy

Table 2. Establishing the maximum overshoot MO [Hz] and time T_S[s] comparison when combined with solar energy

	The suggested approach		The new approach in [27]	
Measurement	$T_s[s]$	MO [Hz]	$T_s[s]$	MO [Hz]
Δf_1	8.2	0.009	78	0.170
Δf_2	9.0	0.010	75	0.173
Δf_3	9.5	0.013	75	0.137

5 Conclusion

In this paper, we develop an extended state observer (ESO) to control the load frequency of a multi-area power plant (MAPP). We obtain matrix equations of MAPP combined with solar energy. The three-area non-reheat, reheat, and hydro turbine power plant is the model used to assess the viability of the built ESO. The MAPP model also takes the situation's uncertainty and related parameters into account. It is this new linear matrix inequality that provides MAPP stability. Simulation results from some recent methods are compared with the areas where the ESO performs better. In terms of MAPP performance, the suggested ESO performs better than the recently mentioned techniques. The resilience of the ESO was further demonstrated by the fact that it was unaffected by random demand disruptions, variations in subsystem variables, and unpredictability of variables in the connected structure and state. As a result, the AGC of MAPP is greatly improved by the proposed ESO.

References

1. Fini, M.H., Yousefi, G.R., Alhelou, H.H.: Comparative study on the performance of many-objective and single-objective optimisation algorithms in tuning load frequency controllers of multi-area power systems. IET Gener., Transmiss. Distrib. **10**(12), 2915–2923 (2016)
2. Kayalvizhi, S., Vinod Kumar, D.M.: Load frequency control of an isolated micro grid using fuzzy adaptive model predictive control. IEEE Access **5**, 16241–16251 (2017)
3. Polajzer, B., Petrun, M., Ritonja, J.: Adaptation of load-frequency control target values based on the covariances between area-control errors. IEEE Trans. Power Syst. **33**(6), 5865–5874 (2018)
4. Xu, Y., Li, C., Wang, Z., Zhang, N., Peng, B.: Load frequency control of a novel renewable energy integrated micro-grid containing pumped hydropower energy storage. IEEE Access **6**, 29067–29077 (2018)
5. Chen, C., Cui, M., Wang, X., Zhang, K., Yin, S.: An investigation of coordinated attack on load frequency control. IEEE Access **6**, 30414–30423 (2018)
6. Huynh, V.V., Tran, P.T., Minh, B.L.N., Tran, A.T., Tuan, D.H., Nguyen, T.M., Vu, P.T.: New second-order sliding mode control design for load frequency control of a power system. Energies 13(24) (2020). https://doi.org/10.3390/en13246509
7. Rosyiana, F.I., Kim, J.S., Song, H.: High-gain disturbance observer-based robust load frequency control of power systems with multiple areas. Energies **10**(5), 595 (2017)

8. Dei, G., et al.: Improved squirrel search algorithm driven cascaded 2DOF-PID-FOI controller for load frequency control of renewable energy based hybrid power system. IEEE Access **10**, 46372–46391 (2022). https://doi.org/10.1109/ACCESS.2022.3169749
9. Yakout, A.H., Kotb, H., Hasanien, H.M., Aboras, K.M.: Optimal fuzzy PIDF load frequency controller for hybrid microgrid system using marine predator algorithm. IEEE Access **9**, 54220–54232 (2021). https://doi.org/10.1109/ACCESS.2021.3070076
10. Haroun, A.G., Li, Y.Y.: A novel optimized hybrid fuzzy logic intelligent PID controller for an interconnected multi-area power system with physical constraints and boiler dynamics. ISA Trans. **71**, 364–379 (2017)
11. Ahmadi Kamarposhti, M., Shokouhandeh, H., Alipur, M., Colak, I., Zare, H., Eguchi, K.: Optimal designing of fuzzy-PID controller in the load-frequency control loop of hydro-thermal power system connected to wind farm by HVDC lines. IEEE Access **10**, 63812–63822 (2022). https://doi.org/10.1109/ACCESS.2022.3183155
12. Gheisarnejad, M., Khooban, M.H.: Design an optimal fuzzy fractional proportional integral derivative controller with derivative filter for load frequency control in power systems. Trans. Inst. Meas. Control. **41**(9), 2563–2581 (2019)
13. Saxena, S., Hote, Y.V.: Decentralized PID load frequency control for perturbed multi-area power systems. Int. J. Electr. Power Energy Syst. **81**, 405–415 (2016)
14. Barakat, M.: Novel chaos game optimization tuned-fractional-order PID fractional-order PI controller for load-frequency control of interconnected power systems. Prot Control Mod Power Syst **7**(2), 1–20 (2022). https://doi.org/10.1186/s41601-022-00238-x
15. Hote, Y.V., Jain, S.: PID controller design for load frequency control: past, present and future challenges. IFAC-PapersOnLine **51**(4), 604–609 (2018). https://doi.org/10.1016/j.ifacol.2018.06.16
16. Gupta, N.K., Kar, M.K., Singh, A.K.: Design of a 2-DOF-PID controller using an improved sine-cosine algorithm for load frequency control of a three-area system with nonlinearities. Prot Control Modern Power Syst **7**(3), 1–18 (2022). https://doi.org/10.1186/s41601-022-00255-w
17. Yang, B., et al.: Passivity-based sliding-mode control design for optimal power extraction of a PMSG based variable speed wind turbine. Renew. Energy **119**, 577–589 (2018)
18. Polyakov, A., Poznyak, A.: Reaching time estimation for supert wisting second order sliding mode controller via Lyapunov function designing. IEEE Trans. Autom. Control **54**(8), 1951–1955 (2009)
19. Basin, M., Panathula, C.B., Shtessel, Y.: Continuous second-order sliding mode control: convergence time estimation. In Proceedings 54th IEEE Conference Decision and Control, pp. 5408–5413. Osaka, Japan, (2015)
20. Guo, J.: Application of a novel adaptive sliding mode control method to the load frequency control. Eur. J. Control. **57**, 172–178 (2021)
21. Ning, B., Han, Q.-L., Ding, L.: Distributed secondary control of AC microgrids with external disturbances and directed communication topologies: a full-order sliding-mode approach. IEEE/CAA J. Autom. Sinica **8**(3), 554–564 (2021)
22. Prasad, S., Purwar, S., Kishor, N.: H-infinity based non-linear sliding mode controller for frequency regulation in interconnected power systems with constant and time-varying delays. IET Gener. Transm. Distrib. **10**(11), 2771–2784 (2016)
23. Guo, J.: Application of full order sliding mode control based on different areas power system with load frequency control. ISA Trans. **92**, 23–34 (2019)
24. Rinaldi, G., Menon, P.P., Edwards, C., Ferrara, A.: Design and validation of a distributed observer-based estimation scheme for power grids. IEEE Trans. Control Syst. Technol. (2018)
25. Cui, Y., Xu, L., Fei, M., Shen, Y.: Observer based robust integral sliding mode load frequency control for wind power systems. Control. Eng. Pract. **65**, 1–10 (2017)

26. Yang, B., Yu, T., Shu, H., Dong, J., Jiang, L.: Robust sliding-mode control of wind energy conversion systems for optimal power extraction via nonlinear perturbation observers. Appl. Energy **210**, 711–723 (2018)
27. Wang, N., et al.: Load-frequency control of multi-area power system based on the improved weighted fruit fly optimization algorithm. Energies **13**(2), 437 (2020)

An Approach to Analyze Factors Affecting the Quality of ICT Teaching Based on Data Mining

Bao The Nguyen and Hung Ho-Dac(✉)

Thu Dau Mot University, Binh Duong, Vietnam
{baont,hunghd}@tdmu.edu.vn

Abstract. Data mining is a powerful instrument in the revolution of artificial intelligence. Data contains important information that is rarely exposed or appropriately utilized. Mining data to elucidate patterns, and intelligent data analysis is an extremely valuable resource. It could result in meaningful perspectives and insightful data. Applying data mining to improve higher education is essential as Vietnam is now revolving its higher education system. In higher education, instructors play a key role in all courses so the quality of the courses heavily relies on the instructors. To aid in the assessment process, ultilize the data mining techniques to analyze and extract factors impacting instructors' performance. To predict teacher performance, we first performed a survey of Information Technology students at Thu Dau Mot University. We then employ multiple classifiers, compare their accuracy, and identify significant characteristics of instructor performance. We also provide our open-source model and source code on GitHub https://bitbucket.org/tdmudc/vdim/src/main/.

Keywords: ICT teaching · Instructors' performance · Undergraduate program · Data mining

1 Introduction

Ensuring and improving quality is a vital factor for any organization, higher education institutions are no exception [1]. In order to survive and develop, higher education institutions need to affirm the quality of training. Quality assurance is the top concern of higher education institutions today. Identifying important factors that affect student learning outcomes contributes to the quality improvement process. From there, coming up with appropriate implementation solutions is necessary to improve training quality. Therefore, surveying the opinions of stakeholders and then using data mining techniques to analyze and evaluate to find useful information will help support decision making [2].

Data mining [3] is a powerful tool in the artificial intelligence revolution. While the amount of data is constantly increasing, increasingly overwhelming us with the vast amounts of data being generated, our ability to understand it is decreasing. Hidden within data is useful information that is rarely revealed or used properly. Mining data to

M. Pagac et al. (Eds.): SCFF 2024, LNNS 1062, pp. 256–264, 2024.
https://doi.org/10.1007/978-3-031-65656-9_26

elucidate patterns, intelligent data analysis is an extremely valuable resource. It can lead to new insights and valuable information. Data mining is solving problems by analyzing existing data. For example, supermarkets can collect customer purchasing information about purchasing habits and habits when combining types of goods; From there, analyze to be able to arrange and arrange related items near each other to create convenience and attract customers' attention to the items, thereby improving sales ability. In today's highly competitive, customer-focused, service-oriented economy, properly analyzing customer data will drive business growth.

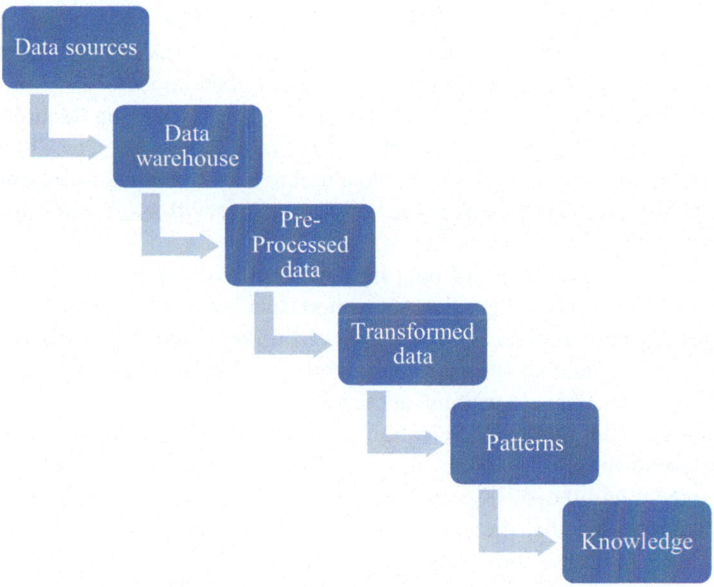

Fig. 1. General process of data mining

Data mining is defined as the process of automatically discovering data patterns (Fig. 1). The discovered patterns must carry useful information that can create an advantage, usually a business advantage [4]. Rely on useful patterns to make predictions about new data.

Data mining is used in a wide range of industries. A lot of educational data has been produced recently as a result of the expansion of electronic learning resources, learning support tools, and internet access in the classroom. These data are very important resources for data mining that can be used. With such a big amount of data, educational institutions are faced with a challenge: how to use this data source to their advantage and raise the caliber of their programs.

In this study, we combine feature selection methods and data mining methods to analyze and extract factors that affect instructors' performance.

2 Related Work

Research, development, and use of computer approaches are all part of educational data mining (EDM) [5], which aims to extract significant patterns from vast amounts of educational data. Starting with studies on intelligent tutoring systems and artificial intelligence in education, EDM has become a stand-alone research area. In 2005, "Educational Data Mining" was the name of the inaugural conference, which in 2008 was changed to "Annual International Conference on Education." The Conferences have been held annually ever since.

Alshareef et al. [6] introduced applications of EDM including but not limited to: study on whether attendance is statistically significant for the performance of students, prediction of student performance, grouping students for placements. Jacob et al. [7] also reviewed and applications of EDM with popular techniques.

Nguyen Thai Nghe et al. employed the Matrix Decomposition Technique in their studies [8] and [9] to forecast the learning outcomes of students at Can Tho University. The authors combine their ideas and create a tool to assist students in creating effective study schedules. Establishing a suitable study schedule will assist students, as well as schools, in avoiding waste in many areas.

In the study [10], a variety of data mining techniques were employed to forecast student learning outcomes using a data set gathered at Gazi University in Ankara, Turkey. These techniques included Decision Tree (J48), Multilayer Perceptron, Naive Bayes, and SMO algorithm (Sequential Minimal Optimization). When using the gathered data set, the authors compared the algorithms' accuracy.

Data mining relies on feature selection because not all of the features in the original data are pertinent to mining. In the study [11], feature selection methods were used to analyze the data, and the WEKA tool was then used to compare the results [12]. The outcomes demonstrate that the classification method will produce superior outcomes when used in conjunction with feature selection.

3 Proposed Approach

To build the model, we carry out a process of steps (Fig. 2): design survey questions, survey students, process data, apply data mining algorithms, evaluate results. Results, provide rcommendations.

The survey questionnaire was created to include questions broken down into categories pertaining to the lecturers' teaching, assessment, and skill-building activities, as well as surveys on the learning environment. The original data set can then be obtained by removing invalid data. The feature ranking algorithms built into the Weka toolbox will be applied to this data set to rank the features. The aspects that have the greatest impact on the outcomes will be kept, while the features with the least impact will be dropped. Models are created using the initial data set and the data set after removing less significant elements using classification methods from data mining. Compare and assess the outcomes based on the model's construction findings. There are some tips and advice provided for enhancing the caliber of instruction.

A component of assessing and enhancing training programs and teaching quality includes surveying stakeholders. Survey-related activities are frequently and continually

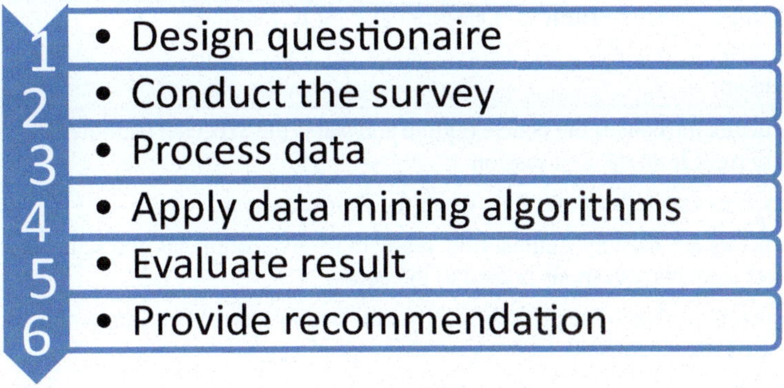

Fig. 2. Model building process

carried out at Thu Dau Mot University. Implementation strategies for conducting surveys through the School Leaders' individual plans. One of the plans is School Leaders Plan No. 64/KH-HTDM dated October 4, 2018, which is about "Getting feedback from students on the teaching activities of lecturers in the 2018–2019 school year." Students will be surveyed as part of this strategy regarding the lecturers' teaching activities, learning facilities, how lecturers are evaluated, etc. The inquiries will be answered by students-five level agreement range from 1 to 5 (Table 1).

A large number of features does not always translate into great accuracy in machine learning approaches. On the other hand, if the number of features is excessive, it will make training and creating the model much more difficult. The data must therefore be processed. Feature selection is a crucial component of data processing [11, 13]. A data collection that is appropriate for data analysis will be produced through feature selection. When assessing the data, feature selection must ensure the process' correctness, save training time, and prevent over-fitting.

The methods used in this work to rate the features of the obtained data include Correlation Attribute Evaluation, Gain Ratio Attribute Evaluation, Information Gain Attribute Evaluation, and OneR [14].

With feature selection, you can receive a reduced data set while still getting accurate results. Additionally, it will aid in identifying the key traits and influences that have a significant impact on raising standards at educational institutions.

The research's objective is to pinpoint crucial variables influencing student learning outcomes so that we may provide remedies to raise the standard of instruction. As a result, the dataset will be used in this study's data mining techniques, including Decision Table (DT), J48, Naive Bayes (NB), Multilayer Perceptron (MP), K-Nearest Neighbor (k-NN), and Sequential Minimal Optimization (SMO).

To determine which algorithm produces the best results and to pinpoint the influencing variables, compare the accuracy and execution times of various algorithms when applied to the data set (both before and after feature selection). Significant impact on raising the caliber of instruction.

Table 1. Questionnaire for ICT students

#	Question
Q1	Instructors introduces the course outline and states the expected learning outcomes of the course right from the first session
Q2	Instructors ensure time and fully teach the course content
Q3	Instructors are friendly, enthusiastic, ready to receive comments and satisfactorily answer students' questions related to the course
Q4	Instructors explain issues closely, easily, and step by step to help students achieve expected learning results
Q5	Instructors always motivated and encouraged students to develop their abilities related to the module
Q6	Instructors always discuss, listen and know the difficulties of students to have timely help methods
Q7	Instructors use many teaching methods to promote students' positivity and initiative
Q8	Instructors design many activities to help students develop their ability to self-study and research
Q9	The lessons conducted by the instructors create excitement/attractiveness/vibrancy/attractiveness for the students
Q10	Instructors guide students to plan their study work
Q11	Instructors guide and ask students to read and research documents related to the module
Q12	All information of the course (syllabi, lectures, practice exercises, etc.) has been provided by the instructors on the E-learning system
Q13	Instructors interact regularly with students through the E-learning system
Q14	Practical content helps students clearly understand theoretical knowledge
Q15	Instructors clearly announce the form and method of testing and evaluating the course (time, submission deadline, evaluation criteria)
Q16	Students understand the course evaluation criteria set by the instructor
Q17	Exercises, tests, and exams are consistent with the content and goals that the instructor initially set
Q18	The workload (exercises, projects, exams) of the course is appropriate and suitable for students
Q19	Instructors took the time to comment, evaluate, and respond to assigned student assignments in a detailed and timely manner
Q20	This module helps students develop specialized skills
Q21	The labs at the faculty has helped students develop their skills and support them to learn this module better
Q22	The module helped students develop their ability to work in groups
Q23	The module helped students develop problem-solving skills

(*continued*)

Table 1. (*continued*)

#	Question
Q24	The tasks (exercises, practice exercises, etc.) help students develop higher-order thinking skills (analysis, synthesis, evaluation, creativity)
Q25	Classrooms and labs are arranged reasonably, and safe
Q26	Teaching and learning equipment (computers, projectors, etc.) are of good quality, meeting the learning needs of this module
Q27	The information system (Wifi, Edusoft, E-learning) effectively supports students in this module
Q28	Documents related to the module are abundant and can be found easily

4 Experiment

To conduct experiments and evaluate the results, this thesis builds an application in Java combined with the Weka library [12]. All experiments were performed on Asus K55A Laptop, Intel(R) Core i7-3630QM processor, 2.4GHz CPU, 8GB RAM installed with Windows 10 Education 64bit operating system, data is divided into two The part includes district training (80%) and testing (20%).

To determine the appropriate data mining algorithm, and at the same time identify important features. We perform the experiments follow the process described in the Fig. 3.

Based on the designed questionnaire, the survey was conducted at the Institute of Engineering and Technology, Thu Dau Mot University. The subjects surveyed are the basic subjects of students majoring in Information Technology. The resulting data set includes 551 rows (corresponding to 551 student survey forms), 28 columns (corresponding to questions 1 to 28), the class column is the course score that the student has received (Table 2).

Table 2. Dataset overview

Lecture	No. of courses	No. of students
Lecture #1	3	189
Lecture#2	3	196
Lecture#3	2	166

With the number of selected features being six, less important features will be eliminated. In this experimental part, we build a model from a feature-reduced data set. Table 3 shows the accuracy when building a model with a combination of data mining algorithms and feature ranking algorithms. Table 4 compares the execution time between data mining algorithms when combined with the feature ranking algorithm.

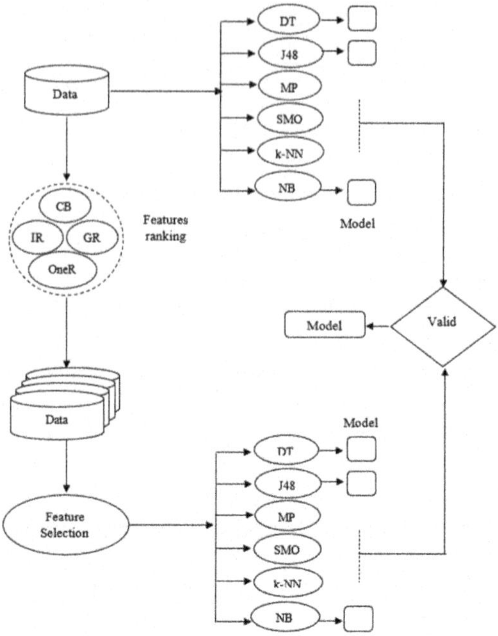

Fig. 3. Experiments process

From the experimental results, it shows that the important features in this data set are Q9 (The lessons taught by the lecturer are interesting/attractive/lively/charismatic for the students), Q10 (Lecturer guides students to plan their study work), Q11 (Lecturer guides, asks students to read and research documents related to the module), Q12 (All module information including syllabi, lectures, practice exercises, etc.) has been provided by the instructor on the E-learning system), Q27 (The information system of university including Wifi, Edusoft, E-learning, etc.) supports effective student support in this module) and Q28 (Documents related to the module are rich and can be found easily) are important influencing factors in improving the quality of teaching at the ICT program.

Table 3. The accuracy of experimental algorithms

Data mining algorithm Feature selection algorithm	DT	J48	NB	MP	k-NN	SMO
Correlation	85.3	85.3	88.2	76.5	79.4	88.2
Info gain	85.3	82.4	88.2	88.2	85.3	88.2
Gain ratio	88.2	85.3	88.2	88.2	85.3	88.2
OneR	85.3	88.2	88.2	82.4	79.4	88.2

Table 4. The time consumption of experimental algorithms

Data mining algorithm / Feature selection algorithm	DT	J48	NB	MP	k-NN	SMO
Correlation	16	16	≈ 0	147	≈ 0	15
Info gain	16	≈ 0	≈ 0	160	2	10
Gain ratio	7	≈ 0	≈ 0	185	≈ 0	16
OneR	10	10	≈ 0	161	≈ 0	≈ 0

5 Conclusion

The study presented an approach to evaluate and predict the performance of instructors at the ICT program of Thu Dau Mot University. On the other hand, the study also shows factors that strongly impact the performance of instructors including but not limited to "The lessons taught by the lecturer are interesting/attractive/lively/charismatic for the students", "Lecturer guides students to plan their study work", "Lecturer guides, asks students to read and research documents related to the module", "The information system of university including Wifi, Edusoft, E-learning, etc.", "Documents related to the module are rich and can be found easily". In future studies, we will propose a universal approach for multiple programs in different fields.

References

1. Frazer, M.: Quality Assurance in Higher Education pp. 17–34 (2003)
2. Blackmur, D.: Issues in higher education quality assurance. Aust. J. Public Adm. **63**(2), 105–116 (2004)
3. Cleve, J., Lämmel, U.: Data Mining. Walter de Gruyter GmbH & Co KG, (2020)
4. Ian, H.W., Eibe, F.: Data mining—Practical Machine Learning Tools and Techniques. Morgan Kaufman, (2005)
5. Romero, C., Ventura, S.: Data mining in education. WIREs Data Min Knowl Discov **3**, 12–27 (2013)
6. Alshareef, F., et al.: Educational data mining applications and techniques. Inter. J. Adv. Comput. Sci. Appl. 11(4) (2020)
7. Jacob, J., et al.: Educational data mining techniques and their applications. 2015 International Conference on Green Computing and Internet of Things (ICGCIoT). IEEE, (2015)
8. Thanh-Nhan, H.L., Thai-Nghe, N.: A system that predicts learning outcomes and suggests subject selection. Proceedings of the 16th National Conference: Some Selected Issues of ICT (@2013), pp. 110–118. Science and Technology Publishing House. ISBN: 987-604-67-0251-1 (2013a)
9. Thanh-Nhan, H.L., Thai-Nghe, N.: A system that predicts student learning outcomes using the open-source recommendation system library MYMEDIALITE. Proceedings of the 2013 National Conference on IT, pp. 192–201. Can Tho University Publishing House. ISBN: 978-604-919-012-4 (2013b)
10. Ahmed, A.M., Rizaner, A., Ulusoy, A.H.: Using data mining to predict instructor performance, international conference on application of fuzzy systems and soft computing, ICAFS 2016. Procedia Comput. Sci. **102**, 137–142 (2016)

11. Gnanambal, S., Thangaraj, M., Meenatchi, V.T., Gayathri, V., Classification algorithms with attribute selection: an evaluation study using WEKA. Int. J. Adv. Network. Appl. **09**(06), 3640–3644 (2018). ISSN: 0975-0290
12. Weka 3.: https://www.cs.waikato.ac.nz/ml. Accessed 1 Sept, 2023
13. Sutha, K., Tamilselvi, J.J.: A review of feature selection algorithms for data mining techniques. Int. J. Comput. Sci. Eng. (IJCSE) 63–67 (2015). ISSN: 0975-3397.B
14. Robert, C.: Holte, very simple classification rules perform well on most commonly used datasets. Mach. Learn. **11**, 63–91 (1993)

An Experimental Comparison Between Tree-Based and Machine Learning Approaches for Predicting an Oil Temperature in an Electricity Transformer Dataset

Anh Kim Thi Do[1] and Vinh Dinh Nguyen[2(✉)]

[1] School of Computing and Information Technology, Eastern International University, Thủ Dầu Một, Vietnam

[2] Department of Software Engineering, FPT University, Can Tho City, Vietnam

VinhND29@fe.edu.vn

Abstract. Time series data forecasting remains a challenging problem within research communities. A plethora of machine learning and deep learning architectures have been devised to tackle the wide array of time series datasets across different domains. Recent advancements in machine learning algorithms have demonstrated impressive performance. However, there remains a limitation regarding the time-consuming nature of both training and testing phases. Hence, this study endeavors to address whether any methods can achieve comparable performance to recent machine learning algorithms without requiring extensive training time. Through our research, we have identified XGBoost as the optimal solution for addressing this issue. Through comprehensive experiments across three study cases encompassing multivariate to multivariate prediction, multivariate to univariate prediction, and univariate to univariate prediction scenarios, we discovered that XGBoost outperformed other state-of-the-art machine learning techniques in the multivariate-to-multivariate prediction case under ETD dataset.

Keywords: tree-based method · machine learning · time series forecasting

1 Introduction

Time-series forecasting plays a crucial role in various domains such as sensor network monitoring, energy and smart grid management, economics, finance, and disease propagation analysis. In these contexts, a significant amount of historical time-series data can be utilized to forecast long-term behavior, known as long sequence time-series forecasting (LSTF). However, existing methods are predominantly designed for short-term prediction tasks, typically forecasting 48 points or fewer. The growing demand for forecasting increasingly longer sequences challenges the predictive capacity of existing models, thus hindering progress in LSTF research [1, 2].

This research aims to investigate the benefits of Multilayer perceptron (MLP), Recurrent Neural Network (RNN), Long short-term memory networks (LSTM), Bidirectional

M. Pagac et al. (Eds.): SCFF 2024, LNNS 1062, pp. 265–274, 2024.
https://doi.org/10.1007/978-3-031-65656-9_27

LSTM (BiLSTM) and XGBoost for predicting an oil temperature in Electricity Transformer Dataset [3]. There are three cases for studying the performance of MLP, RNN, LSTM, BiLSTM and Xboost: (1) Multivariate to Multivariate (MTM), (2) Multivariate to Univariate, (3) Univariate to Univariate. Our research is the first one that deeply evaluate the performance of MLP, RNN, LSTM, BiLSTM and Xboost on above three case studies. Experimental result found that Xboost obtained the best performance with mean absolute error(MAE) 2.02 in the MTM test. This yield interesting conclustion that performance of Xboost is better than other dominant machine learning methods, such as LSTM and BiLSTM. In addition, for MTU test, RNN achieved the best performance with MAE 2.63. For UTU test, MLP obtained the best MAE 2.29. The organization of this paper is presented as follows: Sect. 2 offers a review of existing machine learning algorithms. Section 3 present our approach for preparing dataset, system configuration, how to train the proposed system. Section 4 outlines the presents experimental findings. Finally, Sect. 5 concludes the paper, discussing findings and proposing potential directions for future research.

2 Related Works

2.1 Multilayer Perception

The multilayer perceptron (MLP) is widely recognized as the most commonly used type of neural network [4]. MLP stands as one of the most commonly implemented neural network architectures. MLPs are typically trained using the backpropagation algorithm. This algorithm propagates errors backward through the network, enabling the adjustment of the hidden processing elements (PEs). Two significant characteristics of the multilayer perceptron (MLP) include its nonlinear processing elements (PEs) with smooth nonlinearity, typically implemented using functions like the logistic function or hyperbolic tangent. Additionally, MLPs exhibit extensive interconnectivity, where each element in a given layer connects to all elements in the subsequent layer. Training of the multilayer perceptron is typically achieved through error correction learning, requiring knowledge of the desired system response. Figure 1 show sample model of multilayer perception network.

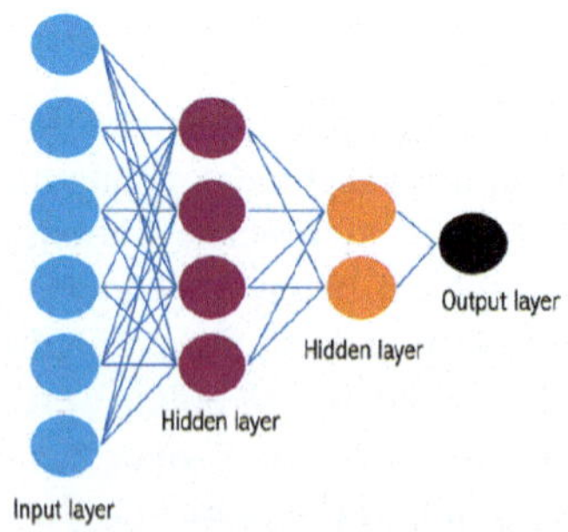

Fig. 1. Sample model of multiplayer perceptron.

2.2 Recurrent Neural Network (RNN)

Recurrent Neural Networks (RNNs) stand out as the predominant neural network architecture for sequence prediction tasks, with notable prevalence in the field of natural language processing [5]. Like Artificial Neural Networks (ANNs), RNNs also function as universal approximators. However, what sets RNNs apart is their utilization of feedback loops within recurrent cells, enabling them to inherently capture both temporal order and dependencies within sequences. Every Recurrent Neural Network (RNN) is composed of multiple RNN units. Elman (1990) introduced the fundamental recurrent unit, known as the Elman Recurrent Neural Network (ERNN) cell. The structure of the basic ERNN cell is illustrated in Fig. 2.

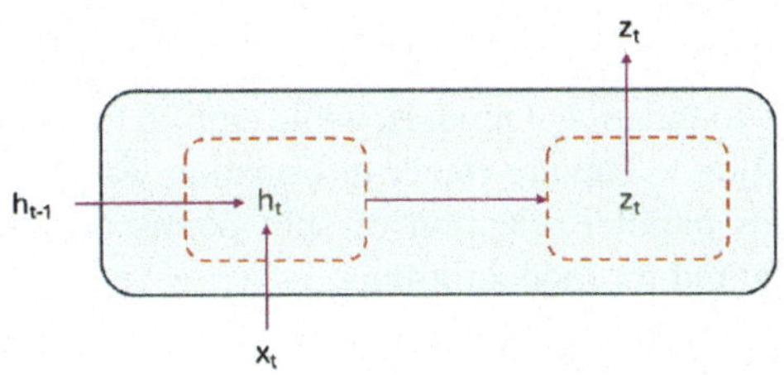

$$h_t = \sigma(W_i \cdot h_{t-1} + V_i \cdot x_t + b_i)$$
$$z_t = tanh(W_o \cdot h_t + b_o)$$

Fig. 2. Sample of recurrent unit

2.3 Long Short Time Memory (LSTM)

The RNN Cell is plagued by the well-known challenges of vanishing and exploding gradients, particularly over extended sequences. This means that simple RNN cells struggle to maintain long-term dependencies into the future. As sequences lengthen, backpropagated gradients often diminish (vanish), leading to inadequate weight updates. Conversely, gradients can become excessively large (explode) over prolonged sequences, resulting in unstable weight matrices. Both of these issues stem from the gradients becoming intractable, hindering the ability of RNN cells to effectively capture long-term dependencies. Over time, various modifications have been proposed to improve upon the base recurrent unit and address its limitations. One notable advancement is the Long Short-Term Memory (LSTM) cell introduced by Hochreiter and Schmidhuber [5]. It has become particularly popular for natural language processing tasks because it effectively captures long-term dependencies in sequences while mitigating issues related to vanishing gradients. In contrast to the basic RNN cell, the LSTM cell incorporates two distinct components for its state: the hidden state and the internal cell state. The hidden state corresponds to the short-term memory aspect, while the cell state represents the long-term memory. Additionally, LSTM introduces a gating mechanism comprising the input gate, forget gate, and output gate. These gates regulate the flow of information within the cell, allowing LSTM to selectively retain or discard information over time, thus facilitating the retention of long-term dependencies while preventing the vanishing or exploding gradient problems.

2.4 Bidirectional LSTM

Deep-bidirectional LSTMs represent an extension of the conventional LSTM models [6]. In this approach, two LSTM layers are applied to the input data. Initially, an LSTM is applied to the input sequence in its original order (referred to as the forward layer). Subsequently, the input sequence in its reverse form is fed into another LSTM layer (known as the backward layer). By applying the LSTM model twice, this technique enhances the learning of long-term dependencies, consequently improving the accuracy of the model.

2.5 XGBoost

XGBoost represents an adaptation of Gradient Boosting, where numerous feeble classifiers are amalgamated into a potent classifier in a linear fashion [7]. It accommodates both CART and linear classifiers as foundational models, while employing second-order Taylor expansion on the cost function to capture richer information. Regarding operational efficiency, XGBoost enables parallel selection of split points during training, significantly reducing the time required for model training.

3 Approach for Time Series Prediction

3.1 Dataset Preparation

The issue of distributing electric power involves delivering electricity to various regions based on their sequential usage patterns. However, accurately predicting the future demand in specific areas is challenging due to fluctuations influenced by factors like weekdays, holidays, seasons, weather, and temperatures. Current methods struggle to provide long-term predictions using extensive real-world data with high precision, as an inaccurate forecast could harm electrical transformers. Consequently, managers often rely on empirical data, leading to overestimations compared to actual demands, resulting in unnecessary electricity wastage and equipment depreciation. Alternatively, monitoring oil temperatures in transformers can indicate their condition. Thus, an effective approach involves forecasting safe oil temperatures to prevent unnecessary waste. To tackle this issue, Haoyi et al. develop a real-world platform and gather two years of data called as Electricity Transformer Dataset (ETDataset) [3]. They aim to predict electrical transformers' oil temperatures and assess extreme load capacity. Figure 3 shows example of ETD dataset.

3.2 Dataset Analysis

To deeply understand the ETD dataset, we setup an experiment to visualize the distribution of data of each feature in the dataset as shown in Fig. 4 (Fig. 5).

To further study more the relationship between each feature in the dataset, we use Person correlation coefficient method. Figure 6 show the result of the analysis. Experiment results reveal that HUFL and MUFL are highly correlation, while LULL and MUFL are less correlation.

date	HUFL	HULL	MUFL	MULL	LUFL	LULL	OT
str	f64	f64	f64	f64	f64	f64	f64
"2016-07-01 00:...	5.827	2.009	1.599	0.462	4.203	1.34	30.531
"2016-07-01 01:...	5.693	2.076	1.492	0.426	4.142	1.371	27.787001
"2016-07-01 02:...	5.157	1.741	1.279	0.355	3.777	1.218	27.787001
"2016-07-01 03:...	5.09	1.942	1.279	0.391	3.807	1.279	25.044001
"2016-07-01 04:...	5.358	1.942	1.492	0.462	3.868	1.279	21.948

Fig. 3. Sample data from ETD dataset

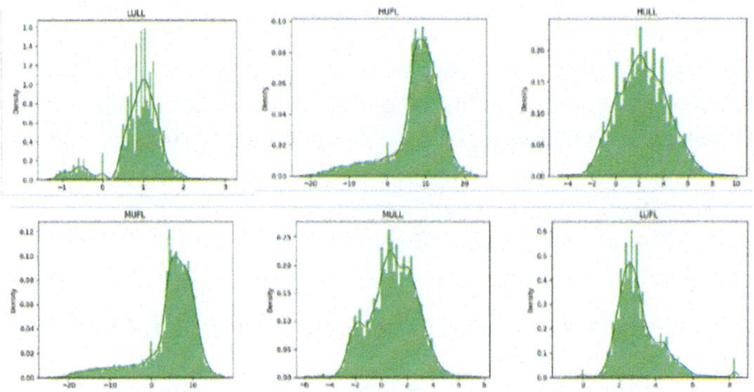

Fig. 4. Data distribution of each feature in ETD dataset

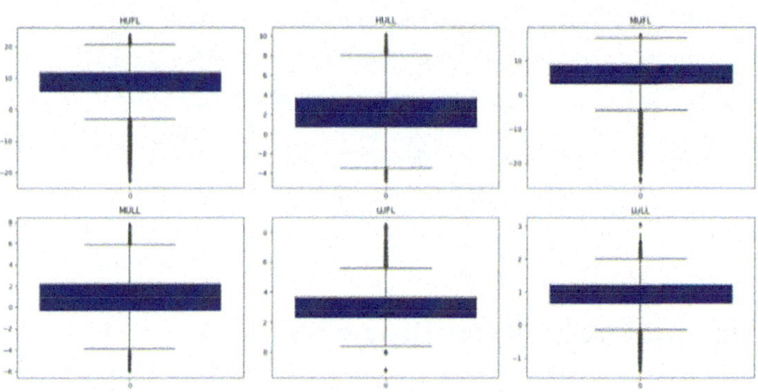

Fig. 5. Data distribution of each feature in ETD dataset

3.3 Methods for Predicting Oil Temperature

For mutltivariate to multivariate case, we designed a system with 336 × 7 input units corresponding to 336 interval time in the dataset. 96 × 7 output units.

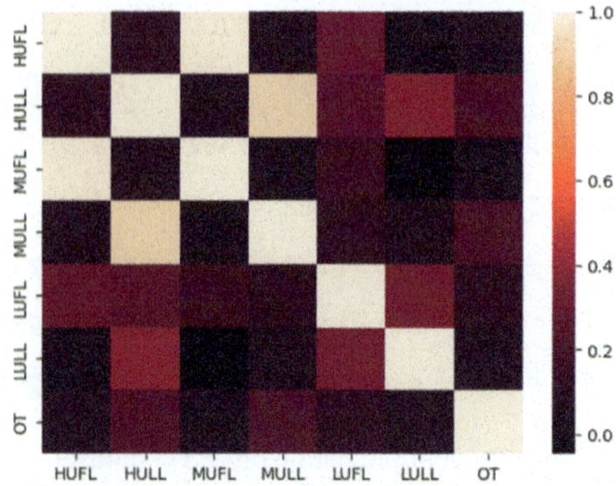

Fig. 6. Feature relationship by using person correlation coefficient

For mutltivariate to univariate case, we designed a system with 336 × 7 input units corresponding to 336 interval time in the dataset. 96 × 1 output units.

For mutltivariate to univariate case, we designed a system with 336 × 1 input units corresponding to 336 interval time in the dataset. 96 × 1 output units. The following shows information of train, validate and testing set:

Table 1 shows detail configurations for MLP, RCNN, LSTM, BiLSTM and XGBoost.

Table 1. The proposed configuration for MLP, RCNN, LSTM, BiLSTM, and XGBoost

Method	Parameter
MLP, RNN, LSTM and BiLSTM	Input layer: 336 ×7 units for MTM, MTU. 336 units for UTU. Hidden layer: 64 units Output layer: 96 ×7 units for MTM, 96 ×1 units for MTU and UTU. Learning rate: 0.01
XGBoost	'objective': 'reg:squarederror', 'eval_metric': 'mae', 'booster': 'gbtree', 'n_estimators': num_epochs, 'max_depth': 3, 'learning_rate': 0.1, 'early_stopping_rounds': patience

4 Experimental Results and Discussion

4.1 Metric for Performance Evaluation

To evaluate the performance of the proposed system, we use mean absolute (MAE) error as follows:

$$\begin{aligned} \text{MAE} &= \frac{1}{n}\sum_{i=1}^{n}\text{AE}_i \\ \text{AE}_i &= \left|\text{actual}_i - \text{predicted}_i\right| \end{aligned} \tag{1}$$

4.2 Results

We evaluate the performance of MLP, RCNN, LSTM, BiLSTM, and XGBoost on MTU case under ETD dataset. Experimental results are shown in Figs. 7, 8, 9, 10 and 11. The results are quite interesting because the XGBoost obtained better performance than other methods. MLP, RNN, LSTM, BiLSTM, and XGBoost obtained the MAE of 2.47, 2.21, 2.29, 2.56, and 2.02, respectively.

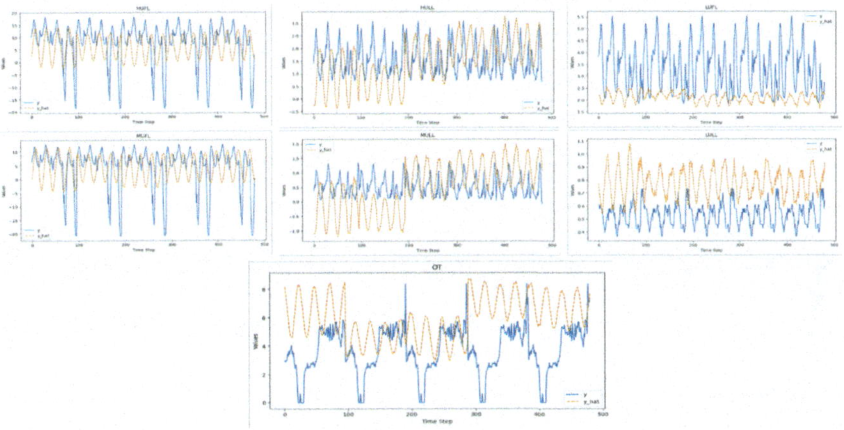

Fig. 7. Experimental results of MLP on multivariate-to-multivariate prediction

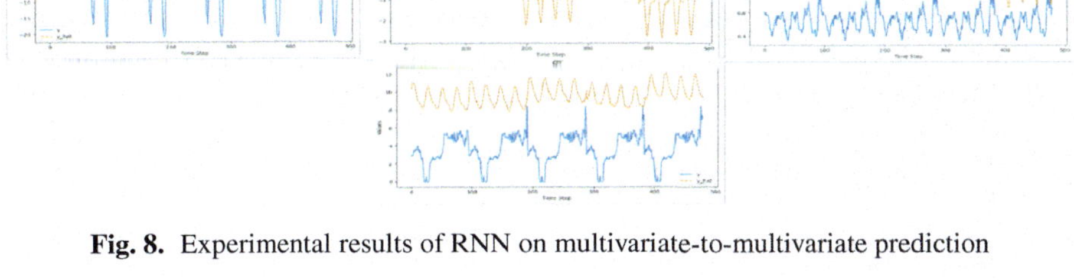

Fig. 8. Experimental results of RNN on multivariate-to-multivariate prediction

Figure 12 shows a comprehensive the performance of the MLP, RNN, LSTM, BiLSTM, and XGBoost on MTM, MTU, and UTU cases. Experimental results described that RNN achieved the lowest MAE (2.62) in compare to MLP (MAE: 3.08), LSTM (MAE: 3.18), BiLSTM (MAE: 2.74) and XGBoost (MAE: 5.01) under MTU test as shown in Fig. 12 (b). Another experiment on UTU test also shows that MLP got the highest performance (MAE: 2.29) where MAEs of RNN, LSTM, BiLSTM, and XGBoost are

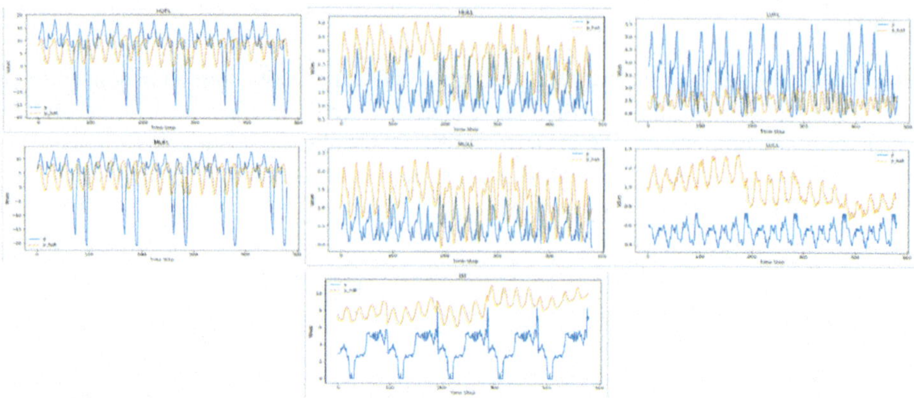

Fig. 9. Experimental results of LSTM on multivariate-to-multivariate prediction

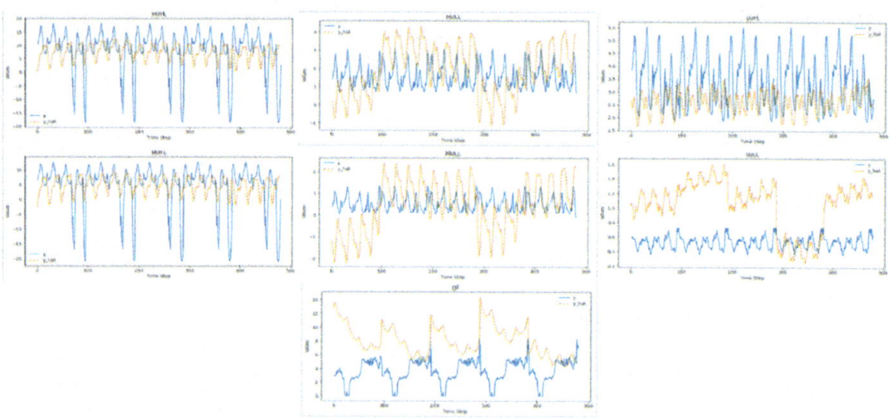

Fig. 10. Experimental results of BiLSML on multivariate-to-multivariate prediction

2.8, 2.56, 3.0, and 5.01, respectively. In summary, we found that the XGBoost are good under MTM test. The RNN are the best under the MTU test, while MLP obtained the best result under the UTU test.

5 Conclusion

Our research is the pioneering investigation into the advantages of employing XGBoost, MLP, RNN, LSTM, and BiLSTM models on the ETD dataset. Through comprehensive experiments across three study cases encompassing multivariate to multivariate prediction, multivariate to univariate prediction, and univariate to univariate prediction scenarios, we discovered that XGBoost outperformed other state-of-the-art machine learning techniques in the multivariate-to-multivariate prediction case. These findings offer an alternative approach for readers seeking solutions for time series data forecasting beyond traditional machine learning methods. In future endeavors, we aim to explore

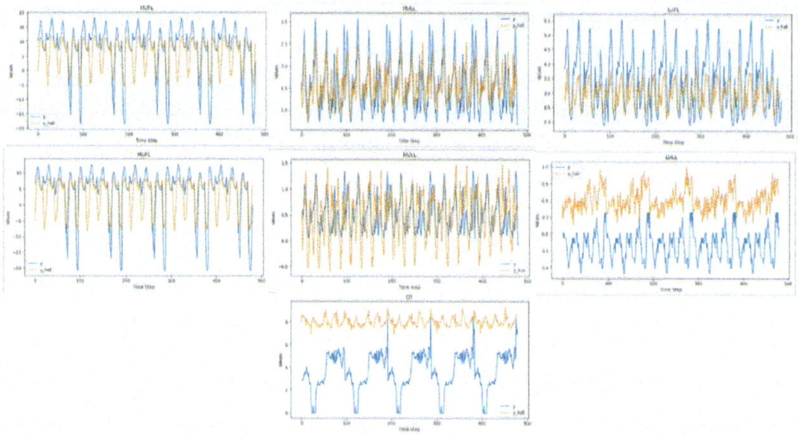

Fig. 11. Experimental results of XGboost on multivariate-to-multivariate prediction

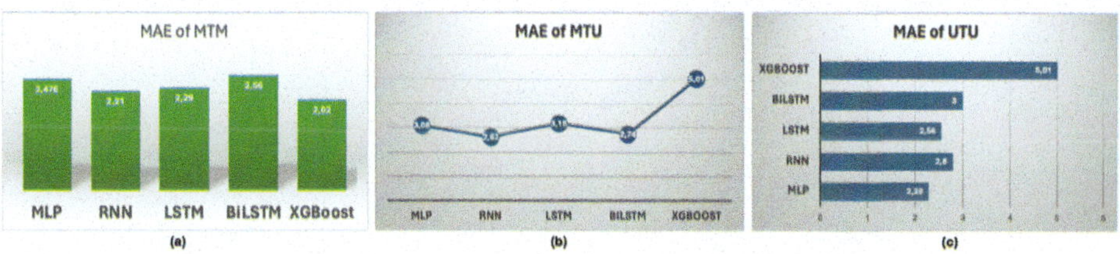

Fig. 12. Experimental results of MTM, MTU, UTU on ETD dataset. (a) Resulst on MTM case. (b) Results on MTU case. (c) Results on UTU case

more recent state-of-the-art methods, such as transformers, for enhancing time series data prediction tasks.

Acknowledgments. The authors would like to thank FPT University Can Tho City for funding this research.

References

1. Mudelsee, M.: Trend analysis of climate time series: a review of methods. Earth Sci. Rev. **190**, 310–322 (2019). https://doi.org/10.1016/j.earscirev.2018.12.005
2. Topol, E.J.: High-performance medicine: the convergence of human and artificial intelligence. Nat. Med. **25**, 44–56 (2019). https://doi.org/10.1038/s41591-018-0300-7
3. Haoyi, Z., Shanghang, Z., Jieqi, P., Shuai, Z., Jianxin, L., Hui, X., Wang, Z: Informer: beyond efficient transformer for long sequence time-series forecasting. AAAI, 11106–11115 (2021)
4. Popescu, M.-C., Balas, V., Perescu-Popescu, L., Mastorakis, N.: Multilayer perceptron and neural networks. WSEAS Trans. Circ. Syst. **8** (2009)
5. Ansika, H., Christoph, B., Kasun, B.: Recurrent neural networks for time series forecasting: current status and future directions. Int. J. Forecast. **37**(1), 388–427 (2021). https://doi.org/10.1016/j.ijforecast.2020.06.008

6. Schuster, M., Paliwal, K.K.: Bidirectional recurrent neural networks. IEEE Trans. Signal Process. **45**(11), 2673–2681 (1997). https://doi.org/10.1109/78.650093
7. Chen, T., Guestrin, C.: XGboost: a scalable tree boosting system. In: Proceedings of the 22nd ACM SIGKDD International Conference on Knowledge Discovery and Data Mining, pp. 785–794 (2016)

Binary-SegNet: Efficient Convolutional Architecture for Semantic Segmentation Based on Monocular Camera

Thai-Viet Dang[1(✉)], Nhu-Nghia Bui[1], and Ngoc-Tam Bui[2]

[1] Hanoi University of Science and Technology, Hanoi 10000, Vietnam
viet.dangthai@hust.edu.vn

[2] Shibaura Institute of Technology, Tokyo 135-8548, Japan

Abstract. The paper utilizes a semantic segmentation model for images obtained from a monocular camera based on Binary Neural Networks (BNN). The proposed model uses Resnet18 as the backbone along with a decoder based on the PSP-Net architecture. By decreasing the magnitude of processing parameters and propagation within the network, this approach facilitates enhanced training efficiency, rapid inference speed, and reduced hardware demands. When implemented, our methodology produces superior outcomes in comparison to conventional CNN architectures. More precisely, the model's inference pace experiences a 3.11-fold increase, whereas the accuracy remains untouched at 6.21%. Moreover, BNNs enable the implementation of semantic segmentation models on hardware that is resource-efficient or has a limited physical capacity. In conclusion, this architectural design presents prospects for integration into intelligent mechatronic systems, particularly in the development of knowledge systems for robotics traversing diverse environments, by maximizing costs and optimizing the monocular camera's operational capabilities.

Keywords: Semantic segmentation · computer vision · convolution neural network · binary convolution · binary classification

1 Introduction

In the domain of artificial intelligence, deep neural networks (DNNs) are prevalent and hold significant importance in the era of technology networks 4.0. DNNs significantly advance the frontiers of computer vision, automation, the Internet of Things (IoT), robotics, and other related fields [1, 2]. An increasing number of computer vision problems and future directions based on deep learning are being investigated, and solutions are being proposed. These encompass models for semi-surveillance object detection that rely on training image sets [3] or point clouds gathered by Lidar sensors for three-dimensional object detection. Nonetheless, these methods have overall limitations. Particularly the calculations and cost specifications for the hardware.

M. Pagac et al. (Eds.): SCFF 2024, LNNS 1062, pp. 275–285, 2024.
https://doi.org/10.1007/978-3-031-65656-9_28

Because of the efficacy and cost-effectiveness, semantic segmentation for monocular camera images is a challenging problem. The accuracy and complexity of neural networks regarding segmentation have been progressively enhanced in recent years. On the contrary, the implementation and training of these convolutional networks pose significant challenges, particularly when applied to intelligent embedded devices, such as excessive computational equipment requirements and frequent overfitting. Approximately speaking, this impacts the system's processing speed, precision, and efficiency [4]. The magnitude and intricacy of the weighting and activation in the model, in addition to the configuration of the convolutional layers, are fundamental causes of these issues. Including the calculation, accumulation, and propagation of parameters, this has a substantial impact on the operation of the network. To fulfill the specified criteria of optimal structure, training efficiency, rapid inference speed, high accuracy, and ease of effective deployment on peripheral devices, neural networks must be enhanced.

In order to address the aforementioned challenges, a binary neural network architecture is employed, featuring parameters (weights and triggers), which are utilized to convert the output to $+1$ or -1 via binary mathematical operations based on the decoder-encoder architecture. By adhering to the framework, the model acquires the ability to proficiently comprehend and depict image information, subsequently employing this knowledge to generate a precise semantic segment map. Encoders, more precisely, are convolutional classes tasked with deriving attributes from the input image via a sequence of filters that produce a characteristic map. Decoder operates in the same manner as Encoder; it resizes the featured map from Encoder to its original dimensions. During this procedure, a semantic segmentation map [5] is generated by predicting the semantic label for every pixel. We present a semantic segmentation model with the architecture in this paper. More specifically, the initial stratum comprises the Convolutional backbone classes, which employ the Resnet18 network architecture [6] to extract the attributes of the input data. After undergoing a pyramid aggregation module, the features are subsequently grouped at various dimensions. This facilitates the model's ability to acquire data at various magnitudes. The design of this structure drew inspiration from Pyramid Scene Parsing Network (PSP-Net), an image segmentation model that comprehends visual landscapes through the utilization of pyramid-based network architecture [7]. To specifically minimize the size of the data and models, we substitute binary convolutional layers for the customary 2D convolutional layers in the decoder. The map illustrates the binary convolution that will be applied to the output prior to joining and transmitting it to the final layer (Binary conv). Alterations will also be made to the activation functions to incorporate the binary convolution layer. In conclusion, the prediction layer generates a final segment map through synthesis.

Based on the outcomes derived from the implementation of the suggested BNNs architecture, the viability of this binary deep neural network technique is evident. The authors examine the training and performance of the model using Cityscape and CIFAR10, two well-known datasets in the field of semantic partitioning. The training procedure is reasonably optimized, and evaluation metrics indicate that our method achieves an accuracy comparable to that of conventional DNNs (deviation of 5–8%). Accompanied by a two to fourfold increase in reasoning efficiency. By optimizing the model's structure and operation, BNNs enable the deployment of artificial intelligence models on a variety of devices, including embedded lightweight processors. In conclusion, the efficacy of artificial neurotransmitter data segmentation based on neural networks for monocular camera data is enhanced.

The rest of the paper is organized as follows. In Part II, the author presents the proposed methodology around the binary of parameters and describes the operation of binary neural networks. The results from training and network operations are presented in Part III. Finally, Part IV includes the conclusions from which the authors draw.

2 Related Work

The proposed transition from DNNs to BNNs comes from the binary of parameters including weight and activation to + 1 or − 1 values. To reduce the volume of computations and improve the propagation process of data in the network [8], traditional 2D convolution classes are replaced by binary convolution layers based on the semantic segmentation model [9].

2.1 Convolution Traditional 2D

In traditional 2D convolution, the convolutional filters have real-valued weights, and the operation involves sliding these filters over the input data (also known as the input feature map) to compute a set of output values. This operation is fundamental in convolutional neural networks (CNNs) for tasks like image processing, computer vision, nature language processing, and other spatial data analysis, in (1).

$$Y(i,j) = \sum_{m=0}^{M-1} \sum_{n=0}^{N-1} X(i+m, j+n) \cdot \mathrm{W}(m,n) + b, \tag{1}$$

where $Y(i,j)$: the value at the position (i,j); $X(i+m, j+n)$: the input value at the position $(i+m, j+n)$; $\mathrm{W}(m,n)$: the weight at the position (m,n); and b is the bias term.

The convolution operation entails calculating the weighted sum at each position after traversing the input data with the filter in increments (step sizes). Adding the bias term to every calculated sum. Frequently, an activation function (such as ReLU) is applied to the convolution output to introduce nonlinearity. The weights W and the bias b are considered learnable parameters, and they are modified via backpropagation throughout the training procedure. The objective is to train the network to recognize significant features in the input data using filters. Indeed, the intricacy associated with parameter calculation renders the model susceptible to overfitting, thereby resulting in suboptimal training and operation of convolutional networks. It is imperative to identify strategies for optimizing the performance of convolutional layers to ensure their compatibility with the computational system's resources.

2.2 Binary Convolution 2D

From the above disadvantages, to replace 2D convolutional layers to binary convolutional layers. Binary convolution 2D refers to a type of convolutional operation where the convolutional filters have binary weights. In traditional convolutional layers, the weights are real-valued parameters that are learned during the training process. In binary convolution 2D, however, the weights are binary, meaning they can only take on two possible values, often − 1 and + 1. The binary nature of the weights in binary convolution brings several advantages, particularly in terms of computational efficiency and reduced memory requirements. Binary convolutional layers are often used in scenarios where there are constraints on computational resources, such as deploying models on edge devices or embedded systems. The parameters in the network are binary in following (2):

$$x^b = Sign(x) = \begin{cases} +1 \text{ if x} \geq 0 \\ -1 \text{ otherwise,} \end{cases} \tag{1}$$

where x^b is the binarized variable (weight or activation) and x the real-valued variable.

Given an input feature map X and a binary filter W, the binary convolution operation is calculated as follows:

$$Y(i, j) = Sign \sum_{m=0}^{M-1} \sum_{n=0}^{N-1} X(i + m, j + n).\mathrm{W}(m, n), \tag{2}$$

where M and N are the dimensions of the binary filter, and Sign(.) is the sign function.

Comparing with traditional 2D convolutional layers [8, 9], we draw several conclusions as follows:

- Fix the value of weight representation instead of any continuous value in traditional convolution.
- Reduce the memory storage and computational requirements by using the binary weight with only two values (usually − 1 and 1).
- Binary convolution is more computationally efficient than a traditional convolution because of simpler mathematical operations using multiplying and plus binary weighting.
- Reduce the model size because binary weights require less memory than real-valued weights.

2.3 Proposed Architecture

The deep learning model is improved with the foundation of encoder–decoder for sematic segmentation problem along with using integrated bar binary convolution layers for traditional 2D convolution layers. The encoder-decoder both helps the network learn the objects of the image and reduces the computational complexity caused by reusing the weight from the encoder without using an additional decoding layer. The structure of the model is depicted as schematic in Fig. 1.

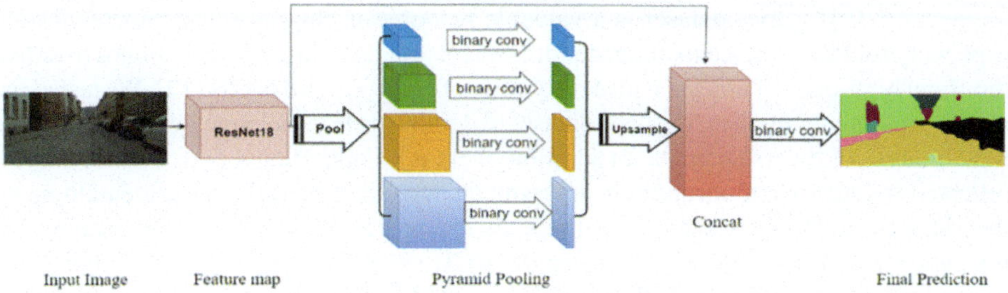

Fig. 1. BNN based PSP-Net using binary convolution, and Resnet18 for extracting the object's characteristics.

Our model uses the convolutional layers of the Resnet18 network as blocks responsible for extracting and encoding features. Resnet18 is designed to consist of blocks with a gradual increase in filters. Each block uses a max pooling layer to reduce the size of the photo and increase the characteristic size. Decoders are designed as pyramid blocks that help the model capture scales. The model uses different kernel sizes in pyramid pooling to collect information from different image sizes. From there, features at different scales are captured and deduced with high accuracy. One of the drawbacks of PSPNet is the complexity of the architecture being built. This leads to inefficiency due to high memory consumption for parameters and slow reasoning speed. The improvement is the use of binary convolutional layers that reduce the size of the model and computational data. The map features the binary convolution of which output will be joined and which it will be transmitted to the final layer (Binary conv). The trigger function in the network will also be changed accordingly. Specifically, due to the use of binary convolution, combining trigger functions requires special consideration due to the binary nature of the weight. The most common approach is to use differentiable approximations of binary step functions, often referred to as sign functions. The model has changed the initial trigger function (RELU) to the TANH function [10].

3 Results and Discussion

3.1 Datasets

First, CIFAR-10 dataset is a popular dataset in the field of machine learning and computer vision. This dataset includes sixty thousand 32 × 32-pixel images with ten classes [11]. Through this dataset, the authors assess BNN's ability to capture characteristics for a large and diverse dataset. This promises the generality of using deep learning models with weight and activation in binary form. The augmentations use on the dataset including randomly flipping the image horizontally, randomly changing the RGB values of the image, normalizing the image based on the mean and standard deviation of each color channel, and converting the image from a numpy array to a torch tensor.

Second, Cityscapes dataset is a valuable resource in the field of computer vision, especially for tasks related to understanding street images. This dataset includes over five thousand high-quality images with a resolution of 1024 × 2048 pixels [12]. We tested the model presented with the Cityscape dataset to demonstrate its accuracy with informative input images of diverse objects. In addition, there is an operational performance survey, reflected in latency and throughput assessments. To perform various augmentations on this dataset, including resizing the image to 256 × 256 pixels, randomly flipping the image horizontally, randomly changing the RGB values of the image, normalizing the image based on the mean and standard deviation of each color channel, and converting the image from a numpy array to a torch tensor.

3.2 Training

Firstly, authors evaluate BNNs with the basic models of Resnet18 and PSP networks. From there, to extract the operating parameters of the model. Secondly, from the training results, our proposed methodology gives relatively positive results. The error values undergoing training all converge and give sufficient results for use in image recognition and classification tasks. In addition, the training indicators (train loss) are less subject to fluctuations in the process of extracting and learning the patterns of the model.

3.3 Testing

To test the inference of BNNs, we ran several resulting models of different binary models on the Cityspaces dataset [9]. The results obtained are visualized as shown in Fig. 2.

The segmentation results are obtained at a relative level, suitable for integration into tasks in computer vision and related environments.

3.4 Benmark Result

The authors begin by comparing BNNs to the fundamental models of Resnet18 and PSP networks. Following this, the operational parameters of the model are obtained. The parameters that have been chosen to assess the outcomes of the model comprise:

- Train-loss and Val-loss: The loss function is calculated on the training set and validation set. The loss function depends on the model and the specific problem. To solve binary classification problems, the authors use the cross-entropy loss function (CELF) in (4).

$$J(\mathrm{w}) = \frac{1}{N}\sum_{n=1}^{N} H(p_n, q_n) = -\frac{1}{N}\sum_{n=1}^{N}\left[y_n \log \hat{y}_n + (1 - y_n)\log(1 - \hat{y}_n)\right] \quad (4)$$

- Train error and validation error: to indicate the error rate of the model's predictions, showing the model's deviation. They both use CELF, like train loss and validation loss.

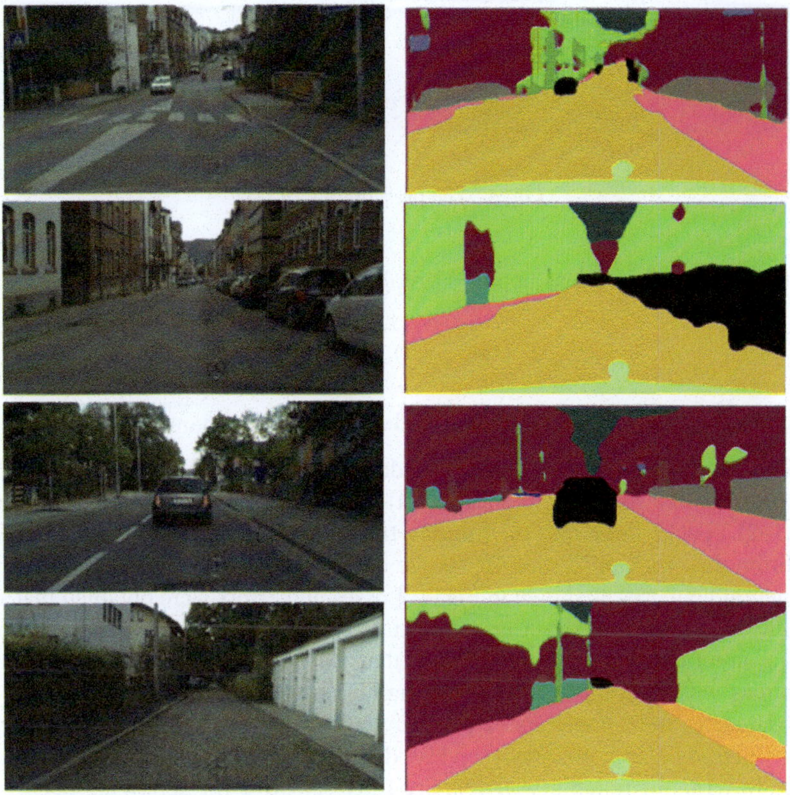

Fig. 2. Semantic segmentation results of BNNs on the Cityscape dataset.

- Accuracy: a metric used in machine learning and statistics to evaluate the performance of a trained model. The correct predictions are proportional to the total number of samples.

$$Accuracy = \frac{TP + TN}{TP + TN + FP + FN}, \tag{5}$$

where TP: True positive; FP: False positive; TN: True negative; and FN: False negative, respectively.

- IoU (Intersection over Union): By way of quantifying the extent of overlap between two images or regions. IoU is a metric utilized in object detection problems to assess the precision of the predicted bounding box in relation to the actual bounding box, as illustrated in Fig. 3.
- Dice Coefficient: The similarity between two samples is quantified by this metric. Frequently employed in the domain of machine learning, this metric assesses the precision of the predicted segmentation in relation to the true segmentation in image segmentation challenges. Dice is more responsive to variations in the overlapping

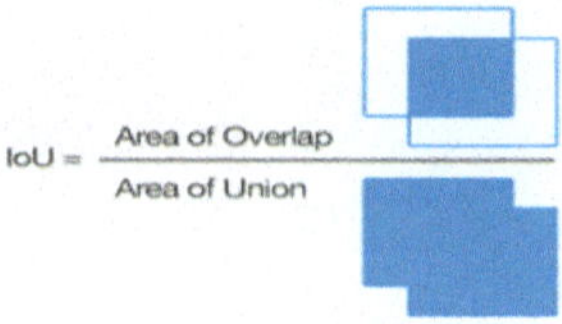

Fig. 3. The presentation of IoU

region expressed in (6), as opposed to IoU.

$$DSC(A, B) := \frac{2|A \cap B|}{|A| + |B|}, \tag{6}$$

where A is the set of true pixels and B the set of predicted ones.

The trained results of the proposed models are represented by two charts, in Figs. 4 and 5. Our training method gives the exact same value as the original models that cost only 15–21 epochs more, but the training time of one epoch is reduced by more than 70%. From this, we can see that BNNs are clearly effective in the training of convolutional neural networks as shown in Fig. 4. Losses and errors were obtained that showed the ability to efficiently extract image features. Along with that is the effectiveness of coaching in terms of time. The reason is that the model tested only performs simple calculations with optimized parameters (mainly at + 1 and − 1). In addition, as shown in Fig. 5 results, the parameters of accuracy, loss, error, IoU, and Dice when conducting tests give the same efficiency or less difference compared to traditional network models. In terms of reliability, BNNs are fully selective for deploying deep learning models in intelligent systems. Finally, authors measure the throughput and latency when implementing a binary approach that implements binary convolutional classes on other semantic segmentation networks. The obtained results are shown in Table 1.

BNN shows great efficiency in reducing latency and increasing the throughput of models that perform various tasks such as classification, object detection, sematic segmentation, instance segmentation, etc. The comparisons show the potential to deploy semantic segmentation models to work in real time. Fast processing speed, high performance while only significantly affecting accuracy as the tests the authors conducted combined with the light weight of the network using binary weight to provide effective deployment on resource-limited hardware devices such as mobile robots, IoT systems, etc.

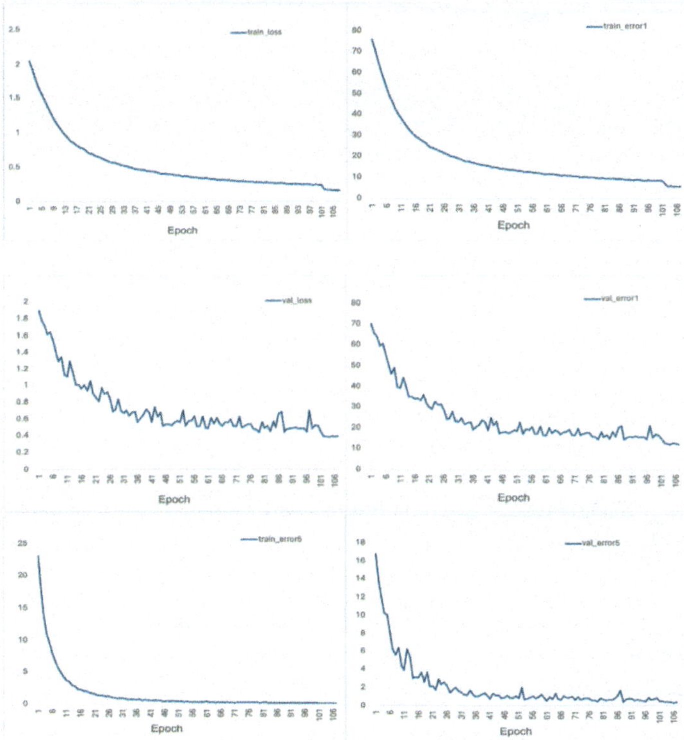

Fig. 4. Evaluation indicators are obtained from the Resnet18 model in combination with the binary module. This model was trained on CIFAR10 dataset.

4 Conclusions

By augmenting a deep neural network model with binary convolutional layers, authors decrease the parameter and model sizes. The bifurcation of the Encoder-Decoder structure, more precisely Resnet18-PSP, has been implemented to enhance the performance of the model during training and operation. Concrete evidence derived from concluded testing datasets (CIFAR10, Cityscape): The duration of training is optimized, and performance is enhanced. Utilizing metrics such as loss, error, accuracy, and IoU to assess outcomes that are comparable or deviate marginally from those of conventional network models. Conversely, the calculation speed and capability of model parameters are network-driven, as evidenced by the model's latency and throughput values. This presents an opportunity to optimize system resources or implement pentatonic segmentation models on devices featuring limited hardware capacity. Convolutional classes find applications beyond semantic segmentation for images and are utilized in numerous domains that incorporate deep learning. Hence, the author intends to broaden the scope of their research beyond semantic segmentation for images to encompass convolutional neural network-based deep learning models. Besides, it is solving the problem of weight fluctuations and optimizing the level of energy used to further improve the accuracy and

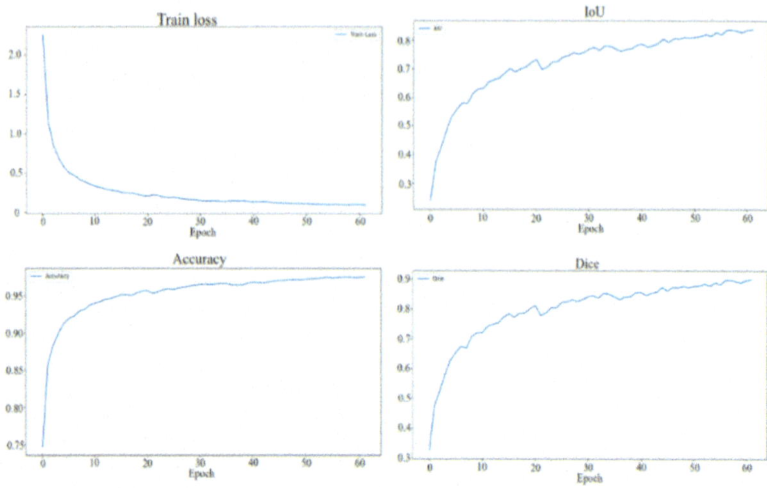

Fig. 5. Inference training results based on PSP network combined with parameter binarization based on Cityscape dataset.

Table 1. Measured result of throughput and latency in the comparison BNN between state-of-the-art methods [6, 7, 12–14].

Model	CPU Latency (s)	CPU Throughput (images/s)	GPU Latency (s)	GPU Throughput (images/s)
ResNet-binary	0.019028	80.156	0.009245	166.25
ResNet [6]	0.0315	54.2514782	0.0721	78.32
PSPNet-binary	0.06054	50.251	0.02359	180.605
PSPNet [7]	0.217	14.98612	0.1198	64.258
FCN-binary	0.01598	70.2984	0.00884	100.7236
FCN [13,14]	0.06689	79.52889	0.028332	94.3268781
Unet-binary	0.091882	20.9924	0.035017	76.00684
Unet [12]	0.4357612	18.11398	0.11157805	41.202

performance of the model. The authors then intend to develop solutions for deploying models that optimize the resource system across various environments, including control systems, mobile robotics, and so forth.

Acknowledgements. This research is funded by Hanoi University of Science and Technology (HUST) under project number T2023-PC-008. Lab-506: Computer Vision and Autonomous

Mobile Robot (CVAMS), VJIIST, HUST is gratefully acknowledged for providing work location, guidance, and expertise.

References

1. Dang, T.V.: Smart home management system with face recognition based on ArcFace model in deep convolutional neural network. J. Robot. Control **3**(6), 754–761 (2022)
2. Dang, T.V.: Smart attendance system based on improved facial recognition. J. Robot. Control **4**(1), 46–53 (2023)
3. Shao, Y., et al.: Semi-supervised object detection with multi-scale regularization and bounding box re-prediction. Electronics **13**(1), 221 (2024)
4. Matthieu, C., Itay, H., Daniel, S., El-Yaniv, R, Yoshua, B.: Binarized neural networks: training neural networks with weights and activations constrained to +1 or −1 (2016). arXiv:1602.02830v3
5. Badrinarayanan, V., Kendall, A., Cipolla, R.: SegNet: a deep convolutional encoder-decoder architecture for image segmentation. IEEE Trans. Pattern Anal. Mach. Intell. **39**(12), 2481–2495 (2017)
6. Yuzhi, C.: Application of Resnet18-Unet in separating tumors from brain MRI images. J. Phys. Conf. Ser. **2580**(1), 012057 (2023)
7. Zhao, H., Shi, J., Qi, X., Wang, Q., Jia J.: Pyramid scene parsing network. In: 2017 IEEE conference on computer vision and pattern recognition (CVPR), pp. 6230–6239 (2017)
8. Constantinides, G.A.: Rethinking arithmetic for deep neural networks (2020). arXiv:1905.02438v2
9. Tran, H.L., Dang, T.V.: Ultrafast semantic segmentation model for road planning. J. Robot. Control **4**(3), 424–430 (2023)
10. Kumar, A., Sodhi, S.S.: Some modified activation functions of hyperbolic tangent (TanH) activation function for artificial neural networks. In: International Conference on Innovations in Data Analytics, vol. 1442, pp. 369–392 (2023)
11. Li, H., Liu, H., Ji, X., Li, G., Shi, L.: Cifar10-dvs: an event-stream dataset for object classification. Front. Neurosci. **11**, 309 (2017)
12. Dang, T.V., Tran, D.M.C., Tan, P.X.: IRDC-Net: lightweight semantic segmentation network based on monocular camera for mobile robot navigation. Sensors **23**(15), 6907 (2023)
13. Dang, T.V., Bui, N.T.: Multi-scale fully convolutional network-based semantic segmentation for mobile robot navigation. Electronics **12**(3), 533 (2023)
14. Dang, T.V., Bui, N.T.: Obstacle avoidance strategy for mobile robot based on monocular camera. Electronics **12**(8), 1932 (2023)

Classification Disease in Tomato Leaf Based on the Mix Convolution Neural Network

Ngo Thi Hoa(✉) and Tang Ha Minh Quan

Faculty of Mechanical - Electrical and Computer Engineering, Van Lang University, Ho Chi Minh City, Vietnam
hoa.ngo@vlu.edu.vn , quan.thm@vlu.edu.vn

Abstract. Tomatoes are among the most common crops we can find everywhere in all culinary backgrounds. Tomatoes diseases can seriously affect crop output and productivity. Therefore, diagnosing and detecting diseases is crucial to help farmers improve tomato crop productivity. The article uses deep learning, specifically convolutional neural networks, to detect and identify leaf diseases. We designed a MIX-CNN network with a mixed network structure and a small-size model, only 62.4K with 29 layers. The network design includes enhancing two consecutive maxpooling layers at the input and two fully connected layers at the output to mitigate overfitting. The results demonstrate that the proposed network effectively classified the leaf disease dataset acquired from the PlantVillage dataset, achieving an accuracy of 97.6%.

Keywords: Deep learning · Convolution neural network · MIX-CNN · Tomato leaf disease

1 Introduction

Tomatoes are one of the most popular and consumed crops in the world [1]. They are rich in vital nutrients, antioxidants, minerals, vitamins, proteins, and critical amino acids. Consuming lycopene, found abundantly in tomatoes, has been reported to have a protective effect against cancer and several other diseases [2]. However, various leaf diseases may occur during tomato cultivation, which is also the main reason for product quality and quantity decline. In severe cases, it may even lead to plant death. Therefore, the detection of leaf diseases is one of the critical factors in the cultivation process.

The most common leaf diseases of tomato plants include Bacterial spot, Septoria leaf spot, Target spot, Mosaic virus, Early blight, Leaf mold, Yellow leaf curl virus, etc. Therefore, diagnosing and identifying diseases early is indispensable in improving productivity. Leaf diseases on tomato plants can be determined using conventional manual methods or diagnostic techniques assisted by artificial intelligence. The manual diagnosis of leaf diseases, if not accurate or timely, can

M. Pagac et al. (Eds.): SCFF 2024, LNNS 1062, pp. 286–295, 2024.
https://doi.org/10.1007/978-3-031-65656-9_29

result in a worsening of disease severity and a reduction in yield. Moreover, common natural conditions such as temperature, wind, humidity, light exposure, and other measurement phenomena can affect the symptoms of leaf diseases. These factors can lead to changes in leaf shape, colour, and the size of the affected area, which also impact the accuracy of disease diagnosis.

Therefore, we detect leaf diseases by convolutional neural networks (CNNs) to see early and accurately diagnose leaf diseases. Detecting diseases early will help farmers apply timely care and prevention measures to minimize disease occurrence and is also an important measure to improve crop productivity. One of the benefits of deep learning is that it can automatically extract image features due to the network can learn features during training [3].

This paper proposes a mix convolutional neural network to classify nine types of leaf diseases and healthy leaves. One notable advantage of convolutional neural networks is that they classify diseases without relying on the position of the symptoms on the leaf. Each disease symptom will appear in different positions on the leaf, and there will be many characteristic shapes for each type of disease. This is highly suitable for the tomato leaf disease problem that we are addressing.

The paper is organized as follows: The background of the convolutional neural networks is described in Sect. 2. Next, Sect. 3 describes the dataset and experimental, and Sect. 4 is results and analysis. Finally, Sect. 5 is conclusions and future works.

2 Methodology

2.1 Convolution Neural Network

CNN is a neural network commonly used to train image data today [1]. CNN network is becoming the most representative neural network in the field of deep learning. Currently, people have done things that were almost impossible to do in the past thanks to using the CNN network to process images, such as self-driving cars, facial recognition, self-service supermarkets, etc. [4].

To increase accuracy during training, many researchers have found solutions such as enhancing and normalizing data and improving network structure [5]. CNN usually includes three main layers: convolution and pooling layers perform data extraction and fully connected are the layer map extraction features for such output as classification [6]. Figure 1 depicts the basic structure of a CNN network.

2.2 Mix Convolution Neural Network (MIX-CNN)

The overall classification process for tomato leaf diseases is shown in Fig. 2. First, the data is processed and pre-labeled for each disease that needs to be classified. The MIX-CNN model is used for the training and validation sets. Figure 3 shows the model structure. The MIX-CNN model is a mixed network integrated with two maxpooling layers after the activation function ReLU instead of a maxpooling; this helps reduce overfitting. At the network output, we combine the

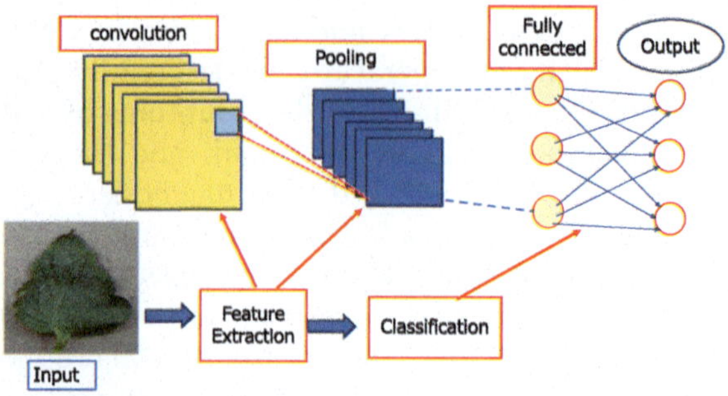

Fig. 1. General structure of CNN model.

dropout before two fully connected layers and connect the class output by one softmax. The parameters during the training process are constantly changed and updated based on the evaluation results of the model. The best results are used to train tomato classification leaf disease dataset.

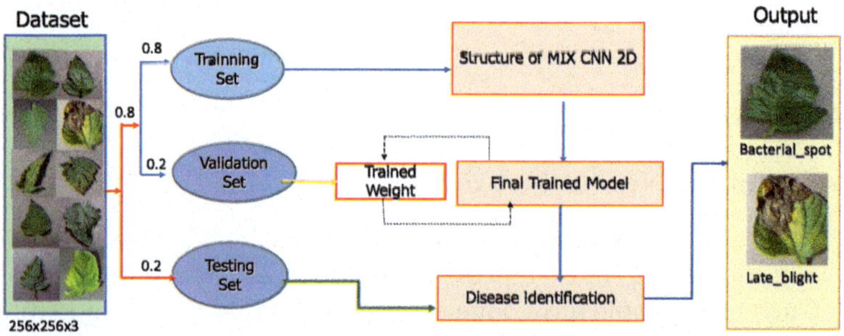

Fig. 2. Overview of tomato leaf disease detection.

Convolution layer is used to blur, sharpen, and define characteristic lines of data throughout the process. Each convolution layer includes different filters that are weights learned during training. The convolution process is performed by sliding the filters and input with the Stride sliding coefficient and convolution between the filter and a portion of the input at each location. The result of the convolution process is a feature map shown in Fig. 4 with the size of the feature map $[a \times a]$ shown in Eq. 1.

$$a = \frac{(n + 2p - f)}{s} + 1, \tag{1}$$

where: n is input size $(n \times n)$, p is padding, f is filter size $(f \times f)$, s is stride.

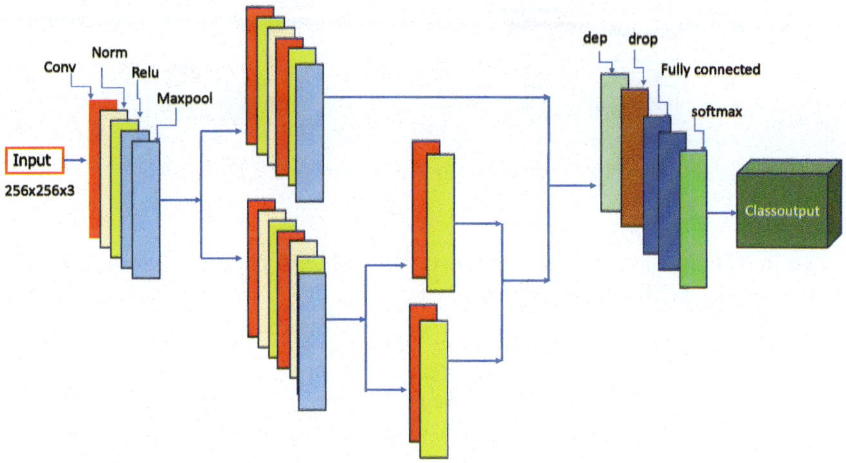

Fig. 3. The MIX-CNN model architecture.

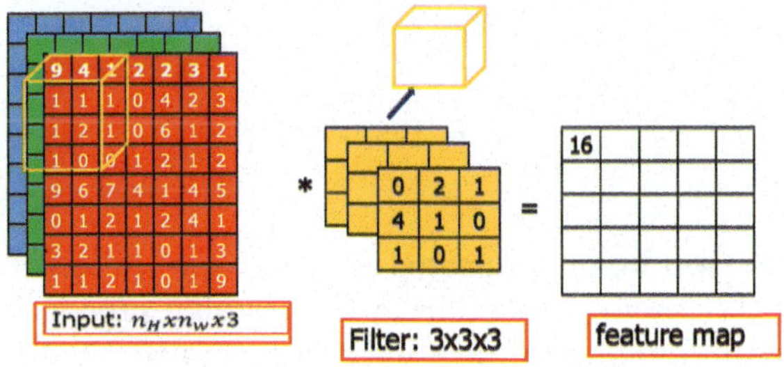

Fig. 4. Convolution with input dimension of 7×7 with zero padding, a filter size 3×3, stride 1.

Pooling layer CNN frequently uses pooling layers to boost computation performance, decrease representation size, and amplify some detection features. In CNN networks, average pooling and maxpooling are often used. Maxpooling represents essential features in a sliding window by selecting the maximum value and putting it in the output as shown in Fig. 5. Therefore, basic features will be retained, and unimportant and duplicate features will be eliminated. Another essential feature of maxpooling is that it can reduce overfitting by reducing the dependence on specific features in the input and helping the model learn more general features instead of just remembering particular features in the input—training data.

ReLu layer is a layer that applies an activation function to apply a threshold to each element of the input, helping to introduce nonlinearity into the model, fast convergence speed, and quicker calculation than other activation functions due to the simple calculation formula. Any x value less than 0 or equal to 0 will

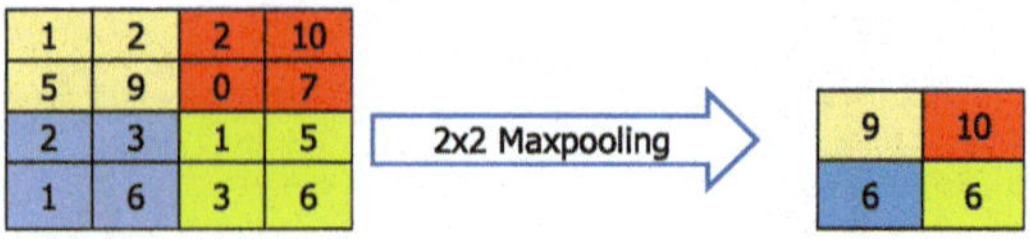

Fig. 5. Maxpooling with stride 2 × 2.

be set to 0. When x is greater than zero, the value of the function will be similar to that x value, as shown in Fig. 6.

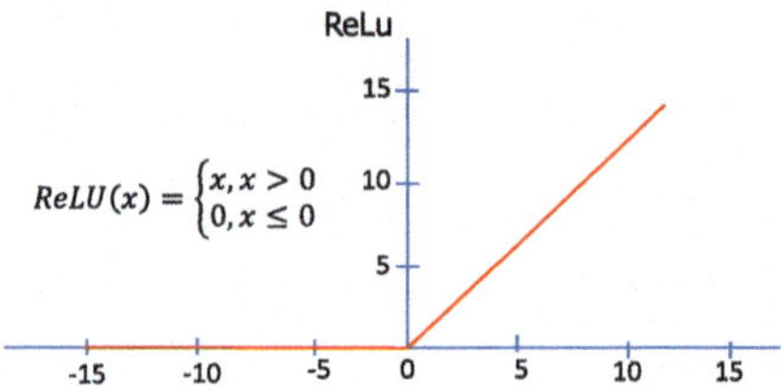

Fig. 6. ReLu layer.

Batch normalization layer connected after the convolution layer has the effect of standardizing the layers' input during the training process, which also helps reduce overfitting and improves the ability to generalize the model.

Dropout using Dropout helps reduce over-fitting during the training process.

Fully connected layer learns the input features, evaluates the feature taken out of the convolution layer in-depth, and produces a vector with n. To classify inputs accurately, fully connected layers often use a softmax activation function, which generates a probability between 0 and 1.

SoftMax is the final layer in the CNN network structure to calculate classification probabilities for different classes, as defined in Eq. 2.

$$a_i = \frac{e^{z_i}}{\sum_{j=1}^{C} e^{z_j}} \tag{2}$$

where: z_i is the ith output in the last layer with C outputs.

3 Experimental Setup

In our research, the dataset named Plant Village Database was used [7]. The dataset includes many different types of plants. However, we only took the dataset on tomato leaf diseases. The dataset is separated into a set of healthy leaves and nine various leaf diseases, including Bacterial spot, Septoria leaf spot, Leaf mold, Early blight, Spider mites: two-spotted spider mite, Target spot, Late blight, Mosaic virus, Yellow leaf curl virus.

Images of tomato leaf diseases used to conduct training are described in Fig. 7. There are a total of 22930 images with the size of 256 × 256, of which the lowest set is the Bacterial spot disease dataset with 2127 images and the highest set is 2451 corresponding to the Yellow leaf curl virus disease dataset. The detailed dataset used during the training process is shown in Table 1.

Fig. 7. Images of leaf diseases tomatoes with (a) Bacterial spot; (b) Early blight; (c) Healthy; (d) Late blight; (e) Leaf mold; (f) Septoria leaf spot; (g) Spider mites two spotted spider mite; (h) Target spot; (i) Mosaic virus; (j) Yellow leaf curl virus

Table 1. Tomato leaf disease dataset.

Category	Train	Test	Total number
Leaf mold	1882	470	2352
Bacterial spot	1702	425	2127
Late blight	1851	463	2314
Early blight	1920	480	2400
Spider mites two spotted spider mite	1741	435	2176
Mosaic virus	1790	448	2238
Target spot	1827	457	2284
Septoria leaf spot	1745	436	2181
Yellow leaf curl virus	1961	490	2451
Healthy	1926	481	2400
Total	18345	4585	22930

Using random sampling, we split the data set into training: testing data at an 8:2 ratio to conduct training. MATLAB R2023a is used to create the 2D MIX-CNN network that trained the model; the ADAM model is utilized for optimization—size used for the maximum training of 100 epochs in the model with a batch size of 128. The model's specific parameters are displayed in Table 2. All experiment was run on a computer with with configuration: Intel(R) Core (TM) i9-12900 RAM 32G GPU.

Table 2. Details of parameter for MIX-CNN model.

Parameter	Description
Number of convolution layers (2D-convolution 3 × 3 × 16)	7.0
Number of max pooling layers (stride 2 × 2)	4.0
Activation function	Relu
Dropout rate	0.2
Fully connected layer	2.0
Learning rate	0.003
Optimizer	ADAM
Number of epoch	100
Batch size	128

4 Result and Discussion

The classification accuracy result of the MIX-CNN model is 97.6% on the test set and 99.83% on the training set. Images of similar-shaped infected tomato leaves can, at different levels, contribute to a specific level of classification complexity that impacts the performance of samples shown in the same class.

A classification model's visual correctness is investigated using a confusion matrix, as shown in Fig. 8.

The accuracy of the class model increases with the number of white cells in the matrix. On the opposite diagonal, all of the successful forecasts are false predictions. These data are used to evaluate the classification's accuracy visually. We use accuracy, precision, recall, and the F1-score to assess the model. Table 3 displays the results of the model performance.

Table 4 compares the proposed MIX-CNN model and several standard models. The findings demonstrate that the MIX-CNN model, which combines many straight networks and integrates two max pooling layers at the input and two fully connected layers at the end, outperforms the previous model despite having a more straightforward structure.

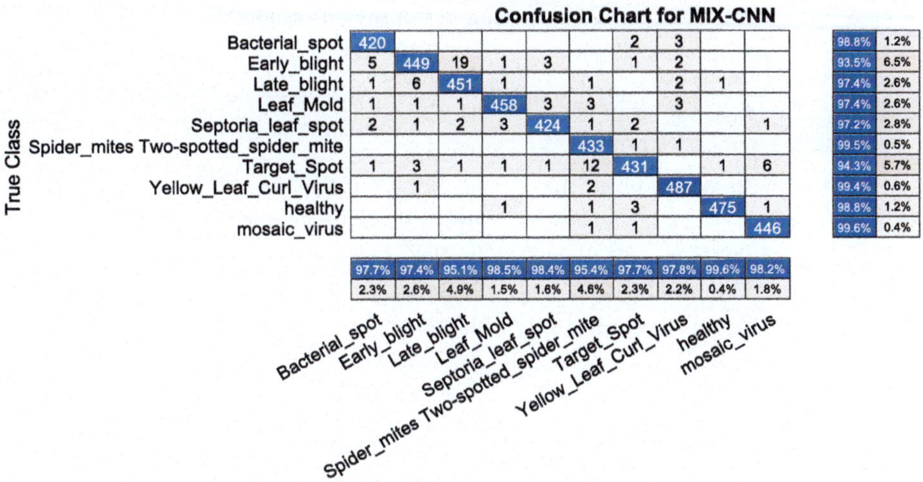

Fig. 8. Confusion matrix of the prediction leaf disease.

Table 3. Metrics of performance in validation.

Leaf disease	Precision %	Recall %	F1-score %
Bacterial spot	98.8	97.7	98.3
Early blight	93.5	97.4	95.4
Late blight	97.4	95.1	96.2
Leaf mold	97.4	98.5	98.0
Septoria leaf spot	97.2	98.4	97.8
Spider mites two potted spider mite	99.5	95.4	97.4
Target spot	94.3	97.7	95.8
Yellow leaf curl virus	98.8	97.6	98.3
Healthy	98.8	99.6	99.2
Mosaic virus	99.6	98.2	98.9
Average	97.6	97.6	97.6

5 Conclusion

The paper proposed the CNN network named MIX-CNN to detect diseases on tomato leaves. This network is a combination of many straight networks. We have combined two maxpooling to reduce overfitting. Experimental results have proven that this model successfully classified 22,930 images with an average accuracy of 97.6%. It is a small size model, only 62.4K with 29 layers, much smaller than some current models such as ResNet, AlexNet, GoogLeNet, etc.

Table 4. Comparing with previous studies.

Reference	Accuracy
Agarwal et al. [8]	91.2%
Khan et al. [9]	93.12%
Suryawati et al. [10]	95.25% VGGNet; 91.52% AlexNet; 89.68% GoogleNet
Tm et al. [11]	94–95% LeNet
Salih et al. [12]	96.34%
Coulibaly et al. [13]	95.0%
Sakkarvarthi et al. [14]	Transfer learning 70.14% VGG 19; 69.58% ResNet 152; 80.24% Inception V3; 88.17% proposed CNN model
Durmus et al. [15]	95.65% AlexNet; 94.3% SqueezeNet
MIX-CNN	97.6%

The classification results show that the classification efficiency is higher. Applying a small model helps reduce time and costs significantly. The proposed model allows farmers to detect and classify diseases on tomato leaves early, helping to improve crop productivity. We will continue to study and improve the MIX-CNN model to obtain better prediction accuracy. Furthermore, we will also use the models on robots or mobile devices to identify and detect tomato diseases.

References

1. Trivedi, N.K., et al.: Early detection and classification of tomato leaf disease using high-performance deep neural network. Sensors **21**(23), 7987 (2021)
2. Ali, M.Y., et al.: Nutritional composition and bioactive compounds in tomatoes and their impact on human health and disease: a review. Foods **10**(1), 45 (2020)
3. Sardogan, M., Tuncer, A., Ozen, Y.: Plant leaf disease detection and classification based on CNN with LVQ algorithm. In: 2018 3rd International Conference on Computer Science and Engineering (UBMK), pp. 382–385. IEEE (2018)
4. Li, Z., Liu, F., Yang, W., Peng, S., Zhou, J.: A survey of convolutional neural networks: analysis, applications, and prospects. IEEE Trans. Neural Netw. Learn. Syst. (2021)
5. Zhu, Q., Zu, X.: Fully convolutional neural network structure and its loss function for image classification. IEEE Access **10**, 35541–35549 (2022)
6. Yamashita, R., Nishio, M., Do, R.K.G., Togashi, K.: Convolutional neural networks: an overview and application in radiology. Insights Imaging **9**, 611–629 (2018)
7. Geetharamani, G., Pandian, A.: Identification of plant leaf diseases using a nine-layer deep convolutional neural network. Comput. Electr. Eng. **76**, 323–338 (2019)

8. Agarwal, M., Singh, A., Arjaria, S., Sinha, A., Gupta, S.: ToLeD: tomato leaf disease detection using convolution neural network. Procedia Comput. Sci. **167**, 293–301 (2020)
9. Khan, S., Narvekar, M.: Novel fusion of color balancing and superpixel based approach for detection of tomato plant diseases in natural complex environment. J. King Saud Univ.-Comput. Inf. Sci. **34**(6), 3506–3516 (2022)
10. Suryawati, E., Sustika, R., Yuwana, R.S., Subekti, A., Pardede, H.F.: Deep structured convolutional neural network for tomato diseases detection. In: 2018 International Conference on Advanced Computer Science and Information Systems (ICACSIS), pp. 385–390. IEEE (2018)
11. Tm, P., Pranathi, A., SaiAshritha, K., Chittaragi, N.B., Koolagudi, S.G.: Tomato leaf disease detection using convolutional neural networks. In: 2018 Eleventh International Conference on Contemporary Computing (IC3), pp. 1–5. IEEE (2018)
12. Salih, T.A., et al.: Deep learning convolution neural network to detect and classify tomato plant leaf diseases. Open Access Libr. J. **7**(05), 1 (2020)
13. Coulibaly, S., Kamsu-Foguem, B., Kamissoko, D., Traore, D.: Deep neural networks with transfer learning in millet crop images. Comput. Ind. **108**, 115–120 (2019)
14. Sakkarvarthi, G., Sathianesan, G.W., Murugan, V.S., Reddy, A.J., Jayagopal, P., Elsisi, M.: Detection and classification of tomato crop disease using convolutional neural network. Electronics **11**(21), 3618 (2022)
15. Durmuş, H., Güneş, E.O., Kırcı, M.: Disease detection on the leaves of the tomato plants by using deep learning. In: 2017 6th International Conference on Agro-Geoinformatics, pp. 1–5. IEEE (2017)

Designing a Smart Lighting System for Illuminating Learning Experiences

Phan Van Vinh[1](✉) and Phan Xuan Dung[2]

[1] School of Computing and Information Technology, Eastern International University, Binh Duong City, Vietnam
vinh.phan@eiu.edu.vn

[2] School of Engineering, Eastern International University, Binh Duong City, Vietnam
dung.phan@eiu.edu.vn

Abstract. The aim of this study is to introduce an educational smart lighting system, offering a practical learning platform for university students across all levels. The proposed system integrates Philips Lighting's Dynalite solution and Home Assistant platform on an Internet of Things (IoT) platform. Simulation results from DIALux software indicate that daylight harvesting and occupancy control can achieve energy savings of over 70%. Furthermore, the implemented results recommend the use of light-emitting diode (LED) lamps and Home Assistant platform as optimal choices for indoor applications, suggesting their practical deployment.

Keywords: Smart lighting · Home Assistant · IoT devices · DIALux

1 Introduction

In the era of rapid technological advancements and a growing awareness of environmental concerns, smart lighting has emerged as a significant aspect of the Internet of Things (IoT). Its multiple benefits, including energy efficiency, improved visual comfort, and flexibility, have not only revolutionized the lighting industry but have also become an integral part of our daily lives. Energy efficiency in lighting aims to deliver appropriate lighting conditions in working areas while utilizing an optimal amount of energy. Visual comfort, on the other hand, aims to provide lighting that is well-suited for various tasks within a specific space. Lighting flexibility involves the ability to adjust in room layout or network connectivity. For the energy efficiency, Sathya et al. designed a lighting system for a laboratory through the use of LED technology [1]. However, they did not address the aspects of smart lighting and energy harvesting, which could further reduce energy consumption. Bannamas et al. introduced an intelligent lighting energy management system that incorporates six lighting control functions and operates in both automatic and manual control modes [2]. Users have the flexibility to control the system via the lighting management program and an internet-based web server. Cheng et al. introduced an approach to decrease energy consumption by deploying a smart lighting system that incorporates sensor devices based on the ZigBee protocol,

M. Pagac et al. (Eds.): SCFF 2024, LNNS 1062, pp. 296–305, 2024.
https://doi.org/10.1007/978-3-031-65656-9_30

and employs illumination control guidelines [3]. Floris et al. introduced a smart building solution focused on IoT-based indoor management technique, such as monitoring room parameters, occupancy, utilizing cloud platforms for data aggregation, and integrating a control dashboard for building management [4]. Burunkaya et al. introduced a smart classroom as an IoT-based solution to mitigate the negative effects of environmental factors on learning by enhancing the accuracy of controlling essential physical parameters through the utilization of cutting-edge technology and methodologies [5]. Saranya et al. proposed a versatile device aimed at energy-efficient and intelligent lighting control, focusing on features like human detection, dynamic lighting adjustments, and wireless compatibility for smartphone control, aiming to promote sustainability and address challenges in lighting systems [6]. Typically, the core component of an IoT system is a control software responsible for collecting, processing, and sharing data. Therefore, in order to create a smart lighting system, choosing the right software platform is very important. While there are ready-made solutions available, it is also possible to develop your own software. Commercial systems like Philips Dynalite solution [7], however, can be expensive and often focus on their own specific devices, which can make it challenging from an economic perspective. Therefore, Home Assistant (HA) platform is suitable due to its open-source nature, support for multiple protocols and device interaction [8]. In fact, Zharikov et al. [9] introduced the use of HA to operate the smart home system. Additionally, TouchGFX framework was used to enable the visualization of system components.

Due to the increasing integration of digital technology with physical environment, there is a need for a comprehensive understanding and practical study of smart lighting, especially in educational and research contexts to reduce the gap between theory and hands-on experience. Therefore, Duijnhoven et al. introduced the smart lighting project on education and development at six universities in Vietnam and Myanmar, aimed at promoting energy-efficient and smart lighting solutions to address the challenges of limited access to daylight and prevalent energy inefficient electric lighting in developing countries [10]. To overcome challenges and improve training and research quality in smart lighting, we introduce a smart lighting system serving as an educational and research framework. Our system utilizes latest technologies, including Philips Lighting's Dynalite solution, the DIALux program, and Home Assistant platform running on the flexible and cost-effective IoT device (like Raspberry Pi). With the goal of enhancing comprehension of lighting technology and its educational and research applications, we introduce the main design of smart lighting system and provide the analysis of energy efficiency with DIALux. In summary, the primary contributions of this work are as follows: (1) to design the lighting system with specific hardware and software components, particularly focusing on cost-effectiveness and educational impact and (2) to implement a smart lighting application with an interactive web-based and mobile-based interface framework for system control and administration. The remainder of the paper is structured as follows: Sect. 2 details the main design of the smart lighting system, Sect. 3 presents the implementation results, and Sect. 4 offers our conclusions.

2 Smart Lighting Laboratory Design

2.1 Smart Lighting System Architecture

The proposed smart lighting system architecture is composed of four individual layers, each one with its own unique design and functions, as shown in Fig. 1:

- **The Sensors/Actuators layer** is composed of various input and output devices, such as sensors, different types of LED lamps, cameras and vacuum robot. The sensors, including light sensor and motion detection sensor (PIR), collect data on the ambient conditions in the room, which is then transmitted to the controller or gateway for further analysis and processing.
- The primary layer of the suggested smart lighting system is the **Controller/Gateway layer**, which employs IoT edge devices to control the system. Raspberry Pi is used as the primary controller due to its powerful performance and reasonable price compared to other IoT-edge devices. Additionally, load controller and gateway devices from Philips Lighting Company are utilized to connect and manage luminaires in the network.
- The **Analytic/Processing layer** is used to connect all components in the lighting system and provides centralized control services. Additionally, a logging system is implemented to store all activities occurring in the room for future analysis.
- The **User/Application layer** enables the owner or admin user to monitor and manage all room activities using either a web or mobile application. In case of abnormal or critical situations, the administrators will receive a notification message to take the right actions.

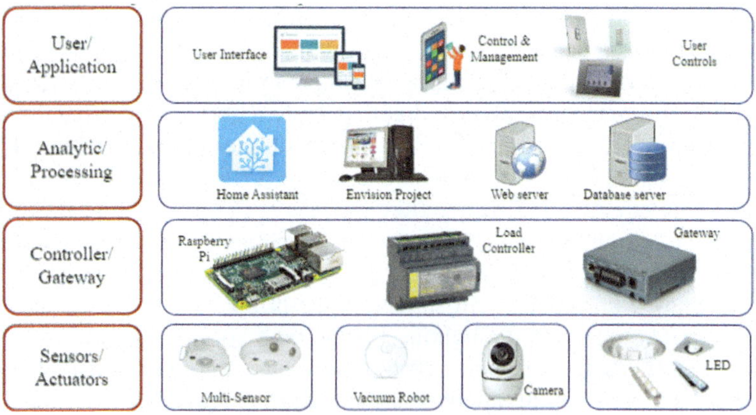

Fig. 1. The proposed system architecture

In this project, Home Assistant serves as the primary user application for smart lighting control, instead of other commercial products in the areas of smart lighting due to its numerous advantages. These include its open-source nature, support for multiple protocols and devices, flexibility in customization, and cost-effectiveness. In fact, Home

Assistant allows users to freely access, modify, and distribute the code, while supporting a wide range of protocols and devices, such as Zigbee, Z-Wave, Wi-Fi, Bluetooth. This enables the integration of various smart lighting products from different manufacturers, providing users with greater flexibility and choice.

2.2 Lighting Laboratory Physical Design

With the aim at providing an open and active learning environment, our educational laboratory was designed with a rectangular room measuring 10.7 m by 9 m and a ceiling height of 4 m from the ground. The room features standard four walls and a concrete ceiling with reflectance properties of 50% and 70%, respectively. On two sides of the walls, there are 10 windows covered by glass panes, and two doors is located on another side of a wall. The floor is tiled, with a reflectance property of 20%. The laboratory contains eight work tables and chairs, and a lighting system mounted on the roof is designed to produce a luminous flux of 60000 lumens, with measurements taken from a work plane positioned at 0.76 m from the ground. Within the rectangular room, we have placed 12 LED Tube lights in a grid 3 × 4, each providing a luminous flux of 3687 lm with a luminous efficiency of 142 lm/W, alongside 14 LED Downlights, each offering a luminous flux of 1124 lm with a luminous efficiency of 125 lm/W. This arrangement ensures the lighting coverage throughout the entire room, with the objective of achieving a luminance level of 500 lx. The actual layout of our smart lighting laboratory is presented in Fig. 2.

(a) The floor plan view (b) The actual design of a smart lighting Lab

Fig. 2. The layout of our designed smart lighting laboratory.

The main hardware components of the smart lighting system are shown in Table 1. Home Assistant platform is running on a Raspberry Pi which is chosen as a primary controller with Wifi & Bluetooth enabled, and General Purpose Input Output (GPIO) support can be one of the best solutions for IoT applications. The Philips Dynalite DDBC320-DALI (Digital Addressable Lighting Interface) is used as a load controller which operates seamlessly with all other Dynalite controllers, sensors, user interfaces.

For the purpose of smart lighting training program, we have also developed eight lighting panels equipped with training modules including light sources, controllers, gateways, sensors, dimmers, actuators, and measurement equipment like power meters, power analyzers, and illuminance meters, as shown in Fig. 3.

Table 1. The main hardware components of the proposed smart lighting system

Hardware Components	Manufacturer	Description
Raspberry Pi 4B	Raspberry	Main controller
Cleaning Robot	Ecovacs	To clean the lighting lab
Load Controller	Philips Lighting	To connect luminaires with DALI standard
Gateway	Philips Lighting	To connect lighting system to outside network
Multi-purpose Sensor	Philips Lighting	Integrated module with light and motion sensor
LED Tube	Philips Lighting	DALI LED tube
LED Downlight	Philips Lighting	DALI Downlight LED
Camera	YOOSEE	WiFi Camera to monitor the lab operations

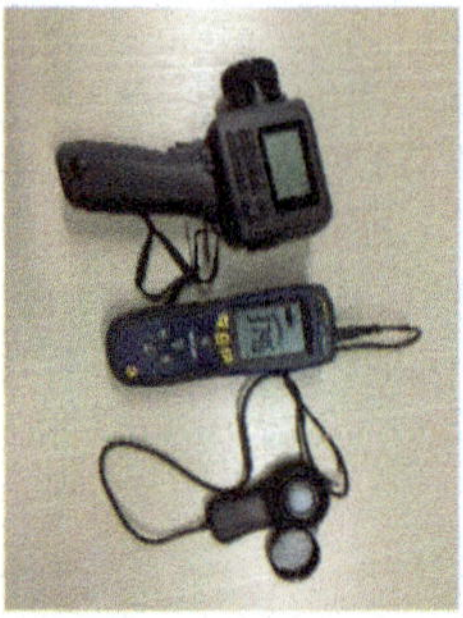

(a) A sample of illuminance meter

(b) A sample of lighting experimental panel

Fig. 3. An illustration of lighting experimental panel and equipment.

3 Implementation Results and Discussions

3.1 Smart Lighting Management with Home Assistant

We present the implementation for both web-based and mobile-based application interfaces which can be utilized to interact with the smart lighting system.

(1) **Web-based Application Interface**

To create a smart lighting system, it is essential to integrate all the required devices and components into the system. There are two ways to add devices: (1) manually add using a configuration file or (2) add devices using the integration feature.

The integration page offers users the capability to incorporate devices or products from other brands that are compatible with HA. Figure 4 illustrates the integration of Philips Dynalite solutions, demonstrating the connectivity between Philips Lighting luminaires and other products. After successfully adding the products, the devices and their components will be listed on the device page. Users can view all the LED luminaires and other devices that have already been added to the system.

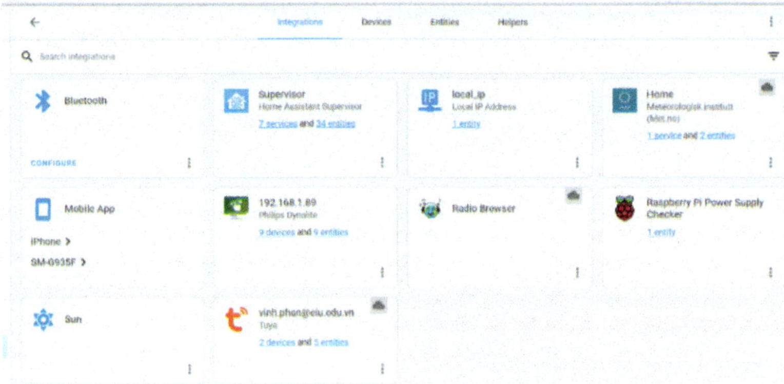

Fig. 4. A screenshot of device integration to HA

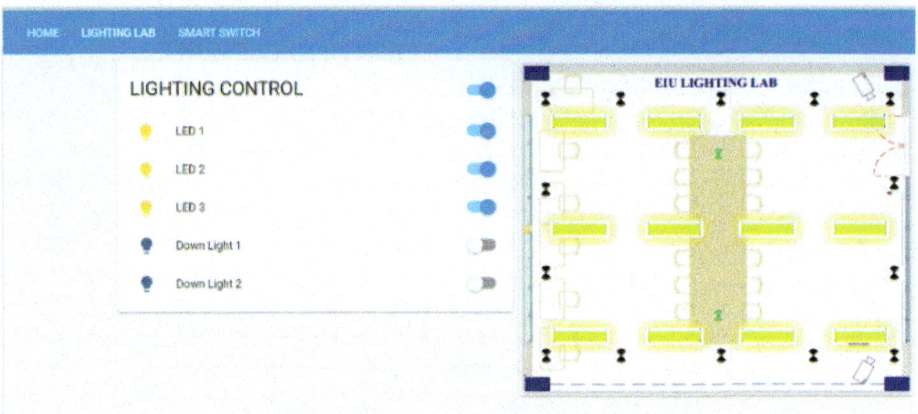

Fig. 5. The smart lighting control dashboard

At the user application level, we provide the control dashboard for managing the smart lighting system, as depicted in Fig. 5. This dashboard enables administrators to create lighting scenarios using the devices listed, which align with the lighting layout design. Each luminaire in the lighting system can be operated independently by the user. The logging features of this system enable users to access a record of all previous activities (Fig. 6).

2) **Mobile-based Application Interface**

An alternative method is to create a mobile application interface to enable users to interact with the lighting system. The structure of the mobile-based application interface is identical to the web-based interface, as depicted in Fig. 7. After successfully logging into the system, the authorized users can access the dashboard on their screen, which allows them to manually interact with the system and control individual appliances.

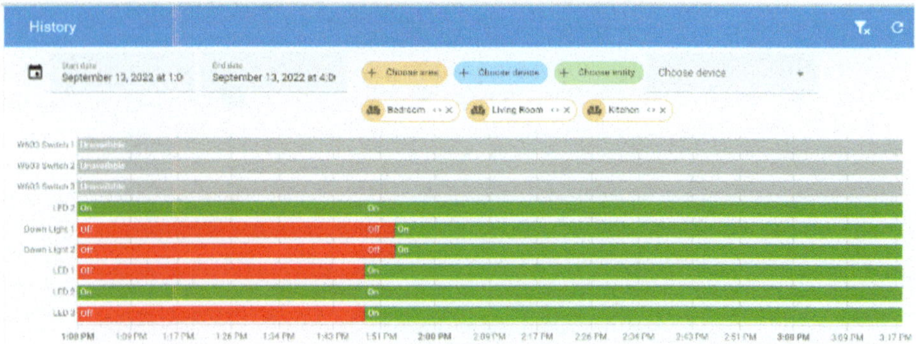

Fig. 6. The logging records of the lighting device operations

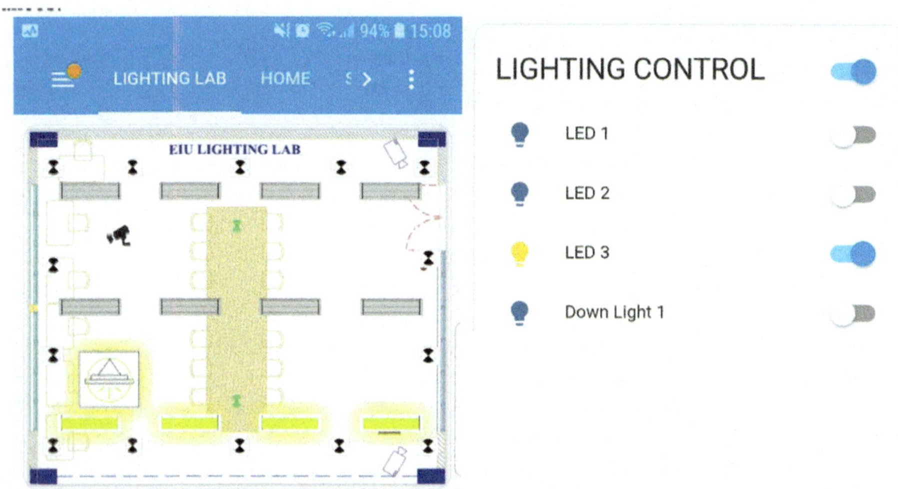

(a) Lighting System Topology (b) Control menu

Fig. 7. The mobile-based application interface for lighting control

3.2 Lighting Design with DIALux

To evaluate the energy efficiency of the lighting design, we performed simulations using DIALux as shown in Fig. 8. It is important to note that the simulation design aligns with the configuration parameters of the physical layout design.

We employed 12 LED Tube lights, each offering a luminous flux of 3687 lm and consuming 26 W, alongside 14 LED Downlights with a luminous flux of 1124 lm and a power rating of 9 W. Consequently, the cumulative power consumption for generating a luminous flux of 59980 lm amounted to 438 W. The maximum energy savings achieved were approximately 36.8%, primarily through the implementation of daylight control methods. This lighting system configuration can maintain an average illuminance of 535 lx. The floor plan view, which displays light distribution with photometric data,

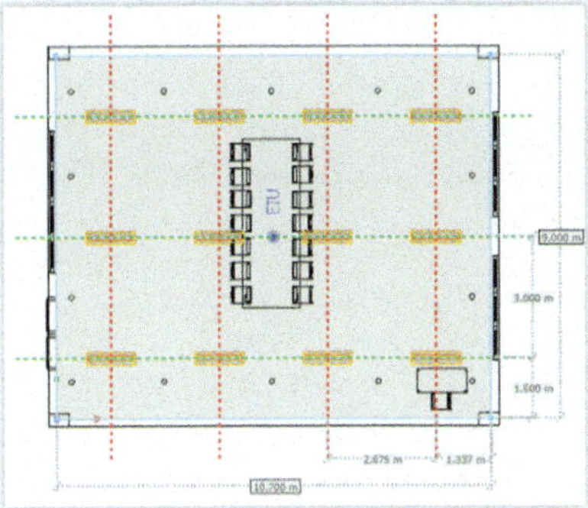

(a) 2D view

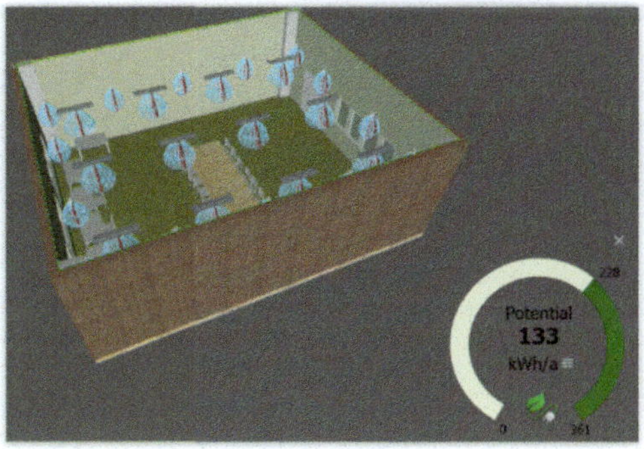

(b) 3D view

Fig. 8. The floor plan view design of the lighting laboratory using DIALux.

is presented in Fig. 9. The visual task area (ET1) is defined as the space required for performing visual tasks, situated at a height of 0.720 m and with a surrounding area of 0.5 m. Adjacent to this area is the surrounding area (ES1), which directly borders the visual task area and should have a minimum width of 0.5 m. The background area (EB1) extends from the direct ambient area to the room borders, with a width of at least 3 m. This area is situated horizontally at floor level.

Table 2 presents the annual energy consumption of the smart lighting system configured with a target illuminance of 500 lx, operating from 8:00 to 17:00, five days a week. The implementation of daylight harvesting and occupancy control solutions results in approximately over 70% energy savings, leading to significant reductions in energy costs and CO_2 emissions.

4 Concluding Remarks

In this study, we have presented the design and implementation of a system for smart lighting intended for both research and educational purposes. The system integrates Philips Dynalite and Home Assistant on a Raspberry IoT platform, with subsequent simulation using DIALux to evaluate energy efficiency. The results have indicated that LED lamps and Home Assistant are the most suitable choice for indoor applications and are therefore recommended for practical deployment. Future works include expanding data collection, exploring correlations between environmental parameters and occupancy, and incorporating additional electrical appliances to assess real-world effectiveness.

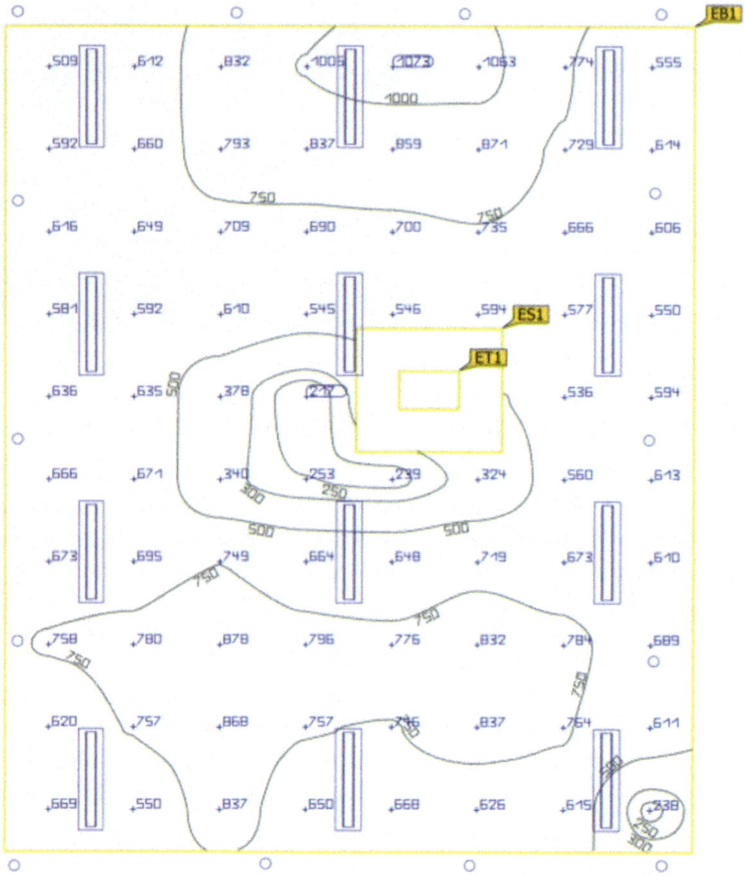

Fig. 9. The light distribution with photometric data.

Table 2. Annual energy consumption of the proposed lighting system

	Uncontrolled	Controlled	Saving
Energy consumption (kWh/a)	102	29.5	73.0
LENI (kWh/(m^2 * a))	1.06	0.31	0.76
Costs (€/a)	30.75	8.85	21.90
CO_2 (kg/a)	41.1	11.8	29.3

LENI: *Lighting energy numeric indicator*

Acknowledgement. This research is financially supported by Eastern International University, Binh Duong Province, Vietnam.

References

1. Sathya, P., Natarajan, R.: Design of energy efficient lighting system for educational laboratory. In: International Conference on Green Computing, Communication and Conservation of Energy (ICGCE), pp. 428–432 (2013)
2. Bannamas, S., Jirapong, P.: An intelligent lighting energy management system for commercial and residential buildings. In: IEEE Innovative Smart Grid Technologies - Asia (ISGT ASIA), pp. 1–6 (2015)
3. Cheng, Y., Fang, C., Yuan, J., Zhu, L.: Design and application of a smart lighting system based on distributed wireless sensor networks. Appl. Sci. **10**, 8545 (2020)
4. Floris, A., Porcu, S., Girau, R., Atzori, L.: An IoT-based smart building solution for indoor environment management and occupants prediction. Energies **14**, 2959 (2021)
5. Burunkaya, M., Duraklar, K.: Design and implementation of an IoT-based smart classroom incubator. Appl. Sci. **12**, 2233 (2022)
6. Saranya, G., Duraiarasu, E., Manoj, S.: Universal smart lighting system. In: IEEE Sustainable Smart Lighting World Conference & Expo (LS18), pp. 1–5 (2023)
7. Philips Dynalite. https://www.dynalite.com/
8. Home Assistant. https://www.home-assistant.io
9. Zharikov, A.M., Kozin, D.A., Nekrasov, P.V.: Design and implementation of Home Assistant and TouchGFX interaction based on STM32 microcontroller. In: Moscow Workshop on Electronic and Networking Technologies (MWENT), pp. 1–3 (2022)
10. Duijnhoven, J.V., Bhusal, P., Bizjak, G., Kobav, M.B., Aarts, M.P.J.: Developing energy efficient and smart lighting education in Vietnam and Myanmar. In: IEEE Sustainable Smart Lighting World Conference & Expo (LS18), pp. 1–4 (2023)

Designing a Student Administration System Utilizing the YOLO-Based Model

Phan Van Vinh(✉), Truong Le My Thanh, and Huynh Thanh Dat

School of Computing and Information Technology, Eastern International University, Binh Duong City, Vietnam
{vinh.phan,thanh.truonglemy,dat.huynh.set16}@eiu.edu.vn

Abstract. The objective of this paper is to develop and implement a student administration system utilizing the YOLO model for information extraction from student cards, integrated into a web application interface using the FastAPI framework. Despite the relatively small size of our student card dataset and the presence of numerous small objects in close proximity within the student card images, we achieved a training mAP of over 83% and an inference of over 89% accuracy with 8-labeled classes. The positive testing results show how crucial it is to use object detection and recognition in a student administration system, providing the groundwork for implementing an automatic attendance system in schools later on.

Keywords: YOLO · object detection · student card · administration system

1 Introduction

Identifying and recognizing objects within images or videos, known as object detection and recognition, is a widely practiced task in computer vision. This involves identifying and localizing specific areas of interest in an image or a video while also categorizing the objects found within those areas. This task becomes more complex when dealing with images containing multiple objects, making it more advanced than standard image classification. For example, a computer examines an image, not just identifying its contents but also outlining each recognized object with boxes, providing precise location information for each one. Fundamentally, deep learning is widely used in object detection to detect and precisely locate objects within images and videos. This technology is widely used, from self-driving cars knowing where pedestrians are, to security cameras spotting intruders, and even a smartphone recognizing owner's face to unlock itself. A typical model of object detection process includes three important parts: Backbone layer, Neck layer and Head layer [1].

First, a backbone layer is typically a convolutional neural network (CNN) acts as a feature extractor in the object detection process (Fig. 1). The backbone layer processes the input image and gradually reduces its spatial dimensions while increasing the depth of feature maps. This allows the model to capture both low-level and high-level features, which are essential for accurate object detection. All of the backbone models are basically classification models, such as VGG [2], ResNet [3], Darknet, MobileNet [4]. Second,

M. Pagac et al. (Eds.): SCFF 2024, LNNS 1062, pp. 306–316, 2024.
https://doi.org/10.1007/978-3-031-65656-9_31

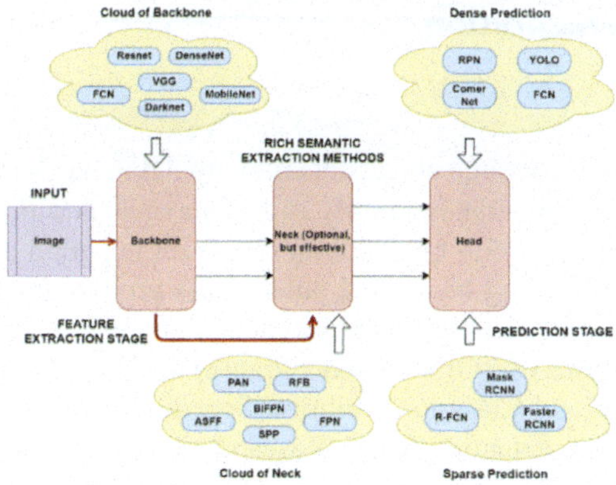

Fig. 1. A typical model of object detection with deep learning.

the neck layer is a critical component that comes after the backbone network. Its primary purpose is to generate feature maps at multiple scales, which are essential for detecting objects of various sizes within an image. Finally, the head layer, often referred to as the object detector, plays a vital role in generating predictions related to detected objects. It relies on the feature maps generated by the backbone and neck layers. There are two primary categories: two-stage detectors and one-stage detectors which further branch into anchor-based and anchor-free detectors. Moreover, specific types of heads are suited for distinct networks. For instance, well-known heads for two-stage sparse object detectors are RCNN [5], Fast RCNN [6], MSCNN [7], Faster RCNN [8]. Similarly, various dense detector heads include YOLO [9], RPN [4]. Among them, YOLO is a highly popular object detection model for its impressive fusion of speed and accuracy. First introduced in 2015, YOLO has evolved over time, with the latest version of YOLOv8 [10] introduced in 2023. A comparative study on YOLO models was introduced in [11]. Its applications span across diverse fields in our daily lives, including real-time object detection [12], surveillance system [13], hand gesture recognition [14], face recognition, self-driving cars, speech detection, disease detection, card detection recognition, etc. However, it is important to note that these applications can impose significant demands on processing time, particularly on devices with constrained resources.

In this project, our focus will involve the utilization of YOLO for the extraction of data from student cards. The acquired information will then be presented through a web application interface. Furthermore, this application has been deployed to enhance and simplify the management of students, leading to better efficiency and improved user experience. In the subsequent sections, we outline the project design and implementation of the proposed method in Sect. 2. Section 3 presents the outcomes of the implementation. Lastly, in Sect. 4, we provide concluding remarks.

2 Project Implementation

2.1 YOLO Model

YOLO is a well-known real-time object detection model in the field of computer vision. It revolutionized the way machines identify and locate objects within images and videos. YOLO takes a unique approach by handling the entire image in a single pass, making predictions for both bounding boxes and object class probabilities simultaneously [9]. This approach significantly accelerates object detection, making it suitable for real-time applications. Key features of YOLO include: Single Pass Processing, Grid-Based Detection, Bounding Boxes and Class Probabilities, Speed and Efficiency. YOLO is widely used for various object detection tasks. Researchers and developers have continued to advance the YOLO algorithm, creating different versions, introducing architectural improvements and innovations. The key models in the YOLO family are presented in Table 1.

For our real-time applications, we are implementing both YOLOv4 [15] and the newest YOLOv8 to see how they compare. We are also using YOLOv4-tiny [9] because it is designed for faster real-time object detection and quicker training, even though it might not be as accurate.

2.2 Project Implementation

Our project implementation is structured into three stages: creating a custom dataset, training the dataset, converting and deploying it within a web application (Fig. 2).

1) **Phase 1: Creating the custom dataset**

In preparation for the training dataset, we gathered over one hundred images of EIU student cards (Fig. 3). When using Darknet to train models, the size of input image is not a concern, and we can retain the original dimensions. However, the crucial aspect is the input resolution, as it determines the number of pixels that the model will use for learning and predictions. A higher resolution can enhance accuracy by aiding the model in detecting smaller objects, but it may lead to slower training and inference times as a trade-off.

We utilize the Roboflow Annotation application to label the appropriate objects in the custom dataset (Fig. 4). Subsequently, the input dataset is divided into the train set and test set. An important step in this stage involves applying pre-processing and augmentation to our labeled images. This process serves two main purposes: reducing training time and enhancing performance. The pre-processing involves applying image transformations to all images, while the augmentation generates new training examples for the model to learn from.

2) **Phase 2: Training the custom dataset**

Once the initial phase is completed, we proceed to train the model using the custom dataset on Collab. This platform is chosen because it offers the advantage of convenient model saving and loading, making the process smoother and less time-consuming. It is

Table 1. The YOLO family models.

Model	Description
YOLOv1 (2015)	– Process image at once to predict bounding boxes and class probabilities – Predict two bounding boxes per grid cell – Real-time performance but had limitations with small objects
YOLOv2 (2016)	– Improve accuracy with anchor boxes for precise bounding box predictions – Maintain two bounding boxes per grid cell – Support detection of over 9000 object categories
YOLOv3 (2018)	– Enhance accuracy further with multiple detection scales – Predict three bounding boxes per grid cell – Achieve a balance of speed and accuracy on smaller object
YOLOv4 (2020)	– Introduce architectural enhancements like CSP Darknet53 and PANet for improved performance – Maintain three predicted bounding boxes per grid cell – Optimize speed and accuracy
YOLOv5 (2020)	– Use the EfficientDet architecture – Focus on lightweight models for real-time applications – Achieve impressive speed and accuracy
YOLOv6 (2022)	– Employ a novel CNN architecture known as SPP-Net (Spatial Pyramid Pooling Network) – Ability to manage objects with diverse sizes and aspect ratios
YOLOv7 (2022)	– Introduce a new multi-scale training strategy, where the model is trained on images at various scales and subsequently merges the predictions – Handle objects with diverse sizes and configurations more effectively
YOLOv8 (2023)	– Two primary modifications have been implemented: + Anchor-free Detection: Instead of calculating the offset from a predefined anchor box, the model now directly predicts an object's center + Mosaic Augmentation: Four distinct images are combined and used as input for the model

important to note that this stage of training might take some time before producing the desired result, which is the weights model.

3) **Phase 3: Convert and Deploy the trained model into the website**

The outcome of training the custom dataset is a trained model in the Darknet weight format. However, for easier deployment, we must convert it into other formats such as PyTorch, tflite, and TensorFlow. These alternative formats facilitate seamless integration and deployment of the model in different environments and frameworks. In the final step, we take the trained model in TensorFlow format and deploy it within a web application. This web application is designed to extract relevant information from the EIU student card. The extracted information serves a valuable purpose in the student administration

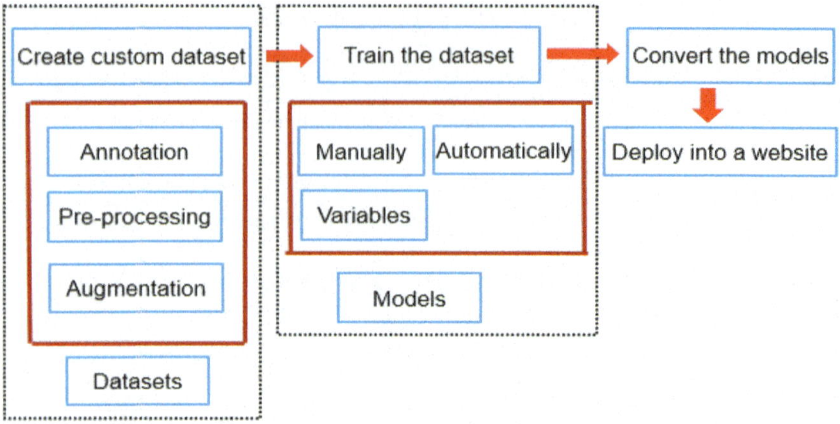

Fig. 2. Project implementation process.

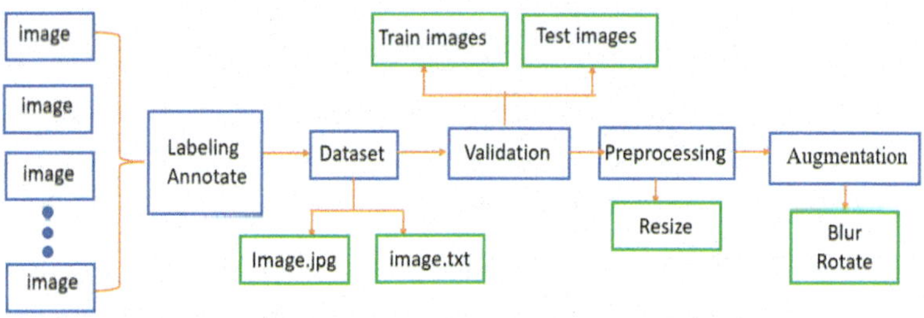

Fig. 3. The procedure of creating the customize dataset.

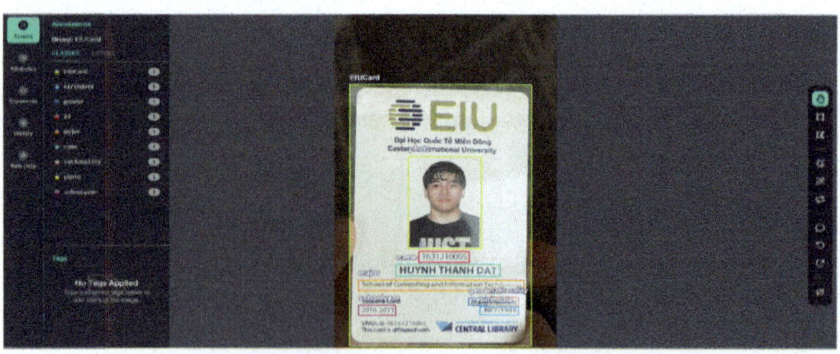

Fig. 4. Roboflow annotation.

system, particularly when adding new students to the student list. By utilizing this system, the process of managing student data becomes more efficient and streamlined.

3 Implementation Results and Discussion

3.1 The Web Application Interface

Application Programming Interfaces (APIs), are like the building blocks of modern software. They make different programs talk to each other, share information, and work together smoothly. APIs also make it easier for developers to create software and keep it up to date. In Python, there are several frameworks available for constructing APIs, each with its own strengths, such as Django, Flask, and FastAPI. In this project, FastAPI is used to build the web application interface for our student administration system. FastAPI is a modern, high-performance, and easy-to-use web framework for building APIs with Python. It is designed to streamline the development of web-based applications and services, particularly RESTful APIs, by providing developers with powerful tools and features. FastAPI has gained popularity within the Python community due to its speed, simplicity, and automatic generation of interactive API documentation. Using FastAPI to build a web application involves several steps, from setting up the development environment to defining routes and handling requests.

As a fundamental component, the website provides a login page which serves as the entry point for users to access the website functionalities. Upon accessing the Homepage, users are presented with two options:

- **EIU Card Extractor**: This option allows users to test the project and perform card extraction.
- **View Result**: By selecting this option, users can view the extracted data that has been processed and obtained from the EIU student cards.

3.2 Student Card Extraction

To perform the student card extraction feature, complete the following procedures:

- *Step 1*: To initiate data extraction, simply drag the EIU Card image to the left side of the interface.
- *Step 2*: Click the *Extract* button to start the extraction process.
- *Step 3*: Allow some time for the extractor to process the information, and the results will be presented once the extraction is complete. Please be patient during this process (Fig. 5).

3.3 Results of Student Card Extraction

The result of EIU student card extraction is illustrated in Fig. 6.

Figure 7 shows the list of students whose information is extracted using card detection method. Moreover, the student information can be modified manually as shown in Fig. 8.

3.4 Discussion

The EIU student card contains multiple valuable pieces of information that can be extracted, including student ID, student name, birthdate, gender, major, student photo,

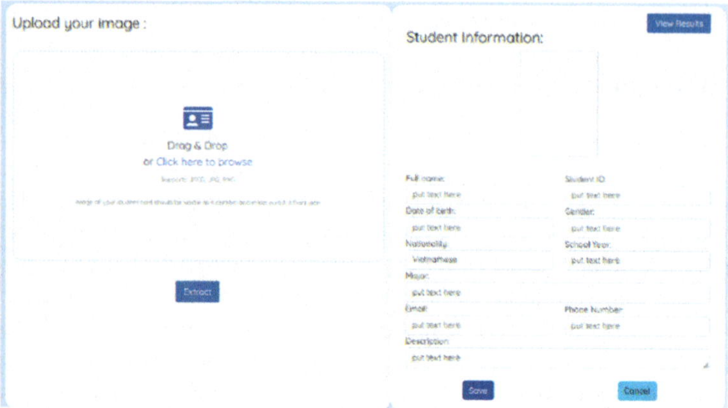

Fig. 5. An example of student card extraction interface.

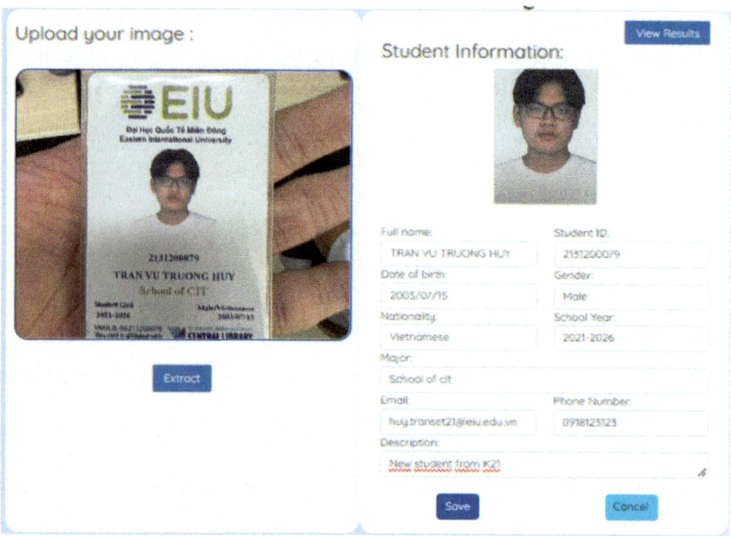

Fig. 6. The result of student card extraction.

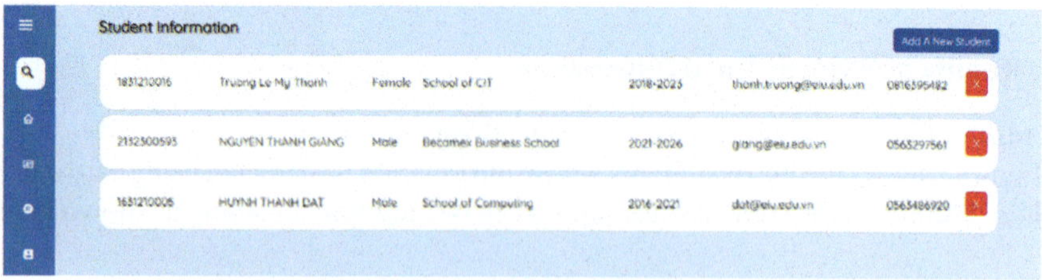

Fig. 7. List of student entries.

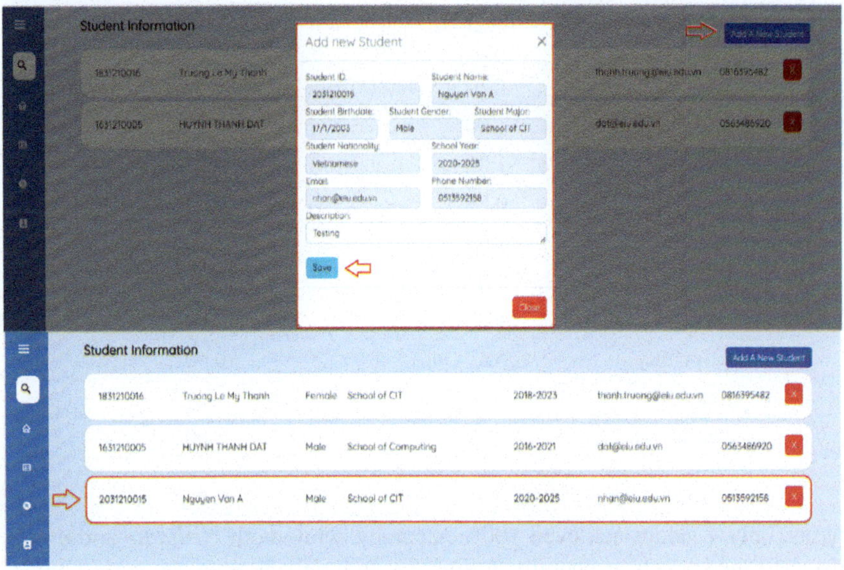

Fig. 8. Adding a student entry manually.

school year, and nationality. However, it is important to carefully determine which specific information needs to be extracted, as it can significantly impact the overall performance. Obtaining more information from the EIU student card can lead to slower inference times and potentially reduced accuracy. Therefore, a balance between the required information and performance considerations is essential.

1) **Case study 1: Training EIU student card dataset with YOLOv4**

In the first study, each student card image is labeled with 9 classes, which comprise EIUCard, birthdate, gender, id, major, name, photo, school year, and nationality. However, upon training, we have observed that the overall performance for the dataset with 9 classes is not satisfactory, particularly with the nationality label achieving only 50.46% accuracy, as depicted in Table 2.

The reason behind this can be attributed to the spatial proximity of the *nationality* field and the *birthdate* field on the EIU student card. Due to their close proximity, it becomes challenging to precisely detect both fields during the inference phase if they are labeled together. The close proximity of these fields creates ambiguity and makes it difficult for the model to distinguish between them accurately. As a result, the model performance in detecting both the *nationality* and *birthdate* fields simultaneously is affected, leading to the observed decrease in accuracy for the *nationality* label during training.

In the next study, we have made a modification by labeling the EIU student cards with only 8 classes, and specifically, we have removed the *nationality* field from the training data. The outcomes of the trained model are illustrated in Table 2. After conducting further analysis, we discovered a significant improvement in the extractor's accuracy when training with only 8 classes, excluding the nationality field. In the previous training with

Table 2. Result of training EIU student card dataset with YOLOv4

#	Class ID	mAP (8 classes)	mAP (9 classes)
1	EIUCard	1.00	1.00
2	birthdate	1.00	0.73
3	gender	0.83	0.83
4	id	1.00	1.00
5	major	0.91	0.92
6	name	1.00	1.00
7	nationality	*NA*	0.50
8	photo	1.00	1.00
9	schoolyear	1.00	1.00

9 classes, only 4 values achieved 100% accuracy, while both *birthdate* and *nationality* labels performed poorly, especially *nationality* with only 50.46% accuracy. Upon closer examination, we realized that the *nationality* value was consistently Vietnamese since the majority of EIU students are Vietnamese, and very few are foreigners. As a result, by removing the *nationality* field, the accuracy of the *birthdate* field dramatically increased to 100% (previously at 73.15%).

2) **Case study 2: Training EIU student card dataset with YOLOv8**

To evaluate the YOLOv8 performance on our custom datasets, we conducted training using the EIU student card dataset for a more meaningful comparison. The results of this training are presented in Table 3.

Table 3. Result of training EIU student card dataset with YOLOv8

#	Class ID	mAP (8 classes)	mAP (9 classes)
1	EIUCard	1.00	1.00
2	birthdate	0.75	0.76
3	gender	0.7	0.68
4	id	0.81	0.79
5	major	0.78	0.78
6	name	0.88	0.85
7	nationality	*NA*	0.76
8	photo	0.97	0.97
9	schoolyear	0.86	0.82

It is evident that training the dataset with 8 classes achieves better results compared to the 9-class training. Specifically, when training with only 8 classes, excluding the nationality field, we achieved better performance in key labeled objects such as gender (70%), id (81%), name (88%), and school year (86%). This improvement may be attributed to the close proximity of labeled objects in our student card dataset. For instance, objects like id, major, and name are grouped closely together, while gender, nationality, and birthdate are small and closely positioned, making them challenging to detect accurately. When objects are smaller and positioned close to each other, their training accuracy tends to decrease. It is worth noting that training our dataset with YOLOv8 did not achieve better results compared to YOLOv4. This may be due to the dataset's small size and the presence of numerous small objects in close proximity within our student card dataset.

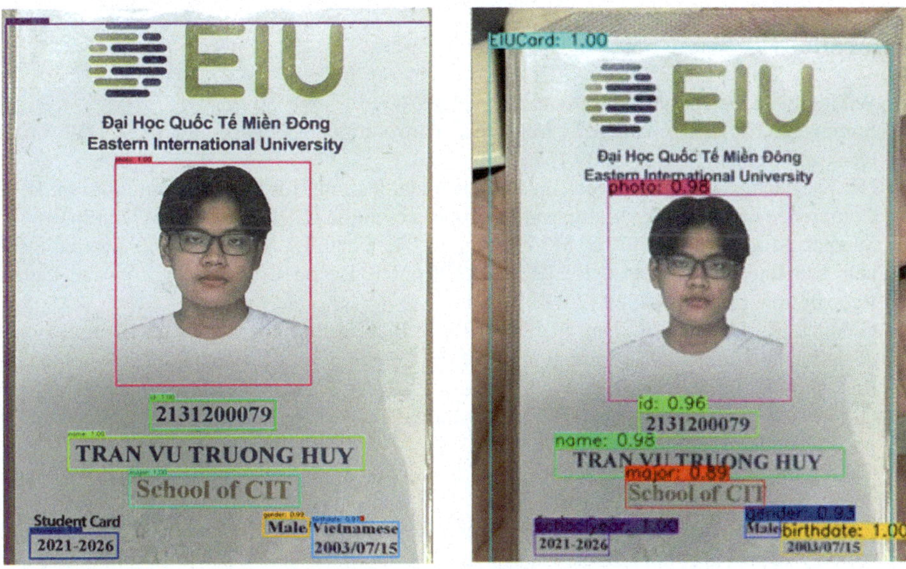

(a) Detection with 9-class trained models (b) Detection with 8-class trained models

Fig. 9. The result of student card inference using the trained models with two different classes.

In Fig. 9, we can see the performance comparison between the two models during inference across different labeled classes. It is clear that the model trained with 8 classes achieves better performance, especially in accurately recognizing the birthdate object. In Fig. 9(a), despite each object being detected with a high score, there was an overlap between the nationality and birthdate fields, resulting in the inability to extract information correctly with the text recognition module. Conversely, with the 8-class training model, detecting and extracting information from the birthdate field is straightforward, as depicted in Fig. 9(b).

4 Concluding Remarks

Within this research paper, we have introduced our methodology which focuses on customizing a trained model for detecting and recognizing the important information of a student card based on YOLO model. Additionally, we have effectively implemented the project as a web application, streamlining the extraction of data from these student cards. However, to enhance the speed and accuracy of the extracted information for practical applications, it is essential to expand the size of the student card dataset and explore implementing with other deep learning models to achieve better performance.

Acknowledgment. This research is financially supported by Eastern International University, Binh Duong Province, Vietnam.

References

1. Wiki: Object Detection. https://wikidocs.net/167699
2. Simonyan, K., Zisserman, A.J.: Very deep convolutional networks for large-scale image recognition (2014). arXiv:1409.1556
3. He, K., Zhang, X., Ren, S., Sun, J.: Deep residual learning for image recognition. In: IEEE Conference on Computer Vision and Pattern Recognition (CVPR), pp. 770–778 (2016)
4. Sandler, M., Howard, A., Zhu, M., Zhmoginov, A., Chen, L.: MobileNetV2: inverted residuals and linear bottlenecks. In: IEEE/CVF Conference on Computer Vision and Pattern Recognition, pp. 4510–4520 (2018)
5. Girshick, R., Donahue, J., Darrell, T., Malik, J.: Rich feature hierarchies for accurate object detection and semantic segmentation. In: IEEE Conference on Computer Vision and Pattern Recognition, pp. 580–587 (2014)
6. Girshick, R.: Fast R-CNN. In: IEEE International Conference on Computer Vision (ICCV), pp. 1440–1448 (2015)
7. Cai, Z., Fan, Q., Feris, R., Vasconcelos, N.: A unified multi-scale deep convolutional neural network for fast object detection (2016)
8. Ren, S., He, K., Girshick, R., Sun, J.: Faster R-CNN: towards real-time object detection with region proposal networks. IEEE Trans. Pattern Anal. Mach. Intell. **39**, 1137–1149 (2017)
9. Redmon, J., Divvala, S., Girshick, R., Farhadi, A.: You only look once: unified, real-time object detection. In: IEEE Conference on Computer Vision and Pattern Recognition (CVPR), pp. 779–788 (2016)
10. Ultralytics: YOLOv8. https://docs.ultralytics.com/
11. Balakrishnan, B., Chelliah, R., Venkatesan, M., Sah, C.: Comparative study on various architectures of YOLO models used in object recognition. In: International Conference on Computing, Communication, and Intelligent Systems (ICCCIS), pp. 685–690 (2022)
12. Lavanya, G., Pande, S.: Enhancing real-time object detection with YOLO algorithm. EAI Endorsed Trans. Internet Things **10** (2023)
13. Anish, A., Sharan, R., Hema Malini, A., Archana, T.: Enhancing surveillance systems with YOLO algorithm for real-time object detection and tracking. In: 2nd International Conference on Automation, Computing and Renewable Systems (ICACRS), pp. 1254–1257 (2023)
14. Herbaz, N., Idrissi, H.E., Badri, A.: Deep learning empowered hand gesture recognition: using YOLO techniques. In: 14th International Conference on Intelligent Systems: Theories and Applications (SITA), pp. 1–7 (2023)
15. Bochkovskiy, A., Wang, C.-Y., Liao, H.: YOLOv4: Optimal Speed and Accuracy of Object Detection (2020)

Enhancing Medical Insurance Pricing Prediction with SHAP-XGBoost for Informed Decision-Making

Danh Hong Le(✉)

Van Hien University, Ho Chi Minh City, Vietnam
danhlh@vhu.edu.vn

Abstract. This research explores the evolving field of predictive modeling in healthcare highlighting the continued interest of insurance companies in using Machine Learning (ML) techniques to improve operational efficiency. The author uses a set of regression based ML models incorporating different versions of Extreme Gradient Boosting (XGBoost) methods to predict medical insurance costs. Furthermore the study utilizes Explainable Artificial Intelligence (XAI) methods, Shapley Additive Explanations (SHAP) to identify and explain the key factors influencing medical insurance premium prices within the dataset. The dataset consists of 986 records from the KAGGLE repository and the models effectiveness is thoroughly assessed using various performance evaluation metrics such as R squared (R^2), Mean Absolute Error (MAE), Root Mean Squared Error (RMSE), and Mean Absolute Percentage Error (MAPE). Additionally a comparison is made between the results generated by XGBoost and Random Forest (RF) models, in determining the features affecting Premium Prices. Despite requiring computational resources the XGBoost model stands out as the top performer overall. The authors aim to offer insights to help policymakers, insurance providers and individuals looking for medical coverage make informed decisions. This will assist them in choosing policies that best suit their needs and preferences.

Keywords: Medical insurance pricing · Predictive · SHAP · XGBoost

1 Introduction

In times there has been a growing emphasis on developing effective premium structures in the health insurance sector through actuarial modeling of insurance claims. This focus is crucial for attracting and retaining policyholders and managing existing plan members efficiently. However creating a model for medical insurance costs faces significant challenges due to the complex interplay of various influencing factors. Factors such as characteristics, health status, geographical accessibility, lifestyle preferences and provider attributes play a substantial role in determining medical insurance expenses [1]. Additionally key elements like coverage extent, plan type, deductible amounts and customer enrollment age greatly impact the costs associated with medical insurance coverage.

M. Pagac et al. (Eds.): SCFF 2024, LNNS 1062, pp. 317–328, 2024.
https://doi.org/10.1007/978-3-031-65656-9_32

Health insurance serves as a mechanism for pre-payment and risk sharing to cover expenses related to illnesses including hospitalizations, medications and consultations with healthcare providers. The introduction of national health insurance programs has improved equitable access, to healthcare services and helps individuals mitigate financial burdens from illnesses [2]. Currently most health insurance systems operate under multiple financing structures. In a single payer system there is a health insurance pool while the multiple payer model consists of several separate pools. Both funding methods have differences. The single payer system usually involves increased government supervision of healthcare delivery focusing on fairness concerns. Conversely, the multi-payer model affords consumers the freedom to select from available insurance providers, fostering innovation and competition within the industry [3].

In countries health insurance systems often blend elements from both payer and multi payer models creating various hybrid setups. These setups can generally be categorized into three types [4]: i). Health Maintenance Organization (HMO): This model lets insured individuals choose from a group of doctors affiliated with the HMO or contracted by it giving them the freedom to select their healthcare providers; ii). Preferred Provider Organization (PPO): PPO plans offer a list of contracted healthcare providers. Typically reimbursement follows a 60/40 split, where the insurer covers 60% of expenses and the insured pays the remaining 40%; iii) High Deductible Health Plan (HDHP) With a Health Savings Account (HSA): HDHPs involve policyholders setting up a health savings account from which treatment costs are deducted based on a percentage. This setup often leads to lower premiums. Additionally the text discusses two methods used by medical insurance companies to reimburse policyholders [5]: i). Cashless Treatment: In this system the insurance company directly handles payments, with hospitals removing the need for payment, by policyholders; ii). Reimbursement: Here policyholders typically pay for medical expenses upfront. Then request reimbursement from their insurance provider.

Machine learning (ML) models, often referred to as "black boxes," in the field of intelligence (AI) focusing on its ability to create systems that can learn independently from large datasets. These "black box" models pose a significant challenge, as their decisions are often difficult to comprehend, leading to concerns about trust, accountability, and ethical implications [6]. While AI includes technologies like expert systems, deep learning and robotics ML specifically concentrates on learning from data driven approaches as supported by research studies [7]. The widespread adoption of ML techniques across industries is credited to advancements in methodologies improved computing power of GPUs and the availability of diverse datasets according to experts [8,9]. Despite these progressions the text highlights that there is still more to explore in the potential of AI and ML leading to research in this area [10]. ML tools aid decision making by making predictions based on data and improving performance as they receive data, evident in medical applications as mentioned in a specific source [11]. In years health insurance companies have increasingly utilized AI and ML to better identify individuals who need protection and streamline their insurance processes. The text emphasizes that one key strength of ML in healthcare management is its skill, in reasoning and quick trend analysis [12]. By utilizing intelligence to create accurate risk assessments and pinpoint clients requiring tailored attention insurance companies can allocate resources, towards policyholders instead of bureaucratic tasks. Systems that integrate data analysis enhance assessments and provide recommendations can significantly reduce the need, for human analysts and administrative costs. Moreover, the text highlights extensive research conducted on utilizing ML systems to forecast medical insurance costs, which employed various regression techniques and machine learning algorithms [13].

The authors in [14] introduced the SHapley Additive exPlanations (SHAP) method as a way to understand how different features contribute to an instance. This approach has been widely used to interpret social phenomena [15]. Using principles, from game theory SHAP assigns zero importance to features that don't impact the models prediction [16]. By analyzing SHAP values researchers can determine the relationship between each variable and the target variable, whether it's positive or negative. Moreover SHAP offers interpretability by assigning a value to all features [17]. Recent studies have shown an uptick in the use of eXplainable Artificial Intelligence (XAI) methods in fields, such as predicting medical insurance costs. In a study how Machine Learning models can predict costs for employer groups during health insurance renewals [18]. They focused on groups that could benefit from cost saving measures and developed models, at both patient and employer group levels. Based on a study involving data, from 14 million patients researchers found a 20% improvement in the effectiveness of insurance pricing models compared to existing ones. They successfully identified 84% of cost saving opportunities showcasing how machine learning systems can enhance the accuracy and fairness of health insurance pricing. Furthermore they used the SHAP XAI method with the LightGBM model to provide explanations for rate adjustments at the individual member level making the model more reliable [19]. The authors in [20] introduced a machine learning approach for predicting health insurance prices using regression models like linear regression and random forest regression. Researchers improved medical insurance pricing models by incorporating XAI techniques like Microsoft InterpretML, LIME, and SHAP, and demonstrated superior performance using machine learning methods such as Random Forest (RF), Gradient Boosting Machine (GBM), Artificial Neural Network (ANN), and Logistic Regression (LR) in predicting high-cost patients from healthcare claims data [21]. The study particularly emphasized the utilization of the SHAP XAI method exclusively for the RF model to assess variable influences on predicted class probabilities.

In a study, the author in [22] compared tree based classifiers to improve risk assessment for life insurance companies using predictive analytics. Their findings indicated that XGBoost outperformed other methods, underscoring the significance of model interpretability for stakeholders in the insurance sector. Using the SHAP method, they further analysed how dataset features influenced overall model performance. The comparison of various XAI methods underscored machine learning's potential in addressing healthcare challenges, particularly in health insurance, by utilising Kaggle datasets to provide cost-effective solutions amidst growing demands and technological complexities. This research offers contributions:

- It forecasts medical insurance expenses using an ensemble learning approach with the XGBoost model.
- It evaluates the effectiveness of two XAI techniques, XGBoost and RF, in explaining how a black box model operates.

The papers structure is organized as follows; Sect. 2 covers medical insurance concepts and relevant literature reviews. Section 3 details the methodology used in the research. Section 4 presents outcomes and determinant analyses. Finally Sect. 5 summarizes the findings.

2 Methodology

2.1 Dataset and Preprocessing

The dataset utilized for medical insurance cost analysis was obtained from KAGGLE's repository in 2021 [23]. This dataset contains 986 entries, with 11 attributes. Following this data preprocessing steps were taken, which are essential in any data focused project seeking to uncover insights. During this phase checks were carried out to address missing values and ensure dataset uniqueness. Moreover the data was standardized using the Standard Scalar method.

Table 1. Statistical summary of the features

Features	Description	Mean	STD	Min.	25%	50%	75%	Max.
Age	Years old at the time of data collection	41.75	13.96	18.00	30.00	42.00	53.00	66.00
Diabetes	Whether the customer has been diagnosed with diabetes	0.42	0.49	0.00	0.00	0.00	1.00	1.00
BloodPressureProblems	Categorization of the customer's blood pressure based on pre-defined thresholds	0.47	0.50	0.00	0.00	0.00	1.00	1.00
AnyTransplants	Presence or absence of any major organ transplants in the customer's medical history	0.06	0.23	0.00	0.00	0.00	0.00	1.00
AnyChronicDiseases	Presence or absence of specific chronic illnesses (e.g., asthma, heart disease)	0.18	0.38	0.00	0.00	0.00	0.00	1.00
Height	Customer's height	168.18	10.10	145.00	161.00	168.00	176.00	188.0
Weight	Customer's weight	76.95	14.27	51.00	67.00	75.00	87.00	132.00
KnownAllergies	Presence or absence of any known allergies	0.22	0.41	0.00	0.00	0.00	0.00	1.00
HistoryOfCancerInFamily	Whether any blood relative of the customer has been diagnosed with cancer	0.12	0.32	0.00	0.00	0.00	0.00	1.00
NumberOfMajorSurgeries	Total number of major surgeries the customer has undergone	0.67	0.75	0.00	0.00	1.00	1.00	3.00
Premium Price	Cost of the insurance premium based on various risk factors, including the listed elements	24336.71	6248.18	15000.00	21000.00	23000.00	28000.00	40000.00

2.2 Exploratory Data Analysis

Exploratory Data Analysis (EDA) was performed to explore the dataset aiming to uncover hidden patterns identify anomalies, assumptions and guide the selection of suitable ML techniques for solving the specific problem. The Statistical summary table provided insights into the distribution of features within the dataset. Additionally a Pearson correlation Heatmap (referenced as Fig. 1) was utilized to evaluate relationships among features and their correlations. The Heatmap indicated that aside from age there were no correlations, among variables. Such discoveries are quite usual; as, per the World Health Organization (WHO) getting older often leads to increased use of healthcare services and expenses.

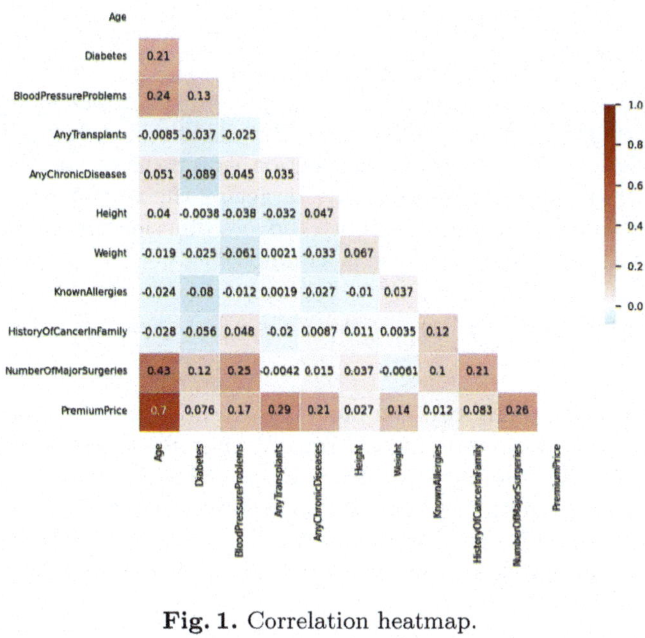

Fig. 1. Correlation heatmap.

2.3 eXtreme Gradient Boosting (XGBoost) Model

After the processing we selected all the features, for our study. Following that we split the dataset into two parts; one for training and the other for testing with 75% of the data used for training and 25% for testing purposes. During the training phase we created models to predict medical insurance costs while we used the testing dataset to evaluate how well the regression model performed. This approach demonstrates how ensemble models are effective in predictive healthcare modeling as discussed in [24]. In the section a brief overview of the XGBoost model that was implemented is presented.

The XGBoost model was designed to provide an scalable implementation of gradient boosting techniques first introduced in [25]. It is widely accepted as a tool among gradient boosted trees algorithms due to its user nature as an open source platform and its effectiveness. Gradient boosting, a type of learning method aims to make predictions of a target variable by combining forecasts from simpler models that are less complex. As mentioned in reference [26], XGBoost generally delivers predictions that outperform those from RF model is comparable to results, from networks. One key feature of XGBoost is its use of threading strategies which optimize CPU core usage to improve overall performance and computational speed compared to traditional gradient boosting methods.

In regression analysis using gradient boosting, regression trees are utilized as fundamental learners, with each tree assigning input data points to leaf nodes holding continuous scores, while the model's objective function combines a convex loss function capturing prediction disparities with a penalty term addressing model complexity. Through iterative training, the algorithm integrates new trees to predict residuals or errors from prior trees. For a given dataset with n examples and m features, denoted $D = \{(x_i, y_i)\}$ $(|D| = n, x_i \in R^m, y_i \in R)$, a tree ensemble model is constructed. Let $\hat{y}_i^{(t)}$ represent the prediction of the i-th instance at the t-th iteration, l is a differentiable convex loss function that quantifies the disparity between the prediction $\hat{y}_i$ and the target y_i, $\Omega()$ penalizes the complexity of the model, specifically targeting the regression tree functions. The objective function (comprising both the loss function and regularization) at iteration t that we aim to minimize is as follows:

$$\mathcal{L}^{(t)} = \sum_{i=1}^{n} l\left(y_i, \hat{y}_i{}^{(t-1)} + f_t\left(\mathbf{x}_i\right)\right) + \Omega\left(f_t\right). \tag{1}$$

In our research we assessed the models using four metrics to evaluate their performance; R^2, Mean Absolute Error (MAE), Root Mean Squared Error (RMSE) and Mean Absolute Percentage Error (MAPE). R^2 also known as the coefficient of determination measures how well a model fits by looking at how much of the predicted price's explained by the features. The MAE provides a reflection of prediction errors while the RMSE indicates how well a regression model predicts the value of a response, by measuring the standard deviation of residuals. The RSME evaluate errors based on the value being predicted such as premium prices in our scenario. The MAPE calculates errors in percentage form showing the percentage difference between predictions and their intended targets in the dataset. It can be thought of as MAE presented as a percentage. As noted in reference [27], a MAPE below 10% suggests quality modeling. Moreover in a situation an R^2 score nearing 100% signals results and signifies a more precise and superior performing model. These metrics can be represented visually as shown below:

$$R^2 = 1 - \frac{\sum_{i=1}^{n} \left(y_i - y_i\right)^2}{\sum_{i=1}^{n} \left(y_i - \bar{y}_i\right)^2}, \tag{2}$$

$$MAE = \frac{1}{n} \sum_{i=1}^{n} \left|\left(y_i - y_i^p\right)\right|, \tag{3}$$

$$RMSE = \sqrt{\frac{1}{n}\sum_{i=1}^{n}(y_i - y_i^p)^2}, \tag{4}$$

$$MAPE = \frac{1}{n}\sum_{i=1}^{n}\left|\left(\frac{y_i - y_i^p}{y_i}\right)\right| * 100, \tag{5}$$

where $\bar{y} = \frac{1}{n}\sum_{i=1}^{n} y_i$, n denotes the sample size, y_i^p represents the prediction for the ith sample, and y_i denotes the actual value corresponding to the mean of the sample.

2.4 Shapley Additive Explanations (SHAP)

Shapley additive explanations (SHAP) serves the purpose of elucidating a model f at a specific individual point x^* through a value function denoted as e_S, $e_S = E\left[f(x) \mid x_S = x_S^*\right]$, where S represents a subset of $S \subseteq \{1, \ldots, p\}$. This approach, called SHAP allows for an examination of how each features impact varies across space and the linear connections, within the dataset on medical insurance costs. By assessing the importance of each factor through changes in variables SHAP offers insights, into feature significance making it easier to generate feature dependency graphs and conduct interaction analyses as mentioned in [28]

$$I_j = \sum_{i=1}^{n}\left|\phi_j^{(i)}\right| \tag{6}$$

where j is denoted by ϕ_j and calculated as the weighted average over all possible subsets S, $\phi_j^{(i)}$ represents the SHAP value of the j-th feature for instance i, $\phi_j(val) = \sum_{S\subseteq\{x_1,\ldots,x_p\backslash\backslash\{x_j\}} \frac{|s|!(p-|s|-1)!}{p!}\left(val\left(S \cup \{x_j\}\right) - val(S)\right)$. Here, p denotes the number of features, S refers to a subset of these features, x represents the feature values of a specific instance in the explained model, and $val(S)$ indicates the prediction for the feature values within set S.

3 Results and Discussions

3.1 The Overview of the Model Outcomes on the Test Dataset

The results, from the models shown in Table 2 demonstrate performance across the models that were evaluated. Specifically the XGBoost model utilized resources compared to the RF models. Notably the XGBoost model outperformed the RF model with a R^2 score of 87.290% and an RMSE of 2229.842 showcasing its superior predictive accuracy. While the RF model showed MAE and MAPE scores than the XGBoost models it's crucial to consider that RMSE penalizes deviations more harshly than MAE does. The high R^2 score achieved by the XGBoost model highlights its enhanced ability to explain variations in data compared to the RF models. With a Premium Price of 2421.753 based on Table 1 data all models displayed predictive capabilities.

Table 2. Overview performance outcomes on the test dataset

Model	MAE	RMSE	R^2 (%)	MAPE (%)	Elapsed time (s)	Memory used (MB)
XGBoost	1439.805	2229.842	87.290	4.975	5267.084	6.760
RF	1381.870	2421.753	83.863	5.951	54.978	0.892

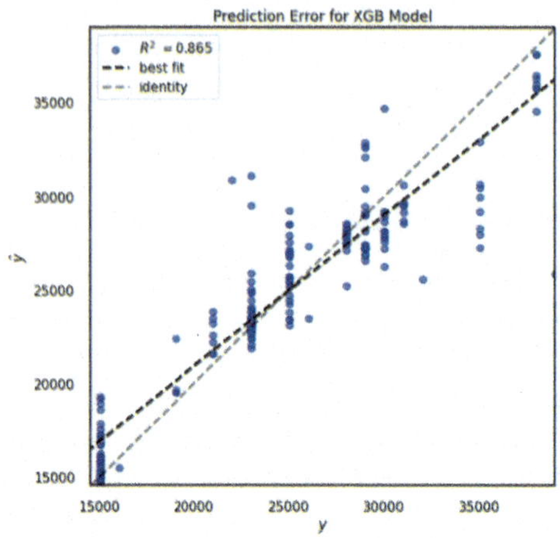

Fig. 2. Prediction error for XGBoost

Furthermore using Yellowbricks Prediction Error Visualizer we plotted targets from the dataset against predicted values. The plot shown in Figs. 2 and 3 illustrates how well each models predictions align with performance, along a 45° line.

3.2 The Feature Importance Analysis Conducted Using SHAP

The SHAP summary plot is a tool, for visualizing how various features impact the models predictions especially when it comes to predicting PremiumPrice. It combines SHAPley values. Feature importance to show the contributions of features, across the models output range. In Figs. 4 and 5, we start to see some insights, about these connections. The summary plot of the XGBoost model shows that when features like "Age", "BMI", "AnyTransplant", "AnyChronicDiseases", "HistoryOfCancerInFamily" and "BloodPressureProblems" have values the SHAP value also increases. This pattern suggests that higher values in these features are linked to premium prices. The SHAP analysis conducted on the RF model revealed that the impact of the "BloodPressureProblems" feature on premium prices was more pronounced compared to its influence in the XGBoost model. On the hand a high value for the "NumberOfMajorSurgeries" feature seems to have an effect on premium prices while a lower value has a positive impact. In addition the features "KnownAllergies" and "Diabetes" seem to have influence on premium prices based on the XGBoost models analysis.

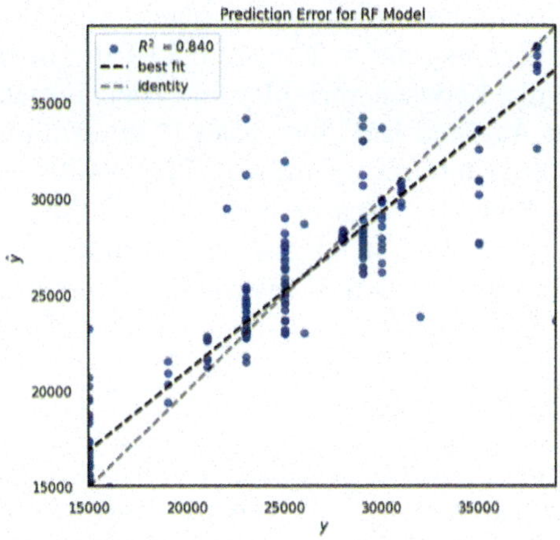

Fig. 3. Prediction error for RF

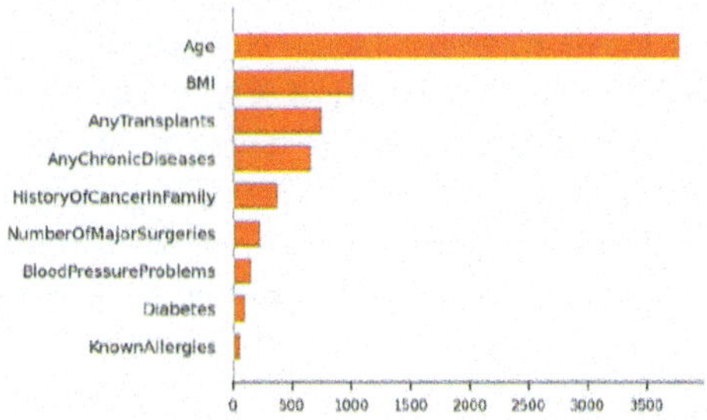

Fig. 4. SHAP summary plot (XGBoost)

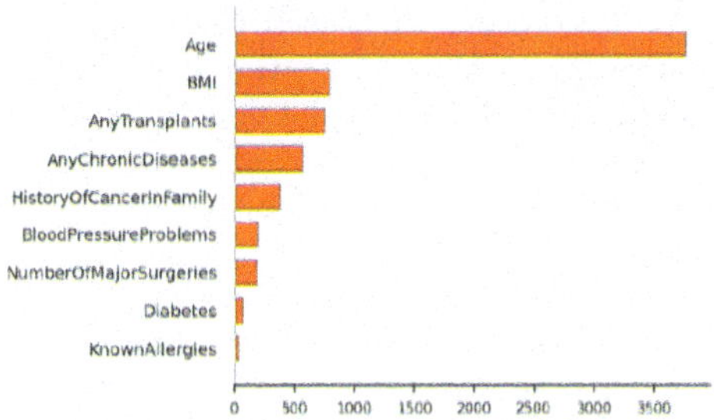

Fig. 5. SHAP summary plot (RF)

This study corresponds well with research by [19,21] which emphasize the effectiveness of XAI techniques in clarifying black box ML models used for predicting medical insurance expenses. Through XAI methods this research sheds light on the relationships between variables and their impact, on premium costs. Notably it shows that Age and BMI play roles in determining medical insurance expenses across all three models examined. The results align, with studies in Actuarial analysis as shown by research from [29,30]. This study emphasizes the importance of XAI techniques, in improving the understandability of machine learning models for predicting medical insurance costs through an investigation.

4 Conclusion

In this study the application of modeling, in the healthcare sector continues to be a focus of actuarial research, driven by insurance companies increasing interest in utilizing ML methods to enhance efficiency and productivity. By using the XGBoost model predictions were made on medical insurance costs based on a dataset from KAGGLEs database showing performance, across models. Notably the XGBoost model achieved a R^2 score of 86.470% and an RMSE of 2231.524 while the RF model excelled in terms of MAE and MAPE with values of 1379.960 and 5.831% respectively. Additionally the RF model demonstrated construction time and lower memory usage compared to XGBoost. To improve model interpretability the SHAP method of XAI was utilized to pinpoint factors influencing price predictions among the features analyzed. This research holds significance in offering decision support to overwhelmed prospective buyers in the medical insurance field by enabling insurers to streamline policy selection through a feature screening process that empowers buyers to find customized policies that suit their needs and financial circumstances.

References

1. Duncan, I., et al.: Testing alternative regression frameworks for predictive modeling of health care costs. N. Am. Actuar. J. **20**(1), 65–87 (2016). https://doi.org/10.1080/10920277.2015.1110491
2. Hartman, B., et al.: Predicting high-cost health insurance members through boosted trees and oversampling: an application using the HCCI database. N. Am. Actuar. J. **25**(1), 53–61 (2020). https://doi.org/10.1080/10920277.2020.1754242
3. Improving health insurance systems, coverage, and service quality. [Online]. https://ww1.issa.int/analysis/improving-health-insurance-systems-coverage-and-service-quality. Accessed 01/02/2023
4. Health Insurance: Definition, How It Works. Investopedia. [Online]. https://www.investopedia.com/terms/h/healthinsurance.asp. Accessed 01/02/2023
5. What Is Health Insurance: Meaning, Benefits & Types. Forbes Advisor INDIA. [Online]. https://www.forbes.com/advisor/in/health-insurance/what-is-health-insurance/. Accessed 01/02/2023
6. Carvalho, D.V., et al.: Machine learning interpretability: a survey on methods and metrics. Electronics **8**(8), 832 (2019). https://doi.org/10.3390/electronics8080832

7. Akter, S., et al.: Transforming business using digital innovations: the application of AI, blockchain, cloud and data analytics. Ann. Oper. Res. **308**(1–2), 7–39 (2020). https://doi.org/10.1007/s10479-020-03620-w
8. Nguyen, H.-S., et al.: Deep reinforcement learning autoencoder with RA-GAN and GAN. Int. J. Adv. Intell. Inform. **8**(3), 313 (2022). https://doi.org/10.26555/ijain.v8i3.896
9. Sánchez Fernández, I., Peters, J.M.: Machine learning and deep learning in medicine and neuroimaging. Ann. Child Neurol. Soc. **1**(2), 102–122 (2023). https://doi.org/10.1002/cns3.5
10. Nguyen, H.-S., et al.: Digital transformation for shipping container terminals using automated container code recognition. TELKOMNIKA (Telecommun. Comput. Electron. Control) **21**(3), 535 (2023). https://doi.org/10.12928/telkomnika.v21i3.24137
11. Ngiam, K.Y., Khor, I.W.: Big data and machine learning algorithms for health-care delivery. Lancet Oncol. **20**(5), e262–e273 (2019). https://doi.org/10.1016/s1470-2045(19)30149-4
12. Using AI and Machine Learning to Improve the Health Insurance Process. Forbes. [Online]. https://www.forbes.com/sites/forbesbusinesscouncil/2022/01/10/using-ai-and-machine-learning-to-improve-the-health-insurance-process/?sh=47ed47de42b1. Accessed 03/03/2023
13. ul Hassan, Ch.A., et al.: A computational intelligence approach for predicting medical insurance cost. Math. Probl. Eng. **2021**, 1–13 (2021). https://doi.org/10.1155/2021/1162553
14. Lundberg, S.M., et al.: A unified approach to interpreting model predictions. In: Proceedings of the 31st International Conference on Neural Information Processing Systems (NIPS'17), pp. 4768–4777. Curran Associates Inc., Red Hook, NY (2017)
15. Janizek, J.D., et al.: Explainable machine learning prediction of synergistic drug combinations for precision cancer medicine, May 2018. https://doi.org/10.1101/331769
16. Stojić, A., et al.: Explainable extreme gradient boosting tree-based prediction of toluene, ethylbenzene and xylene wet deposition. Sci. Total Environ. **653**, 140–147 (2019). https://doi.org/10.1016/j.scitotenv.2018.10.368
17. Stiglic, G., et al.: Interpretability of machine learning-based prediction models in healthcare. WIREs Data Min. Knowl. Discov. **10**(5) (2020). https://doi.org/10.1002/widm.1379
18. Kshirsagar, R.: Accurate and interpretable machine learning for transparent pricing of health insurance plans. Proc. AAAI Conf. Artif. Intell. **35**(17), 15127–15136 (2021)
19. Bora, A., et al.: Interpretation of machine learning models using XAI - a study on health insurance dataset. In: 2022 10th International Conference on Reliability, Infocom Technologies and Optimization (Trends and Future Directions) (ICRITO), Oct 2022. https://doi.org/10.1109/icrito56286.2022.9964649
20. Gaurav, D., Tiwari, S.: Interpretability vs explainability: the black box of machine learning. In: 2023 International Conference on Computer Science, Information Technology and Engineering (ICCoSITE), Feb 2023. https://doi.org/10.1109/iccosite57641.2023.10127717
21. Langenberger, B., et al.: The application of machine learning to predict high-cost patients: a performance-comparison of different models using healthcare claims data. PLoS ONE **18**(1), e0279540 (2023). https://doi.org/10.1371/journal.pone.0279540

22. Sahai, R., et al.: Insurance risk prediction using machine learning. In: Lecture Notes on Data Engineering and Communications Technologies, pp. 419–433 (2023). https://doi.org/10.1007/978-981-99-0741-0_30
23. Medical Insurance Premium Prediction. [Online]. https://www.kaggle.com/datasets/tejashvi14/medical-insurance-premium-prediction. Accessed 01/03/2023
24. Hartman, B., et al.: Predicting high-cost health insurance members through boosted trees and oversampling: an application using the HCCI database. N. Am. Actuar. J. **25**(1), 53–61 (2020). https://doi.org/10.1080/10920277.2020.1754242
25. Friedman, J., et al.: Additive logistic regression: a statistical view of boosting. Ann. Statis. **28**(2) (2000). https://doi.org/10.1214/aos/1016218223
26. Sheridan, R.P., et al.: Extreme gradient boosting as a method for quantitative structure-activity relationships. J. Chem. Inf. Model. **56**(12), 2353–2360 (2016). https://doi.org/10.1021/acs.jcim.6b00591
27. What is a good MAPE score and how do I calculate it? [Online]. https://stephenallwright.com/good-mape-score/. Accessed 01/02/2023
28. Teoh, E.Z., et al.: Explainable housing price prediction with determinant analysis. Int. J. Hous. Mark. Anal. **16**(5), 1021–1045 (2022). https://doi.org/10.1108/ijhma-02-2022-0025
29. Mendelson, D.N., et al.: The effects of aging and population growth on health care costs. Health Aff. **12**(1), 119–125 (1993). https://doi.org/10.1377/hlthaff.12.1.119
30. Kamble, P.S., et al.: Association of obesity with healthcare resource utilization and costs in a commercial population. Curr. Med. Res. Opin. **34**(7), 1335–1343 (2018). https://doi.org/10.1080/03007995.2018.1464435

Fair Allocation Game Approach for Cloud-Hosted Applications by Ant Colony Optimization Algorithm

Thanh-Khiet Bui(✉)

Thu Dau Mot University, Thu Dau Mot City, Binh Duong Province, Vietnam
khietbt@tdmu.edu.vn

Abstract. Service providers and end-users have significant challenges in managing resources in cloud computing. Resource allocation will be developed by leveraging the user's resource needs and the level of service quality. The available physical resources in a cloud computing environment are diverse, and the resource demands from users likewise vary in terms of resource types. Allocating resources can be challenging, especially in situations when resources are limited. This study involved the construction of allocation model that aimed to achieve fairness across physical servers and maximize resource utilization using game theory principles. An Ant Colony Optimization algorithm will be utilized to develop a fair resource allocation approach. The simulation results demonstrate the utilization of parameters to achieve equitable virtual machine allocation and benchmarking using Round-Robin, Ant system, and Max-Min ant system.

Keywords: Fair allocation · Cloud-hosted applications · ACO algorithm

1 Introduction

The way that people and businesses access and use data, apps, and services has completely changed as a result of cloud computing [1]. There has been significant acceptance of technology paradigms that allows users to store, analyze, and retrieve information online, reducing the need for local hardware and infrastructure. Cloud computing infrastructure services involve the transmission of resource demands, such as CPU, memory, and storage space, from customers to cloud computing service providers in order to construct virtual machines (VMs). Cloud computing service providers typically construct systems with a substantial quantity of physical servers (PMs) and impose charges on consumers based on the duration of service utilization.

Service providers have the ability to adapt resource allocation in response to the evolving resource use requirements of users, therefore ensuring both service quality and profitability. The successful utilization of resources in cloud computing necessitates the implementation of optimal resource allocation strategies. The issue of fairness is analyzed in the context of resource allocation, ensuring that no one is allotted significantly more resources than others. Users with different needs require resources of

M. Pagac et al. (Eds.): SCFF 2024, LNNS 1062, pp. 329–339, 2024.
https://doi.org/10.1007/978-3-031-65656-9_33

multiple sorts, including CPU, memory, and storage. In this situation, fair allocation aims to equalize the maximum resource share of total availability allocated to each user. Another objective of resource allocation is to ensure the optimal utilization of resources. Due to the heterogeneity of different VM types, numerous of resource pieces on PMs could be produced during the VMs deployment. Therefore, an efficient VM allocation technique should limit the amount of resources fragments. This particular optimization issue typically falls into the category of NP-Hard or NP-Complete [2]. The solution of VM allocation is typically dependent on the individual features of each situation, from which methods such as exhaustive algorithms, deterministic algorithms or deterministic algorithms or metaheuristics [3]. In practice, deterministic algorithms are usually always greater than exhaustive algorithms. However, deterministic algorithms are unsuccessful in distributed environments, consequently leaving them unfit for resource allocation difficulties in scaled systems. Meanwhile, cloud computing is an environment with distributed data, requiring scalability and the ability to meet high user requirements. Therefore, the problem of VM allocation on cloud computing may be handled in a metaheuristic manner is viable although metaheuristic algorithms can produce near-optimal solutions in acceptable time [4]. This study presents a VM fair allocation strategy based on game theory described as a finite game to ensure a fair distribution among users while keeping a high resource utilization level. Each PM giving resources is considered as a game participant. The Ant Colony Optimization (ACO) algorithm has been applied to identify ways to obtain optimum or near optimal VM allocation based on Nash equilibrium [5, 6]. The remaining part presents fair allocation model based on game theory in Sect. 2. The ACO algorithm for the VM allocation problem is presented in Sect. 3. Section 4 presents the experimental results of the model. Section 5 presents the conclusions.

2 Fair Allocation Game Approach

Suppose the cloud system has p PM availability and PM is denoted by i, where $1 \leq i \leq p$. In this study, different k types of PM resources are considered, and the resource availability of a PM is i denoted as a vector $\vec{\mathrm{N}}^{(i)} = \left(\mathrm{N}_1^{(i)}, \mathrm{N}_2^{(i)}, \ldots, \mathrm{N}_k^{(i)}\right)$. Each PM resource type is denoted by j, where $1 \leq j \leq k$. User-submitted jobs are n denoted by $\mathcal{L}_n$, where $n \in \{1, 2, \ldots, u\}$. Multiple VM types are predefined by the cloud provider and a VM type is represented as a vector $\underset{i}{\vec{v}} = (v_{i1}, v_{i1}, \ldots, v_{ik})$.

The user's resource requirements are defined as a matrix $\mathcal{R}^{u\times k}$, where the horizontal row of the matrix depicts the number of resource types of the user. VM for each user, the vertical row depicts the number of different requests of each resource type by the users.

$$\mathcal{R} = \begin{pmatrix} \underset{1}{\vec{v}} \\ \underset{2}{\vec{v}} \\ \vdots \\ \underset{u}{\vec{v}} \end{pmatrix} = \begin{pmatrix} v_{11} & v_{12} & \cdots & v_{1k} \\ v_{21} & v_{22} & \cdots & v_{2k} \\ \vdots & \vdots & \ddots & \vdots \\ v_{u1} & v_{u2} & \cdots & v_{uk} \end{pmatrix} \tag{1}$$

Each resource allocation decision of the control center is represented as a matrix $\mathcal{A}$ as follows.

$$\mathcal{A}^{(i)} = \begin{pmatrix} \vec{\rho}_1^{(i)} \\ \vec{\rho}_2^{(i)} \\ \vdots \\ \vec{\rho}_u^{(i)} \end{pmatrix} = \begin{pmatrix} \rho_{11}^{(i)} & \rho_{12}^{(i)} & \cdots & \rho_{1k}^{(i)} \\ \rho_{21}^{(i)} & \rho_{22}^{(i)} & \cdots & \rho_{2k}^{(i)} \\ \vdots & \vdots & \ddots & \vdots \\ \rho_{u1}^{(i)} & \rho_{u2}^{(i)} & \cdots & \rho_{uk}^{(i)} \end{pmatrix} \tag{2}$$

in there $\rho_{uk}^{(i)}$ is the number of resources k on the PM i allocated to the user u.

In this work, the subject of fair allocation is explored for numerous sorts of resources. To facilitate effective and flexible resource management, this study explores the ideal trade-off between fair and efficiency.

2.1 Fair Allocation

Fair allocation refers to the equitable distribution of a single type of resource, ensuring that each user receives an equal portion of the resource. In a multi-resource environment, it is imperative to allocate resources to users based on their diverse requests for various types of resources. The dominant share refers to the greatest allocation of the total capacity among various resources for each user. The primary objective of equitable allocation is to achieve a balanced distribution of dominance among users. The assessment of equity in the provision of numerous resources is conducted through the expansion of the dominant resource fairness (DRF) method [7].

The normalized resource requirements matrix is as follows:

$$\mathcal{H} = \begin{pmatrix} h_{11} & h_{12} & \cdots & h_{1k} \\ h_{21} & h_{22} & \cdots & h_{2k} \\ \vdots & \vdots & \ddots & \vdots \\ h_{u1} & h_{u2} & \cdots & h_{uk} \end{pmatrix} = \begin{pmatrix} \frac{v_{11}}{\mathcal{L}_1} & \frac{v_{12}}{\mathcal{L}_2} & \cdots & \frac{v_{1k}}{\mathcal{L}_k} \\ \frac{v_{21}}{\mathcal{L}_1} & \frac{v_{22}}{\mathcal{L}_2} & \cdots & \frac{v_{1k}}{\mathcal{L}_k} \\ \vdots & \vdots & \ddots & \vdots \\ \frac{v_{u1}}{\mathcal{L}_1} & \frac{v_{u2}}{\mathcal{L}_2} & \cdots & \frac{v_{uk}}{\mathcal{L}_k} \end{pmatrix} \tag{3}$$

where, $\mathcal{L}_k = \sum_{i=1}^{p} \mathrm{N}_k^{(i)}$ is the total amount of type resources k of all PMs.

A user's dominant share is the largest share of the resource types allocated to that user. The dominance sharing index is the largest proportion of a type of resource allocated to a user, calculated as follows:

$$d = \frac{1}{\max\limits_{j} \sum_{i=1}^{u} g_{ij}} \tag{4}$$

which $g_{ij} = \frac{h_{ij}}{\max\limits_{j} h_{ij}}$ is standardizing user requirements.

The fairness variance is defined to measure the fairness of resource allocation. Each resource allocation decision has a deviation from the reasonable resource allocation called the equity variance $\omega(\mathcal{A})$, calculated according to the following formula:

$$\omega(\mathcal{A}) = \left(\sum_{i=1}^{u} \sum_{j=1}^{k} \left(\left| \frac{x_{ij}}{\mathcal{L}_k} - d * g_{ij} \right| \right)^{\alpha-1} \right)^{\frac{1}{\alpha}} \tag{5}$$

which represents the amount of resource type x_{ij} allocated to the user i and $\alpha \in R$.

2.2 Resource Utilization

Given the collection of VMs coupled with their resource configuration, the most effective technique to package these VMs on PM to avoid resource wastage in terms of space will be addressed in this section. In a multi-resource context, to increase resource utilization, resource consumption on each resource dimension needs to be considered. It is vital to examine the resource consumption of the bottleneck between multiple resource kinds. Deviation is recommended to quantify the unevenness in the consumption of diverse resources. Reducing skew might be positive for improved resource mix and higher utilization. The waste of resources is determined according to the following formula

$$\varphi(p) = \frac{1}{\sqrt{\sum_{j=1}^{k} \left(\frac{u_j^{(p)}}{\overline{u^{(p)}}} - 1 \right)^2}} \tag{6}$$

in there $\overline{u^{(p)}}$ is the average of the used resources of the th PM p and $u_j^{(p)}$ is the level of category resource utilization j at the th PM p.

2.3 Fairness-Utilization Tradeoff Game

To achieve maximum use of computer resources, cloud providers aim to substitute virtual machines on available machines such that the resource requirements on a server complement each other. In this part, a game theoretic method to resource distribution is provided, which tries to preserve fair allocation as well as limit the quantity of fragmented resources. The resource allocation problem is described as a finite expansion game with perfect knowledge. Physical servers with idle resources are portrayed as selfish players, and each player has a limited amount of alternative allocation matrices. Accordingly, PMs with idle resources are considered players in the game. A resource allocation game is conventionally referred to as a vector $\mathcal{G} = (\mathcal{P}, \mathcal{R}, \mathcal{A}, \mathcal{F})$ where $\mathcal{P}$ is the player, $\mathcal{R}$ is the user's resource request matri, $\mathcal{A}$ is the set of player strategies, and $\mathcal{F}$ is the player's utility function.

Overall goal of this allocation game is to share resources fairly. Furthermore, based on the principle of efficiency, each player tries to minimize the waste of their resources, meaning they prioritize choosing combinations with high utilization. To exploit fair resource sharing and also take into account the maximization of resource utilization rate, the fair utilization balance utility function is designed as follows:

$$\mathcal{F}^{(p)}(\mathcal{A}) = \text{sgn}(1 - \gamma) \cdot \omega(\mathcal{A}) + \varphi(p) \tag{7}$$

where is γ the coefficient influencing the weight of equity and utilization, $\omega(\mathcal{A})$ is the fairness variance, and $\varphi(p)$ is a deviation that reflects the uneven use of different resources. Allocation decisions $\mathcal{A}$ receive the less equity variance, the more utility players receive. Likewise, each PM prefer to choose the less biased combination to optimize its own utility.

Each player in this game will choose a strategy to maximize their own harvest. Therefore, it is necessary to maximize the benefit function of each player. Each player will choose the strategy that maximizes his payoff $\max_{\mathcal{A}} \mathcal{F}^{(p)}$.

$$\max_{\mathcal{A}} \mathcal{F}^{(p)} \tag{8}$$

$$\text{subject to } \sum_{i=1}^{n}\sum_{j=1}^{k} x_{ij}^{(p)} \le c_j^{(p)} \tag{9}$$

$$x_{ij}^{(p)} > 0 \tag{10}$$

The player's strategy is a resource allocation decision of each PM based on the user's resource requirements matrix. The players know each other's information, strategies, and decision points, so we can set up a perfectly informed, cooperative game. From there we approach the concept of Pareto efficient Nash equilibrium of the game which is the point where no player receives more harvest by changing his strategy. The Nash equilibrium of a game is a strategy where no one player can increase profits while the other players have fixed strategies. Then, if the strategy of the second player p is the optimal strategy denoted by $\mathcal{A}^{(p)*}p_i^*$,he optimal strategy of the other players is denoted by $\mathcal{A}^{(-p)*}p_{-i}^*$, then the Nash equilibrium of the strategy $\mathcal{A}^{(p)*}p_i^*$ will comply with the following condition:

$$\mathcal{F}\left(\mathcal{A}^{(-p)*}, \mathcal{A}^{(p)*}\right) \ge \mathcal{F}\left(\mathcal{A}^{(-p)*}, \mathcal{A}^{(p)}\right) \tag{11}$$

3 Fair Allocation Algorithm

The reinforcement learning process has the impact of boosting the effectiveness of the Ant Colony Optimization (ACO) algorithm in the process of ants discovering solutions [5]. One of the first crucial things in implementing ACO algorithm is the process of finding reinforcement learning information through trails, in other words, recognizing information that trails represent (Fig. 1).

To handle VM allocation problem, as illustrated Fig. 2, an allocation solution is constructed utilizing an array encoding approach which reflects the relationship between the VMs and PMs. The first dimension of ant^k solution is a set VMs which each VM corresponds an element of array. The second dimension of ant solution is a collection of available PMs which each element value of array represents a PM. For example, VM1 is allocated to PM2, VM2 is allocated to PM1, VM3 is allocated to PM3, … and VMi is allocated to PMj.

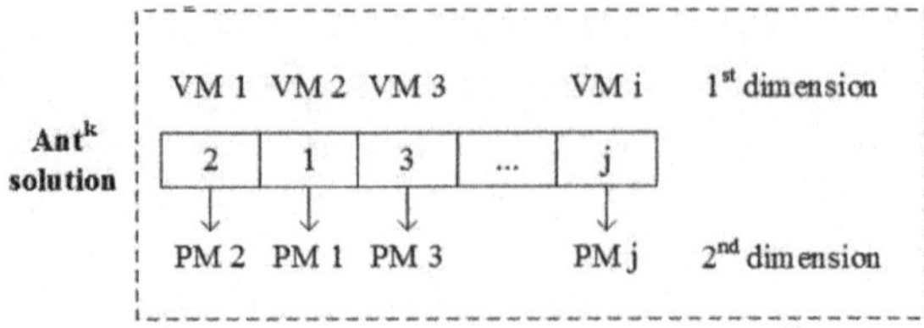

Fig. 1. Ant solution encoding scheme

For each VM, the software will compute to extract valid PMs that are enough circumstances to allocate that virtual machine and calculate the selected probability of each PM. The likelihood that each available PM is picked to assign a VM which is determined according to the formula

$$p_{ij}^{k} = \begin{cases} \dfrac{\left([\tau_{ij}]^{\alpha}.[\eta_{ij}]^{\beta}\right)}{\sum_{j\in N_i^k)}\left([\tau_{ij}]^{\alpha}.[\eta_{ij}]^{\beta}\right)} & if\ j \in N_j^k \\ 0 & otherwise \end{cases} \tag{12}$$

where p_{ij}^{k} is probability of ant k to select PM j for VM i, N_j^k is a set of available PMs candidate sub-solution that can meet the requirement of the VM k^{th}; the parameter α, β are used to determine the influence of odor and heuristic information η_i. The heuristic information will be computed after each allocation activity as the configuration information of PM will alter after each successful allocation event. The recalculation will aid make the heuristic information more correct for the upcoming allocation step. The heuristic information calculated by the formula $\eta_i = \frac{1}{u^{(j)}}$.

iACO fair allocation algorithm

```
setupAnts()
clearTrails()
WHILE (¬StopCondition(ε)) DO
    FOR (i = 1 To Ants_size)
        FOR (k = 1 To n)
            //Move ant colonies to build solutions to allocate VMs
            MoveAnt(Φ_k, trails, probabilities)
        END
    END
    updateTrails()
    updateBest()
    //Inversing the best ant solution
    FOR (k = 1 to Ants_size) DO
        FOR (j = 1 to VM_size) DO
            antSolution[j] = bestAntSolution[VM_size - j]
        END
    END
END
```

After each algorithm loop, the trail value is updated by all ants that have built the solution in its loop. The τ_{ij} trail value on the edge of (i, j) is calculated as follows

$$\tau_{ij} \leftarrow (1-\rho).\tau_{ij} + \sum_{k=1}^{m} \Delta\tau_{ij}^{k} \tag{13}$$

in which, ρ is the evaporation rate of trail. The parameter evaporates to avoid excessive accumulation of trails and eliminates previously selected inefficient PMs. $\Delta\tau_i^k$ is the trail quality of the nth ant on the edge of (i, j) on the graph is calculated as follows

$$\Delta\tau_i^k = \begin{cases} \frac{Q}{L_k} \ \textit{if the ant chooses } (i, j) \\ 0 \ \ \textit{otherwise} \end{cases} \tag{14}$$

where Q is a constant, L_k is the cost of ant k edge (i, j).

Building a new solution can be done by constructive or perturbative approaches [8]. Accordingly, we propose introducing an inverse operation from line 13 to line 17 in the iACO algorithm to initialize solutions for the ants for the following loop. In a multi-agent system setting, the equilibrium point may not be. The stopping condition of iACO algorithm is computed according to [9] as follows:

$$\sum_{i=1}^{n} \left(F^{itr} - F^{itr-1}\right)^2 < \varepsilon \tag{15}$$

in there, ε is a constant, F_k^{itr} is the benefit of the second player k at iteration step itr, F_k^{itr-1} is the benefit of the second player k at the iteration $itr - 1$.

Table 1. VM and PM configurations

Type	VM			PM		
	CPU (core)	RAM (GB)	Disk (GB)	CPU (core)	RAM (GB)	Disk (GB)
Tiny	1	1	5	8	16	128
Small	1	3	15	16	64	512
Medium	2	6	30	64	128	1024
Large	4	12	60	64	256	2048
X Large	8	24	80	128	1024	5096

4 Evaluation

This part gives a complete examination of the resource allocation method provided in the preceding section. The fairness evaluation is accomplished by an iACO algorithm prototype on CloudSim [10]. To simplify the complexity of the simulations, the following assumptions are made three categories of resources (i.e., CPU, RAM, DISK) are considered in our simulations. The entire quantity of resources allocated for each time slot is already calculated by the cloud provider. We focus on VMs allocation to satisfy dominant resource fairness in Eq. (5) and waste of resources in Eq. (6). The operational parameters of all infrastructure services are created randomly, whereas values of the configuration of VM and PM are shown in Table 1.

4.1 Study Parameters

The experiment was done performed on a simulation of a data center with 50 PMs with diverse configurations, deploying 30 three-tier applications. First, we choose $\rho = 0.8$, ant of number is 15, and then change ε in Eq. (15) from 0.01 to 0.1. As shown in Fig. 2, the iACO algorithm remains stable after a small number of iterations when the value of ε is changed from 0.04 to 0.1. However, the value of benefit function tends to decrease when ε decreases in Fig. 3. This shows that the iACO algorithm will be more accurate with increasing number of loops.

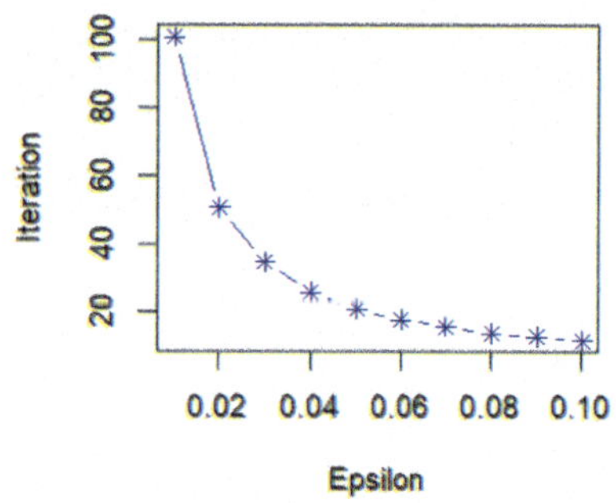

Fig. 2. Iteration with ε

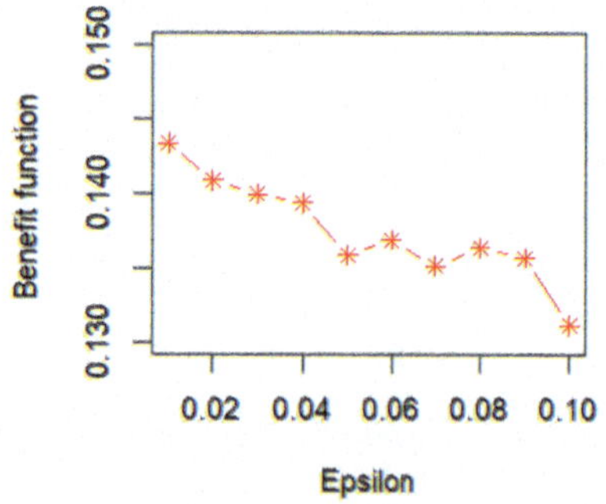

Fig. 3. Benefit function with ε

Next step, we choose ε = 0.03, ant of number is 15, and change coefficient of discovery ρ in Eq. (13) from 0.1 to 0.9. In Fig. 4, with small coefficient of discovery, almost target function is small. Final, we choose ε = 0.03, $\rho = 0.8$, and change ant of number from 10 to 100. In Fig. 5, with the number of ants from 10 to 100, the results of iACO are feasible with more of ant number.

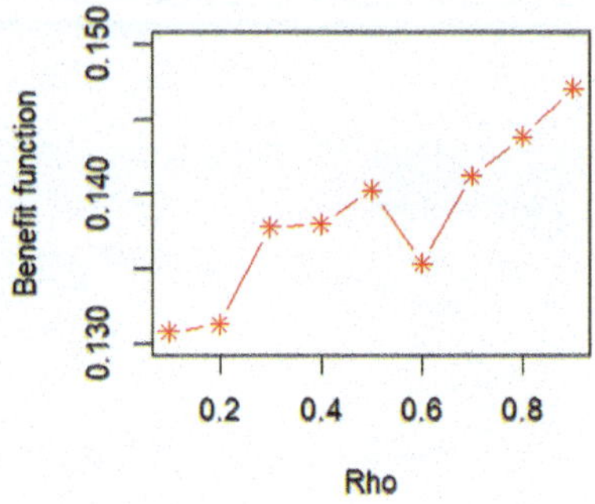

Fig. 4. Benefit function with ρ

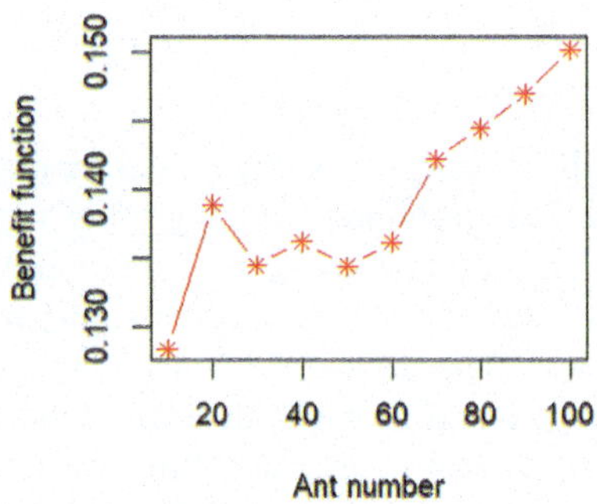

Fig. 5. Benefit function with ant number

4.2 Benchmarking

This section presents the results of evaluating the effectiveness of VM allocation based on fair and resource wastage of Round-Robin (RR), Ant System (AS), Max-Min Ant System (MMAS), and iACO algorithms. We choose $\varepsilon = 0.03$, $\rho = 0.8$, the number of ants is 80. We made comparisons on the three data sets including Dataset 1 has 50 PMs, 30 Apps, have 13 PMs are fault and 77 VMs need to be allocated and Dataset 2 has 450 PMs, 200 Apps, have 630 VMs need to be allocated. Figure 6 shows the benefit value of iACO algorithm is better than AS and MMAS algorithm with Dataset 1 and Dataset 2.

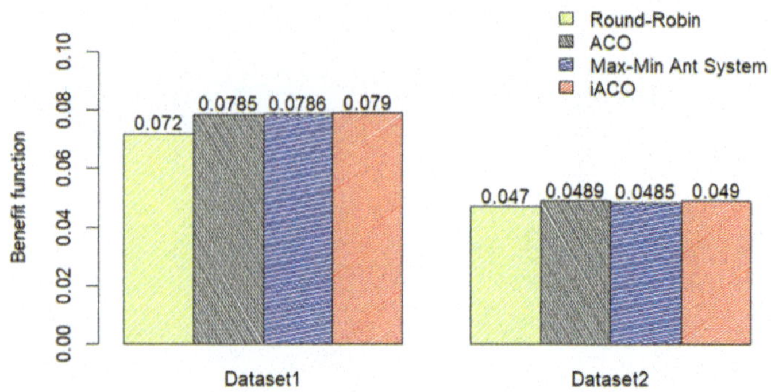

Fig. 6. Benchmarking of RR, AS, MMAS, iACO

5 Conclusion

In this work, we accomplish fair using the game model for players in the VM allocation. The outcome of the game is determined by integrating the two variables, dominant resource fairness and resources utilization. Ant colony optimization, which is based on Nash equilibrium, is used to estimate the allocation solution. Using the coefficients to accomplish dominant resource fairness in virtual machine provisioning is demonstrated in the tests. The goals of the cloud provider determine these coefficients. Optimal or near-optimal solutions are the goal of the next time we look at this problem, so we investigate several optimization techniques and how to apply the coefficients.

References

1. Namasudra, S., Roy, P., Balusamy, B.: Cloud computing: fundamentals and research issues. In: 2017 Second International Conference on Recent Trends and Challenges in Computational Models (ICRTCCM), pp. 7–12. IEEE (2017)
2. Gary, M.R., Johnson, D.S.: Computers and Intractability: A Guide to the Theory of NP-Completeness. WH Freeman and Company, New York (1979)
3. Tsai, C.-W., Rodrigues, J.J.: Metaheuristic scheduling for cloud: a survey. IEEE Syst. J. **8**, 279–291 (2014)
4. Thanh Bui, K., Dac Ho, H., Pham, T.V., Tran, H.C.: Virtual machines migration game approach for multi-tier application in infrastructure as a service cloud computing. IET Netw. **9**, 326–337 (2020)
5. Asghari, S., Jafari Navimipour, N.: The role of an ant colony optimisation algorithm in solving the major issues of the cloud computing. J. Exp. Theoret. Artif. Intell. **35**, 755–790 (2023)
6. Bui, K.T., Pham, T.V., Tran, H.C.: A load balancing game approach for VM provision cloud computing based on ant colony optimization. In: Context-Aware Systems and Applications: 5th International Conference, ICCASA 2016, Thu Dau Mot, Vietnam, November 24–25, 2016, Proceedings 5, pp. 52–63. Springer (2017)
7. Xu, X., Yu, H.: A game theory approach to fair and efficient resource allocation in cloud computing. Math. Probl. Eng. **2014** (2014)
8. Hoos, H.H., Stützle, T.: Stochastic Local Search: Foundations and Applications. Elsevier (2004)

9. Siar, H., Kiani, K., Chronopoulos, A.T.: An effective game theoretic static load balancing applied to distributed computing. Cluster Comput., 1–15 (2015)
10. Bouhouch, L., Zbakh, M., Tadonki, C.J.I.J.o.C., Applications: DFMCloudsim: an extension of cloudsim for modeling and simulation of data fragments migration over distributed data centers. Int. J. Comput. Appl. **46**, 1–20 (2024)

Insights from Traveler Feedback: Topic Analysis of Short-Day Tours in Southern Vietnam

T. H. Giang Tran(✉) and T. Doan Dang

Eastern International University, Thu Dau Mot City, Binh Duong Province, Vietnam
{giang.tranthi,doan.dang}@eiu.edu.vn

Abstract. This study utilizes an innovative approach combining natural language processing (NLP) techniques and Non-negative Matrix Factorization (NMF) to extract and analyze topics from TripAdvisor reviews. Focusing on short-day tours in Southern Vietnam, the research identifies key topics that represent travelers' experiences and expectations. By mapping these topics into the push and pull theory, it delves into the underlying motivations driving tourist behaviors and preferences. The findings reveal that travelers are influenced by a blend of push factors, such as the desire for knowledge and facility convenience, and pull factors, including unique activities and cultural immersion. These insights offer valuable implications for the tourism industry in Southern Vietnam, highlighting the importance of curating experiences that cater to both the intellectual and experiential desires of tourists. The study presents actionable recommendations for tour operators and the broader tourism sector to enhance the quality and appeal of their offerings, thereby enriching the customer experience and fostering a deeper connection with the destination.

Keywords: User-generated content · topic modeling · customer experience · short day tours · Non-negative Matrix Factorization

1 Introduction

Tourist attractions are important components for numerous fields of study about domestic and international tourism [1]. Traveler reviews are one of the noticeable resources for managers, researchers and travelers used to evaluate the quality of a tour management and understand customer decision-making [2]. In the dynamic context of the tourism industry, the ability to understand and adapt to traveler behaviors and preferences is crucial for destinations aiming to enhance their appeal and competitiveness. The advent of digital platforms, especially review sites like TripAdvisor, has opened a treasure trove of user-generated content, offering deep insights into tourists' experiences and expectations [3]. Against this backdrop, this study employs topic modeling—a method in text analysis for identifying themes within large textual datasets—to distill the essence of tourists' experiences on short-day tours in Southern Vietnam. Utilizing NLP techniques and NMF, it aims to systematically dissect TripAdvisor reviews to unearth core topics reflective of tourist experiences [4, 5]. Southern Vietnam's tourist attractions, with their rich cultural

M. Pagac et al. (Eds.): SCFF 2024, LNNS 1062, pp. 340–349, 2024.
https://doi.org/10.1007/978-3-031-65656-9_34

heritage, historical significance, and natural beauty, attract a global audience [6]. To maintain and increase this allure, the local tourism industry must remain attuned to the evolving desires of tourists.

This necessitates an understanding of the motivations behind travel behaviors, well-captured by the push and pull theory [7, 8]. This theory distinguishes between push factors—internal motivations that compel individuals towards travel, such as the quest for knowledge, relaxation, or personal growth—and pull factors, which are the destination's attributes that draw visitors, including cultural sites, natural landscapes, and recreational opportunities [9]. Given this framework, this research seeks to address two pivotal questions:

(1) What key topics emerge from TripAdvisor reviews of Southern Vietnam's short-day tours, and what do they reveal about travelers' experiences?
(2) How do these topics align with push and pull factors, shedding light on the motivations and satisfaction of travelers?

By answering these questions, the study aims to offer actionable insights for tour operators and the tourism sector to refine and enhance the quality of tourist experiences. This effort not only seeks to elevate the attractiveness of Southern Vietnam as a travel destination but also to contribute to the sustainable growth of the tourism industry in the region.

2 Literature Review

2.1 User-Generated Content

In the dynamic landscape of modern consumer behavior, user-generated content (UGC) presents a potent tool for researchers delving into traveler expectations and evolving tourism strategies [10]. UGC, encompassing online reviews, social media posts, and other digital artifacts created by consumers themselves, offers a real-time window into the dynamic world of traveler sentiment, desires, and experiences [11]. By using the power of UGC, The Southern Vietnam tours through the lens of traveler expectations and current strategies from Vietnam Adventure Tours can provide an updated understanding about consumer behaviors [6]. By combining UGC with TripAdvisor, related information can be the foundation for optimizing operational and marketing strategies in Southern tourism. Additionally, identifying thematic threads within reviews can unlock opportunities for niche marketing campaigns that resonate with specific traveler segments, fostering a more tailored and effective approach to attracting and engaging visitors [5].

2.2 Topic Modeling

Topic modeling is an analytical technique that mines underlying themes from extensive text data, with various algorithms offering unique advantages. Latent Semantic Analysis (LSA) is an early technique that uses singular value decomposition to identify relationships between words and topics [12]. Probabilistic Latent Semantic Analysis (PLSA), an advancement over LSA, incorporates a probabilistic framework for topic discovery [13]. Latent Dirichlet Allocation (LDA) further refines this process by assigning probability distributions to both documents and topics [14]. Each method offers different pathways to extract thematic structures within text corpora, useful for analyzing complex datasets like customer reviews.

By employing algorithms like NMF, this method identifies patterns and clusters of words that represent key topics within the data. This approach is particularly useful for businesses and researchers to distill insights from customer feedback, online reviews, or any extensive textual datasets. In topic modeling, NMF identifies hidden themes and topics within reviews, even if they are not explicitly mentioned. This can reveal deeper insights into what truly matters to travelers and the hidden factors that influence their experiences [11]. Therefore, using NLP techniques and NMF to analyze online feedback of travelers in Tripadvisor can be helpful in understanding feelings and evaluations of customers about services and products of the Southern Vietnam Tours without missing valuable information [11, 15].

2.3 Push and Pull Theory in Tourism Traveler Experiences

The push and pull theory were introduced by Dann in 1977 to understand consumer motivation and their decision in the travel industry [8]. The study articulated travel motivation as a profound mental disposition prompting individuals or groups to embark on travel. This concept was clarified through the distinction between two motivational forces, termed push and pull factors [7]. From this perspective on travel motivation, push and pull factors elucidate the reasons behind individuals' decisions to travel and their selection of specific destinations over others [16, 17].

- Push factors are also known as psychographic motivations [9]. Based on the research of Crompton [18], push factors related to relaxation, health and fitness, or personal growth that propel individuals towards travel experiences. Based on these specific factors, direction of travelers' behaviors in specific tourism destinations are evaluated [7, 18].
- Pull factors, on the other hand, reside within the inherent characteristics of a destination. These elements attract people by natural beauty, cultural attractions, diverse entertainment options, comfortable accommodations, or the promise of unique experiences [7, 19]. The differences in promotions, accessibility and service quality in different destinations can affect the foreign tourists' preferences. This can be one of the factors for tours to create unique experiences in each destination for tourists.

Travel choices are not solely influenced by push or pull factors; rather, they interact in complex ways to shape travelers' decisions. Push factors spark the desire to travel, while pull factors determine the choice of destination [20]. Beyond guiding destination selection, the push-pull framework highlights travel's role in fostering personal growth and social connections, with shared experiences often leading to self-discovery and a sense of well-being. Table 1 outlines key push and pull factors in tourism [17]. Table 1 below shows some main push and pull factors in the tourism industry.

Table 1. Push and pull factors in tourism industry.

Push factors	Pull factors
Cultural values; Family togetherness; Knowledge Interest; Relaxation	Safety; Budget; Convenience of facilities; Activity; Nature; Religion

3 Research Method and Dataset

3.1 Dataset Description

This research conducted a comprehensive evaluation of 17,337 TripAdvisor reviews concerning Vietnam Adventure Tours, collected in November 2023. The extraction process utilized the powerful programming language Python, specifically employing Selenium and Beautiful Soup for web scraping, alongside relevant libraries [21]. These tools were crucial for collecting and parsing extensive web data into a dataset rich in traveler feedback, laying the groundwork for our analysis of customer experiences on short-day tours in Southern Vietnam. The comprehensive range of reviews forms a solid foundation for using advanced NLP techniques to reveal insights into traveler sentiments and preferences, as detailed in Fig. 1.

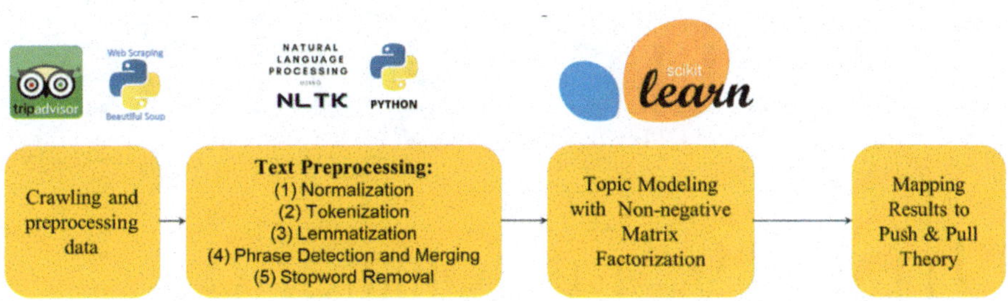

Fig. 1. The process of analyzing data

The dataset includes TripAdvisor review attributes such as reviewer IDs ('NameReviewer'), origins ('ReviewerCountry'), travel types ('TypeTravellers'), likes ('Like'), review dates ('date_review'), detailed reviews ('FullReview'), ratings ('Rating'), sentiments ('Sentiment'), and review lengths ('ReviewLength'), as shown in Table 2. It focuses on 'FullReview' for topic modeling, offering a detailed insight into customer feedback.

Some popular words that appear in the travelers' feedback are shown in Fig. 3.

3.2 Analyses Method

The study uses topic modeling, particularly the NMF algorithm, to analyze TripAdvisor reviews for its ability to efficiently distill large texts into distinct, interpretable topics by identifying themes with non-negative weights [4]. This feature makes NMF ideal

Table 2. Summary of dataset

Attributes	Data types	Range/sample
NameReviewer	object	[Explore56044221797, Martin S, Ariel I, Aidan...]
ReviewerCountry	object	[Unknown, UK, Canada, Italy, NY]
TypeTravellers	object	[Family, Couples, Friends, unknown, Solo]
DateReview	Int64	[2014, 2023] [1, 12]
Like	int64	[0, 23]
FullReview	object	[We enjoyed the trip. Theo was an informative...]
Rating	float64	[1.0, 5.0]
Sentiment	int64	[0, 12]
ReviewLength	int64	[95, 15577]

Fig. 3. The word cloud from tourists' reviews

for analyzing TripAdvisor reviews, allowing for accurate alignment with the push and pull framework. Through this method, the research successfully extracted 10 distinct topics that represent the core themes discussed by travelers in their reviews of Vietnam Adventure Tours [22].

Following the extraction of these topics, the study applied the push and pull theory as a theoretical framework to map and interpret the topics. This dual approach enables a nuanced understanding of the motivations behind travel decisions and overall satisfaction with the travel experiences provided by Vietnam Adventure Tours. By linking the identified topics to push factors (internal motivations to travel) and pull factors (destination's attractions that draw visitors) [7, 8], the research offers valuable insights into the dynamics of traveler behavior and destination choice [21].

4 Experiment Results

4.1 Topic Modeling with NMF

By using NMF analysis, the present study revealed a spectrum of latent topics. These 10 distinct categories encompass elements of visitor experience on short trips organized by Vietnam Adventure Tours. These themes highlight the diverse aspects of visitor enjoyment, revealing the key factors contributing to positive sentiment and satisfaction. A detailed examination of these themes is presented below, drawing upon insights gleaned from visitor reviews:

Topic #1: tour, recommend, highly, guide, informative, amazing, knowledgeable, definitely, friendly, funny
Topic #2: war, jackie, vietnam, tuan, experience, story, veteran, tour, guide, personal
Topic #3: mekong, delta, day, boat, local, river, lunch, ride, coconut, island
Topic #4: good, english, nice, guide, ele, food, tour, time, friendly, spoke
Topic #5: great, day, time, experience, sense, guide, humour, thanks, knowledge, kai
Topic #6: trip, guide, enjoyable, ele, amazing, day, thank, excellent, fun, friendly
Topic #7: tunnel, chi, history, half, visit, chu, went, visiting, interesting, bus
Topic #8: lan, unicorn, highly, recommended, recommend, funny, experience, guy, knowledgeable, guide
Topic #9: really, enjoyed, nice, funny, interesting, tour, john, guide, helpful, explained
Topic #10: lot, information, interesting, history, fun, vietnam, learned, gave, culture, tour

4.2 Mapping Non-negative Matrix Factorization Topics to Push and Pull Theory

Through this innovative approach, ten topics are grouped into a push and pull framework. These factors reveal some main reasons that influence tourist decision-making and travelers' satisfaction.

Push Factors - Knowledge: Among the identified topics, topics #1, #8, #9 highlight the inherent human desire for intellectual enrichment and cultural understanding. These topics resonate with tourists seeking to delve into the rich tapestry of Vietnamese history and customs, suggesting a strong intrinsic motivator beyond mere sightseeing.

Pull Factors - Facilities: Topics #2, #4, #7 focusing on the importance of tangible conveniences. These topics reveal the critical role that reliable transportation, comfortable accommodation, and well-maintained facilities play in shaping a positive travel experience, emphasizing the role of perceived risk reduction in influencing travel decisions.

Pull Factors - Activities: Topics #3, #5, #6 show a preference for enriching activities that promote cultural immersion and personal growth, reflecting a shift towards authentic, educational travel experiences in Vietnam. Push-pull analysis underscores a demand for knowledge and high-quality experiences, highlighted by the influence of informed guides (Fig. 4) and superior facilities (Fig. 5) on tourist satisfaction. These points stress the value of focusing on educational content and excellence in operations to meet and exceed short-day tour expectations, as shown in Fig. 6.

Oct 2023 • Friends

We had the best time with Duc Pham. Duc is so friendly so professional so on time with great sense of humor too. He had a cut during coconut island tour while demonstrating how to cut open a coconut but he just kept going, that's some professionalism. He went above and beyond for tour mates who had different needs as well. Overall it was an excellent experience, definitely book with Duc. Thanks again Duc for a great time!

Fig. 4. A traveler's review about professional attitude of a tour guide [26]

Mar 2023 • Friends

we spent a whole day with a this company and our gorgeous tour guide Van. the bus was very comfortable and air condition was our salvation during hot weather. Our guide was very nice, positive and active. the trip was very ballanced and Van told us many interesting stories. the river trip, folk music and traditional, special snake wine were the most amazing 😊

Fig. 5. A traveler's review about the facilities of the tour [26]

May 2023 • Family

We had a fantastic experience on the day tour to chuchi tunnels and mekong delta. Our guide Kevin is a history student and was so knowledgeable about the history of Vietnam. He gave us an excellent summary of the history of the war. This tour is a must for all visitors to Vietnam.

After chuchi tunnels, we went to mekong delta which was also an amazing experience. When i first read the tour description, i was not very sure about it. But the tour included so many different experiences like seeing the coconut candy making, trying the fruits and different local village experience was just too good.

I would highly recommend this tour and the Vietnam Adventure tour group to all our friends.

Fig. 6. A traveler's review their positive experience about the knowledge they have from a tour [26]

Beyond the push factors, the research illuminates the potent pull factors that draw tourists to heritage experiences. The analysis reveals a strong desire for meaningful activities directly related to Vietnamese culture and history (#3, #5, #6). These topics showcase the growing preference for experiential tourism, where travelers seek to engage in activities that offer deeper understanding and a transformative connection with the local culture (Fig. 7). This trend creates great potential for tourism stakeholders to construct unique and real experiences that connect with today's sensitive visitors.

5 Discussion and Implication

5.1 Discussion

The extensive analysis of TripAdvisor reviews has yielded critical insights into the experiences of travelers using Vietnam Adventure Tours. By applying NMF topic modeling to the review content, this study not only uncovers the core themes within traveler feedback but also provides definitive answers to the research questions initially posed. The

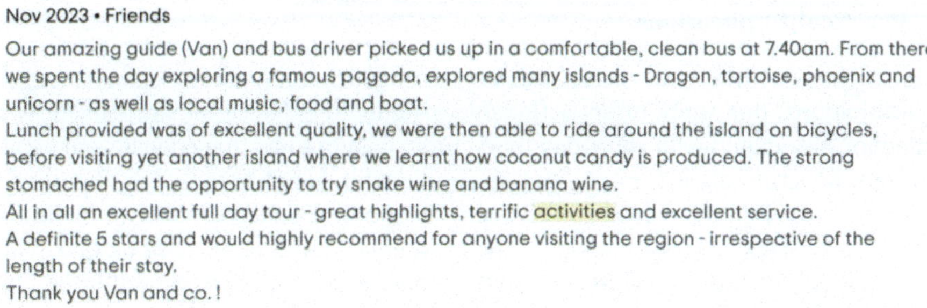

Nov 2023 • Friends

Our amazing guide (Van) and bus driver picked us up in a comfortable, clean bus at 7.40am. From there we spent the day exploring a famous pagoda, explored many islands - Dragon, tortoise, phoenix and unicorn - as well as local music, food and boat.

Lunch provided was of excellent quality, we were then able to ride around the island on bicycles, before visiting yet another island where we learnt how coconut candy is produced. The strong stomached had the opportunity to try snake wine and banana wine.

All in all an excellent full day tour - great highlights, terrific activities and excellent service.

A definite 5 stars and would highly recommend for anyone visiting the region - irrespective of the length of their stay.

Thank you Van and co. !

Fig. 7. A traveler's review about the interesting activities of the tour [26]

extracted topics from the reviews shed light on the multifaceted nature of traveler experiences [23]. The findings illustrate that travelers' discussions revolve around a desire for enriching experiences (topics #1, #8, #9), comfort and reliability of facilities (topics #2, #4, #7), and engaging activities (topics #3, #5, #6). The alignment of these topics with push and pull factors illuminates the motivations behind travelers' decisions and their satisfaction levels [24]. The intrinsic push factors, such as knowledge acquisition and quality facilities, drive travelers toward seeking out these tours. Simultaneously, the extrinsic pull factors, like unique activities and the opportunity for cultural immersion, play a crucial role in destination choice, confirming the theoretical framework's applicability in this digital era of travel reviews.

Based on the exploration, the suggestions are to increase the activities related to Vietnamese cultures and histories. Travel companies can organize more traditional and unique activities from the Southeast areas such as producing steamed cassava with coconut milk and expanding the visiting historical areas of the war. These activities can strengthen knowledge and the intellectual and emotional engagement with the destination. In addition, Khare et al. shared useful information that travels tend to choose homestays to experience the culture exchange [27]. Expanding the quality of homestays not only support the pull factors but also ensure the sustainable tourism in the Southern of Vietnam.

5.2 Theoretical Implications

The application of push and pull factors in understanding tourist behavior presents a nuanced framework that enriches the theoretical discourse on travel motivation [24]. The study's findings affirm the relevance of Dann's push and pull theory within the context of modern tourism, particularly in the realm of digital consumer feedback. By demonstrating how distinct topics extracted from online reviews map onto push and pull factors, this research contributes to a deeper understanding of the dynamic interplay between a traveler's internal desires and the external attributes of a destination [25]. It underscores the theory's adaptability in deciphering the complex motivations behind why people travel and how they select destinations, enriching the academic dialogue on tourism studies.

5.3 Practical Implications

For practitioners within the tourism industry, particularly in Southern Vietnam and similar destinations, this study offers actionable insights to enhance the attractiveness and satisfaction associated with short-day tours. First, design tours that educate and engage, incorporating cultural, historical, and environmental insights to satisfy knowledge-seeking tourists. Second, focus on the quality and convenience of tourism facilities to meet tourists' expectations for comfort and reliability. Third, emphasize unique, hands-on activities in marketing efforts to appeal to tourists seeking experiential travel. Last, use insights from this study for marketing strategies that resonate with specific traveler segments, highlighting enriching travel experiences.

6 Limitations and Future Works

This study on Vietnam Adventure Tours, while insightful, is limited by the length of the research. Therefore, its reliance on TripAdvisor reviews and NLP techniques, potentially missing the full spectrum of tourist experiences and subtler insights. It also overlooks broader aspects of tourist behavior like decision-making and satisfaction.

Future research should expand using other techniques such as LDA and LSA and employ advanced analytical methods for a more thorough understanding of Southern Vietnam tourism, improving destination management.

Acknowledgements. The authors would like to thank Eastern International University, Binh Duong Province, Vietnam for funding this research.

References

1. Fernández, J.A.S., Azevedo, P.S., Martín, J.M.M., Martín, J.A.R.: Determinants of tourism destination competitiveness in the countries most visited by international tourists: proposal of a synthetic index. Tour. Manage. Perspect. **33**, 100582 (2020)
2. Mardani, A., Jusoh, A., Zavadskas, E.K., Khalifah, Z., Nor, K.M.: Application of multiple-criteria decision-making techniques and approaches to evaluating of service quality: a systematic review of the literature. J. Bus. Econ. Manag. **16**(5), 1034–1068 (2015)
3. Nelson, V.: Liminality and difficult heritage in tourism. In: Liminality in Tourism, pp. 80–100. Routledge (2021)
4. Lee, D., Seung, H.S.: Algorithms for non-negative matrix factorization. In: Advances in Neural Information Processing Systems, vol. 13 (2000)
5. Albalawi, R., Yeap, T.H., Benyoucef, M.: Using topic modeling methods for short-text data: a comparative analysis. Front. Artif. Intell. **3**, 42 (2020)
6. Tien, N.H., Ngoc, N.M., Anh, D.B.H.: Sustainability of coastal tourism development: comparative analysis of Vietnam's Northern and Southern provinces. J. Southwest Jiaotong Univ. **55**(6) (2020)
7. Dann, G.M.: Tourist motivation an appraisal. Ann. Tour. Res. **8**(2), 187–219 (1981)
8. Dann, G.M.: Anomie, ego-enhancement and tourism. Ann. Tour. Res. **4**(4), 184–194 (1977)
9. Getz, D., Page, S.J.: Progress and prospects for event tourism research. Tour. Manage. **52**, 593–631 (2016)

10. Garner, B., Kim, D.: Analyzing user-generated content to improve customer satisfaction at local wine tourism destinations: an analysis of Yelp and TripAdvisor reviews. Consum. Behav. Tour. Hosp. **17**(4), 413–435 (2022)
11. Mishra, R.K., Jothi, J.A.A., Urolagin, S., Irani, K.: Knowledge based topic retrieval for recommendations and tourism promotions. Int. J. Inf. Manage. Data Insights **3**(1), 100145 (2023)
12. Deerwester, S., Dumais, S.T., Furnas, G.W., Landauer, T.K., Harshman, R.: Indexing by latent semantic analysis. J. Am. Soc. Inf. Sci. **41**(6), 391–407 (1990)
13. Hofmann, T.: Probabilistic latent semantic indexing. In: Proceedings of the 22nd Annual International ACM SIGIR Conference on Research and Development in Information Retrieval, pp. 50–57 (1999)
14. Blei, D.M., Ng, A.Y., Jordan, M.I.: Latent Dirichlet allocation. J. Mach. Learn. Res. **3**(Jan), 993–1022 (2003)
15. Nithin Goud, K., Ramanjaneyulu, Y., Sathya Babu, K., Patra, B.K.: User preference multi-criteria recommendations using neural collaborative filtering methods. In: Proceedings of the Sixth International Conference on Mathematics and Computing: ICMC 2020, pp. 55–64. Springer (2021)
16. Bright, A.D.: Motivations, attitudes, and beliefs. In: Handbook of Hospitality Marketing Management, pp. 239–265. Routledge (2009)
17. Correia, A., do Valle, P.O., Moço, C.: Modeling motivations and perceptions of Portuguese tourists. J. Bus. Res. **60**(1), 76–80 (2007)
18. Crompton, J.L., McKay, S.L.: Motives of visitors attending festival events. Ann. Tour. Res. **24**(2), 425–439 (1997)
19. Erol, G., Deliormanli, A., Düşmezkalender, E.: A tourism accommodation perspective in a post-pandemic nature-based tourism development context. In: The Routledge Handbook of Nature Based Tourism Development, pp. 474–489. Routledge (2023)
20. Cifci, I., Atsiz, O., Kahraman, O.C., Istanbullu Dincer, F.: Recreational fishers' motivations at wetland destinations: the push and pull theory approach. Leisure Sci., 1–23 (2021)
21. Han, S., Anderson, C.K.: Web scraping for hospitality research: overview, opportunities, and implications. Cornell Hosp. Q. **62**(1), 89–104 (2021)
22. Singh, S., Chauhan, T., Wahi, V., Meel, P.: Mining tourists' opinions on popular Indian tourism hotspots using sentiment analysis and topic modeling. In: 2021 5th International Conference on Computing Methodologies and Communication (ICCMC), pp. 1306–1313. IEEE (2021)
23. Sánchez-Franco, M.J., Rey-Tienda, S.: The role of user-generated content in tourism decision-making: an exemplary study of Andalusia, Spain. Manage. Decis. (2023)
24. Sastre, R.P., Phakdee-Auksorn, P.: Examining tourists' push and pull travel motivations and behavioral intentions: the case of British outbound tourists to Phuket, Thailand. J. Qual. Assur. Hosp. Tour. **18**(4), 437–464 (2017)
25. Carvache-Franco, M., Solis-Radilla, M.M., Carvache-Franco, W., Carvache-Franco, O.: Push and pull motivations as demand predictors of coastal and marine destinations: a study in Acapulco, Mexico. Tour. Plan. Dev. **20**(6), 1000–1018 (2023)
26. Tripadvisor (2023). Vietnam adventure tours. https://www.tripadvisor.com/Attraction_Review-g293925-d7171779-Reviews-Vietnam_Adventure_Tours-Ho_Chi_Minh_City.html
27. Bi, J., Wang, Y., Han, T., Zhang, K.: A study of tourists perception of homestays as a sustainable accommodation choice. J. Asiatic Soc. Mumbai **XCVI** (2023)

Performance Analysis of Cognitive Underlay NOMA Random Networks Under Nakagami-*m* Channel

Pham Thi Dan Ngoc[1], Thi-Tuyet-Hai Nguyen[2], Tan N. Nguyen[3], and Lam-Thanh Tu[3(✉)]

[1] Faculty of Electronics Engineering, Posts and Telecommunications Institute of Technology, Ho Chi Minh City, Vietnam
ngocptd@ptit.edu.vn

[2] Faculty of Information Technology, Posts and Telecommunications Institute of Technology, Ho Chi Minh City, Vietnam
ntthai@ptit.edu.vn

[3] Communication and Signal Processing Research Group, Faculty of Electrical and Electronics Engineering, Ton Duc Thang University, Ho Chi Minh City, Vietnam
{nguyennhattan,tulamthanh}@tdtu.edu.vn

Abstract. This study addresses the performance evaluation of cognitive non-orthogonal multiple access (CONOMA) networks. Specifically, we focus on cognitive underlay networks employing non-orthogonal multiple access (NOMA), where all secondary users (SUs) are characterized using tools from stochastic geometry. In addition, we utilize the Nakagami-m distribution, which offers a more versatile model compared to the conventional Rayleigh or Rician distributions, applicable across various transmission environments. In this context, we derive closed-form expressions for the outage probability of both favorable and unfavorable users. The selection criteria for identifying favorable and unfavorable users are based on large-scale path-loss considerations. By doing so, we aim to provide a comprehensive understanding of the network performance under different conditions. Finally, we present simulation results obtained through Monte-Carlo simulations to illustrate and validate our findings.

Keywords: Cognitive radio networks · Nakagami-*m* · non-orthogonal multiple access · outage probability · random networks

1 Introduction

Cognitive radio networks (CRNs) is one of the most effective techniques to facilitate the spectrum usage since it permits the secondary users (SUs) to access the licensed spectrum of the primary users (PUs) provided that the quality-of-service (QoS) of the primary networks (PNs) is strictly guaranteed [1]. The practical implementation of the CRNs is done via two well-known approaches namely, underlay and interweave protocols. Both protocols have its pros and cons. For example, the former allows the SUs to

M. Pagac et al. (Eds.): SCFF 2024, LNNS 1062, pp. 350–358, 2024.
https://doi.org/10.1007/978-3-031-65656-9_35

simultaneously operate with the PNs thus possibly applying for the real-time applications. Nonetheless, the QoS of the secondary networks (SNs) is hard to achieve owing to the constraints on the aggregate interference. One of the powerful combinations of cognitive underlay networks is to utilize non-orthogonal multiple access (NOMA) to concurrently serve multiple users thus improving the networks spectral efficiency [2]. Alternatively, stochastic geometry (SGs) is a vital tool to study the spatial distributions and interactions of the wireless nodes. Hence, it can provide a system-level performance instead of link-level [3]. Additionally, system-level performance is a useful reference for the whole networks optimization. As a consequence, in the present work, we study the cognitive underlay NOMA networks (CONOMA) under the impact of the Nakagami-m distribution. It is emphasized that Nakagami-m distribution is considered a general fading modelling which can be adapted to both the Rayleigh and Rician fading.

The performance of the CRNs, SGs, and NOMA in wireless networks was briefly summarized here. Particularly, the cumulative distribution function (CDF) of the needed time-slot to broadcast a common message to multiple users in cognitive radio networks by employing Fountain codes was investigated in [4]. On the other hand, Tin et al. derived the outage probability in energy harvesting enabled cooperative CRNs networks [5]. These works, nonetheless, do not consider both the NOMA and SG. The OP in the cooperative NOMA power splitting was investigated in [6] while the reliability and security trade-off in wireless information and power transfer (SWIPT)-enabled NOMA networks were addressed in [7]. The relay selection in cognitive NOMA networks was studied in [9] while the time switching based NOMA networks was given in [8]. Besides, [10, 11] were derived the coverage probability and energy efficiency of the LoRa and cellular networks via stochastic geometry. The work in [12] also studied CONOMA with SG, however, they investigate the performance under Rayleigh fading instead of Nakagami-*m*.

These aforementioned works, nevertheless, do not simultaneously take into consideration the combination of the cognitive NOMA networks with stochastic geometry. Additionally, they also consider the rich-scatter environment via Rayleigh fading while we consider the general environment via Nakagami-*m* distribution. Particularly, in the present work, we study the OP of the CONOMA networks with Nakagami-*m* distribution. The present manuscript can serve as a reference for network operators and planners aiming to employ CONOMA networks. It also highlights the impacts of key parameters such as the transmit power of secondary transmitters (STx), the network radius, and other relevant factors.

2 System Model

Let us analyze the CONOMA networks illustrated in Fig. 1, which consist of primary networks (PNs) and secondary networks (SNs). The primary networks consist of a transmitter, PTx, and a receiver, PRx. In the secondary networks, there is a transmitter, STx, and multiple secondary receivers (SRx). Each node is equipped with a single antenna. STx is situated at the center of a disk with a radius O, while the secondary receivers are randomly distributed around STx. Assuming that PTx is situated far enough from SRx so that interference from PNs to SNs can be treated as background noise. In this study,

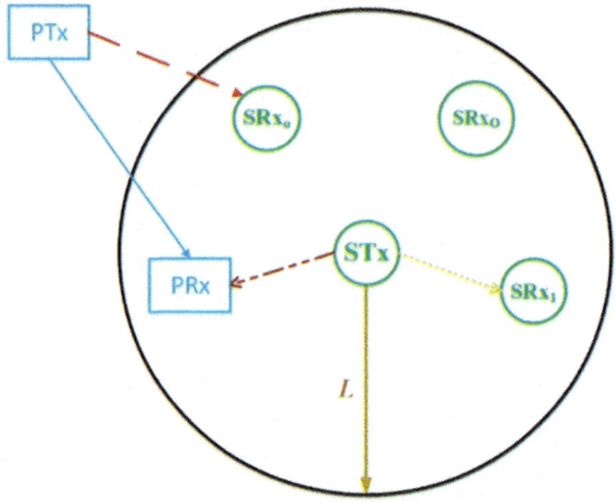

Fig. 1. The considered CONOMA networks

we adopt the underlay protocol, allowing seamless transmissions as long as they do not compromise the quality of service (QoS) for the primary networks. While we discuss the two-user NOMA technique under the power domain in this context, the scenario involving several users is reserved for subsequent investigation due to the significant impact of imperfect successive interference cancellation (SIC) under such conditions [13]. We further assume that perfect CSI at the STx and consider low-mobility users.

2.1 Channel Modelling

For any link between a transmitter and receiver, both large-scale path-loss and small-scale fading are experienced. Shadowing influence is excluded from this study, as they are a well-documented scenario in existing work [11].

Small-Scale fading

Considering the channel coefficient from sender $a \in \{\text{STx}, \text{PTx}\}$ to recipient $b \in \{\text{PRx}\}$ as $h_{a,b}$ and is followed by a Nakagami-m distribution with scale and shape parameters χ and υ, respectively. The channel gain, denoted by $|h_{a,b}|^2$, subsequently follows Gamma distribution with shape and scale parameters χ and υ/χ. Additionally, we assume the system experiences block fading, whereby $h_{a,b}$ remains constant throughout a transmission but differs among transmissions [14].

Remark: In the present manuscript, we adopt Nakagami-*m* fading, as it is considered one of the general distributions capable of modeling other popular fading distributions by properly setting its shape parameter, i.e., *m*.

Large-scale path-loss When transmitting from the sender $a \in \{\text{STx}, \text{PTx}\}$ to the receiver $b \in \{\text{PRx}\}$ the formulation for the large-scale path-loss is as follows:

$$N_{a,b} = T(\max\{1, r_{ab}\})^{\beta} \tag{1}$$

where r_{ab} represents the Euclidean distance from transmitter a and receiver b. The parameter β, , which falls within the range (2, 7), denotes the path-loss exponent, and their values depend greatly on the practical environment. T denotes the path-loss constant at one meter, which is determined by the speed of light, $c = 3 \times 10^8$ m/s, and f_c in (Hz) is the carrier frequency. Specifically, T is calculated as $T = \left(\frac{4\pi}{\theta}\right)^2$ where $\theta = \frac{c}{f_c}$ is wavelength in meters.

2.2 Secondary Receiver Selection

In this work, we adopt the receiver selection according to the path-loss to the STx. The adopted strategy is regarded as an extended-term selection, as changes in the positions of secondary receivers occur less frequently compared to fading. Mathematically, the formulation for the h-th smallest path-loss to the STx is given as [15]:

$$N_{(h)} : N_{(1)} < \cdots < N_{(h-1)} < N_{(h)} < N_{(h+1)} < \cdots < N_{(U)}, \tag{2}$$

where $N_{(h)} = T\left(\max\{1, r_{(h)}\}\right)^{\beta}$ represents the path-loss from the (h)-th SR to the ST, arranged in increasing order, where (1) denotes the smallest path-loss while (U) is the furthest one.

2.3 Signal-to-Interference-Plus-Noise Ratio (SINR) at SRs

The current study employs the two-user power domain NOMA technique. Specifically, the user at a disadvantage, either due to being farther from the secondary transmitter or a weaker channel gain, is assigned higher transmit power to ensure a relatively fair received power compared to the favorable user.

By denoting y_1 and y_2 as the received information at the unfavorable and favorable users, respectively, they are given as follows [16]:

$$\begin{aligned} y_1 &= \sqrt{P_S N_{(n)}^{-1}}\left(\sqrt{a_1}x_1 + \sqrt{a_2}x_2\right)h_{S,(n)} + n_1 \\ y_2 &= \sqrt{P_S N_{(m)}^{-1}}\left(\sqrt{a_1}x_1 + \sqrt{a_2}x_2\right)h_{S,(m)} + n_2 \end{aligned} \tag{3}$$

Here, a_1 and a_2 represent the power allocation coefficients for the unfavorable user denoted by "1" and the favorable user denoted by "2", respectively, satisfying the following properties: $a_1 > a_2$ and $a_1 + a_2 = 1$.[1] n_1 and n_2 represent the additive white Gaussian noise (AWGN) for the far and near users respectively. $h_{S,(n)}$ and $h_{S,(m)}$ are the channel coefficients from STx to SRx_1 and SRx_2 respectively. $N_{(n)}$ and $N_{(m)}$ represent the path-loss from the far and near SRx to STx, where $N_{(n)} > N_{(m)}$. P_S is the transmit power of the secondary transmitter and is given as:

$$P_S = \min\left(P_{max} \frac{I_{agg}}{\left|h_{STx,PRx}\right|^2}\right), \tag{4}$$

[1] Here, the fixed power allocation coefficient is adopted for simplicity.

where P_{max} is the maximal transmitter power that the STx can be emitted owing to the hardware limitation while I_{max} is the aggregate interreference created by the SNs at the PRx.

In this manuscript, we assume that the primary transmitter is situated at a distance from the secondary receivers, enabling us to disregard any interference caused by the PT at the SRs, or treat it as background noise. The signal-to-interference-plus-noise ratio (SINR) is then given as follows:

$$\gamma_1 = \frac{a_1\left(P_S/N_{(n)}\right)\left|h_{S,(n)}\right|^2}{a_2\left(P_S/N_{(n)}\right)\left|h_{S,(n)}\right|^2 + \sigma_1^2}$$
$$\gamma_2 = \frac{a_2\left(P_S/N_{(m)}\right)\left|h_{S,(m)}\right|^2}{\sigma_2^2} \quad (5)$$

In (5), σ_1^2 and σ_2^2 are the variance of the AWGN at the SRx. γ_2 is obtained after employing successive interference cancellation.

2.4 Performance Metrics

In this work, the OP is considered as the sole metric to measure the performance of the considered networks. The OP denotes the likelihood that the instantaneous rate at the receiver falls below the specified threshold. Mathematically, it is expressed as:

$$OP_g = \Pr\left\{\mathrm{BW}\log_2\left(1+\gamma_g\right) < R_{\mathrm{th}}\right\}, g \in \{1, 2\}. \quad (6)$$

Here Pr{.} is the probability operator and R_{th} (in bits per second) is the targeted rate, and BW is the transmission bandwidth. It is noted that other vital metrics such as throughput, fairness, ergodic capacity, and bit error rate will be addressed in future work.

3 Numerical Results

In this section, we present simulation results obtained through the Monte Carlo method. Our primary aim is to address the influence of vital parameters, including the power allocation coefficient, the network radius, and the maximal aggregate interference at the primary receiver, on the performance of outage probability.

Throughout this section, we maintain consistency in the utilization of the following parameters, unless explicitly stated otherwise: $\beta = 3.25$, $f_c = 1.8$ GHz, Bw = 1 MHz, NF = 6 dB, $O = 500$ m, $U = 6$, $\chi = 2.5$, $\upsilon_{\mathrm{STx,SRx}} = 3.1$, $\forall u$, $\upsilon_{\mathrm{STx,PRx}} = 2.75$, $r_{\mathrm{STx,PRx}} = 400$ m, $R_{\mathrm{th}} = 2$ Mbps, $a_1 = 0.55$, $(n) = 2$, $(m) = 4$, $\sigma_1^2 = \sigma_2^2 = -174 + NF + 10\log_{10}(\mathrm{BW})$ [dBm], $P_{max} = 40$ dBm, $I_{agg} = 35$ dBm.

Figure 2 depicts the influence of aggregate interference on outage probability performance. It is evident that increasing aggregate interference, I_{agg}, yields benefit for outage probability. Furthermore, the figure confirms that favorable users in closer proximity to the secondary transmitter achieve superior performance compared to their counterparts. Notably, the gap between their performances is not linear; it initially starts small,

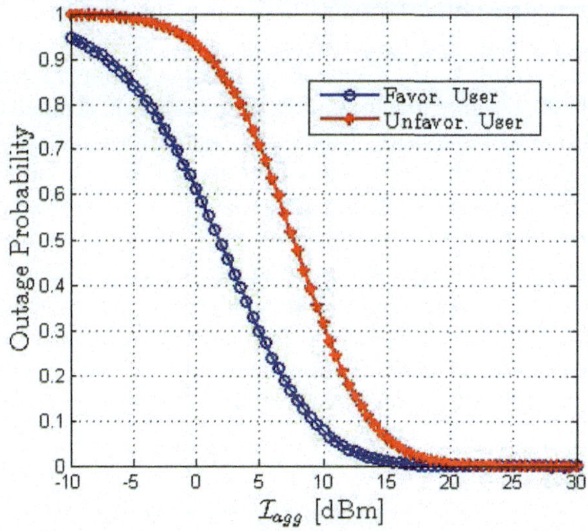

Fig. 2. Outage probability versus the aggregate interference at the primary receiver

expands, and then diminishes as I_{agg} increases sufficiently. Additionally, the figure illustrates that augmenting I_{agg} can mitigate losses attributable to path-loss. Specifically, the outage probability of favorable users reaches 0.1 when I_{agg} is approximately 8 dBm, while unfavorable users achieve similar performance when I_{agg} is not less than 14 dBm.

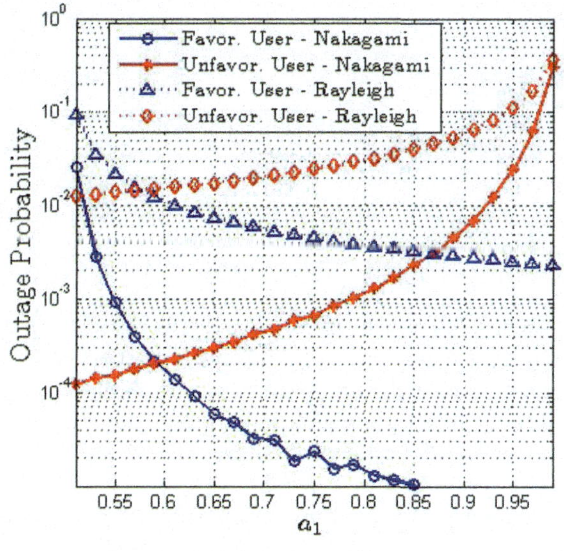

Fig. 3. Outage probability versus power allocation coefficients with different fading distribution

Figure 3 illustrates the OP performance concerning the power allocation coefficients, denoted as a_1, with various fading distributions. The curves labeled 'favorable user - Rayleigh' and 'unfavorable user - Rayleigh' are adapted from [12] to enable comparison with our work. We observe that the curves from our work, denoted as 'favorable user - Nakagami' and 'unfavorable user - Nakagami', outperform those from [12]. Particularly, the best performance of the favorable user under Rayleigh fading solely achieves 0.002, while the best performance of the unfavorable user under Nakagami distribution already attains 10^{-4}. Thus, our work demonstrates superior performance compared to [12]. Moreover, it is evident that increasing a_1 proves advantageous for the favorable user, as it significantly enhances the signal-to-interference-plus-noise ratio. However, this enhancement comes at the expense of decreased performance for the unfavorable user. Consequently, there exists an optimal value of a_1 where the OP of both users is minimized. Notably, this intriguing optimization problem remains unaddressed in the present manuscript.

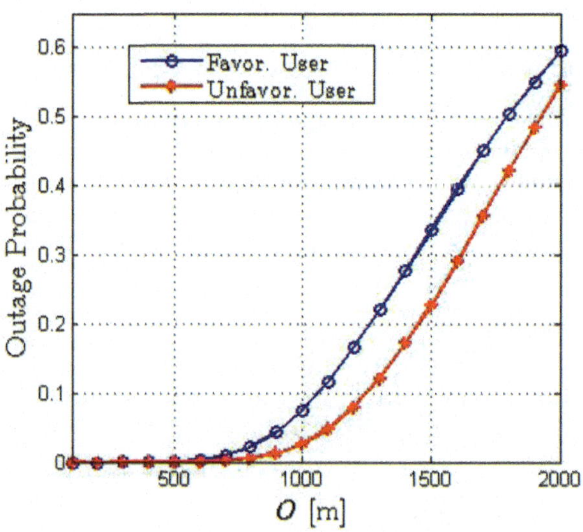

Fig. 4. Outage probability versus the network radius

Figure 4 provides insights into the influence of network radius on OP performance. It is evident that increasing the network radius adversely affects OP. This can be attributed to the larger path-loss associated with greater network radii, leading to higher outage probabilities. Notably, the figure highlights an intriguing observation: closer users experience comparatively worse performance than their counterparts. This can be explained by the increasing path-loss for both the second and fourth closest users. However, while the gap between the second and fourth users diminishes, the interference generated by the further user at the closer user intensifies, thereby degrading the OP of the favorable user.

4 Conclusions

This study delved into the performance analysis of CONOMA networks. Specifically, we derived the OP for both users within cognitive underlay NOMA random networks, incorporating the Nakagami-m distribution. Simulation results were then presented to elucidate the influence of several key parameters, including the maximal aggregate interference at the primary receiver, the network radius, and the power allocation coefficients.

References

1. Mitola, J., Maguire, G.Q.: Cognitive radio: making software radios more personal. IEEE Pers. Commun. **6**(4), 13–18 (1999). https://doi.org/10.1109/98.788210
2. Dai, L., Wang, B., Yuan, Y., Han, S., Chih-lin, I., Wang, Z.: Non-orthogonal multiple access for 5G: solutions, challenges, opportunities, and future research trends. IEEE Commun. Mag. **53**(9), 74–81 (2015). https://doi.org/10.1109/MCOM.2015.7263349
3. Andrews, J.G., Baccelli, F., Ganti, R.K.: A tractable approach to coverage and rate in cellular networks. IEEE Trans. Commun. **59**(11), 3122–3134 (2011). https://doi.org/10.1109/TCOMM.2011.100411.100541
4. Tu, L.-T., Nguyen, T.N., Duy, T.T., Tran, P.T., Voznak, M., Aravanis, A.I.: Broadcasting in cognitive radio networks: a fountain codes approach. IEEE Trans. Veh. Technol. **71**(10), 11289–11294 (2022). https://doi.org/10.1109/TVT.2022.3188969
5. Tin, P.T., et al.: Outage analysis of the power splitting based underlay cooperative cognitive radio networks. Sensors **21**, 7653 (2021). https://doi.org/10.3390/s21227653
6. Ly, T.T.H., Nguyen, H.-S., Nguyen, T.-S., Huynh, V.V., Nguyen, T.-L., Voznak, M.: Outage probability analysis in relaying cooperative systems with NOMA considering power splitting. Symmetry **11**, 72 (2019). https://doi.org/10.3390/sym11010072
7. Nguyen, Q.-S., Nguyen, T.N., Lam-Thanh, T.: On the security and reliability performance of SWIPT-enabled full-duplex relaying in the non-orthogonal multiple access networks. J. Inf. Telecommun. **7**(4), 462–476 (2023). https://doi.org/10.1080/24751839.2023.2218046
8. Van, H., et al.: Outage performance analysis of non-orthogonal multiple access with time-switching energy harvesting. Elektronika ir Elektrotechnika. **25**, 85–91 (2019). https://doi.org/10.5755/j01.eie.25.3.23682
9. Duy, T.T., Ngoc, P.T.D., Tran, P.T.: Performance enhancement for multihop cognitive DF and AF relaying protocols under joint impact of interference and hardware noises: NOMA for primary network and best-path selection for secondary network. Wirel. Commun. Mobile Comput. **2021**, Article ID 8861725, 15 p. (2021). https://doi.org/10.1155/2021/8861725
10. Nguyen, T.-T.-H., et al.: Coverage probability of energy harvesting enabled LoRa networks with stochastic geometry. J. Inf. Telecommun. (2023). https://doi.org/10.1080/24751839.2023.2281144
11. Lam, T.T., Di Renzo, M.: On the energy efficiency of heterogeneous cellular networks with renewable energy sources—a stochastic geometry framework. IEEE Trans. Wireless Commun. **19**(10), 6752–6770 (2020). https://doi.org/10.1109/TWC.2020.3005618
12. Ngoc, P.T.D., Duy, T.T., Nhat, N.L., Hanh, T., Chung, Y.H., Tu, L.-T.: On the performance of outage probability in cognitive NOMA random networks with hardware impairments. J. Inf. Telecommun. (2023). https://doi.org/10.1080/24751839.2023.2294570
13. Jee, A., Prakriya, S.: Novel channel aware power control for a multi-user downlink NOMA network. IEEE Wirel. Commun. Lett. **13**(2), 392–396 (2024). https://doi.org/10.1109/LWC.2023.3330150

14. Mohanty, A.K., Perli, S.B.: Fuzzy based optimal network reconfiguration of distribution system with electric vehicle charging stations, distributed generation, and shunt capacitors. Adv. Electr. Electron. Eng. **21**(2). https://doi.org/10.15598/aeee.v21i2.4599
15. Al-qudah, Z., Darabkh, K.: A simple encoding scheme to achieve the capacity of half-duplex relay channel. Adv. Electr. Electron. Eng. **20**(1) (2022). https://doi.org/10.15598/aeee.v20i1.4347
16. Tripathi, A., Rajawat, A.: Early area and power estimation model for rapid system level design and design space exploration. Adv. Electr. Electron. Eng. **20** (2022). https://doi.org/10.15598/aeee.v20i1.4229

Rice Leaf Diseases Classification Based on RiceNet

Tang Ha Minh Quan(✉) and Ngo Thi Hoa

Faculty of Mechanical–Electrical and Computer Engineering, Van Lang University, Ho Chi Minh City, Vietnam
quan.thm@vlu.edu.vn, hoa.ngo@vlu.edu.vn

Abstract. This paper presents a new model called RiceNet, which is applied to classify rice leaf diseases with compatible and improved accuracy. The RiceNet model is built with 30 layers that are responsible for collecting and extracting features from input diseased leaf images. The research experiment was performed on the Paddy Doctor dataset, which includes 10,407 rice leaf disease images labeled with nine disease types and one standard, respectively. The model's performance was evaluated on the correct type of information classification and implemented for various cases. Research results show that the RiceNet classification model has an accurate classification rate of 93.08%.

Keywords: Deep learning · Artificial intelligence · Data augmentation · Rice disease identification

1 Introduction

Rice is one of the most essential human crops; it can provide food for billions worldwide. Many countries such as China, India, and Southeast Asia have a thriving rice agriculture industry, creating jobs for millions of people and contributing to national GDP. In addition, the rice processing industry, which includes the production of rice, flour, cakes, and other products, also creates meaningful business and export opportunities [1]. However, rice diseases are a critical problem and often cause many losses in the agricultural industry. Diseases that develop on rice leaves can weaken rice plants and reduce the quality and productivity of rice grains, causing significant economic losses to farmers.

To deal with this problem, the classification of rice leaf diseases plays a vital role in general managing, controlling, and researching rice leaf diseases in agriculture [2]. Classifying rice leaf diseases helps determine the exact type of disease the rice plant suffers, thereby providing appropriate control and treatment measures. Farmers and agricultural professionals can make decisions about pesticide application, disease prevention, and other interventions to minimize damage and protect plant health by identifying disease information and characteristics.

M. Pagac et al. (Eds.): SCFF 2024, LNNS 1062, pp. 359–368, 2024.
https://doi.org/10.1007/978-3-031-65656-9_36

Classification of rice leaf diseases is a complex and vital task in agricultural research and management. Classifying rice leaf diseases often involves collecting images and analyzing symptoms, lesions, and changes on rice leaves. The development of artificial intelligence and deep learning has contributed to the automation of rice leaf disease classification, helping to increase the accuracy and efficiency of this process [3]. In particular, the convolutional neural network (CNN) has become a powerful tool to solve this problem.

This paper proposes a convolutional neural network named RiceNet to classify rice leaf diseases. The remainder of the article is organized as follows: Sect. 2 presents the proposed solution, including the RiceNet model and model performance evaluations; Sect. 3 presents the experimental setup; Sect. 4 presents the results and discussion; Sect. 5 concludes the paper.

2 Methods

2.1 Overall Workflow

An overview diagram of the rice leaf disease classification process is presented in Fig. 1. First, image data was collected and labeled with each corresponding rice leaf disease to ensure accurate classification. Next, this dataset is divided into three separate parts: training set (70%), evaluation set (20%), and test set (10%). In addition, several data preprocessing techniques are applied to improve the quality and standardize the data in the training set.

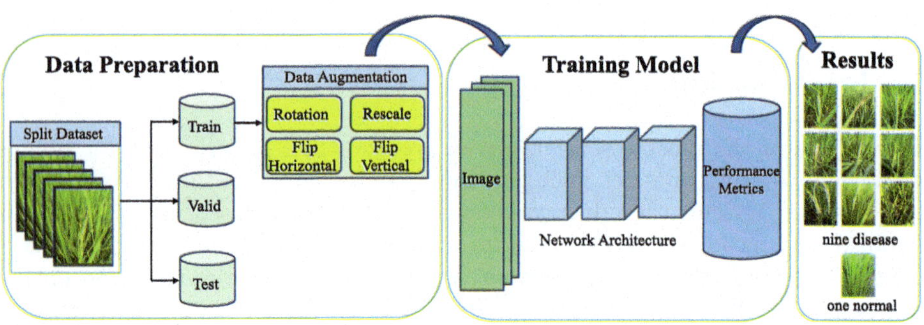

Fig. 1. Overview process for identifying rice leaf diseases.

The next step is to design an improved neural network architecture to extract essential features from images of diseased rice leaves. This model will be trained to detect essential features and classify rice leaf diseases corresponding to the previously assigned labels. The parameters of the RiceNet model are continuously changed and updated during the training process through the evaluation results of the evaluation set.

After the model has been trained, the model will be applied to the rice leaf images in the test set and produce rice leaf disease classification results. The predicted results will be compared with the actual labels to evaluate the model performance.

2.2 RiceNet Model

In this study, we propose a network architecture named RiceNet, as shown in Fig. 2. The RiceNet has 30 layers, and the main goal is to identify rice leaf diseases through visual images. In the proposed model, six essential components comprise the network architecture: Convolutional Layer, Activation Function, Pooling Layer, Fully Connected Layer, Dropout Layer, and Softmax Layer. Combining these layers helps the network learn the features extracted from the input data, perform classification, and produce the desired results. These components will be clarified as follows.

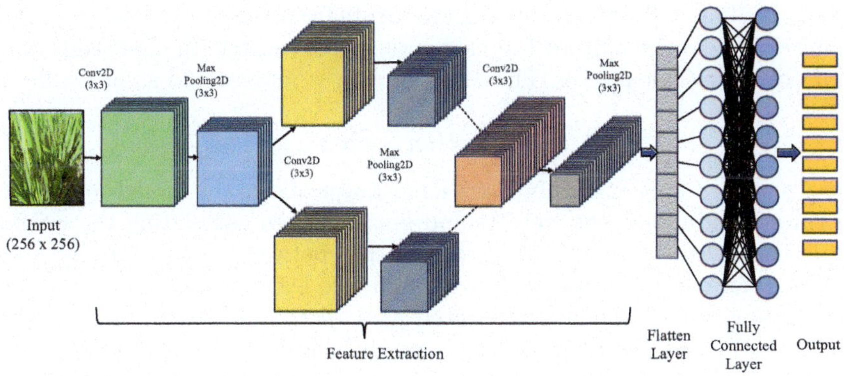

Fig. 2. The architecture of the RiceNet model.

Convolutional Layer: The critical layer often appearing in the first position in CNNs is the Convolutional Layer. The main task of this layer is to extract related features of the input data image. Convolutional captures and maintains relationships between pixels using small plots of input data, allowing it to remove image features. It is a mathematical operation with two inputs: an image matrix and a filter. An image matrix (volume) of dimension is $H * W * D$. A filter is $F_H * F_W * D$.

The outputs a volume dimension is $(\frac{H-F+2P}{S} + 1) * (\frac{W-F+2P}{S} + 1) * K$.

Where:

- H is the height of the image;
- W is the width of the image;
- D the thickness of the image;
- S is stride;
- P is padding;
- K is the number of kernels.

Activation Function: The Rectified Linear Unit (ReLU) is a nonlinear function commonly used in neural networks. In some cases, nonlinear functions such as sigmoid and tanh can replace the ReLU function. However, the ReLU function is often preferred thanks to its good performance. The ReLU activation function converts negative values and retains positive values. So, essential features can be included during the training process, and unnecessary features can be eliminated. The ReLU function for an input value x is represented as in Eq. 1.

$$ReLU(x) = max(0, x). \tag{1}$$

Pooling Layer: For an image that is too large, pooling will have the task of shrinking the spatial size of the input, reducing computational complexity but retaining important information about the image. There are many different pooling options, such as Max Pooling, Average Pooling, and Sum Pooling. In particular, Max Pooling will take the most significant element from the object matrix, Average Pooling will take the average elements from the object matrix, and Sum Pooling will take the sum of the elements in the matrix. In this study, we use Max Pooling for calculation. The Max Pooling is represented as in Eq. 2.

$$MaxPooling = max(x). \tag{2}$$

Fully Connected Layer: A fully Connected Layer allows the model to learn complex rules by adjusting the weights and biases. When calculating the weights, a fully linked layer can capture the relationship between input features, thereby providing more accurate results for different rice leaf diseases. In the Fully Connected Layer, each input is represented as a vector, and each output is also a vector. The Fully Connected Layer for production is defined as in Eq. 3.

$$output = activationfunction(W * input + b). \tag{3}$$

where:

- *input* is the input vector of the Fully Connected Layer;
- W is the weight matrix with size (number of outputs, number of inputs), where each element $W[i, j]$ represents the weight between the jth input and the ith output;
- b is the bias vector with size (number of outputs), representing the translation effect on each output;
- "*" denotes matrix multiplication;
- *activationfunction* is an activation function applied to the output after the computation, such as the ReLU function or the sigmoid function.

Dropout Layer: Dropout Layer is a technique used to limit overfitting, which is common in CNN networks, and helps the model calculate faster and more efficiently. Dropout removes a random number of neurons in the network with a given probability by setting all the neuron weights to 0, which means that the links to that neuron are all invalid. Then, the model must correctly identify while lacking information from the removed neurons. This will help increase the model's recognition rate but not be too dependent on training data.

Softmax Layer: The softmax function calculates the probability that a particular class appears in all possible courses, as in Eq. 4. At that point, this likelihood will be utilized to decide the input's target lesson. If one of the input values is exceptionally low or negative, the softmax work will change to a low likelihood. If an input value is significant, it will be converted into a large likelihood.

$$\sigma(z)_i = \frac{e^{z_i}}{\sum_{j=1}^{K} e^{z_i}}. \tag{4}$$

The result of the Softmax Layer is a vector with values ranging from 0 to 1 and a sum of 1. These values can be interpreted as the probabilities of each class given the corresponding input. The class with the highest probability is often considered as the predicted class.

2.3 Assessing the Performance of Models

Accuracy was used as a measure in this study because it is an essential factor in determining the reliability of research results. Additionally, it evaluates and compares the performance of this model against other models.

True Positive (TP) and True Negative (TN) are used in the classification problem; it is essential for calculating performance metrics of a classification model, such as accuracy, recall, specificity, and F1-score. In particular, TP occurs when the classification model correctly identifies a positive sample; it represents the number of accurate optimistic predictions made by the model. TN occurs when the classification model correctly identifies a negative sample, which is the number of accurate pessimistic predictions made by the model.

Besides TP and TN, the confusion matrix includes FP and FN. FP occurs when the classification model incorrectly identifies a negative sample. It represents the number of wrong optimistic predictions made by the model. FN occurs when the classification model incorrectly identifies a positive sample. It represents the number of wrong pessimistic predictions made by the model.

3 Experimental Setup

3.1 Dataset

In this study, we used the Paddy Doctor [4] to identify rice leaf diseases depicted in Fig. 3. These images were collected from a rice field in India; data collection occurred from February to April 2021, when the rice age was between 40 and 80 days. Agronomists then manually annotated each image according to disease symptoms and labeled them accordingly.

The data set contains 10,407 images of rice leaves with annotations on ten types (9 disease types and regular leaves). The Paddy Doctor annotated public data set has been of great help in developing automated solutions for accurate and effective rice disease diagnosis.

(a) blast (b) brown spot (c) dead heart (d) downy mildew (e) hispa

(f) leaf bligh (g) leaf streak (h) panicle blight (i) tungro (j) normal

Fig. 3. Sample images of the Paddy Doctor dataset.

3.2 Data Acquisition

The data preparation process includes the following steps in the next order: First, the entire data set will be randomly divided into training sets, evaluation sets, and test sets. Details of dividing the data in the image set are presented in Table 1. Second, data augmentation techniques are applied to create new versions of the data by data transformations of different materials. This process involves enlarging, reducing, or rotating random images vertically or horizontally to increase the number and diversify the data set. The main goal of data augmentation is to improve the performance and generalization of the trained model. Finally, the images will be normalized to 256 × 256 pixels before being included in the training model.

3.3 Software Tools

All experiments were conducted on the deep learning toolbox platform MATLAB 2023a. The device we use to train the model is a Dell Vostro 3020 MT Core i7 13th Gen computer - (8 GB/512 GB SSD/Windows 11 Home/4 GB Graphics) equipped with NVIDIA GeForce GTX 1050Ti.

Each classifier's performance was evaluated using the following metrics: accuracy, recall, specificity, precision, and F1 Score. Table 2 presents the parameters used during the experiment, batch size 128 was used with a learning rate of 0.0003, the number of epochs was set to 25, 50, 75, and 100, the dropout rate

Table 1. Allocate the number of details of images after randomly distributed.

Class	Sample	Train	Valid	Test
Blast	1738	1215	348	174
Brown spot	965	675	193	97
Dead heart	1442	1010	288	144
Downy mildew	620	434	124	62
Hispa	1594	1116	319	159
Leaf blight	479	335	96	48
Leaf streak	380	266	76	38
Panicle blight	337	236	67	34
Tungro	1088	762	217	109
Normal	1764	1234	353	177

varied from 0.2 to 0.4, and the optimization algorithm used was Adam. To evaluate the model's performance, we compared it with other models such as DCNN [5], ShuffleNetV2 [6], MobileNetV1 [7], and EfficientNetB4 [8].

Table 2. Model training parameters.

Parameters	Values
Dropout	0.2 to 0.4
Learning rate	0.0003
Batch size	128
Epoch	25 50 75 100
Optimizer	Adam

Choosing parameters to train the model is one of the essential steps; it dramatically affects the model's classification performance. The RiceNet model used different dropout values ranging from 0.2 to 0.4 during training. The experimental process shows that at a dropout value of 0.2, the model's accuracy reaches the most optimal results.

4 Results and Discussion

Table 3 presents the RiceNet model trained from 25 to 100 epochs; the results show that the model accuracy improves after each epoch; it increases significantly from epoch 25 to epoch 100; epoch 100 gives the highest performance in terms of accuracy 93.08%, precision 91,72%, recall 93.21% and F1 score of 92.46%, which shows that the model has improved over time. Furthermore, the performance growth of the model achieved stability from epoch 100 onwards. In this experiment, we stopped the training process to avoid wasting resources.

Furthermore, the RiceNet model consumes fewer resources (memory, computing power) and costs (hardware, energy) per epoch, which helps the model's scalability become more convenient and can to put into practical application to classify diseases on rice leaves.

Table 3. The table compares model performance between epochs.

Epoch	Accuracy	Precision	Recall	F1 Score
25	88.92	87.24	88.37	88.83
50	90.72	89.82	89.76	89.03
75	91.78	90.96	90.97	90.34
100	**93.08**	**91.72**	**93.21**	**92.46**

In Table 4, the performance of the RiceNet model is compared with four previously published models in classification including: DCNN, ShuffleNetV2, MobileNetV1, EfficientNetB4, in which the RiceNet model stands out with its classification accuracy. The DCNN model has the lowest classification performance with 88.84% accuracy, 89.22% precision, 88.84% recall, and 88.81% F1 score. This shows the ability of the RiceNet model to classify images related to diseases of rice leaves. With better accuracy, the RiceNet model can reliably classify different types of rice plants in the most accurate and efficient way.

Table 4. The table compares the performance of the RiceNet model with other models.

S. No	Model	References	Accuracy	Precision	Recall	F1 Score
1	DCNN	[5]	88.84	89.22	88.84	88.81
2	ShuffleNetV2	[6]	91.14	85.83	84.18	84.94
3	MobileNetV1	[7]	92.06	87.74	85.25	86.32
4	EfficientNetB4	[8]	92.59	91.09	91.01	90.99
5	**RiceNet**		**93.08**	**91.72**	**93.21**	**92.46**

Figure 4 shows the confusion matrix of the model's predictions on the experimental data set. The RiceNet model achieved the highest classification results for the hispa disease type, which showed that the model can accurately classify rice leaves infected with hispa disease. During training, the RiceNet model learned the features of this disease and could distinguish them from other diseases most effectively.

The classification of bacterial leaf blight by the RiceNet model has low accuracy. This may be due to confusion about the characteristics of other rice leaf diseases or a lack of input data for this disease. To improve classification performance for bacterial leaf blight, the authors will continue researching to enhance a more diverse training data set for this disease.

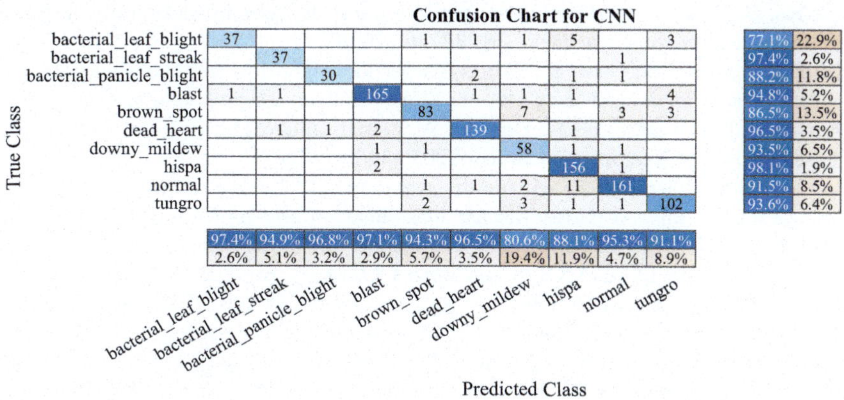

Fig. 4. Confusion matrix of the RiceNet model

5 Conclusion

The application of deep learning models is bringing enormous potential to improve the productivity of the agricultural industry. One of them is the application of deep learning models to identify and classify rice leaf diseases from input images. This helps agriculturalists promptly detect leaf diseases to have more effective prevention and control plans.

In this study, the authors proposed a RiceNet model that can effectively detect rice leaf diseases. Images were collected from infected rice leaves from farmers to create a complete dataset labeled with nine disease types and one regular healthy leaf. Based on that data set, we added several data augmentation techniques to make the data set more prosperous and diverse, from which the model's classification process was significantly improved. After the model training process, the RiceNet model shows more effective classification performance than other deep learning models.

The RiceNet model has demonstrated outstanding performance with an accuracy of 93.08%, 91.72% precision, 93.21% recall, and an F1 score of 92.46%, compared with four other published deep learning models. The next plan of this research is to integrate the model into a mobile application so that farmers can take images of suspected diseased plants. The application will use the trained model to classify. And evaluate diseases on rice leaves, thereby suggesting early prevention and treatment measures to farmers or providing guidance to promptly contact agricultural experts.

References

1. Bishwajit, G., Sarker, S., Kpoghomou, M.A., Gao, H., Jun, L., Yin, D., Ghosh, S.: Self-sufficiency in rice and food security: a South Asian perspective. Agric Food Secur **2**(1), 1–6 (2013)
2. Dordas, C.: Role of nutrients in controlling plant diseases in sustainable agriculture. a review. Agron. Sustain. Dev. **28**, 33–46 (2008)
3. Kaur, S., Pandey, S., Goel, S.: Plants disease identification and classification through leaf images: a survey. Arch. Comput. Methods Eng. **26**, 507–530 (2019)
4. Murugan, D., et al.: Paddy Doctor: A Visual Image Dataset for Paddy Disease Classification. arXiv preprint arXiv:2205.11108 (2022)
5. Petchiammal, Kiruba, B., Murugan, Arjunan, P.: Paddy doctor: a visual image dataset for automated paddy disease classification and benchmarking. In: Proceedings of the 6th Joint International Conference on Data Science & Management of Data (10th ACM IKDD CODS and 28th COMAD), pp. 203–207 (2023)
6. Ma, N., Zhang, X., Zheng, H.T., Sun, J.: Shufflenet v2: Practical guidelines for efficient cnn architecture design. In: Proceedings of the European Conference on Computer Vision (ECCV), pp. 116–131 (2018)
7. Howard, A.G., Zhu, M., Chen, B., Kalenichenko, D., Wang, W., Weyand, T., Andreetto, M., Adam, H.: Mobilenets: Efficient Convolutional Neural Networks for Mobile Vision Applications. arXiv preprint arXiv:1704.04861 (2017)
8. Nigam, S., Jain, R., Marwaha, S., Arora, A., Haque, M.A., Dheeraj, A., Singh, V.K.: Deep transfer learning model for disease identification in wheat crop. Eco. Inform. **75**, 102068 (2023)

Smart Economy, Industrial Management, Mathematical and Simulation Model, Digital Transformation, E-Commerce, Business Intelligence, Economics, Political Economy, Management, Logistics

Analyze the Backward and Forward Linkages of Intermediate Products Network in a Frontier Market

Quoc-Nha Le(✉), Cong-Minh Huynh, and Hoang-Sy Nguyen

Eastern International University, Thu Dau Mot City, Binh Duong, Vietnam
nha.le@eiu.edu.vn

Abstract. This study takes a close look at how different industries in Vietnam are connected to each other, using a method called network analysis and data from Input-Output tables. It aims to figure out which industries are most important in driving the economy and how they're linked to each other. The findings show that Vietnam relies heavily on basic industries like farming and simple manufacturing. If something big happens to one of these industries, it could have a ripple effect throughout the whole economy. The study also points out specific industries that are closely connected to others, both in terms of buying from and selling to them. It emphasizes the crucial role of basic industries in keeping the economy going. Understanding these connections can help government officials make better decisions about how to manage the economy and deal with challenges like global tensions and new technologies.

Keywords: Input-Output · Network analysis · Backward linkages · Forward linkages

1 Introduction

As an emerging approach to research, network theory considers how a set of agents behaves and interacts with the others in a complex system (Phelan 2001). The approach has also become increasingly popular in economics studies that consider intersectoral linkages and strategies to determining the key sectors in the economy. According to Michael and Battiston (2009), the behavior of the economy cannot be examined by the behavior of isolated individuals. Those models excluding the interlink patterns from their analysis may not fully capture the features and properties of certain phenomena either (Jackson 2010). Unlike traditional cause–effect relationships and linear thinking, network-based economists consider the interactions and involve them in the models. Therefore, a growing number of economists have suggested that input-output linkages play a vital role to explain the relationship between sectoral shocks and macro-level volatility (e.g. Horvath 1998, 2000; Gabaix 2011; Acemoglu et al. 2013; Acemoglu et al. 2016; Acemoglu et al. 2017; Jones 2013; Carvalho et al. 2014; Cavallo et al. 2014; Carvalho et al. 2016). Hence, capturing the network structure of industrial linkages can help better explain the origins of aggregate fluctuation as well as the potential channels

M. Pagac et al. (Eds.): SCFF 2024, LNNS 1062, pp. 371–381, 2024.
https://doi.org/10.1007/978-3-031-65656-9_37

of shock diffusion, which is, in turn, useful for policymakers to propose effective policies to mitigate negative shocks such as geopolitical tensions, energy crisis, technology wars, etc. or to adapt with the age of Generative AI and digitalization.

From a developmental standpoint, industries' ability to drive demand for products from other sectors, known as linkage, is crucial. Linkages represent input-output connections between firms or sectors within an economy. Forward linkage occurs when a manufacturing sub-sector supplies output to various sectors of the economy, while backward linkage describes the relationship where the manufacturing sub-sector acts as a consumer of inputs from other sectors. Understanding these linkages is essential for assessing the manufacturing sector's impact on the national economy.

In this paper, we apply network theory to analyse the input–output table to explore the backward and forward linkage of Vietnam's production network. We use the 2022 input-output table of Vietnam economy to firstly investigate whether that the shocks to different sectors in the industrial network can create huge volatility at the macro level through both hierarchical structure and the dense connections of industrial sectors. Second, we identify the specific sectors of Vietnam economy that asymmetrically affect others and so can be considered as "systemically important hubs". The recent economic transformation has propelled Vietnam onto the global stage, attracting foreign investment, fostering industrialization, and lifting millions out of poverty. Looking ahead, Vietnam aims to become a high-income country by 2045 while pursuing sustainable and inclusive growth. However, Vietnam is also facing constraints such as low external reserves, high public debt, recurrent tensions with China, complicated business environment, weak infrastructure. By understanding the interdependencies within the industrial sectors, Vietnam policy makers can better manage the circular flow of the economy, more effectively reallocate public expenditures to the crucial sectors, avoid and mitigate the propagation of shocks from one sector to the whole economy, and formulate economic policies to stimulate and resolve bottlenecks to economic growth.

The remaining part of the paper proceeds as follows. Section 2 data and methodology is presented in Sect. 2. Section 3 provides results and discussions. Concluding remarks and policy implications are indicated in Sect. 4.

2 Data and Methodology

2.1 Network Representations

We represent the IOTs raw data in the forms of standardized matrices for further network analysis procedure. Mathematically, the input-output linkages among sectors are represented by a directed, weighted network with possible self-loops, where vertices represent industrial sectors and links represent the flows of intermediate inputs.

Therefore, the domestic production network of Vietnam (with M industrial sectors indexed by i = 1, 2,…, M) can be represented by the directed weighted matrix W_d, and the adjacency matrix Ad:

$$\boldsymbol{W}^{d} = \left\{w_{ij}^{d}\right\}_{M\times M} = \begin{bmatrix} w_{11}^{d} & \cdots & w_{1M}^{d} \\ \vdots & \ddots & \vdots \\ w_{M1}^{d} & \cdots & w_{MM}^{d} \end{bmatrix} \tag{1}$$

$$A^d = \left\{a^d_{ij}\right\}_{M \times M} = \begin{bmatrix} a^d_{11} & \cdots & a^d_{1M} \\ \vdots & \ddots & \vdots \\ a^d_{M1} & \cdots & a^d_{MM} \end{bmatrix} \tag{2}$$

The weighted matrix W_d represents directed network with possible self-loops, in which each element indicates an economic flow of intermediate goods from one sector to another sector. For the adjacency matrix A_d, $a_{ij} = 1$ if there is a link between nodes i and j and $a_{ij} = 0$ otherwise. With self-loop, $aii = 1\ \forall i = 1, \ldots, M$; and without self-loop, $a_{ii} = 0$.

Besides, the imported production network in of Vietnam (with M industrial sectors indexed by $i = 1, 2, \ldots, M$) can be mathematically represented by a directed weighted matrix WI:

For imported production network, the weighted matrix WI is:

$$W^I = \left\{w^I_{ij}\right\}_{M \times M} = \begin{bmatrix} w^I_{11} & \cdots & w^I_{1M} \\ \vdots & \ddots & \vdots \\ w^I_{M1} & \cdots & w^I_{MM} \end{bmatrix}, \tag{3}$$

and the adjacency matrix A^I is:

$$A^I = \left\{a^I_{ij}\right\}_{M \times M} = \begin{bmatrix} a^I_{11} & \cdots & a^I_{1M} \\ \vdots & \ddots & \vdots \\ a^I_{M1} & \cdots & a^I_{MM} \end{bmatrix}, \tag{4}$$

where a_{ij} represents the flow of imported intermediate goods from sectors i to domestic sector j.

The weighted matrix W^I represents directed network with possible self-loops, in which each element indicates an economic flow of imported intermediate goods to domestic sectors. For the adjacency matrix A^I, $a_{ij} = 1$ if there is a link between nodes i and j and $a_{ij} = 0$ otherwise. With self-loop, $aii = 1\ \forall i = 1, \ldots, M$; and without self-loop, $a_{ii} = 0$.

2.2 Distribution of Degree and Weighted Degree

A sector in a network is considered as a "node". A directed weighted network is mainly characterized by its node strength and centrality properties. The strength of nodes is basically measured by the weighted degree. Weighted degree of a node measures the strength of interaction with all its neighbors. Degree, on the other hand, is the total number of neighbors that one node has connection. Node degree is the simplest measure of network centrality.

Mathematically, for every node i, its in-degree ($k^{in}{}_i$: the total number of in-coming connections) and weighted in-degree ($wd^{in}{}_i$: the total strength of in-coming connections) are respectively defined as the sums of all elements in the column i^{th} of the adjacency and weighted matrices. Similarly, the out-degree ($k^{out}{}_i$: the total number of out-going connections) and weighted out-degree (wdouti: the total strength of out-going connections) are respectively defined as the sums of all elements in the row ith of the adjacency and weighted matrices. Nodes with high degree and weighted degree are basically referred to as hubs. Recall that, in directed network, in-degree and out-degree are given as follows:

$$k_i^{in} = \sum_j A_{ji}, \tag{5}$$

$$k_i^{out} = \sum_j A_{ij}, \tag{6}$$

where $A_{ij} = 1$ indicates a link from i to j.

Meanwhile, the weighted in-degree and weighted out-degree are given by:

$$s_i^{in} = \sum_j W_{ji}, \tag{7}$$

$$s_i^{out} = \sum_j W_{ij}, \tag{8}$$

where W_{ij} is the element of the W_d matrix.

2.3 Eigenvector Centrality

Eigenvector centrality is like a popularity score for a node in a network. It's different from just counting how many friends a node has. It also looks at how popular those friends are. Nodes with high eigenvector centrality are not only popular themselves but are connected to other popular nodes. This helps find important nodes that shape how information flows in the network. It's measured using a mathematical formula. It is measured as follows:

$$C_e(n_i) = \lambda \sum_j x_{ij} e_j, \tag{9}$$

where Xij is the relationship matrix weighted by the degree centrality of node j, and λ is the graph's eigenvalue for the eigenvector of node i.

3 Results and Discussions

3.1 Vietnam Production Network

Based on total output values, the top sectors contributing significantly to the economy, listed in Fig. 1. The total output values of various sectors were aggregated and ranked to determine the top sectors driving economic output. This implies that the economy of Vietnam is still mainly drived by primary and low-tech industries. These low-tech industries also stay within the top sectors which require as well as supply the intermediate goods and services (See Figs. 2 and 3). This reflects that the economy still heavily depends on higher-tech products from foreign countries (Figs. 4, 5 and 6).

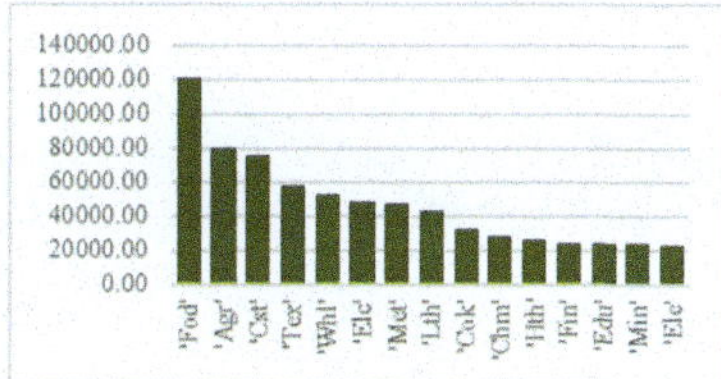

Fig. 1. Total output of industrial sectors in Vietnam, 2022.

Figure 7 demonstrates that the sectors with higher backward linkage, which are measured by weighted in-degree, also tend to have higher value added in the total output. Compared with the 45-degree line, those sectors lying below the lines indicate that intermediate goods play a crucial role in these sectors' output production.

Figure 8 demonstrates that the sectors with higher forward linkage, which are measured by weighted out-degree, also tend to have higher quantity of final demand. Compared with the 45-degree line, those sectors lying above the lines are recognized as industries which mainly serve the final consumers both domestically and internationally rather than provide intermediate goods for domestic production network. The minority number of sectors lie below the 45-degree line are identified as auxiliary industries of Vietnam production network.

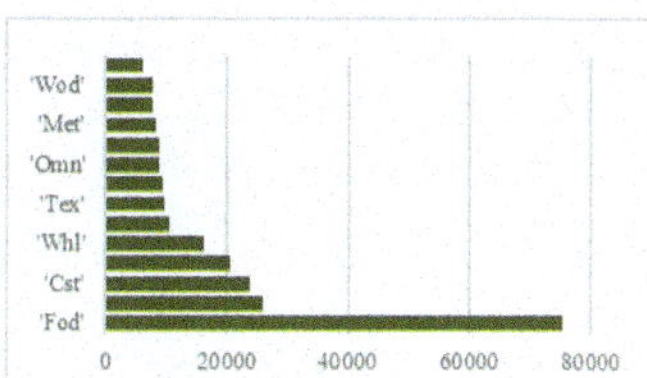

Fig. 2. Weighted in-degree.

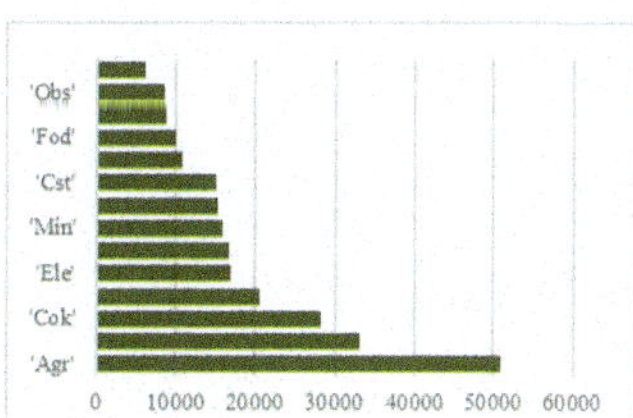

Fig. 3. Weighted out-degree.

Figures 4, 5 and 6 show the ego network of key sectors to better understand the production network connected to a single key industry.

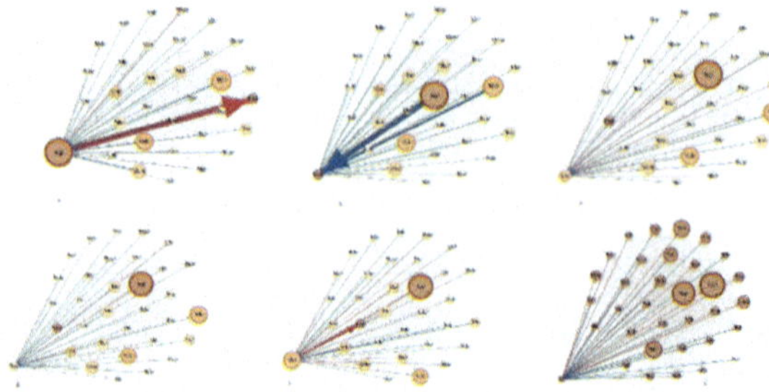

Fig. 4. Ego network of a. Agriculture, hunting, forestry, and fishing; b. Food, beverages, and tobacco; c. Construction; d. Textiles and textile products; e. Wholesale trade and commission trade, except of motor vehicles and motorcycles; f. Electrical and optical equipment.

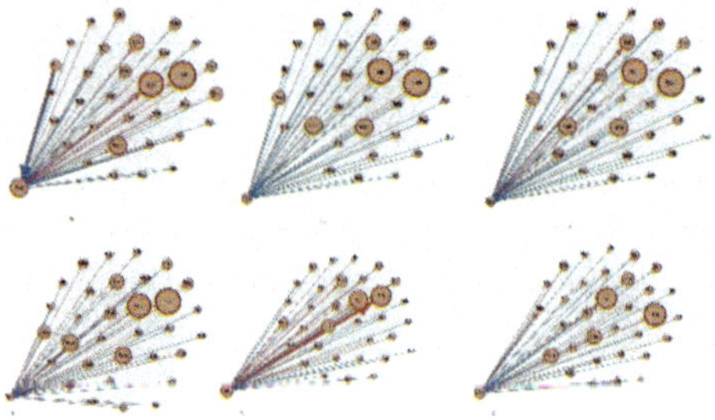

Fig. 5. Ego network of a. Coke, refined petroleum, and nuclear fuel; b. Textiles and textile products; c. Other nonmetallic minerals; d. Manufacturing, nec; recycling; e. Wholesale trade and commission trade, except for motor vehicles and motorcycles; f. Chemicals and chemical products.

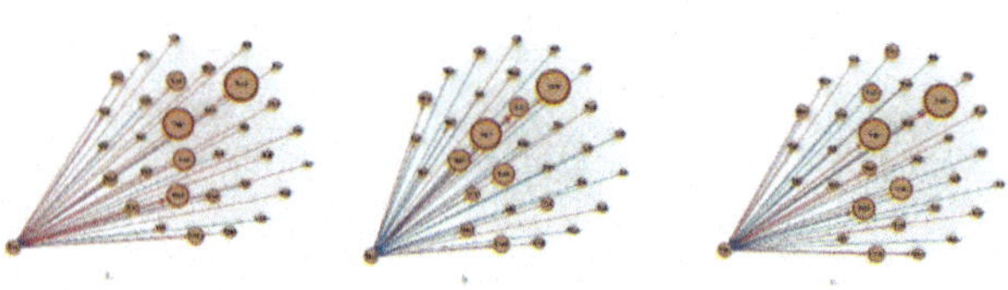

Fig. 6. Ego network of a. financial intermediation; b. Basic metals and fabricated metals; c. Electrical and optical equipment.

3.2 Direct and Indirect Linkage

To discuss the direct and indirect linkage of the production network. We employ the Leontief Inverse matrix, where each node represents an industry, and the weight of the edge corresponds to the strength of the linkage or the size of the Leontief Inverse coefficient.

The histograms of the degrees sectors in Vietnam, 2022 are shown in Fig. 9. We observe that the mass of the distributions of degrees concentrates on the left (positive skewness), and the ranges of the distributions of degrees in the domestic production network are wide. This indicates the presence of hubs in the input-output linkages in

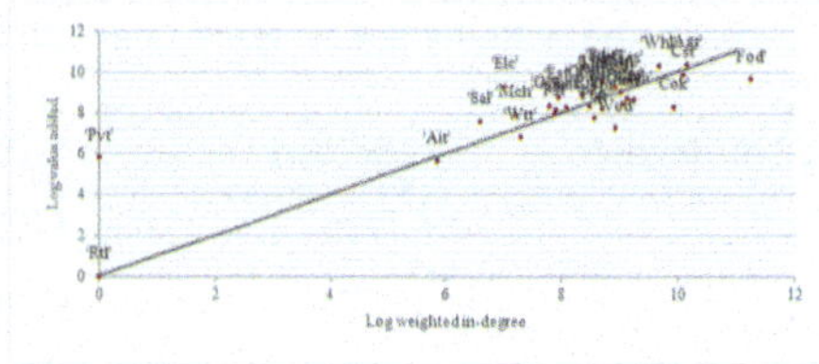

Fig. 7. Correlation between value added and weighted in-degree.

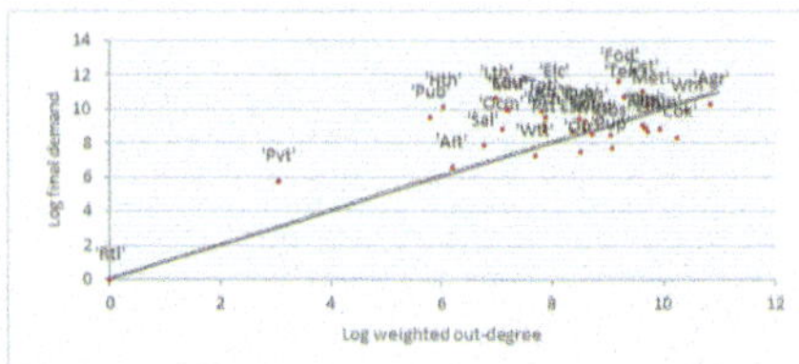

Fig. 8. Correlation between final demand and weighted out-degree.

Vietnam industrial network. In other words, there are key sectors with dominant connection in intermediate products exchange within the industrial network. These sectors are listed in Table 1:

Table 1. Top 16 sectors with highest degree centrality

Sectors	Degree	Sectors	Degree
'Whl'	34	'Met'	14
'Ele'	30	'Agr'	13
'Obs'	23	'Min'	12
'Cok'	22	'Wod'	12
'Fin'	19	'Rub'	12
'Cst'	18	'Est'	12
'Pup'	17	'Pub'	12
'Chm'	17	'Omn'	11

In detail, Fig. 10 with the distribution of weighted in-degree of the Leontief inverse coefficient reflects the backward linkage of the sectors. Furthermore, the positive skewness emphasizes that there are some key sectors with significant demand for intermediate input. These sectors have a significant impact on downstream industries in the economic value chain. In other words, a large proportion of goods and services from other sectors are invested in these sectors (Ahmad and Alvi 2024). See the list of these sectors in Table 2 as follows:

Table 2. Top 16 sectors with highest weighted in-degree centrality

Sectors	Weighted in-degree	Sectors	Weighted in-degree
'Wod'	0.621	'Fin'	0.320
'Cok'	0.597	'Chm'	0.314
'Fod'	0.557	'Htl'	0.296
'Omn'	0.485	'Pub'	0.291
'Pup'	0.422	'Ait'	0.279
'Mnf'	0.370	'Pst'	0.275
'Wtt'	0.358	'Agr'	0.268
'Otr'	0.342	'Obs'	0.261

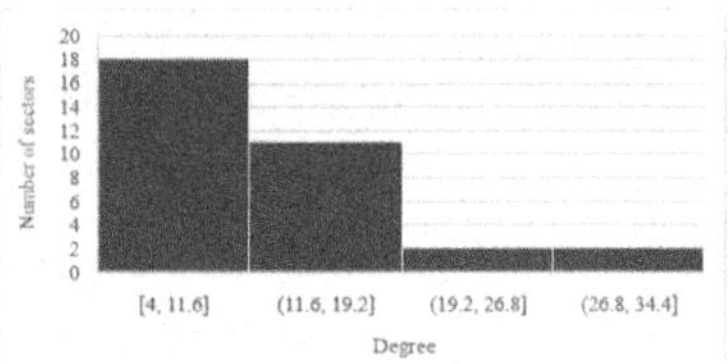

Fig. 9. Degree distribution of Leontief inverse coefficient.

Furthermore, Fig. 11 with the distribution of weighted out-degree of the Leontief inverse coefficient reflects the forward linkage of the sectors. The positive skewness highlights the importance of specific key sectors within the supply chain. These sectors play a vital role by providing essential inputs or components that downstream industries further process or utilize in the production of final goods or services. Thus, these sectors are considered as upstream industries or strategic industries. Table 3 lists the top 16 largest upstream industries as follows:

A node will have high eigenvector centrality if it has high connectivity with high-influencing nodes. As highlighted in Fig. 12, services sectors are the highest eigenvector centrality industries. This result reflects that all the sectors in the product network perceive the spillover effects and are vulnerable to the quality of business and public services.

Figure 13 shows the visualization of the Leontief inverse matrix representing the direct and indirect linkage of production network in Vietnam in 2022. Different colors represent sectors in different linkage communities within the network and the size of each node is proportional to its weighted out-degree. The arrows represent the flow of intermediate inputs. Dense connections and scale-free structure of the network imply that sectoral linkages could be an important transmission path of exogenous shocks. In other words, the shocks to different sectors can create huge volatility at the macro level.

Table 3. Top 16 sectors with highest weighted out-degree centrality

Sectors	Weighted in-degree	Sectors	Weighted in-degree
'Cok'	1.317	'Otr'	0.456
'Whl'	0.814	'Met'	0.433
'Agr'	0.748	'Pup'	0.413
'Chm'	0.639	'Obs'	0.397
'Ele'	0.589	'Wod'	0.364
'Fin'	0.561	'Tex'	0.221
'Min'	0.534	'Est'	0.155
'Cst'	0.493	'Rub'	0.138

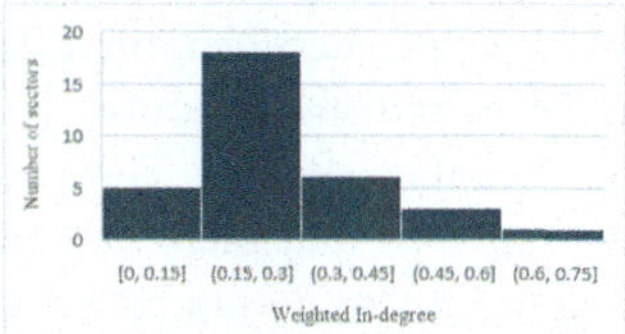

Fig. 10. Weighted in-degree distribution of Leontief inverse coefficient.

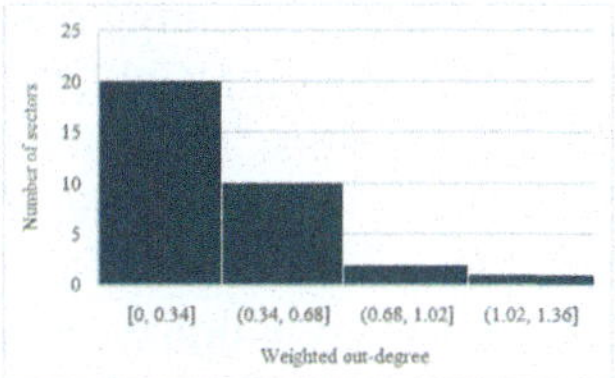

Fig. 11. Weighted out-degree distribution of Leontief inverse coefficient.

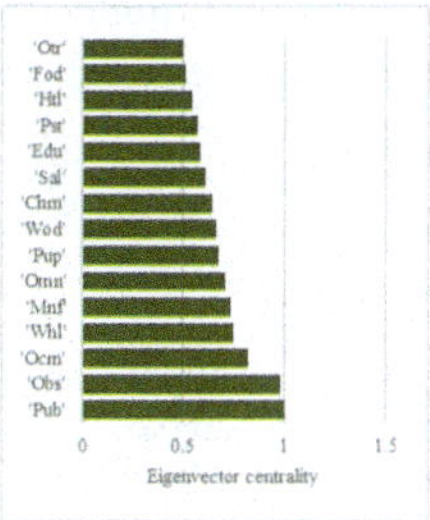

Fig. 12. Top 15 sectors by eigenvector centrality

Fig. 13. Leontief inverse coefficient network.

4 Conclusion and Policy Implications

In general, the economy of Vietnam is still mainly drived by primary and low-tech industries and the economy still heavily depends on higher-tech products from foreign countries. With the results from degree and weighted degree distribution, we realize that if the shocks affect some key industrial sectors, it can create huge volatility at the macro level and propagate to other sectors. The presence of scale-free structure of the Leontief inverse matrix indicates that there are potential channels of the quick transmission of shocks to different sectors within each network community. Moreover, the dense connections among domestic sectors imply that sectoral linkages could be an important transmission path of exogenous shock.

From the perspective of both direct and indirect linkage, the backward linkage of 'Wod', 'Cok', 'Fod', 'Omn', 'Pup' to their in-coming neighbors is high compared to the other sectors. Meanwhile, the forward linkage of 'Cok', 'Whl', 'Agr', 'Chm', 'Ele' is also relatively high compared to the other sectors. The results from eigenvector centralities also highlight that the key sectors that hold crucial roles in the production network flow are mostly low-tech and primary industries.

By understanding the interdependencies within the industrial sectors, Vietnam policy makers can better manage the circular flow of the economy, more effectively reallocate public expenditures to the crucial sectors, avoid and mitigate the propagation of shocks from one sector to the whole economy, and formulate economic policies to stimulate and resolve bottlenecks to economic growth under the current pressure of geopolitical tensions, the rise of generative AI and the age of digital technology.

References

Acemoglu, D., Akcigit, U., Kerr, W.: Networks and the macroeconomy: an empirical exploration. NBER Macroecon. Annu. **30**(1), 273–335 (2016)

Acemoglu, D., Ozdaglar, A., Tahbaz-Salehi, A.:. The network origins of large economic downturns (No. w19230). National Bureau of Economic Research (2013)

Acemoglu, D., Ozdaglar, A., Tahbaz-Salehi, A.: Microeconomic origins of macroeconomic tail risks. Am. Econ. Rev. **107**(1), 54–108 (2017)

Ahmad, I., Alvi, S.: Exploring the backward and forward linkages of production network in a developing country. Hum. Soc. Sci. Commun. **11**(1), 1–13 (2024)

Asian Development Bank: Viet Nam: Input-Output Economic Indicators. ADB's Economic Research and Regional Cooperation Department (ERCD) (2019)
Barabási, A.L.: Network science. Cambridge University Press (2016)
Blondel, V.D., Guillaume, J.L., Lambiotte, R., Lefebvre, E.: Fast unfolding of communities in large networks. J. Stat. Mech: Theory Exp. **2008**(10), P10008 (2008)
Brandes, U.: A faster algorithm for betweenness centrality. J. Math. Sociol. **25**(2), 163–177 (2001)
Brin, S., Page, L.: The anatomy of a large-scale hypertextual web search engine (1998)
Carvalho, V.M., Nirei, M., Saito, Y., Tahbaz-Salehi, A.: Supply Chain Disruptions: Evidence from the Great East Japan Earthquake. Columbia Business School Research Paper (17-5) (2016)
Cavallo, A., Cavallo, E., Rigobon, R.: Prices and supply disruptions during natural disasters. Rev. Income Wealth **60**, S449–S471 (2014)
Gabaix, X.: The granular origins of aggregate fluctuations. Econometrica **79**(3), 733–772 (2011)
Horvath, M.: Cyclicality and sectoral linkages: aggregate fluctuations from independent sectoral shocks. Rev. Econ. Dyn. **1**(4), 781–808 (1998)
Jackson, M.O.: Social and Economic Networks. Princeton University Press (2010)
Jones, C.I.: Misallocation, Input-Output Economics, and Economic Growth. In: Advances in Economics and Econometrics: Tenth World Congress, vol. 2, pp. 419–58. Cambridge University Press Cambridge, UK (2013)
Lucas Jr, R.E.: Understanding business cycles. In: Carnegie-Rochester Conference Series on Public Policy, vol. 5, pp. 7–29. North-Holland (1977
Luu, D.T., Napoletano, M., Fagiolo, G., Roventini, A., Sgrignoli, P.: Uncovering the network complexity in input-output linkages among sectors in European countries. ISI Growth Working Paper (2017)
Michael, D.K., Battiston, S.: From graph theory to models of economic networks. A tutorial. In: Networks, Topology and Dynamics, pp. 23–63. Springer, Berlin, Heidelberg (2009)
Phelan, S.E.: What is complexity science, really? Emergence J. Complexity Issues Organ. Manage. **3**(1), 120–136 (2001)

Analyzing Material Consumption Within a Global Input-Output Framework from 2011 to 2023: A Comprehensive Evaluation

Minh Thanh Nguyen, Thi Hoai Yen Le, and Hoang-Sy Nguyen(✉)

Becamex Business School, Eastern International University, Thu Dau Mot City, Binh Duong Province, Vietnam
sy.nguyen@eiu.edu.vn

Abstract. Interdisciplinary Analysis of Global Material Consumption and The transition from the Millennium Development Goals to the Sustainable Development Goals (SDG) has emphasized the importance of using a variety of approaches to achieve development goals. This paper offers a look at material consumption and trade patterns in six countries from 2011 to 2023 focusing on metallic minerals, biomass and fossil fuels. By using the Global Material Flows Database and multi regional input output analysis (MRIOA) this research uncovers valuable insights into how much material's used the raw materials traded globally and the balance of physical material exchanges. The study sheds light on the nature of material consumption with metallic minerals taking the lead in 2023 as a sign of an increasing global reliance on mineral resources. Biomass and fossil fuels also play roles reflecting trends in energy use and industrial practices. Additionally it delves into trade patterns highlighting differences in trade balances across regions and countries. Overall this interdisciplinary examination enhances our understanding of global resource usage. Emphasizes the importance of development strategies, for reaching the goals set out in the SDG.

Keywords: Material footprint · resource efficiency · interdisciplinary analysis · multi-regional input-output analysis

1 Introduction

In 2015, The 2030 Agenda for Sustainable Development was unanimously adopted by all member states of the United Nations [1]. This comprehensive framework comprises 17 sustainable development goals, offering a collective roadmap towards global peace and prosperity for both present and future generations. The transition from the Millennium Development Goals (MDGs) to the Sustainable Development Goals has introduced novel objectives that necessitate interdisciplinary approaches and tactics for accomplishment [2]. Notably,

M. Pagac et al. (Eds.): SCFF 2024, LNNS 1062, pp. 382–391, 2024.
https://doi.org/10.1007/978-3-031-65656-9_38

the concept of resource efficiency is now explicitly addressed within these goals, underscoring the heightened attention of the United Nations towards global resource utilization.

In 2014, the authors in [5] employed the Global Trade Analysis Project (GTAP) alongside a multi-regional input-output analysis method to analyze global material consumption and trade patterns from 1997 to 2007. They assessed metrics like domestic material consumption (DMC) and raw material consumption (RMC), also known as material footprint (MF), identifying major importers and exporters of embodied materials. This study, the first of its kind to use the GTAP database for such analysis, compared its effectiveness with other global databases. Additionally, they introduced innovative methods to illustrate the results, including disaggregating RMCs by product groups and regions of origin. Multi-regional input-output analysis (MRIOA), pioneered by Isard in [6], extends the principles of input-output analysis and has been recognized as an effective tool for delineating and comprehending supply chains and interrelations among consuming and producing sectors [7].

Since its establishment in 2007, the Global Innovation Index (GII) has significantly influenced the discourse on measuring innovation and has emerged as a fundamental tool in shaping economic policy [8]. Released on a yearly basis, the GII primarily evaluates and ranks 132 economies based on their innovation frameworks. Utilizing a comprehensive dataset comprising 81 indicators sourced from various international public and private outlets, the index extends beyond conventional metrics, reflecting the expanded understanding of innovation.

This study leverages data from the Global Material Flows Database (GMFD) provided by the United Nations Environment Programme (UNEP) and Global Innovation Index from World Intellectual Property Organization (WIPO) [3,4]. Our objective is to re-evaluate the findings put forth in 2014 [5], with the time frame from 2011 to 2023. Additionally, we aim to assess if there exists a comparable trend between the GII and imports serving domestic final demand (RME_IMP), domestically extracted materials embodied in exports (RME_EXP), while also examining the alterations occurring within the specified timeframe. Finally, this study focuses on India, Switzerland, Chile, the United States of America, Israel, and Singapore. These countries were selected due to their consistent ranking within the top tier of the Global Innovation Index (GII) for more than nine years in their respective regions.

2 Methodology

2.1 Advantages and Disadvantages of Global Input-Output Framework and Multi-regional Input-Output Analysis

Simply put, MRIO models connect Input-Output tables (IOTs) from various countries or regions using bilateral trade data. This linkage enables MRIO models to track not only domestic supply chains but also international ones, allowing for consideration of varying resource intensities across different countries [9,10]. MRIO models offer the capability to compute a range of footprint indicators for numerous products and sectors, even those characterized by intricate supply chains [11]. This attribute enables MRIO models to circumvent truncation errors inherent in bottom-up approaches like LCA, which arise due to the inability to fully analyze complex production chains, necessitating the omission of certain upstream chains [12,13]. Additionally, MRIO-based assessments evade ambiguity in defining system boundaries, as they consider the entire global economy as the assessment framework [14]. By adhering to a top-down accounting logic, I-O models also prevent double counting, ensuring that a specific resource input is allocated only once to final demand since all supply chains are comprehensively represented [15]. Notably, the MRIO approach aligns closely with standard economic and environmental accounting frameworks [16,17].

Nevertheless, employing IOA comes with drawbacks. MRIO models operate at the level of economic sectors, assuming uniformity within industries or product groups outlined in the IOTs [18]. This assumption means that various products and processes with potentially distinct material intensities are amalgamated within a single sector, limiting disaggregation levels [19]. Consequently, this amalgamation may distort results, particularly when diverse materials, such as industrial minerals and metal ores, are grouped into one sector. Other disadvantages highlighted in the literature regarding footprint-type indicators based on IOA include significant time lags for the publication of IOTs, the heightened sensitivity of I-O models to minor errors in trade data, especially in cases where a country's imports and exports substantially outweigh domestic production in [20], the utilization of monetary economic structures for allocating physical flows, and the application of the proportionality assumption, which assumes consistent material intensity for flows from a specific sector regardless of the recipient country and sector [21]. Additionally, the results are highly sensitive to the method used to allocate materials use data to sectors.

The Global Innovation Index (GII) operates on the basis of two fundamental sub-indices: the Innovation Input Sub-Index and the Innovation Output Sub-Index. These sub-indices are structured around specific pillars, which serve as foundational components in the assessment of innovation within economies. The Innovation Input Sub-Index, constituting the first aspect of the GII, encompasses five distinct pillars designed to capture elements of the national economy that facilitate and promote innovative activities. These pillars include Institutions, Human capital and research, Infrastructure, Market sophistication, and Business sophistication, collectively defining the conducive environment for innovation within an economy. In contrast, the Innovation Output Sub-Index evaluates the outcomes of innovative endeavors within the economy. Despite consisting of only two pillars-Knowledge and technology outputs, and Creative outputs-the Output Sub-Index holds equal significance in determining the overall GII scores alongside the Input Sub-Index.

The overall GII score is derived as a simple average of the Input and Output Sub-Indices, providing a comprehensive assessment of a country's innovation performance. Additionally, the Innovation Efficiency Ratio, calculated as the ratio of the Output Sub-Index to the Input Sub-Index, offers insights into the efficiency of innovation processes, indicating the level of innovation output relative to the inputs invested. Each pillar is further subdivided into three sub-pillars, with each sub-pillar comprising individual indicators. This detailed breakdown yields a total of 82 indicators for the current year's GII, facilitating a thorough evaluation of various facets of innovation and offering valuable insights for policymakers and stakeholders alike.

2.2 Data Sources

The utilized the GTAP database, curated at Purdue University, for their research, citing its comprehensive compilation of Input-Output Tables (IOTs) and bilateral trade data as the primary rationale for selection [5]. This choice was motivated by the database's expansive temporal coverage and its explicit representation of numerous countries through national tables. However, for the present study, the data source was transitioned to the GMFD, facilitated by the UNEP. The decision was chiefly driven by the availability of annually updated data, crucial for conducting trend analyses on various indicators.

In aligning material extraction data with GMFD, the study aggregated over 90 material extraction categories, also known as Trade Common Compilation Categories (TCCC), into four primary material flow categories: biomass, fossil fuels, metal ores, and non-metallic minerals, alongside some sub-categories.

3 Results

This study shows values for domestic material extraction (DE), Raw Material Equivalents of exports (RME_EXP), and Raw Material Equivalents of imports (RME_IMP). Imports, which are re-exported directly or after further processing, are therefore neither included on the import nor on the export side. The Physical Material Trade Balance (PTB) is calculated as Physical imports (IM) minus Physical exports (EX) and the MF as DE plus RME_IMP minus RME_EXP.

3.1 Continental Perspective

Table 1 presents a summary of the MRIO-MFA results for all continents and selected major countries in 2011 and 2023. Note: DE = domestic material extraction; RME_IMP = Raw Material Equivalents of imports; RME_EXP = Raw Material Equivalents of exports; PTB = Physical material trade balance; MF = Material Footprint = Raw Material Consumption; DMC = Domestic Material Consumption.

Africa experienced a growth of 37.1%, increasing from 5,906 million tonnes in 2011 to 8,098 million tonnes in 2023. Similarly, Asia & Pacific exhibited a substantial rise of 24.3%, escalating from 44,323 million tonnes to 55,111 million tonnes during the same period. Europe also demonstrated growth, albeit at a lower rate of 9.4%, with domestic material extraction increasing from 8,420

Table 1. Material footprints (RMC) of continents and comparison with DMC, in million tonnes, 2011 and 2023

	DE		RME_IMP		RME_EXP		PTB		MF		2*MF (%)	DMC		2* DMC (%)	2*MF/ DMC (2023) (%)
	2011	2023	2011	2023	2011	2023	2011	2023	2011	2023		2011	2023		
Africa	5,906	8,098	1,520	1,465	2,762	2,862	−266	−182	4,616	6,471	40%	5,640	7,915	40%	82%
Asia & Pacific	44,323	55,111	10,761	14,866	9,852	12,332	1,420	2,390	45,017	57,031	27%	45,743	57,500	26%	99%
West Asia	2,628	2,894	1,517	1,336	1,664	1,721	−901	−962	2,481	2,432	−2%	1,726	1,933	12%	126%
Europe	8,420	9,218	7,468	7,750	3,959	4,915	1,027	940	11,929	11,973	0%	9,448	10,158	8%	118%
Latin America & Caribbean	9,934	11,099	1,887	2,012	3,832	4,655	−709	−700	7,893	8,161	3%	9,226	10,399	13%	78%
North America	9,044	10,497	3,912	4,657	3,269	3,569	104	−576	9,687	11,504	19%	9,148	9,922	8%	116%
World	85,142	101,851	28,185	32,945	28,169	32,912	−262	−451	84,798	100,382	18%	84,881	101,400	19%	99%

million tonnes to 9,218 million tonnes. West Asia also saw a growth of 10.1%, increasing from 2,628 million tonnes to 2,894 million tonnes.

The Asia & Pacific region stood out with significant growth rates for both imports and exports between 2011 and 2023. For instance, in 2023, Asia & Pacific experienced a growth rate of 38.2% for imports and 31.5% for exports. Europe also displayed substantial growth rates with a 3.8% increase in imports and a notable 24.5% rise in exports during the same period. Africa, West Asia, Latin America & Caribbean, and North America also showed varying levels of imports and exports.

While Africa and Latin America & Caribbean exhibited negative trade balances in both years, regions like Asia & Pacific, Europe, and North America showcased positive trade balances. Notably, North America's trade balance fluctuated from a positive value of 104 million tonnes in 2011 to a negative value of 576 million tonnes in 2023. The world's physical material trade balance also showed a decline from −262 million tonnes in 2011 to −451 million tonnes in 2023.

Asia & Pacific consistently led in material footprint with notable growth rates over the years. In 2023, the region recorded a growth rate of 26.3% in raw material consumption. Europe followed closely behind with a modest growth rate of 0.4% in the same year. Africa, West Asia, Latin America & Caribbean, and North America also displayed varying levels of raw material consumption.

Asia & Pacific maintained its lead in domestic material consumption with substantial values and a growth rate of 26.3% in 2023. Europe also saw a slight increase with a growth rate of 0.1% in domestic material consumption figures during the same year. Africa, West Asia, Latin America & Caribbean, and North America also demonstrated varying levels of domestic material consumption.

3.2 Material Footprint (MF) of Selected Countries

It is evident overall that all of the chosen countries' overall MF profiles tend to rise over time. The most prominent observation pertains to the abrupt surge in material consumption exhibited by India and the United States, reaching approximately 2000 million tonnes (See Fig. 1). The MF of Switzerland, Chile, Singapore, and Israel exhibit distinct yet nuanced trajectories when analyzed within the contextual framework of temporal evolution. Notably, Fig. 1 displays these

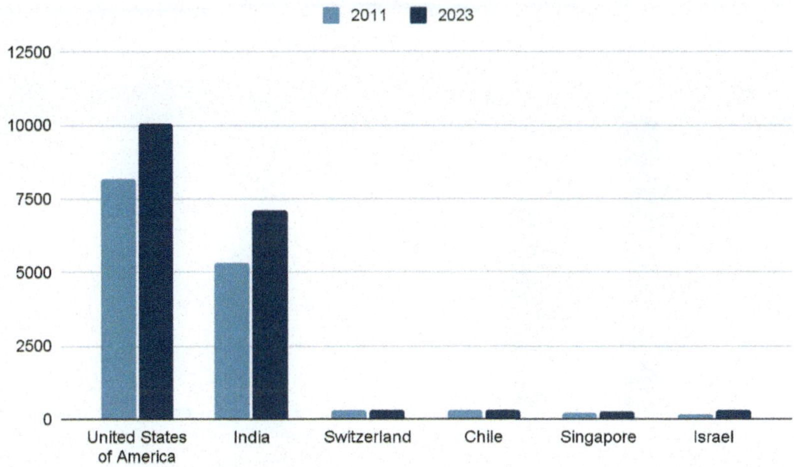

Fig. 1. The total raw material consumption of 4 main categories of selected countries

countries have demonstrated relative stability in their respective MF profiles throughout the period under the data, thereby illustrating concerted endeavors aimed at preserving sustainable resource consumption patterns.

3.3 Material Footprint Versus Domestic Material Consumption

Figure 2 provided on domestic material consumption (DMC) and material footprint (MF) for different categories of materials in 2023 offers valuable insights into the material consumption patterns of six selected countries. One key finding is that non-metallic minerals are the most consumed materials in all countries, followed by biomass and fossil fuels. India demonstrates a substantial dependence on metal ores, substantiated by its status as the leading importer of such materials.

Comparing the countries using Fig. 2, we find that the United States has the highest MF values around 10 billion tonnes, suggesting a more resource-intensive economy. India, despite having a lower MF (approximately 7 billion tonnes), has the highest DMC (more than 8 billion tonnes), indicating a high level of domestic consumption. India's high DMC compared to its MF suggests a high dependence on material resources for economic output, possibly due to more material-intensive industries. A higher DMC could indicate a large amount of imported goods, while a higher MF suggests a high level of domestic production and consumption (European Commission in [22]. Countries like Israel, Singapore, and Switzerland show lower material intensities relative to their populations, economic structures, and more efficient resource use.

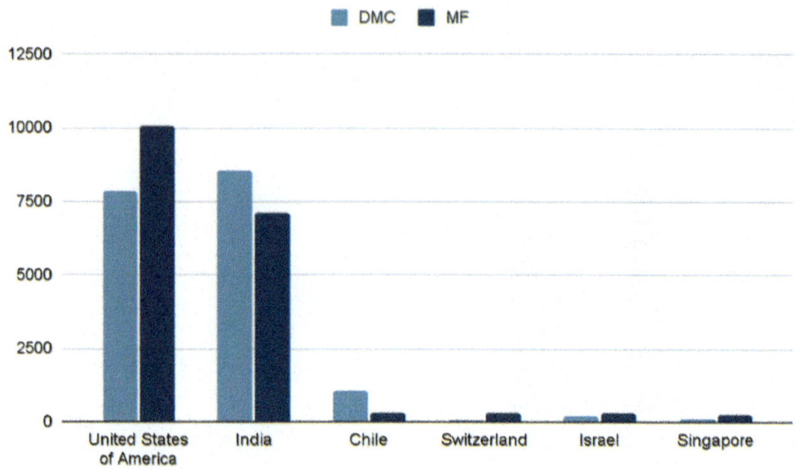

Fig. 2. Comparison of Raw Material Consumption (MF) with Domestic Material Consumption (DMC) of Selected Countries

3.4 Patterns of International Trade

To comprehensively grasp the dynamics of global trade balances, it is imperative to broaden the scope of the Physical Trade Balance (PTB) terminology. This expansion is crucial for capturing a holistic perspective of trade dynamics across diverse nations, thereby mitigating the risk of overlooking countries with unique trade circumstances. By encompassing a wider array of countries, the PTB framework can effectively account for varying trade patterns influenced by factors such as levels of natural resources. Indeed, nations endowed with different resource endowments exhibit distinct trade behaviors, underscoring the necessity of extending the measurement scope. Expanding the PTB concept facilitates a more nuanced understanding of international trade dynamics and fosters inclusive analyses that reflect the diverse economic landscapes across the globe.

Observing Fig. 3, it is evident that in the year 2023, China emerged as the leading nation in Asia with the most significant positive PTB, boasting a deficit of approximately 2,700 million tonnes. Various factors contributed to this deficit, including a trade deficit, escalating demand for services, trends in export prices, and strategic initiatives aimed at fostering income growth and product diversification. Following China, Japan recorded a trade deficit of approximately 570 million tonnes, while South Korea and India demonstrated nearly identical PTBs of 440 million tonnes each. The Russian Federation secured the second position in trade surplus, amounting to 700 million tonnes.

Conversely, Australia faced the highest negative PTB in Asia (Fig. 3), reaching approximately 1,500 million tonnes. This adverse balance was ascribed to the country's heavy reliance on exports of resources such as iron ore, coal, and liquid natural gas, characterized by volatile prices, consequently leading to fluctuations in Australia's trade balance. In Europe, countries like Germany, Spain,

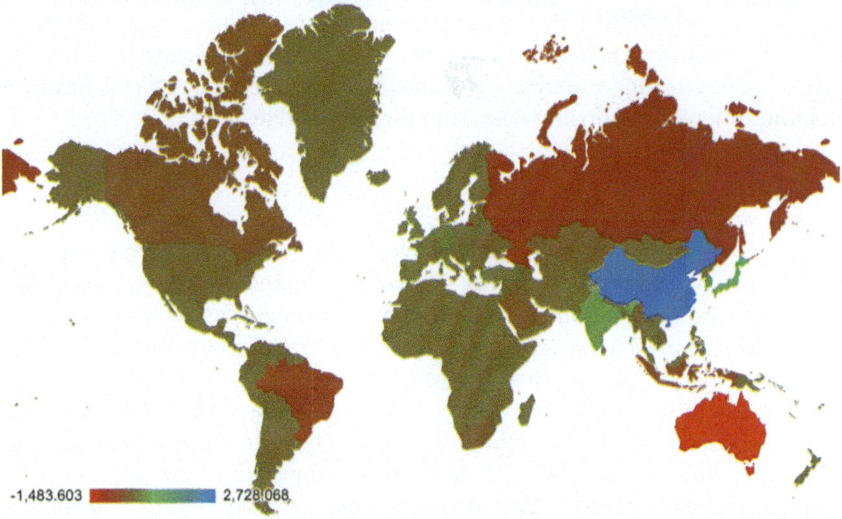

Fig. 3. Patterns of Global Physical Material Trade Balance

and the United Kingdom (the UK) displayed evident trade deficits. Germany notably exhibited the largest export deficit in Europe, amounting to about 200 million tonnes. Spain and the UK, meanwhile, recorded nearly identical export volumes of 100 million tonnes each. Turning to the Americas, a majority of nations experienced negative PTBs, indicating that exports surpassed imports. Brazil in South America and Canada in North America reported the highest negative PTBs on the continent, with figures standing at around 650 million tonnes and 322 million tonnes, respectively, in the year 2023.

3.5 Pattern of GII Versus Pattern of RME_IMP and RME_EXP

This section aims to analyze the correlation between the Global Innovation Index (GII), Raw Material Equivalents Imports (RME_IMP), and Raw Material Equivalents Exports (RME_EXP) to investigate whether the level of innovation aligns with the contribution of nations in raw material trade. Raw Material Equivalents (RME) are coefficients used to quantify the amount of material extraction required globally to produce a unit of traded goods. The GII assesses innovation within a context of economic and geopolitical uncertainty, ranking approximately 132 economies based on their innovation performance and identifying strengths and weaknesses. By examining the trend of these indicators from 2011 to 2023, this analysis seeks to determine if there is a correlation between RME_IMP, RME_EXP, and GII. The trend is derived from the combined RME_IMP and RME_EXP across four main material categories for 107 countries with at least 13 GII rankings for increased accuracy and better illustrated correlation. The findings reveal a 53.5% similarity between RME_IMP and GII, indicating that 53.5% of countries experiencing positive growth in RME_IMP also exhibit positive growth in GII. Similarly, the similarity between RME_EXP and GII is

52.8%. These results suggest ongoing governmental efforts worldwide to foster innovation. However, the escalating utilization of raw materials in traded goods highlights a concerning trend of continued exploitation of natural resources to meet human demands, posing sustainability challenges.

4 Conclusion

The analysis conducted in this paper delves into the intricate dynamics of material consumption and trade patterns across six selected countries, shedding light on the significant role of non-metallic minerals, biomass, and fossil fuels in global resource utilization. The study underscores the transition from the Millennium Development Goals to the Sustainable Development Goals, emphasizing the imperative for interdisciplinary approaches to achieve sustainable development objectives. By leveraging the Global Trade Analysis Project and multi-regional input-output analysis, the research provides a comprehensive evaluation of material consumption trends from 2011 to 2023, offering insights into domestic material consumption, raw material equivalents of exports and imports, and physical material trade balances. Furthermore, the study intertwines the analysis of material footprints with the Global Innovation Index (GII), highlighting the interconnectedness between resource efficiency and innovation capacities within economies. Through a meticulous examination of material flows and innovation metrics, the research aims to discern patterns that elucidate the relationship between resource utilization and innovative performance. By focusing on countries consistently ranked highly in the GII, such as India, Switzerland, Chile, the United States of America, Israel, and Singapore, this study contributes to a nuanced understanding of how innovation and resource consumption intersect on a global scale.

Acknowledgements. The authors would like to thank Eastern International University, Binh Duong Province, Vietnam for funding this research.

References

1. Department of Economic and Social Affairs:. THE 17 GOALS - Sustainable Development. Sustainable Development (n.d.). https://sdgs.un.org/goals
2. The United Nations: United Nations millennium development goals. The United Nations (n.d.). https://www.un.org/millenniumgoals/
3. Australia's National Science Agency: Technical annex for Global Material Flows Database - 2021 edition. International Resource Panel. Available online: https://resourcepanel.org/sites/default/files/technical_annex_for_global_material_flows_database_-_vers_30_aug22.pdf. Last accessed: 30 Aug 2023
4. United Nations Environment Programme: Global Material Flows Database. Resource Panel. International Resource Panel. Available online: https://www.resourcepanel.org/global-material-flows-database. Last accessed: 30 Aug 2023
5. Giljum, S., Bruckner, M., Martinez, A.: Material footprint assessment in a global input-output framework. J. Ind. Ecol. **19**(5), 792–804 (2014). https://doi.org/10.1111/jiec.12214. Dec.

6. Isard, W.: Interregional and regional input-output analysis: a model of a space-economy. Rev. Econ. Stat. **33**(4), 318 (1951). https://doi.org/10.2307/1926459. Nov.
7. Andrew, R., Peters, G.P., Lennox, J.: Approximation and regional aggregation in multi-regional input-output analysis for national carbon footprint accounting. Econ. Syst. Res. **21**(3), 311–335 (2009). https://doi.org/10.1080/09535310903541751. Sep.
8. World Intellectual Property Organization: Global Innovation Index (GII). WIPO (n.d.). https://www.wipo.int/global_innovation_index/en/
9. Feng, K., et al.: Comparison of bottom-up and top-down approaches to calculating the water footprints of nations. Econ. Syst. Res. **23**(4), 371–385 (2011). https://doi.org/10.1080/09535314.2011.638276
10. Tukker, A., Dietzenbacher, E.: Global multiregional input-output frameworks: an introduction and outlook. Econ. Syst. Res. **25**(1), 1–19 (2013). https://doi.org/10.1080/09535314.2012.761179
11. Chen, Z.-M., Chen, G.Q.: Virtual water accounting for the globalized world economy: National water footprint and international virtual water trade. Ecol. Ind. **28**, 142–149 (2013). https://doi.org/10.1016/j.ecolind.2012.07.024
12. Lenzen, M.: Errors in conventional and input-output-based life-cycle inventories. J. Ind. Ecol. **4**(4), 127–148 (2000). https://doi.org/10.1162/10881980052541981
13. Wiedmann, T.: Editorial: Carbon footprint and input-output analysis - an introduction. Econ. Syst. Res. **21**(3), 175–186 (2009). https://doi.org/10.1080/09535310903541256
14. Wiedmann, T.: A review of recent multi-region input-output models used for consumption-based emission and resource accounting. Ecol. Econ. **69**(2), 211–222 (2009). https://doi.org/10.1016/j.ecolecon.2009.08.026
15. Daniels, P.L., Lenzen, M., Kenway, S.J.: The ins and outs of water use - a review of multi-region input-output analysis and water footprints for regional sustainability analysis and policy. Econ. Syst. Res. **23**(4), 353–370 (2011). https://doi.org/10.1080/09535314.2011.633500
16. System of Environmental-Economic Accounting 2012. International Monetary Fund (2017). https://doi.org/10.5089/9789211615630.069.
17. Environmental input-output economics. The Economics of Input-Output Analysis, pp. 139–150 (2006). https://doi.org/10.1017/cbo9780511610783.013
18. Tukker, A., et al.: EXIOPOL - development and illustrative analyses of a detailed global Mr Ee Sut/Iot. Econ. Syst. Res. **25**(1), 50–70 (2013). https://doi.org/10.1080/09535314.2012.761952. Mar.
19. Ewing, B.R., et al.: Integrating ecological and water footprint accounting in a multi-regional input-output framework. Ecol. Ind. **23**, 1–8 (2012). https://doi.org/10.1016/j.ecolind.2012.02.025
20. Mekonnen, M., Hoekstra, A.Y.: National Water Footprint Accounts: The Green, Blue and Grey Water Footprint of Production and Consumption, vol. 50, no. 50. UNESCO-IHE Institute for Water Education, 2021. http://www.unesco-ihe.org/Value-of-Water-Research-Report-Series/Research-Papers
21. Bruckner, M., Giljum, S., Lutz, C., Wiebe, K.S.: Materials embodied in international trade - Global material extraction and consumption between 1995 and 2005. Glob. Environ. Chang. **22**(3), 568–576 (2012). https://doi.org/10.1016/j.gloenvcha.2012.03.011
22. European Commission: Material footprints. European Commission (n.d.). https://ec.europa.eu/eurostat/cache/metadata/en/env_ac_rme_esms.htm

Cybersecurity in Small and Medium-Sized Enterprises: A Bibliometric Analysis

Thanh-Cong Truong(✉) and Huy Khang Nguyen

Faculty of Information Technology, University of Finance-Marketing, Ho Chi Minh City, Vietnam
ttcong@ufm.edu.vn

Abstract. This study delves into the critical domain of cybersecurity within small and medium-sized enterprises (SMEs) through a meticulous bibliometric analysis. Scopus electronic database, renowned for its inclusivity of various publication formats, serves as the primary source for data retrieval. Through a meticulously crafted search strategy, 888 publications spanning the years 2014–2023 were unearthed, encompassing a spectrum of scholarly outputs such as journal articles, conference papers, and contributions from a staggering 112 countries worldwide. The amassed publications command significant attention, garnering over 8000 citations, with an impressive average of nearly 9.4 citations per record. Furthermore, authorship analysis underscores a culture of collaboration within this scholarly domain, with an average of 3.2 authors per publication, highlighting the interdisciplinary nature and collective endeavor inherent in tackling cybersecurity challenges within SMEs. Data cleaning and analysis tools like Microsoft Excel and VOSviewer will be utilized to explore publication trends, prominent research areas, and potential knowledge gaps within the field of SME cybersecurity. Keyword analysis unveils four pivotal emerging themes: integrating digital technologies while managing risk, emphasizing information management for securing cloud-based operations, addressing specific cyber threats and vulnerabilities, and advocating for proactive information security risk management with cost-effective solutions. This study offers a comprehensive picture of emerging trends, key research areas, and potential knowledge gaps within the field. These insights will be valuable for researchers, policymakers, and practitioners working to strengthen the cybersecurity posture of SMEs on a global scale.

Keywords: Cybersecurity · SMEs · SMBs · Bibliometric analysis · Cyber security

1 Introduction

Small and medium-sized enterprises (SMEs) or, in some documents, Small and medium-sized Businesses (SMBs) are vital for innovation, growth, and development of national and international economies, making them a top priority in economic policy [1]. According to the statistics of Statista [2], as of 2021, approximately 332.99 million small and

M. Pagac et al. (Eds.): SCFF 2024, LNNS 1062, pp. 392–402, 2024.
https://doi.org/10.1007/978-3-031-65656-9_39

medium-sized enterprises worldwide. This number represents a slight increase compared to 2019 when there were 328.5 million SMEs/SMBs globally. These businesses play a crucial role in economies worldwide, accounting for over 95% of all firms and contributing to 60–70% of employment [2]. The Asia-Pacific region stands out as the leader in terms of SMEs/SMBs, boasting the highest number of such enterprises. This highlights the region's vibrant and thriving entrepreneurial landscape, where small and medium-sized businesses play a crucial role in driving economic growth. Europe, on the other hand, holds a significant share of SMEs, with a staggering 99% of all businesses falling into this category. This showcases the importance and prevalence of small and medium-sized enterprises in the European business ecosystem. Meanwhile, North America serves as a hub for innovation, with a notable distinction. A remarkable 285,334 entities, equivalent to 97.5% of all exporters in the region, are classified as SMEs/SMBs. This emphasizes the pivotal role played by these enterprises in fostering trade and driving economic development through their innovative endeavors.

Electronic communication networks, interconnected information systems and digital services are an essential part of businesses in the industrial revolution 4.0 [3]. Digital transformation makes SMEs/SMBs more dependent on software and networks to provide services to customers and meet business objectives [4]. In parallel with the benefits of technology adoption, threats to privacy and information security are emerging and becoming obstacles to digital transformation [5]. In addition, the COVID-19 crisis has made more businesses dependent on digital technology than before, creating opportunities for cybercriminals to ramp up attacks [6]. A report highlights that cyberattacks against small businesses have been on the rise, with certain types of attacks like phishing being more commonly aimed at SMEs. It also notes that small businesses are generally not financially prepared for an attack, and most lack cyber insurance. In this context, SMEs need to implement appropriate information security procedures, apply technical mechanisms, and take appropriate organizational measures. Without such safeguards, SMEs can be severely affected by cyber threats or deliberate attacks on their information systems and networks, which can lead to negative business impacts.

The work by Bada and Nurse [7] emphasizes the role of social engineering tactics that exploit human vulnerabilities within SMBs, leading to data breaches and financial losses. They also suggested a cybersecurity education and awareness programme for SMEs/SMBs at a city level, emphasizing the importance of such initiatives in an organization's cybersecurity strategy.

The paper [8] discussed the security divide between SMEs and large companies and how SME characteristics influenced IT security investments. It mentioned constraints such as limited budget planning and resources that hinder IT security investment in SMEs. It offered context-specific insights regarding the effects of identified constraints on IT security investments for all involved stakeholders.

The authors in [9] discuss the challenges that SMBs face in the cyber threat landscape due to limited resources. It emphasizes the need for a more holistic approach to cybersecurity, known as cyber resilience, which includes anticipating, detecting, withstanding, recovering from, and evolving after cyber incidents. The paper presents a framework and implementation order for SMEs to operationalize cyber resilience. The framework aims to guide SME managers in understanding and implementing cyber resilience with concrete actions based on expert experience, thus easing the process for SMEs.

The work in [10] states that SMBs are prime targets for phishing emails due to the potential lack of employee awareness and robust spam filtering mechanisms. They pointed out that authority and urgency are identified as key techniques that increase the likelihood of employees clicking on malicious links in emails.

A study by Chidukwani et al. [11] points out that SMBs often rely on basic security measures like firewalls and antivirus software, which might be insufficient against evolving cyberattacks. They also highlighted the lack of in-house IT expertise and limited budgets as significant barriers for SMBs to implement robust security solutions.

While previous reviews have highlighted vulnerabilities and mitigation strategies for SMEs/SMBs cybersecurity, a critical research gap exists - the lack of extensive bibliometric studies systematically analyzing the existing literature in this domain. To address this gap, this study will conduct a comprehensive bibliometric analysis of research on SMB cybersecurity.

By examining the body of literature through bibliometric techniques, this study aims to provide a quantitative and objective assessment of the current state of SMB cybersecurity research. The bibliometric analysis will involve evaluating various bibliometric indicators, such as publication and citation trends, collaboration networks, and prominent research themes and methodologies.

The bibliometric analysis will serve as a valuable resource for researchers, practitioners, and policymakers by providing a comprehensive overview of the SMB cybersecurity research domain, highlighting its evolution, identifying potential research gaps and underexplored areas, and guiding future efforts toward addressing the most pressing issues and contributing new insights to this critical field.

2 Methodology

This article uses bibliometric analysis to analyze the impact of works in the field of technology on digital transformation at higher education institutions. The steps to carry out the research followed the instructions of Donthu et al. in [12], including:

- Determining the objectives and scope of the research,
- Choosing analytical techniques,
- Collect data,
- Analyze data and report results.

The goal of this study is to understand the characteristics of scientific publications indexed in a database on Cybersecurity in SMEs/SMBs. Within the scope of this study, we use the Scopus database. The reason for choosing Scopus is that Scopus indexes a larger number of documents than Web of Science and Dimensions, which facilitates more comprehensive information searches [13]. Scopus provides a wide range of output types, reflecting the diversity of research work within the science and research community. Therefore, using this database makes bibliometric analysis richer and more multidimensional, helping to better understand trends and developments in specific fields.

For data collection: (a) The keywords "cybersecurity", "cyber security", "Small and medium sized enterprises" and "Small and medium sized Business" were used in a search query on Scopus. (b) Documents published between 2014 and 2023 were considered to ensure continuity (over 10 years) and reduce the delay between publication and indexing in Scopus. (c) Only documents written in English were included for consistency. The data set was cleaned to remove duplicates and correct errors that may appear during indexing. Data analysis tools used: Microsoft Excel (version 2401) and VOSviewer (version 1.6.20). Results from bibliometric analysis will be reported based on the research objectives.

3 Results

A collection of 888 records were gathered from Scopus, comprising 467 journal articles and 427 conference papers, spanning across 589 distinct sources over the period from 2014 to 2023. These records garnered a cumulative total of 8375 citations, averaging 9.4 citations per record. Authored by a collective of 2833 individuals, each record boasted an average of 3.2 authors. These 888 records were contributed by authors hailing from 112 unique countries or territories.

3.1 The Trend of Publications in the Period

The analysis of publications on "Cybersecurity in SMEs" based on the Scopus database data reveals a compelling narrative of growing scholarly interest and engagement in understanding and addressing cybersecurity challenges faced by small and medium-sized enterprises. The trend indicates a steady progression from 2014 onwards, with the number of publications gradually increasing each year. Notably, there is a significant acceleration in research activity observed in 2016, suggesting a heightened awareness and urgency surrounding the topic. Although there are minor fluctuations in publication numbers in certain years, such as a slight decrease in 2017, the overall trend remains positive, reflecting a sustained commitment to advancing knowledge in the field. The substantial increase in publications in 2019, followed by continued growth in subsequent years, underscores the escalating significance of cybersecurity for SMEs in an increasingly digitalized business environment. The peak in research activity reached in 2023 further emphasizes the growing recognition of SMEs as vulnerable targets for cyber threats and the critical need for effective cybersecurity measures tailored to their unique needs and constraints. This trend analysis provides valuable insights into the evolving landscape of cybersecurity research and highlights the pivotal role of academic scholarship in addressing the cybersecurity challenges confronting SMEs. Overall, the trend analysis underscores the escalating scholarly interest and recognition of the critical importance of cybersecurity for SMEs, reflecting a sustained commitment to advancing knowledge and addressing the unique challenges faced by these businesses in an increasingly digitalized landscape (Fig. 1).

3.2 Country-Wise Analysis: Publication and Citation Trends

Table 1 provides information about the country, total publications, and total citations. The table analysis detailing the research output and impact across various countries

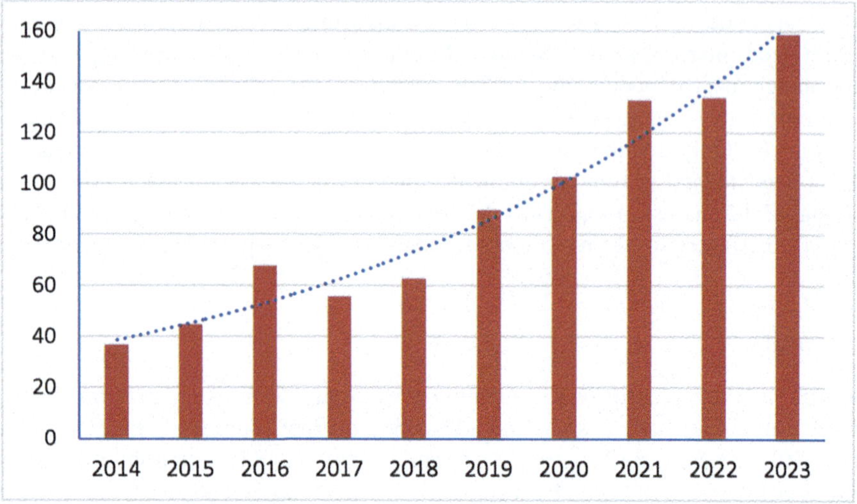

Fig. 1. A time series graph displaying the number of publications per year related to cybersecurity in SMEs, showing trends over the specified period

provides valuable insights into the global landscape of academic publications. Among the nations listed, the United Kingdom is a leader in research productivity, boasting the highest number of publications at 125 and citations at 1598. This indicates a robust research ecosystem within the United Kingdom, characterized by a prolific output of scholarly work across diverse fields.

However, while the United Kingdom leads in sheer volume, China exhibits a remarkable impact per publication despite having fewer overall publications. With 68 publications and 1329 citations, China's research output significantly influences the academic community, suggesting a high degree of relevance and quality in its scholarly contributions. This underscores China's growing prominence in the global research landscape and its ability to produce impactful research despite numerical disparities.

In contrast, larger countries like the United States and Germany, while producing substantial numbers of publications (81 and 89, respectively), do not match China's level of impact per publication. This suggests that while these countries contribute significantly to the overall volume of academic output, the average impact of their research may be lower compared to more focused efforts seen in countries like China.

Moreover, examining countries with smaller research outputs, such as Switzerland and Belgium, sheds light on the varying degrees of research engagement across different regions. While these nations may have fewer publications and citations compared to their larger counterparts, their contributions are nonetheless significant within their respective academic communities, reflecting the diverse and distributed nature of global research endeavors.

Table 1. Top 20 countries based on total publications

Rank	Country	Number of publications	Total citations
1	United Kingdom	125	1598
2	Germany	89	373
3	United States	81	1287
4	Malaysia	77	1041
5	India	76	801
6	China	68	1329
7	Australia	45	616
8	South Africa	41	168
9	Spain	36	236
10	Italy	35	124
11	Indonesia	31	99
12	Netherlands	31	156
13	Portugal	25	101
14	Switzerland	25	97
15	Saudi Arabia	23	506
16	Belgium	21	87
17	Norway	21	81
18	Poland	21	163
19	Sweden	21	173
20	Greece	19	253

3.3 The Landscape of International Collaboration in Cybersecurity Research for SMEs/SMBs

Figure 2 offers a fascinating glimpse into the world of international collaboration for research in cybersecurity within Small and Medium-Sized Enterprises (SMEs). The network visualizes 32 countries, each a significant contributor to the field, boasting at least 10 publications on the topic. The size of each country's corresponding circle serves as a visual indicator of its research output, with larger circles representing a greater number of publications. More importantly, the network reveals collaboration patterns between these nations, depicted by connecting lines. By analyzing these connections, we can group the countries into distinct clusters based on their collaborative tendencies.

The first cluster, boasting 16 members, paints a picture of a dynamic European research hub. Countries like Belgium, the Czech Republic, France, Germany, Greece, Italy, Netherlands, Norway, Poland, Portugal, Romania, Slovakia, Spain, Sweden, Switzerland, and the United Kingdom form this tightly-knit group. The web of connections within this cluster is particularly dense, signifying a high level of collaboration on cybersecurity research specific to SMEs. This suggests a well-established network

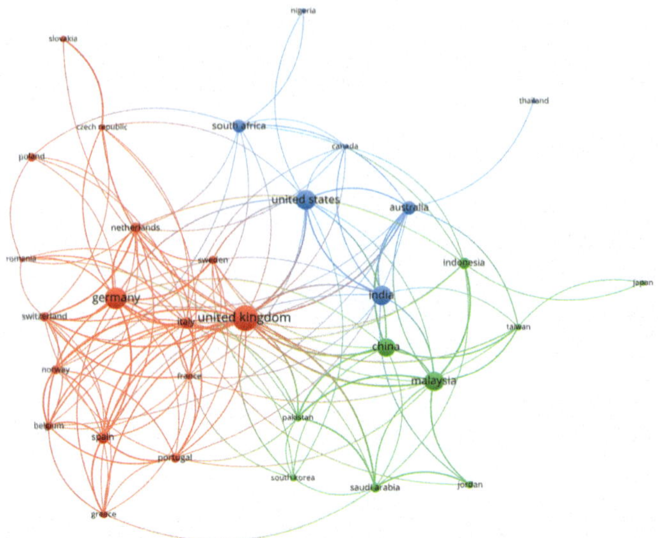

Fig. 2. International collaboration in cybersecurity research for SMEs/SMBs

where knowledge and expertise are readily shared, potentially accelerating advancements in the field.

In contrast, the second cluster, comprising 9 Asian nations, presents a different dynamic. China, Indonesia, Japan, Jordan, Malaysia, Pakistan, Saudi Arabia, South Korea, and Taiwan fall under this category. While all these countries contribute to the research effort, the number of publications from each appears to be lower compared to the European cluster. Furthermore, the sparsity of connecting lines within this cluster suggests a lesser degree of collaboration amongst these Asian nations. This doesn't necessarily negate all collaboration; it simply indicates that it might not be as prevalent as within the European group.

The final cluster, consisting of 7 members, reveals a geographically diverse group. Australia, Canada, India, Nigeria, South Africa, Thailand, and the United States come together in this cluster. Similar to the Asian cluster, the number of publications from each country seems to be lower when compared to the European group. Additionally, the absence of connecting lines suggests a lower level of collaboration amongst these nations in the context of cybersecurity research for SMEs.

3.4 Keyword Co-occurrence Analysis

All the publications in the data set together yielded a total of 5526 keywords; of these, 44 keywords that had featured in at least 20 publications were selected and represented as a network based on the co-occurrence of these keywords. Figure 3 explores the key themes emerging from a dataset of publications on cybersecurity in SMEs/SMBs. The data is visualized as a network of keywords, where connections represent frequent co-occurrences within the publications. By examining these connections and the size of

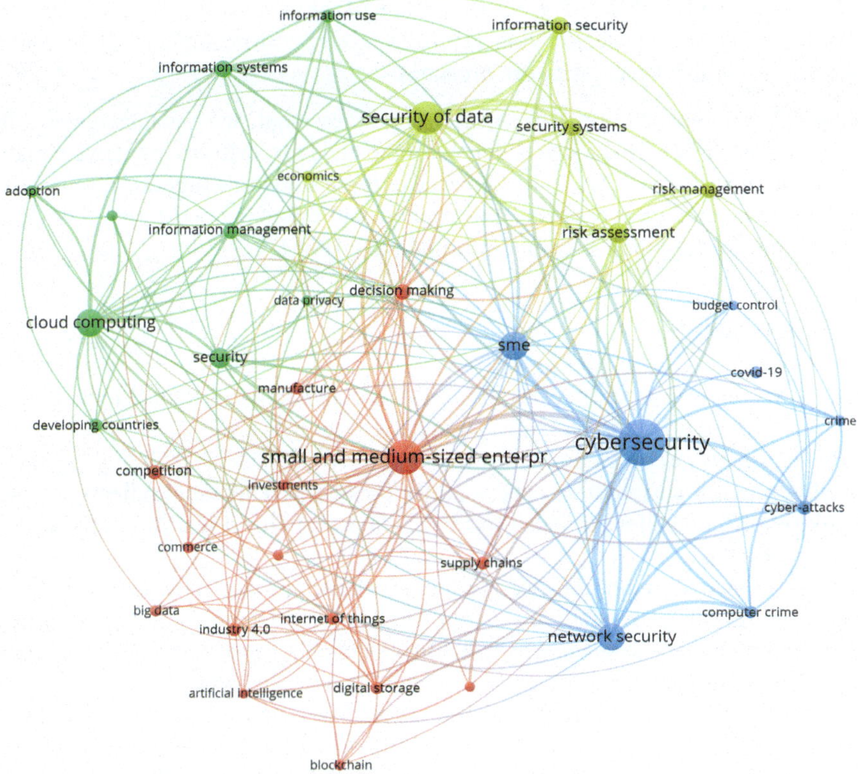

Fig. 3. A network diagram showing the relationships between keywords within the analyzed publications, potentially highlighting clusters of related terms

each keyword, we gain valuable insights into the most pressing cybersecurity concerns for SMEs in today's evolving technological landscape.

The network reveals four distinct clusters of keywords, each highlighting a critical aspect of cybersecurity for SMEs.

Cluster 1: This cluster features keywords like artificial intelligence, big data, blockchain, digital transformation, and industry 4.0. It signifies the growing importance of digital technologies for SMEs. The prominence of these keywords suggests that many publications discuss how SMEs are incorporating these advancements while navigating the associated cybersecurity risks.

Cluster 2: Cloud computing, data privacy, information systems, and information management dominate this cluster. This focus highlights the crucial role of information management in securing SME operations. The co-occurrence of these keywords suggests discussions on how SMEs can leverage cloud computing for efficiency while ensuring data privacy and robust information security measures within their systems.

Cluster 3: This cluster prominently features keywords like computer crime, cyber-attacks, network security, and SME. Its presence emphasizes the significant challenges cyber threats pose for SMEs. The co-occurrence of these keywords suggests that many publications address the specific vulnerabilities faced by SMEs and potential strategies to mitigate cyberattacks and network breaches.

Cluster 4: Economics, information security, risk assessment, and risk management form the core of this cluster. This focus underscores the need for SMEs to proactively manage information security risks. The co-occurrence of these keywords suggests discussions on how SMEs can conduct proper risk assessments and implement cost-effective security solutions to safeguard their data and information systems.

4 Discussion

This analysis of cybersecurity research for SMEs in the Scopus database unveils several key trends and areas for further exploration.

The data showcases a clear rise in publications from 2014 to 2023, highlighting the increasing urgency surrounding cybersecurity for SMEs. This trend reflects a growing recognition of the challenges faced by SMEs in a digitalized world and the need for effective solutions.

The result reveals a multifaceted picture of research productivity and impact across countries. While the UK leads in overall publications, China exhibits a remarkable impact per publication, suggesting high-quality research. The varying degrees of research engagement across regions highlight the need to foster collaboration.

Network visualization unveils distinct clusters based on collaboration patterns. The European cluster stands out for its dense network and prolific research output, suggesting a well-established collaborative environment. In contrast, collaboration appears less frequent in the Asian and African clusters, indicating potential areas for future development.

The keyword analysis pinpoints four prominent clusters reflecting critical aspects of cybersecurity for SMEs.

(i) Integration of Digital Technologies: One prominent trend highlighted in the keyword analysis is the emphasis on integrating digital technologies, such as artificial intelligence and big data analytics, while managing cybersecurity risks for SMEs. With the proliferation of digital transformation initiatives among SMEs, there is a growing recognition of the potential of emerging technologies to enhance cybersecurity capabilities. AI, for instance, offers advanced threat detection and response mechanisms, while big data analytics enable proactive risk assessment and predictive analytics to anticipate cyber threats before they materialize.

(ii) Importance of Information Management in Securing Cloud-Based Operations: As SMEs increasingly rely on cloud computing services for storage, processing, and collaboration, ensuring robust information management practices becomes paramount. The keyword analysis underscores the significance of effective information management in securing cloud-based operations against cyber threats. This includes implementing encryption protocols, access controls, and data loss prevention mechanisms to safeguard sensitive information stored in the cloud and mitigate the risk of unauthorized access or data breaches.

(iii) Addressing Specific Cyber Threats and Vulnerabilities Faced by SMEs/SMBs: SMEs/SMBsoften face unique cyber threats and vulnerabilities due to resource constraints, limited cybersecurity expertise, and reliance on third-party vendors. The analysis highlights the importance of tailoring cybersecurity strategies to address these specific challenges, such as phishing attacks, ransomware, supply chain vulnerabilities, and insider threats. By identifying and prioritizing these threats, SMEs can implement targeted cybersecurity measures to mitigate risks and strengthen their cyber resilience.

(iv) Proactive Risk Assessment and Cost-Effective Security Solutions: In light of the evolving cyber threat landscape, there is a growing recognition of the need for proactive risk assessment and cost-effective security solutions tailored to the needs and constraints of SMEs. This entails adopting risk-based approaches to cybersecurity, conducting regular vulnerability assessments, and investing in scalable and affordable security technologies and practices. By proactively identifying and addressing vulnerabilities, SMEs can minimize the impact of cyber incidents and optimize their cybersecurity investments to maximize efficacy and cost-effectiveness.

This research lays the groundwork for further exploration. Potential areas for future inquiry include:

- Delving deeper into the factors influencing international collaboration in cybersecurity research for SMEs.
- Investigating the specific challenges and best practices for cybersecurity in different industries or regions.
- Analyzing the effectiveness of various cybersecurity solutions tailored for SMEs.
- Exploring the role of government policies and regulations in promoting cybersecurity awareness and preparedness among SMEs.

5 Conclusion

In conclusion, this bibliometric analysis provides a comprehensive overview of the landscape of cybersecurity research within SMEs/SMBs over the past decade. Through meticulous data collection and analysis, we have uncovered emerging trends, identified key research areas, and highlighted potential knowledge gaps in the field.

The findings of this study underscore the critical importance of addressing cybersecurity challenges faced by SMEs/SMBs, considering their significant contribution to economic growth and innovation globally. Collaboration emerges as a central theme, emphasizing the interdisciplinary nature of cybersecurity research and the collective endeavor required to safeguard SMEs/SMBs against cyber threats effectively.

Moreover, the identified themes, including the integration of digital technologies, the importance of information management, addressing specific cyber threats, and advocating for proactive risk management, offer actionable insights for researchers, policymakers, and practitioners. By leveraging these insights, stakeholders can formulate targeted strategies and interventions to enhance the cybersecurity posture of SMEs, thereby mitigating risks and fostering resilience in an increasingly digitized world.

Moving forward, it is imperative to sustain momentum in research and collaboration efforts, continually adapting to the evolving cyber threat landscape and the dynamic needs of SMEs/SMBs. By fostering a culture of innovation, collaboration, and knowledge sharing, we can collectively work towards building a safer and more secure digital ecosystem for SMEs/SMBs worldwide.

In essence, this study serves as a catalyst for advancing cybersecurity research, policy, and practice within SMEs, ultimately contributing to the overarching goal of creating a more secure and resilient digital infrastructure for all.

References

1. Petkovska, T.: Role and importance of innovation in business of small and medium enterprises. Econ. Dev. **17**, 55–74 (2015)
2. Global SMEs (2021) https://www.statista.com/statistics/1261592/global-smes/, last accessed 12 Mar 2024
3. Ardolino, M., et al.: The role of digital technologies for the service transformation of industrial companies. Int. J. Prod. Res. **56**, 2116–2132 (2018). https://doi.org/10.1080/00207543.2017.1324224
4. Garzoni, A., et al.: Fostering digital transformation of SMEs: a four levels approach. Manag. Decis. **58**, 1543–1562 (2020). https://doi.org/10.1108/md-07-2019-0939
5. Lesmana, D., et al.: Challenges and cybersecurity threats in digital economic transformation. Int. J. Human. Educ. Soc. Sci. (IJHESS) (2023). https://doi.org/10.55227/ijhess.v2i6.515
6. OECD: The Digital Transformation of SMEs. OECD (2021). https://doi.org/10.1787/bdb9256a-en
7. Bada, M., Nurse, J.R.C.: Developing cybersecurity education and awareness programmes for small- and medium-sized enterprises (SMEs). Inf. Comput. Secur. **27**(3), 393–410 (2019). https://doi.org/10.1108/ICS-07-2018-0080
8. Heidt, M., et al.: Investigating the security divide between SME and large companies: how SME characteristics influence organizational IT security investments. Inf. Syst. Front. **21**(6), 1285–1305 (2019). https://doi.org/10.1007/s10796-019-09959-1
9. Carías, J.F., et al.: Systematic approach to cyber resilience operationalization in SMEs. IEEE Access **8**, 174200–174221 (2020). https://doi.org/10.1109/ACCESS.2020.3026063
10. Williams, E.J., et al.: Exploring susceptibility to phishing in the workplace. Int. J. Hum. Comput. Stud. **120**, 1–13 (2018). https://doi.org/10.1016/j.ijhcs.2018.06.004
11. Chidukwani, A., et al.: A survey on the cyber security of small-to-medium businesses: challenges, research focus and recommendations. IEEE Access **10**, 85701–85719 (2022). https://doi.org/10.1109/ACCESS.2022.3197899
12. Donthu, N., et al.: How to conduct a bibliometric analysis: an overview and guidelines. J. Bus. Res. **133**, 285–296 (2021). https://doi.org/10.1016/j.jbusres.2021.04.070
13. Visser, M., et al.: Large-scale comparison of bibliographic data sources: scopus, web of science, dimensions, crossref, and microsoft academic. Quant. Sci. Stud. 1–22 (2020). https://doi.org/10.1162/qss_a_00112

Exploring Opportunities and Challenges in Sustainable and Smart Cities Through the Lens of Smart Technologies

Danh Hong Le(✉)

Van Hien University, Ho Chi Minh City, Vietnam
danhlh@vhu.edu.vn

Abstract. This paper offers a tutorial based on a survey, focusing on the technologies that support and the potential applications within smart cities. It begins by providing a thorough definition of smart cities concepts from both industrial and academic viewpoints, then proceeds to investigate various applications within smart urban environments, spanning from the Internet of Everything to intelligent healthcare, cloud manufacturing, smart additive manufacturing, and geospatial intelligence. Emphasizing key-enabling technologies, the paper underscores the importance of achieving consistent harmony between human and machine efficiency. With the aid of emerging applications and supporting technologies, smart technologies have the potential to augment manufacturing output and enhance customer satisfaction. However, to fully realize the concept of smart cities in the foreseeable future, several challenges and open issues such as security, privacy, human-robot collaboration in industrial settings, scalability, and the skilled workforce need to be effectively addressed. This comprehensive approach aims to provide solutions to contemporary urban challenges while advancing the smart city concept.

Keywords: Smart city · Smart additive manufacturing · Geospatial intelligence · Internet of everything · Cloud manufacturing · Security and privacy

1 Introduction

Urbanization is a process fueled by population growth seen through the expansion of cities globally. By 2050 it is projected that 68% of the worlds population will be living in areas highlighting the trend, towards more interconnected city structures [28]. The 11th United Nations Sustainable Development Goal (SDG) focuses on establishing sustainable cities making it crucial for city planning to incorporate these objectives. The concept of technology driven cities aims to transform centers into high tech hubs with fully automated systems impacting not only the economy but also raising environmental concerns. Cities have

M. Pagac et al. (Eds.): SCFF 2024, LNNS 1062, pp. 403–413, 2024.
https://doi.org/10.1007/978-3-031-65656-9_40

become centers of resource consumption and regional dependence leading to challenges such as land use issues, strained infrastructure, transportation problems, environmental risks and improving residents quality of life [15]. Emphasizing the development of systems has been widely advocated as a solution to address these complex issues. Existing infrastructure endeavors to provide solutions for overcoming barriers, in constructing adaptable communities [16]. Cities are often described using a range of adjectives, such, as "intelligent" or "digital" [3]. A city is seen as a system with layers; technology forms the base layer, specific applications make up the second layer and user adoption is key, in the layer. Smart cities fundamentally rely on infrastructure [12].

The incorporation of information technology in governance, as highlighted by references [32,53], aims to enhance infrastructure management and optimize resource utilization, with cities adopting electronic technologies to elevate living standards and cultivate sustainable environments within their boundaries. This thinking urban system prioritizes resident services, sustainability, and innovation, leveraging technological systems such as communication networks, machine learning (ML), big data, and the Internet of Things (IoTs) to improve infrastructure efficiency and address environmental challenges [25].

Smart buildings, incorporating sensors, enhance disaster preparedness and maintenance, prioritizing residents' well-being. These sensors, interconnected and invaluable, improve warning systems and optimize traffic flow, focusing on quality of life rather than ICT [27]. Smart cities leverage data and human insights, exploring blockchain technology's challenges and opportunities to enhance service delivery [38]. Technological solutions support city managers in meeting resource demands effectively, driving technological progress and blurring boundaries between physical, virtual, and biological domains [8]. Cutting-edge technologies play pivotal roles in various smart city domains, including energy, transportation, healthcare, information analytics, and city mobility [31].

Geospatial technology has revolutionized the creation, analysis, and visualization of information, encompassing analytics, business intelligence, and data infrastructure. Advancements in sensing, GIS, GPS, radar, and LiDAR enable the capture, processing, and visualization of data, driving the integration of analytics into business operations and increasing government investments in the geospatial technology market, projected to reach USD 1.44 trillion by 2030 [55]. Global initiatives such as the UN Global Geospatial Information Management (UN GGIM) underscore the importance of geomatics, a multidisciplinary field crucial for collecting geospatial data through survey methods, satellite systems, and geographical information systems [11].

The integration of technologies with various disciplines has significantly bolstered the significance of geomatics. Starting from the early days of cartography, the field has evolved with the integration of satellite imagery, enabling a deeper understanding of dynamics within natural and human systems, facilitated by computational software for extensive spatial data processing. This research aims to explore the challenges and opportunities linked to integrating these enabling technologies into applications for smart cities.

2 Application in Smart City Environments

This section explores various potential applications of key enabling technologies within smart city environments, as illustrated in Fig. 1.

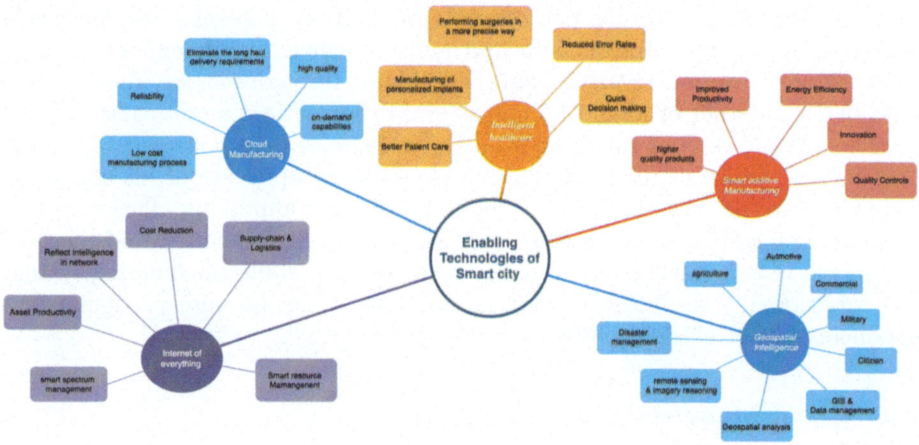

Fig. 1. Key enabling technologies of smart city environments.

Furthermore, the Economist Intelligence Unit, supported by NEC, has seized an opportune moment to introduce technology solutions aimed at enhancing city safety. This initiative involves evaluating 31 segments of industries and economy across 162 indicators, grouped into six pillars, including high tech, intelligent, smart transport, safe cities—digital security, smartest cities, innovative cities, and sustainable cities [10,37,46,48]. Statistical analysis of city performance across various indices in Table 1 reveals that top-ranking cities are predominantly located in Europe, the United States, Southeast Asia, and Australia. Tokyo consistently leads among these cities, with London, New York, and Singapore maintaining their positions in the top five across multiple indices, underscoring the importance of technology in city rankings.

Table 1. Top five cities ranking smart city

Ranking cities	High tech cities	Intelligent cities	Smart transport cities	Safe cities (digital security)	Smartest cities	Innovative cities	Sustainable cities
1	Tokyo	New York City	Singapore	Sydney	London	Tokyo	Oslo
2	London	London	Amsterdam	Singapore	Zurich	London	Stockholm
3	San Francisco	Paris	London	Copenhagen	Taipei City	New York	Tokyo
4	New York	Tokyo	Shanghai	Los Angeles	New York	Paris	Copenhagen
5	Los Angeles	San Francisco	New York	San Francisco	Barcelona	Singapore	Berlin

2.1 Internet of Everything

The Internet of Everything (IoE) acts as a vital link between individuals, processes, data, and objects, offering immense potential for smart city applications [23]. Within industries, IoE advancements can enhance user experiences and provide benefits for businesses and countries, particularly in boosting customer satisfaction and streamlining processes [51]. Despite its potential, IoE faces challenges in optimizing supply chain and logistics efficiency, though it simplifies wireless information exchange among individuals through sensors, such as those used in the Internet of Medical Things (IoMT) for patient monitoring [14].

The effectiveness of IoE hinges on the volume of data collected by sensors and devices, which is analysed through cloud-based applications, as envisioned by initiatives like the UN GGIM [56]. IoE's role in improving efficiency, cost-effectiveness, and security measures is anticipated, with demonstrated successes in wireless sensing technology applied to homes and intelligent building systems [9]. Additionally, the incorporation of IoE has shown promise in significantly enhancing power efficiency [50].

2.2 Cloud Manufacturing

Cloud manufacturing modernizes production by integrating cutting-edge technologies like cloud computing, IoT, and virtualization for efficient and cost-effective collaboration among partners, ensuring reliability, quality, and demand responsiveness while minimizing environmental impact [2]. Control systems oversee operations, IoT sensors gather production data, and studies support the effectiveness of cloud manufacturing as a service-oriented model [43]. Business strategies such as the pay-as-you-go model are explored, anticipating cloud manufacturing's evolution to meet complex demands across engineering, production, and logistics sectors, driven by advancements in intelligence, ML, edge computing (EC), and green communication networks [54].

Furthermore, cognitive computing addresses flexibility challenges in the cognitive Internet of Things (CIoT) by integrating diverse sensory devices, as seen in [5]. Additionally, the anticipated AI-based semantic Internet of Everything (AI-SIoE) hybrid service architecture is expected to cater to various device types across different scenarios [13].

2.3 Smart Additive Manufacturing

Sustainable manufacturing, recognised as a cost-effective approach in today's industries, prioritises efficient production, pollution reduction, and optimal resource utilisation throughout product life-cycles [41]. Additive manufacturing, building product components layer by layer, offers lighter and more durable parts, with smart additive manufacturing (SAM) utilising AI and computer vision to enhance precision and visual representation in 3D printing. Businesses and researchers are increasingly embracing manufacturing solutions, leveraging technologies like AI, IoT, cloud computing, Big Data, and CPS to achieve sustainability, profitability, and productivity [34]. The SAM has emerged as a pivotal technology in the manufacturing sector, enhancing sustainability and customer satisfaction through integrated product features and services [30]. By integrating SAM with automated functions, supply chain processes can minimise product delivery times, maximising benefits in smart city environments.

Moreover, in [1], practical insights into smart system realization and productivity impact were provided, highlighting enabling technologies like networked sensor data interoperability and machine cognition. Additionally, [17] identified key indicators for smart industry in manufacturing, emphasizing its potential to enhance company value and customer satisfaction. Furthermore, the application of smart industry in the textile sector was explored, suggesting its adoption in fibre computing solutions in the near future [36].

2.4 Intelligent Healthcare

Doctors are increasingly using ML models for disease diagnosis, improving accuracy and benefiting patients by saving time and resources [6]. However, there is a growing demand for personalised technology to monitor health indicators such as blood pressure and blood sugar levels [47]. Wearable devices like smartwatches and sensors, supported by cities, offer real-time health data monitoring, stored in the cloud for ML analysis to detect conditions and notify healthcare providers when intervention is required. Collaborative robots (cobots) can aid surgeons by facilitating communication and automating routine tasks, transforming healthcare in smart city environments [4].

The authors in reference [33] propose expanding the use of technology, such as 3D printing and 4D imaging scans, to enhance the capabilities and accessibility of wearable devices. Furthermore, technologies like digital twins (DT) can assist doctors in offering personalised drug prescriptions to patients [35]. Subsequent sections explore recent research on the potential applications of smart healthcare.

2.5 Geospatial Intelligence (GEOINT)

The term GEOINT, coined by Clapper at the National Geospatial-Intelligence Agency, involves extracting geospatial information within a space-time framework, integrating imagery, imagery intelligence, and geospatial data for comprehensive analysis, including crime pattern analysis using methods like hotspot mapping [7]. Satellite data is crucial for monitoring infrastructure, resources, disasters, and health, while geophysical phenomena studies are conducted using various methodologies [22]. A common intelligence framework incorporating data analytics, sensor fusion, augmented reality, and predictive modeling enhances situational awareness in megacities, with the IoT, AI, Cloud, and Big Data playing significant roles in the geospatial ecosystem, enabling technology-based solutions for sustainable transport and demand forecasting [26]. GEOINT significantly contributes to technology connectivity, locational data, and intelligence solutions for smart city development.

2.6 Other Enabling Technologies

Cutting-edge technologies such as Network Slicing (NS), eXtended Reality (XR), and Private Mobile Networks (PMNs) are pivotal in advancing technology-based development and supporting its applications. NS enables the creation of tailored networks on a single physical infrastructure, addressing the diverse needs of smart cities efficiently [52]. XR technologies, encompassing augmented and virtual reality, are already being applied in various smart city projects, including health education, remote healthcare services, and maintenance tasks [19].

Future developments in XR technologies will rely on advancements in networking, EC, virtual communication technologies, and computational capabilities. Additionally, upcoming cognitive networks are expected to align with the requirements of a connected society, offering fast data speeds, minimal delays, high reliability, energy efficiency, and sufficient traffic capacity.

3 Future Research Challenges

Insights from cutting-edge sources reveal that the concept of smart cities is already being implemented across various sectors, including healthcare, cloud manufacturing, advanced AM and GEOINT, integrating technologies such as AI, IoT, EC, and advanced connectivity like 6G and beyond. Challenges in city development include handling diverse data types, effective resource allocation, managing vast data volumes efficiently, and addressing delays, requiring attention for its complete realization. Moreover, discussions have centred on leveraging smart city concepts to address challenges posed by the COVID-19 crisis, employing advanced technologies like holography systems, 4D scans, humanoid robotic assistants, telemedicine, and smart inhaler devices to provide tailor-made treatments and diagnoses, optimize supply chains, digitalize medical services, improve pharmaceutical production methods, raise public awareness about health concerns, and enhance crowd surveillance measures.

Following COVID-19, advanced technologies like cobots, CHIPBOT microchips for COVID patients, and currency-less CURBOT systems may revolutionise touchless transactions, healthcare, and security surveillance. Smart city projects offer potential for emerging sectors, integrating human-machine collaboration and tackling ethical issues, while smart industry developments enable customisable production and sustainable practices, though addressing security, privacy, and ethical concerns is crucial for successful deployment [20].

3.1 Security and Privacy

Security concerns in smart cities mirror those in traditional Cyber-Physical Systems (CPSs), requiring scalable authentication mechanisms to accommodate numerous stakeholders, while integrity and access control mechanisms are essential to secure data transfers and restrict resource access [44]. Transparency through log management systems is crucial for assessing service conformity and resolving conflicts, particularly with the incorporation of AI and automation, necessitating trusted execution and data integrity safeguards [45].

Privacy concerns in smart cities necessitate robust safeguards to protect sensitive data and promote trust within cloud manufacturing environments, while addressing ethical considerations in AI integration is crucial for fostering seamless collaboration between humans and cobots in industrial processes [39]. Despite the potential of blockchain to address security issues, its resource-intensive nature and scalability challenges highlight the need for lightweight frameworks and exploration of alternative security measures like quantum computing for securing Cyber-Physical Production Systems (CPPS) without downtime [49].

3.2 Scalability

Scalability in smart city technologies is vital for adapting to changing workloads, particularly in integrating robots and humans across factories, and can be monitored through service level indicators and agreements given the unpredictable growth of data and human involvement [18]. Technology providers must offer flexible services capable of handling increasing workloads and delivering low-latency responses, while AI-based cobots need to process queries efficiently from dynamic cloud and edge servers without delays.

The scalability of AM poses challenges in integrating its outcomes into production, navigating market trends, and refining industry standards [29]. Solutions may involve leveraging advanced printing capabilities and integrating EC with AI to reduce processing delays and improve predictive capabilities at the network edge.

3.3 Skilled Workforce

Addressing challenges in technology, society, and management demands skilled workers, underscoring the necessity for standardised policies guiding workforce development and training initiatives [21]. Overcoming obstacles like trainer shortages and financial limitations necessitates comprehensive training programmes and public-private partnerships to prepare the workforce for evolving technological demands. Furthermore, successful smart city transition mandates management comprehension of technologies, infrastructure upgrades, and business culture openness to support workforce training and human-cobot collaboration.

Smart manufacturing seeks to integrate workers with cobots and intelligent machines, necessitating regulations to ensure workplace safety and productivity, covering aspects such as defining cobots and guiding AI usage [24]. Establishing standards, laws, and guidelines can expedite and effectively manage the adoption of smart city technologies. Leveraging blockchain and quantum computing for transactions and data preservation, alongside integrating cognitive EC for efficient data processing, is crucial for establishing the foundation of smart systems. Despite advancements, scaling AM still poses challenges, as highlighted in Table 2 regarding the role of enabling technologies in smart city applications.

Table 2. Role of enabling technologies in smart city applications

Enabling technologies	Smart factory	Innovation ecosystem	Intelligent healthcare	Cloud manufacturing	Smart additive manufacturing	Geospatial intelligence
Edge computing	High utilization (H)	H	H	H	H	H
Digital twins	H	Medium utilization (M)	M	H	H	M
Collaborative robots	H	H	Low utilization (L)	H	H	M
Internet of everything	H	H	H	H	H	H
AI	H	H	H	H	H	H
Big data	H	H	H	H	H	H
Blockchain	H	H	H	H	H	H
Network slicing	H	L	H	H	H	M
Extended reality	H	H	H	H	H	M
Private mobile networks	H	L	M	M	H	M
6G and beyond	H	H	H	H	H	M

4 Conclusion

In conclusion, smart technologies offer significant opportunities for creating sustainable and intelligent urban environments through diverse applications such as the Internet of Everything, intelligent healthcare, cloud and smart manufacturing, and geospatial intelligence. However, realizing the full potential of the smart city paradigm faces key challenges, including security, privacy, human-robot collaboration, scalability, and skilled workforce development. Overcoming these hurdles requires collaborative, interdisciplinary efforts from stakeholders, supported by sustainable practices balancing economic, environmental, and social factors. Addressing the risks and challenges of smart technologies is crucial for cities to unlock their transformative potential, leading to an era of innovation, prosperity, and resilience in addressing urbanization issues.

References

1. Adel, A.: Future of industry 5.0 in society: human-centric solutions, challenges and prospective research areas. J . Cloud Comput. **11**(1), Sep 2022. https://doi.org/10.1186/s13677-022-00314-5
2. Akbaripour, H., et al.: Cloud manufacturing service selection optimization and scheduling with transportation considerations: mixed-integer programming models. Int. J. Adv. Manuf. Technol. **95**(1–4), 43–70 (2017). https://doi.org/10.1007/s00170-017-1167-3. Oct
3. Aldinhas Ferreira, M.I. (ed) How Smart Is Your City? Springer International Publishing (2021). https://doi.org/10.1007/978-3-030-56926-6
4. Basulo-Ribeiro, J., Teixeira, L.: The future of healthcare with industry 5.0: preliminary interview-based qualitative analysis. Future Int. **16**(3), 68, Feb (2024). https://doi.org/10.3390/fi16030068
5. Braten, A.E., Kraemer, F.A.: Towards cognitive IoT: autonomous prediction model selection for solar-powered nodes. IEEE Int. Congr. Int. Things (ICIOT). San Francisco, CA, USA **2018**, 118–125 (2018). https://doi.org/10.1109/ICIOT.2018.00023

6. Deepa, N., et al.: An AI-based intelligent system for healthcare analysis using Ridge-Adaline Stochastic Gradient Descent Classifier. J. Supercomput. **77**(2), 1998–2017 (2020). https://doi.org/10.1007/s11227-020-03347-2. May
7. Dold, J., Groopman, J.: The future of geospatial intelligence. Geo-spat. Inf. Sci. **20**(2), 151–162 (2017). https://doi.org/10.1080/10095020.2017.1337318. Apr
8. El-Hallaq, M.A., et al.: Enhancing sustainable development through web based 3D Smart City model using GIS and BIM. Case study: Sheikh Hamad City. J. Geogr. Inf. Syst. **11**(03), 321–330 (2019). https://doi.org/10.4236/jgis.2019.113019
9. Ghayvat, H., et al.: WSN- and IOT-based smart homes and their extension to smart buildings. Sensors **15**, 10350–10379 (2015). https://doi.org/10.3390/s150510350
10. Global 500 City Rankings Results for 2022-2023. Available online: https://innovation-cities.com/worlds-most-innovative-cities-2022-2023-city-rankings/26453/. Last accessed on 10 November 2023
11. Goodchild, M.F.: Citizens as sensors: the world of volunteered geography. Geo J **69**(4), 211–221 (2007). https://doi.org/10.1007/s10708-007-9111-y. Nov
12. Gracias, J.S., et al.: Smart cities-a structured literature review. Smart Cities **6**(4), 1719–1743 (2023). https://doi.org/10.3390/smartcities6040080. Jul
13. Guo, K., et al.: Artificial intelligence-based semantic Internet of Things in a user-centric smart city. Sensors **18**(5), 1341 (2018). https://doi.org/10.3390/s18051341. Apr
14. Higginbotham, S.: What 5G hype gets wrong—[Internet of everything]. IEEE Spectr. **57**(3), 22–22 (2020). https://doi.org/10.1109/MSPEC.2020.9014454. March
15. Huang, R.: SDG-oriented sustainability assessment for Central and Eastern European countries. Environ. Sustain. Ind. **19**, 100268 (2023). https://doi.org/10.1016/j.indic.2023.100268. Sep
16. Innovation and Skill 2013. Available online: https://www.gov.uk/government/publications/bis-annual-report-and-accounts-2013-to-2014. Accessed on 6 September 2023
17. Javaid, M., et al.: Industry 5.0: potential applications in COVID-19. J. Ind. Integr. Manage. **05**(04), 507–530 (2020). https://doi.org/10.1142/s2424862220500220. Nov
18. Javed, A.R., et al.: Future smart cities: requirements, emerging technologies, applications, challenges, and future aspects. Cities **129**, 103794 (2022). https://doi.org/10.1016/j.cities.2022.103794. Oct
19. Kandil, A., et al.: Enhancing user experience of interior design mobile augmented reality applications. In: Proceedings of the 5th International Conference on Computer-Human Interaction Research and Applications (2021). https://doi.org/10.5220/0010630400003060
20. Klein, C., Kaefer, G.: From smart homes to smart cities: opportunities and challenges from an industrial perspective. Lect. Notes Comput. Sci. 260–260. https://doi.org/10.1007/978-3-540-85500-2_24
21. Kumar, K.: ITIs / ITCs: industrial training institutes/industrial training centres. In: India: Preparation for the World of Work, pp. 65–80 (2016). https://doi.org/10.1007/978-3-658-08502-5_5
22. Lechner, A.M., et al.: Applications in remote sensing to forest ecology and management. One Earth **2**(5), 405–412 (2020). https://doi.org/10.1016/j.oneear.2020.05.001. May
23. Li X., Xu, L.D.: A review of internet of things-resource allocation. IEEE Int Things J 8(11):8657–8666, 1 June 2021. https://doi.org/10.1109/JIOT.2020.3035542

24. Lima, F.: Digital manufacturing tools in the simulation of collaborative robots: towards industry 4.0. Braz. J. Oper. Prod. Manage. **16**(2), 261–280 (2019). https://doi.org/10.14488/BJOPM.2019.v16.n2.a8. May
25. Lin, S.-H., et al.: Evaluating smart office buildings from a sustainability perspective: a model of hybrid multi-attribute decision-making. Technol. Soc. **68**, 101824 (2022). https://doi.org/10.1016/j.techsoc.2021.101824. Feb
26. Loper, M.L.: Situational awareness in megacities. Adv. Sci. Technol. Secur. Appl. 205–235 (2018). https://doi.org/10.1007/978-3-319-75232-7_12
27. Manchester, H., Cope, G.: Learning to be a smart citizen. Oxf. Rev. Educ. **45**(2), 224–241 (2019). https://doi.org/10.1080/03054985.2018.1552582. Mar
28. Mori, K., Christodoulou, A.: Review of sustainability indices and indicators: towards a new City sustainability index (CSI). Environ. Impact Assess. Rev. **32**(1), 94–106 (2012). https://doi.org/10.1016/j.eiar.2011.06.001. Jan
29. Mukherjee, A.A., et al.: Identification of barriers and their mitigation strategies for industry 5.0 implementation in emerging economies. Int. J. Prod. Econ. **257**, 108770 (2023). https://doi.org/10.1016/j.ijpe.2023.108770. Mar
30. Nguyen, H.-S. et al.: Copyright protection in additive manufacturing. MM Sci. J. 2023(2) Jun 2023. https://doi.org/10.17973/mmsj.2023_06_2023030
31. Nguyen, H.-S., et al.: A bibliometrics analysis of medical internet of things for modern healthcare. Electronics **12**(22), 4586 (2023). https://doi.org/10.3390/electronics12224586. Nov
32. Nguyen, H.-S., Voznak, M.: A bibliometric analysis of technology in digital health: exploring health metaverse and visualizing emerging healthcare management trends. IEEE Access **12**, 23887–23913 (2024). https://doi.org/10.1109/access.2024.3363165
33. Priadythama, I., et al.: Role of rapid manufacturing technology in wearable customized assistive technology for modern industry. In: AIP Conference Proceedings (2020) https://doi.org/10.1063/5.0000747
34. Qi, Q., et al.: Big data analytics challenges to implementing the intelligent industrial Internet of Things (IIoT) systems in sustainable manufacturing operations. Technol. Forecast. Soc. Chang. **190**, 122401 (2023). https://doi.org/10.1016/j.techfore.2023.122401. May
35. Reinhardt, I.C., et al.: Current perspectives on the development of industry 4.0 in the pharmaceutical sector. J. Ind. Inf. Integr. **18**, 100131 (2020). https://doi.org/10.1016/j.jii.2020.100131. Jun
36. Roy, S., Chowdhury, C.: Remote health monitoring protocols for IoT-enabled healthcare infrastructure. In: Healthcare Paradigms in the Internet of Things Ecosystem, pp. 163–188 (2021). https://doi.org/10.1016/b978-0-12-819664-9.00008-9
37. Safe Cities Index 2021. Available online: https://safecities.economist.com/safe-cities-2021-whitepaper/. Last accessed on 10 November 2023
38. Salha, R.A., et al.: Blockchain in smart cities: exploring possibilities in terms of opportunities and challenges. J. Data Anal. Inf. Process. **07**(03), 118–139 (2019). https://doi.org/10.4236/jdaip.2019.73008
39. Samek, W., et al.: Explaining deep neural networks and beyond: a review of methods and applications. Proc. IEEE **109**(3), 247–278 (2021). https://doi.org/10.1109/JPROC.2021.3060483. March
40. Silver, D., et al.: Mastering the game of Go with deep neural networks and tree search. Nature **529**(7587), 484–489 (2016). https://doi.org/10.1038/nature16961. Jan

41. Sommer, M., et al.: Automated generation of digital twin for a built environment using scan and object detection as input for production planning. J. Ind. Inf. Integr. **33**, 100462 (2023). https://doi.org/10.1016/j.jii.2023.100462. Jun
42. Sportillo, D., et al.: On-road evaluation of autonomous driving training. In: 2019 14th ACM/IEEE International Conference on Human-robot Interaction (HRI), pp. 182–190. Daegu, Korea (South). https://doi.org/10.1109/HRI.2019.8673277
43. Tao, F., et al.: Cloud manufacturing: a computing and service-oriented manufacturing model. Proc. Inst. Mech. Eng. Part B J. Eng. Manuf. **225**(10), 1969–1976 (2011). https://doi.org/10.1177/0954405411405575. Aug
44. Tawalbeh, L., et al.: IoT privacy and security: challenges and solutions. Appl. Sci. **10**(12), 4102 (2020). https://doi.org/10.3390/app10124102. Jun
45. The HIPAA Privacy Rule. Available online: https://www.hhs.gov/hipaa/for-professionals/privacy/index.html. Last accessed on 19 December 2023
46. The top 10 cities in the Arcadis Sustainable Cities Index 2022. Available online: https://www.wbcsd.org/Overview/News-Insights/Member-spotlight/The-Arcadis-Sustainable-Cities-Index-2022. Last accessed on 10 November 2023
47. Thippa Reddy, G., et al.: Antlion re-sampling based deep neural network model for classification of imbalanced multimodal stroke dataset. Multimedia Tools Appl. **81**(29):41429-41453, Oct 2020. https://doi.org/10.1007/s11042-020-09988-y
48. Top 10 Most Futuristic Smart Cities in the World 2023. Available online: https://www.digi.com/blog/post/smart-cities-in-the-world. Last accessed on 10 November 2023
49. Tosh, D.: Towards security of cyber-physical systems using quantum computing algorithms. In: IEEE 15th International Conference of System of Systems Engineering (SoSE), pp. 313–320. Budapest, Hungary (2020). https://doi.org/10.1109/SoSE50414.2020.9130525
50. Van, H.T., et al.: Opportunistic DF-AF selection relaying in hybrid wireless and power line communication for indoor IoT networks. Sensors **21**(16), 5469 (2021). https://doi.org/10.3390/s21165469. Aug
51. Wang Y. et al.: Application of IoT optical sensing image processing equipment based on load balancing algorithm in sports physical fitness detection. Opt. Quant. Electron. **56**(584). https://doi.org/10.1007/s11082-024-06292-z
52. Wijethilaka, Liyanage, M.: Realizing Internet of Things with network slicing: opportunities and challenges. In: IEEE 18th Annual Consumer Communications & Networking Conference (CCNC), pp. 1–6. Las Vegas, NV, USA (2021). https://doi.org/10.1109/CCNC49032.2021.9369637
53. Wu, Z., et al.: Mapping the knowledge domain of smart city development to city sustainability: a scientometric study. J. City Technol. **28**(1–2), 29–53 (2020). https://doi.org/10.1080/10630732.2020.1777045. Jul
54. Xu, X.: From cloud computing to cloud manufacturing. Robot. Comput. Integr. Manuf. **28**(1), 75–86 (2012). https://doi.org/10.1016/j.rcim.2011.07.002. Feb
55. Xue, Y., et al.: Telegeoprocessing: the integration of remote sensing, geographic information system (GIS), global positioning system (GPS) and telecommunication. Int. J. Remote Sens. **23**(9), 1851–1893 (2002). https://doi.org/10.1080/01431160110069953. Jan
56. Zanella, A., et al.: Internet of Things for smart cities. IEEE Int. Things J. **1**(1), 22–32 (2014). https://doi.org/10.1109/JIOT.2014.2306328. Feb

Factors Driving Consumers' Adoption of Internet Banking in an Emerging Market

Bang Nguyen-Viet[1], Ngoc Khanh Huynh[2(✉)], and Thanh Hai Nguyen[3]

[1] University of Economics Ho Chi Minh City, Ho Chi Minh City, Vietnam
[2] Bank for Investment and Development of Vietnam JSC, Hanoi, Vietnam
khanhhn2@bidv.com.vn
[3] University of Finance and Marketing, Ho Chi Minh City, Vietnam

Abstract. The primary objective of this study is to examine how the technology acceptance model (TAM) affects the intention to utilize online banking products and services of Vietnamese customers, incorporating perceived risk and consumer innovativeness. Data were collected from 475 respondents who used internet banking in Vietnam, and CB-SEM method was utilized for analyzing the data. The findings show that consumers' intention to use internet banking services is positively influenced by their perceived usefulness and consumer innovativeness. Perceived ease of use has a significantly positive impact on consumer intention, which is mediated by perceived usefulness, while the impact of perceived risk is adverse. The findings validated the expanded TAM and verified its reliability in forecasting customers' intention to use online banking services. This study adds to the literature by developing and verifying the TAM for predicting internet banking service adoption. The results provide helpful information for bank management in developing internet banking service marketing strategies.

Keywords: Innovativeness · Internet banking · Technology acceptance model · Perceived risk · Adoption · Vietnam

1 Introduction

The banking industry is regarded as essential in any economy's financial system, which is continually evolving at a rapid pace in response to the country's economic and political situations and the rapid rise in information and communication technology (Hanafizadeh and Marjaie 2020). Vietnam, situated in Southeast Asia, stands out as one of the countries heavily dependent on cash, boasting one of the most significant rates of embracing cashless transactions within the region during the year 2020 (Nguyen 2021). Many banks are beginning to see the need to invest in digital technology for customer acquisition and satisfaction. Consequently, they strive to use authorised electronic devices to offer more services (Sedaghatparast 2019). Internet banking (IB) holds a significant position as a crucial and impactful component within the expanding global economy and banking industry (Akhter et al. 2022). IB denotes the systems that facilitate customers of financial institutions to retrieve information regarding their accounts and various bank offerings via online platforms, eliminating the need for traditional methods such as letters, faxes, signatures, or phone calls (Alkhowaiter 2020).

M. Pagac et al. (Eds.): SCFF 2024, LNNS 1062, pp. 414–425, 2024.
https://doi.org/10.1007/978-3-031-65656-9_41

IB provides several benefits to both banks and customers (Anouze and Alamro 2020). IB has made banks more competitive by allowing them to access and service a wider spectrum of customers at a cheaper cost (Alhassany and Faisal 2018). Banks can achieve a reduction in costs by around 20–25%, consequently enhancing their competitive advantage (Ananda et al. 2020), due to the decreased necessity for in-person interactions with customers (Namahoot and Laohavichien 2018). Clients have the ability to engage in financial transactions, such as purchasing and transferring funds, with efficiency, simplicity, and convenience (Ananda et al. 2020). E-banking services, specifically, provide advantages to customers as they enable the execution of transactions and various financial tasks without the need for a physical bank branch (Anouze and Alamro 2020).

Most IB research has been conducted in wealthy countries (Sharma et al. 2020) and only a few empirical studies were carried out in developing nations like Vietnam (Nguyen-Viet and Huynh 2021; Lin et al. 2014); however, these investigations have failed to make theoretical or practical advances. Additionally, many research studies utilize TAM or UTAUT models without focusing on innovation characteristics in mobility and banking services (Ho et al. 2020. Many previous studies have identified the crucial roles of perceived risk (Kaur and Arora 2020; Bashir and Madhavaiah 2015) and innovation (Chauhan et al. 2019, Nguyen-Viet and Huynh 2021) in IB adoption. However, the existing literature has an unclear depiction of the relationship between perceived risk and innovativeness.

This research addresses the gap by investigating the factors that drive IB adoption in Vietnam using Davis's (1989) TAM and including new variables (customer innovativeness and perceived risk). It highlights the significant aspects that would benefit researchers in investigating the concerns surrounding IB adoption and approaches in Vietnam. Using the TAM, the findings provide vital insight into the implications of consumer perceived risk and innovativeness of IB adoption in Vietnam.

2 Literature Review and Hypotheses Development

2.1 Technology Acceptance Model (TAM)

This study uses the TAM by Davis (1989) as the theoretical background. The validity of TAM has been demonstrated to have generalizability by examining the adoption of various forms of information technology by a broad set of people several times in time (Alhassany and Faisal 2018).

The primary benefit of this model is that it can measure consumers' willingness to adopt various forms of information technology and/or systems (Singh and Srivastava 2020). Additionally, the TAM supports senior managements' responsibility for offering and developing banking products online and developers of information systems in predicting users' behavioral intentions (Yiu et al. 2007).

It is regarded as the most comprehensive research model for forecasting consumer behaviors in the context of IB over time due to its wide theoretical and empirical support (Chauhan et al. 2019; Nguyen-Viet and Huynh 2021). As a result, TAM was selected for this study as the research framework. It will also be integrated into a model that combines the two most important constructs suggested as single or bilateral in this research (perceived risk and consumer innovativeness). Perceived risk and consumer innovativeness were the two most influential constructs examined in this study.

2.2 Perceived Usefulness

Perceived usefulness relates to individual belief on how technology enhances their job efficiency (PU) (Davis 1989). PU in the Internet context is the consumer's belief in IB services enhancing banking performance. (Bashir and Madhavaiah 2015).

The PU elucidates the beliefs held by consumers and the mechanisms that expedite adoption of information technology (Kaur and Malik 2019). The PU factor within the IB framework pertains to the perceptions held by consumers that the utilization of internet services could enhance their efficiency and effectiveness (Alhassany and Faisal 2018). Furthermore, PU is considered as the customers' perception regarding the utilization of IB services in their financial transactions that satisfy their financial needs, objectives, and dealings (Akhter et al. 2022).

Several IB adoption studies have continuously presented empirical proof of PU's strong favourable influence on user intentions (Akhter et al. 2022; Nguyen-Viet and Huynh 2021). Therefore, H1 is proposed as follows:

H1: PU positively affects consumers' intention to use IB services.

2.3 Perceived Ease of Use

The level to which a consumer perceives the ease of utilizing IB services is defined as the perceived ease of use (PEU) (Davis 1989). The PEU factor in the IB context relates to users' perceptions and evaluations of the difficulty of IB usage, which determine their willingness to use such services. Customers have a tendency to use services that are simple to interact with and do not cause any confusion (Alhassany and Faisal 2018). The importance of PEU in influencing IB adoption is well acknowledged. Many studies have experimentally explored this connection and found that PEU plays a key role in predicting IB adoption (Chauhan et al. 2019; Davis 1989). According to the literature, PEU favourably impacts PU in IB adoption (Chauhan et al. 2019; Nguyen-Viet and Huynh 2021). Therefore, H2 is proposed as follows.

H2: PEU positively impacts PU.

2.4 Attitude

The concept of 'attitude' refers to how individuals evaluate specific behaviors as positive or negative (Ajzen 1991). Because IB services directly include monetary transactions, ATT is more frequent in the IB adoption setting; hence, fundamental consumer views on risk and safety impact customer ATTs (Bashir and Madhavaiah 2015). According to previous research, consumer ATT impact their propensity to employ IB services (Chauhan et al. 2019; Nguyen-Viet and Huynh 2021). Furthermore, there is abundant evidence that the TAM constructs, PEU and PU, significantly influence consumer ATT toward IB adoption (Nguyen-Viet and Huynh 2021; Anouze and Alamro 2020; Chauhan et al. 2019). Therefore, H3, H4, and H5 are proposed as follows.

H3: ATT has a positive effect on consumers' intention to use IB services.
H4: PU has a positive effect on ATT.
H5: PEU has a positive effect on ATT.

2.5 Perceived Risk

Individuals' perceived risk is the uncertainty they feel, which may result in undesirable repercussions when using a service or product (Bauer 1960). According to consumer behavior research, consumers tend to weigh all uncertainties and potential negative effects when making product/service decisions (Mitchell 1999). Consumer reluctance to adopt technology is influenced by perceived risk in various contexts, particularly in IB services (Bashir and Madhavaiah 2015). Customers are worried about the security, privacy, and confidentiality of the services (Chauhan et al, 2019). According to previous research, perceived risk negatively impacts PU (Bashir and Madhavaiah 2015; Marakarkandy et al. 2017) and customer attitudes toward using IB services (Bashir and Madhavaiah 2015). Furthermore, multiple studies have found that perceived risk has a direct negative impact on customers' intentions to use innovative financial services such as IB (Bashir and Madhavaiah 2015; Marakarkandy et al. 2017). Therefore, H6, H7, and H8 are proposed as follows.

H6: Perceived risk negatively impacts consumers' intention to use IB services.
H7: Perceived risk negatively impacts ATT.
H8: Perceived risk negatively impacts PU.

2.6 Consumer Innovativeness

Consumer innovativeness, as a concept, was developed in the early 1970s and has since become increasingly popular due to its significant impact on the adoption and diffusion of innovative products and services. Consumer innovativeness is a new product or service adoption behavior that refers to the degree to which individuals in a social system adopt innovative items ahead of others (Rogers and Shoemaker 1971). Midgley and Dowling (1978) may have been the first to distinguish between actual and potential innovative behavior. Midgley and Dowling (1978) describe consumer innovativeness as the extent to which individuals engage in innovative decision-making processes without seeking input or sharing their experiences with others. They further categorise consumer innovation as actualised and intrinsic. Innovativeness can be described as the capacity and creativity of an individual that allows them to embrace novel concepts, techniques, or innovations (Zhang et al. 2020).

Consumer innovativeness is considered a crucial factor in determining the level of IB adoption according to a previous study (Akhter et al. 2022; Alhassany and Faisal 2018) and plays a role in reducing threats related to using technology in banking channels, affecting the adoption of IB services (Akhter et al. 2022). Moreover, Nguyen-Viet and Huynh (2021) and Chauhan et al. (2019) found that consumer innovativeness is a significant predictor of PEU's ability to support IB. Furthermore, Chauhan et al. (2019) find a negative association between innovative behaviour and perceived risk. Therefore, H9, H10, and H11 can be expressed as

H9: Consumer innovativeness positively impacts consumers' intention to use IB services.

H10: Consumer innovativeness positively impacts PEU.
H11: Consumer innovativeness negatively impacts perceived risk.

3 Methodology

3.1 Measurement

The constructs utilized in this research were assessed using items validated in prior studies. This study encompasses six constructs, with all items evaluated using a five-point Likert scale, ranging from 1 (indicating strong disagreement) to 5 (indicating strong agreement). The questionnaire was disseminated among professionals and scholars specializing in IB and marketing domains to gauge their perspectives regarding the study's subject matter, as well as its evaluative criteria and contextual relevance. The questionnaire was also pre-tested with responses from five IB and five non-IB customers to confirm accuracy in wording, sequence, and completion. The questionnaire sequencing and wording were adjusted based on feedback from respondents; ambiguous questions were removed, and some wordings were changed. As a result, the feedback enhanced the clarity, relevance, and consistency of the questionnaire.

3.2 Sampling and Data Collection

As the sampling frame of the IB user population was unavailable, nonprobability sampling was used. The hypotheses were examined through the utilization of a convenience sample of individuals using banking services for transactions at the bank. Six bank brands were carefully selected to represent the three types of banks in Vietnam: two branches of foreign banks (Citibank Vietnam and HSBC Vietnam), two state-owned (Bank for Investment and Development of Vietnam and Vietnam Joint Stock Commercial Bank for Industry and Trade), two private joint stock commercial banks (DongA Commercial Joint Stock, Asia Commercial Bank). A survey questionnaire was created based on concepts and items found in existing literature. The initial questionnaire was designed in English. A technique of back-translation was employed to ensure a precise rendition of the questionnaire content in the desired language (Tyupa 2011). Hence, several bilingual researchers were requested to convert the questionnaire from English to Vietnamese and then back to English. The questionnaire in Vietnamese was developed then distributed to the participants. All surveys included cover letters explaining the study's aim and ensuring confidentiality and anonymity for responses. Finally, 621 responses were gathered over three months, specifically from the months of June to August in the year 2020. After removing invalid and incomplete entries, 475 completed responses were used for further analyses. The demographic profiles of the participants are displayed in Table 1. Out of the total 475 respondents, 56.6% were male and 43.4% were female, 48.8% of respondents were within the age group of below 30, 32.2% were in the 30–40 age group, and 18.9% were in the age group of above 40 (mean = 32.01). Educated respondents were selected for the study, and 85.3% had graduate degrees or higher. Approximately 67.6% of respondents were from Vietnamese banks.

Table 1. Demographic characteristics (N = 475)

Demographic profile		Frequency	Percent (%)
Gender	Male	269	57
	Female	206	43
Age (in years)	Below 30	232	48
	30–40	153	33
	Above 40	90	19
Education level	Undergraduate	70	14
	Graduate	189	40
	Postgraduate	152	32
	Doctorate	64	14
Type of banks	Vietnamese banks	321	68
	Foreign banks	154	32

3.3 Data Tools and Analytical Techniques

Statistical data analysis results were employed for the assessment of measurement and structural models, as well as for the identification of hypothetical relationships. Evaluation of the measurement model was conducted through the consideration of reliability, convergent validity, and discriminant validity (Hair et al. 2006). The examination of the structural model and the proposed relationships was carried out using Covariance-based structural equation modeling (CB-SEM).

Assessment of the Measurement Model

The reliability and validity of the investigated constructs were used to evaluate the test scale; details are provided in Table 2. Composite reliability was used to assess construct reliability. The value ranges between 0.767 and 0.898, which is higher than the suggested criteria (equal to or greater than 0.6) (Hair et al. 2010). Cronbach's alpha was employed for the computation of the internal consistency of every construct. This value is greater than 0.6, which is considered suitable ensuring for item reliability and internal consistency (Hair et al. 2010). VIF values below 3 suggest no multicollinearity among the items.

Convergent validity was assessed through the utilization of standardized factor loading and the computation of the average variance. The standardised factor loadings of all items range from 0.693 to 0.838, exceeding the recommended criterion of 0.5. (Hair et al. 2010). The extracted variance value ranges from 0.521 to 0.670, which also meets the criterion of 0.5 and higher (Hair et al. 2010). Moreover, Table 3 demonstrates that the correlation among the variables is lower than the square root of the average variance extracted for each variable, thereby validating the discriminant validity adequately (Fornell and Larcker 1981).

Table 2. Validation of the measurement model, reliability and convergent validity

Constructs	SFL	CA	CR	AVE
Consumer innovativeness (CI) (adopted from Alhassany and Faisal 2018)				
CI1	0.760	0.848	0.849	0.584
CI2	0.778			
CI3	0.785			
CI4	0.733			
Perceived usefulness (PU) (adopted from Anouze and Alamro 2020)				
PU1	0.703	0.813	0.813	0.522
PU2	0.747			
PU3	0.708			
PU4	0.730			
Perceived ease of use (PEU) (adopted from Anouze and Alamro 2020)				
PEU1	0.838	0.890	0.890	0.670
PEU2	0.791			
PEU3	0.848			
PEU4	0.796			
Attitude (ATT) (adopted from Anouze and Alamro 2020)				
ATT1	0.761	0.858	0.859	0.603
ATT2	0.823			
ATT3	0.732			
ATT4	0.789			
Perceived risk (PR) (adopted from Chauhan et al. 2019)				
PR1	0.744	0.766	0.767	0.523
PR2	0.733			
PR3	0.693			
Intention to use (IT) (adopted from Yaseen and Qirem 2018)				
IT1	0.784	0.806	0.806	0.580
IT2	0.761			
IT3	0.740			

Assessment of the Structural Model

The outcomes of the experimental model are displayed in Fig. 1: Chi2 = 263.282, df = 198, Cmin/df = 1.330, Tucker–Lewis fit index = 0.984, comparative fit index = 0.986, and root mean square error of approximation = 0.026. Consequently, the data matches the conceptual model. The results of the model in Table 4 demonstrate PU positively influences consumers' intention to use IB services (β = 0.271, p = 0.000), ATT (β =

Table 3. Discriminant validity measurement matrix

	Mean	CI	PU	PEU	ATT	PR	IT
CI	2.847	**0.765**					
PU	2.576	0.159	**0.722**				
PEU	3.626	0.155	0.069	**0.819**			
ATT	2.773	0.294	0.136	0.120	**0.777**		
PR	3.222	– 0.117	– 0.064	– 0.057	– 0.136	**0.724**	
IT	2.700	0.364	0.188	0.205	0.382	– 0.176	**0.762**

Bold significance denotes the square root of the average variance extracted for each variable

0.249, p = 0.006), and consumer innovativeness (β = 0.176, p = 0.002). In contrast, perceived risk (β = – 0.251, p = 0.002) negatively influences IB service adoption. Thus, H1, H3, H6, and H9 are supported. In addition, H2 and H4 are supported, because PU (β = 0.261, p = 0.000) has a positive effect on ATT. The findings suggest a significant positive impact of PEU (β = 0.134, p = 0.006) on ATT, mediated by PU. They also indicate a nonsignificant influence of PEU on ATT (β = 0.059, p = 0.198). Thus, H5 is not supported. The results support H7 and H8, indicating that perceived risk has a significant effect on ATT (β = – 0.424, p = 0.000) and PU (β = – 0.569, p = 0.000). These findings support that consumer innovativeness has a direct and positive effect on PEU (β = 0.247, p = 0.000) and negative effects on perceived risk (β = – 0.553, p = 0.000). Therefore, H10 and H11 are supported.

Table 4. The values obtained from structural models

Hypothesis	Path			Standardized estimate	P	Supported?
H1	PU	→	IT	0.271	0.000	Yes
H2	PEU	→	PU	0.134	0.006	Yes
H3	ATT	→	IT	0.249	0.000	Yes
H4	PU	→	ATT	0.261	0.000	Yes
H5	PEU	→	ATT	0.059	0.198	No
H6	PR	→	IT	– 0.251	0.002	Yes
H7	PR	→	ATT	– 0.424	0.000	Yes
H8	PR	→	PU	– 0.569	0.000	Yes
H9	CI	→	IT	0.176	0.002	Yes
H10	CI	→	PEU	0.247	0.000	Yes
H11	CI	→	PR	– 0.553	0.000	Yes

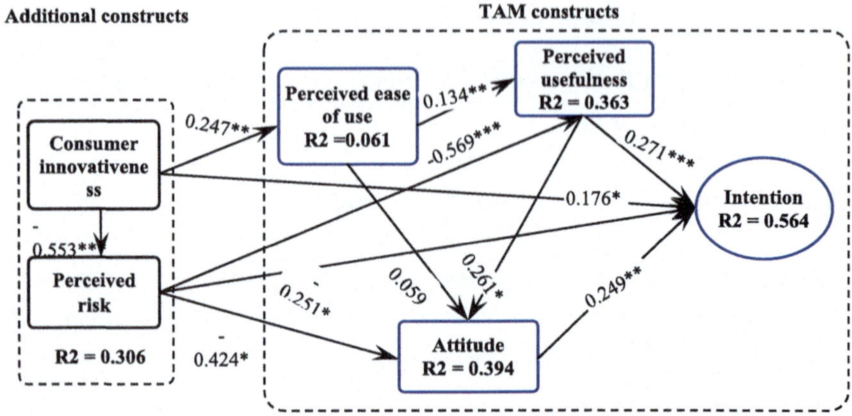

Note: *Significant at 10% level; **Significant at 5% level; ***Significant at 1% level.

Fig. 1. Results of model testing

4 Discussion

By examining the use of the TAM and combining additional factors, the findings offer insight into the crucial concerns surrounding customers' intention to use IB services (consumer innovativeness and perceived risk). The expanded TAM used in this study aids in better understanding and explaining the aspects that lead to success. As a result of this study, the effectiveness of the expanded TAM as a research model to explain customer adoption of IB was confirmed. This implies that authorities should try to build an environment conducive to IB transactions. This study introduces two variables (perceived risk and innovativeness) into the well-accepted TAM and applies it to the IB context in Vietnam and highlights the role of perceived risk in the adoption of IB. The results also indicate that the more customers perceived risks in the operation, the lower their intention to use IB. This study also provides a deep understanding of the role of individual innovativeness in the adoption of IB. Moreover, this study explores the relationship between perceived risk and innovation in the context of Vietnamese IB. This research is among the initial efforts to include innovativeness and perceived risk in a model to assess its relevance in predicting consumers' intention to adopt IB in Vietnam.

5 Managerial Implications

According to the TAM, PU is one of the most significant indicators of consumers' future ATT intentions and IB service adoption, this finding is supported by prior IB research by Chauhan et al. (2019) and Nguyen-Viet and Huynh (2021). Therefore, it is imperative for financial institutions to contemplate enhancing their existing IB platforms in order to provide enhanced accessibility, efficiency, and performance capabilities to their customers.More crucially, this study linked perceived risk and the TAM, providing a unique insight into consumer PU, ATT, and adoption behavior prediction. Consumers who migrate from conventional offline to online banking can save money and time, but also incur the risk of fraud and identity theft (Chauhan et al. 2019). Consequently, banks should strengthen their systems' security features and emphasise their security and preventive functions. Additionally, this study found that ATT affects customers' propensity to use IB services in the suggested model. Consequently, banks must place an even higher premium on fostering positive attitudes toward IB among potential customers by presenting them with all necessary IB information. The findings from this research suggest that PEU does not have a significant impact on ATT. This conclusion contradicts the results of Chauhan et al. (2019) and Nguyen-Viet and Huynh (2021). This means that when individuals perceive and understand that IB technology is easy to use, they are more inclined to adopt online banking services. The results also indicate that consumer innovativeness is a key predictor of customers' behavioral intention to use IB services, which is consistent with earlier research (Alhassany and Faisal 2018; Nguyen-Viet and Huynh 2021). Furthermore, this study's findings indicate that consumer innovativeness has a considerable beneficial influence on PEU, aligning with Chauhan et al. (2019) and Nguyen-Viet and Huynh (2021). They are also related to a decreased risk perception of individuals adopting IB behavior, which is similar to previous research (Chauhan et al. 2019).

6 Limitations and Future Research Directions

This study, like other empirical research, has limitations that can be further examined in future studies. First, the study was limited to a particular site (Ho Chi Minh City, Vietnam), thereby limiting the generalizability of the results. Second, rather than actual usage or behavior, this study focuses on IB service adoption aspirations. Third, due to the utilization of convenience sampling by the researchers, the findings of this study could potentially exhibit bias. Finally, the respondents were restricted to the banking industry, which may have impaired the generalisability of the findings.

Acknowledgements. This research is funded by University of Economics Ho Chi Minh city (UEH).

References

Akhter, A., Karim, M.M., Jannat, S., Islam, K.M.A.: Determining factors of intention to adopt internet banking services: a study on commercial bank users in Bangladesh. Banks Bank Syst. **17**(1), 125–136 (2022). https://doi.org/10.21511/bbs.17(1).2022.11

Alhassany, H., Faisal, F.: Factors influencing the internet banking adoption decision in North Cyprus: an evidence from the partial least square approach of the structural equation modeling. Financ. Innov. **4**(1), 1–21 (2018)

Alkhowaiter, W.A.: Digital payment and banking adoption research in Gulf countries: a systematic literature review. Int. J. Inf. Manage. **53**, 102102 (2020). https://doi.org/10.1016/j.ijinfomgt.2020.102102

Ananda, S., Devesh, S., Al Lawati, A.M.: What factors drive the adoption of digital banking? An empirical study from the perspective of Omani retail banking. J. Financ. Serv. Mark. **25**(1), 14–24 (2020)

Anouze, A.L.M., Alamro, A.S.: Factors affecting intention to use e-banking in Jordan. Int. J. Bank Mark. **38**(1), 86–112 (2020)

Ajzen, I.: The theory of planned behavior. Organ. Behav. Human Decis. Process **50**(2), 179–211 (1991)

Bashir, I., Madhavaiah, C.: Consumer attitude and behavioural intention towards Internet banking adoption in India. J. Indian Bus. Res. **7**(1), 67–102 (2015)

Bauer, R.A.: Consumer behavior as risk taking. In: Hancock, R.S. (eds.) Dynamic Marketing for a Changing World. Proceedings of the 43rd American Marketing Association Conference, pp. 389–398. Chicago, IL (1960)

Chauhan, V., Yadav, R., Choudhary, V.: Analyzing the impact of consumer innovativeness and perceived risk in internet banking adoption: a study of Indian consumers. Int. J. Bank Mark. **37**(1), 323–339 (2019)

Davis, F.D.: Perceived usefulness, perceived ease of use, and user acceptance of information technology. MIS Q. **13**(3), 319–339 (1989)

Fornell, C., Larcker, D.F.: Evaluating structural equation models with unobservable variables and measurement error. J. Mark. Res. **18**(1), 39–50 (1981)

Hair, J., Black, W., Babin, B., Anderson, R., Tatham, R. L.: Mutivariate Data Analysis. vol 31 (2006)

Hair, J.F., Black, W.C., Babin, B.J., Anderson, R.E.: Multivariate Data Analysis, 7th edn. Prentice Hall, Englewood Cliffs (2010)

Hanafizadeh, P., Marjaie, S.: Trends and turning points of banking: a timespan view. Rev. Manag. Sci. **14**(6), 1183–1219 (2020)

Ho, J.C., Wu, C.-G., Lee, C.-S., Pham, T.-T.T.: Factors affecting the behavioral intention to adopt mobile banking: an international comparison. Technol. Soc. **63**, 101360 (2020)

Kaur, S., Arora, S.: Role of perceived risk in online banking and its impact on behavioral intention: trust as a moderator. J. Asia Bus. Stud. **15**(1), 1–30 (2020)

Kaur, A., Malik, G.: Examining factors influencing Indian customers' intentions and adoption of Internet banking: extending TAM with electronic service quality. Innov. Mark. **15**(2), 42–57 (2019)

Lin, F.T., Wu, H.Y., Tran, T.N.N.: Internet banking adoption in a developing country: an empirical study in Vietnam. Inf. Syst. e-Bus. Manag. **13**(2), 267–287 (2014)

Marakarkandy, B., Yajnik, N., Dasgupta, C.: Enabling internet banking adoption: an empirical examination with an augmented technology acceptance model (TAM). J. Enterp. Inf. Manag. **30**(2), 263–294 (2017)

Midgley, D.F., Dowling, G.R.: Innovativeness: the concept and its measurement. J. Consum. Res. **4**(4), 229–242 (1978)

Mitchell, V.: Consumer perceived risk: conceptualisations and models. **33**(1/2), 163–195 (1999). https://doi.org/10.1108/03090569910249229

Namahoot, K.S., Laohavichien, T.: Assessing the intentions to use internet banking. Int. J. Bank Mark. **36**(2), 256–276 (2018)

Nguyen-Viet, B., Huynh, N.K.: Empirical analysis of internet banking adoption in Vietnam. Cogent Econ. Financ. **9**(1), 1992876 (2021)

Nguyen, M.-N.: E-wallets in Vietnam—statistics and facts, 28 Oct 2021. Retrieved from statista: https://www.statista.com/topics/8647/e-wallets-in-vietnam/#topicHeader__wrapper

Rogers, E.M., Shoemaker, F.F.: Communication of Innovations. The Free Press, New York (1971)

Sedaghatparast, E.: A meta-synthesis approach to specify components of future banking. Foresight **21**(4), 482–496 (2019)

Sharma, R., Singh, G., Sharma, S.: Modelling internet banking adoption in Fiji: a developing country perspective. Int. J. Inf. Manage. **53**, 102116 (2020)

Singh, S., Srivastava, R.K.: Understanding the intention to use mobile banking by existing online banking customers: an empirical study. J. Financ. Serv. Mark. **25**, 86–96 (2020)

Tyupa, S.: A theoretical framework for back-translation as a quality assessment tool. New Voices Transl. Stud. **7**(1), 35–46 (2011)

Yaseen, S.G., Qirem, I.A.E.: Intention to use e-banking services in the Jordanian commercial banks. Int. J. Bank Mark. **36**(3), 557–571 (2018)

Yiu, C.S., Grant, K., Edgar, D.: Factors affecting the adoption of internet banking in Hong Kong – Implications for the banking sector. Int. J. Inf. Manage. **27**(5), 336–351 (2007)

Zhang, F., Sun, S., Liu, C., Chang, V.: Consumer innovativeness, product innovation and smart toys. Electron. Commer. Res. Appl. **41**, 100974 (2020)

Foreign Ownership and Corporate Financial Performance in Vietnam: New Insights from the Generalised Method of Moments

Danh Hong Le[1], Phuong Thanh Bui[2](✉), Tra Thi Huong Lam[2], Trang Thi Thu Nguyen[3], and Thang Cong Nguyen[2]

[1] Van Hien University, Ho Chi Minh City, Vietnam
[2] Becamex Business School, Eastern International University, Thu Dau Mot, Vietnam
phuong.buithanh@eiu.edu.vn
[3] Faculty of Public Relations—Communications, Van Lang University, Ho Chi Minh City, Vietnam

Abstract. This study seeks to explore the influence of foreign ownership on the financial performance of corporations. Using panel data of 73 firms from 2009 to 2018 and the Generalised Method of Moments (GMM), our results show that foreign ownership positively correlates with return on assets. Moreover, we document that financial leverage hinders firm performance. Based on the empirical findings, this study suggested various practical implications for corporate managers and policymakers, such as considering the potential benefits of liberalizing foreign investment policies to promote economic growth and development.

Keywords: Foreign ownership · Corporate financial performance · GMM · Vietnam

1 Introduction

Despite some initial setbacks and challenges, foreign investment has generally played a significant role in the development of transition economies over the past few decades. Although several studies have been conducted to interpret the impact of foreign ownership (FO) on corporate financial performance (CFP), achieving a consensus has proven difficult. This presents a significant challenge for companies in developing countries, as foreign investment is essential for financial growth and business expansion. Notably, the relationship between FO and CFP is influenced by endogeneity and the measurement of CFP [1]. In layman's terms, both the mutual connection between the two variables and the proxy of CFP can lead to inconsistent estimates and incorrect conclusions. While addressing endogeneity bias has gained significant attention in applied economics, it is not yet widely explored in management disciplines. Such observations reinforce the call for taking endogeneity into consideration to understand the nexus between FO and CFP [2].

M. Pagac et al. (Eds.): SCFF 2024, LNNS 1062, pp. 426–437, 2024.
https://doi.org/10.1007/978-3-031-65656-9_42

Vietnam is a compelling location to examine the link between FO and CFP. Between 2016 and 2019, Vietnam experienced an average growth rate of approximately 7.1% [3], which can be partially attributed to an increasing inflow of foreign direct investment (FDI) [4, 5]. FDI inflows have risen from US$2.4 billion in 2000 to US$20 billion by 2020, and the M&A activities, which are a form of FDI, have increased significantly for the same period [3]. This has positively impacted firms' operational performance [6].

As discussed, Vietnam presents an interesting setting to study the FO and CFP nexus. Various studies on foreign ownership and corporate financial performance in Vietnam were conducted [5, 7–10]. To the best of our knowledge, although the above studies significantly contributed to the relevant literature, the insights on FO and CFP nexus remains unclear. We are aware that past value of CFP can explain its current value [2, 11] and there is two-way linkage between FO and CFP, which results in a dearth of literature on the subject. As such, this study is conducted to solve these problems. Based on the empirical findings, this study suggests various practical implications.

2 Literature Review

Paper [12] asserted that foreign direct investment can be accomplished through greenfield investment or mergers and acquisitions (M&A). In the case of greenfield investments, companies should decide between complete ownership or joint ventures with local companies. Academics have displayed considerable interest in examining the correlation between foreign ownership and firm performance. However, an ongoing debate surrounds this topic.

The impact of foreign ownership on corporate financial performance is a subject of debate in current literature. While some studies suggest a negative effect of foreign ownership on firm performance [9, 13, 14], others suggest a positive effect [7, 15–17], and others suggest a non-linear, inverted U-shaped relationship [8, 18]. Previous research indicates that foreign ownership can hurt firm performance due to agency problems arising from conflicts of interest between foreign investors and managers, lack of accountability, and issues with monitoring of managers. For example, authors in [9] used data from firms listed on the Ho Chi Minh Stock Exchange during the period 2008–2011 and found that foreign ownership has a negative impact on firm performance. The study's findings indicate that foreign ownership cannot effectively monitor corporate governance practices in Vietnam, primarily due to the lack of ownership concentration.

There is a consensus that foreign ownership has a negative impact on firm performance in Vietnam. However, there are contradictory studies in other emerging markets that acknowledge the role of foreign ownership in economic development and corporate financial growth. For instance, a study on the Indian market found that foreign ownership positively impacts firm performance [3]. They argued that their foreign affiliation gives domestic firms relatively easy access to superior technical, managerial, and financial resources. Similarly, paper [19] investigated the Ukrainian market and studied the effect of ownership structure on corporate governance and performance. They found a positive relationship between foreign ownership and corporate financial performance at a certain level. They argued that if outside owners do not have an unambiguous common interest in enforcing a value-maximizing policy of the firm, then the effect of concentrated outside ownership on performance may be insignificant. These points are contradicted by the findings of studies conducted in the context of Vietnam when the authors argued that there is a negative relationship since there is a need for concentrated ownership. As such, we aim to revisit the relationship between foreign ownership and corporate financial performance in Vietnam.

3 Methodology and Data

3.1 Empirical Model

Based on the existing empirical studies relating to the foreign ownership-firm performance nexus, we propose the following model to investigate the effect of foreign ownership on firm performance. The baseline panel data model is as follows:

$$ROA_{it} = \beta_0 + \beta_1 FO_{it} + \beta_2 A_{it} + \beta_3 S_{it} + \beta_4 FL_{it} + \beta_5 K_{it} + \beta_6 G_{it} + \varepsilon_{it}$$

where ROA is return on asset, FO is foreign ownership, A is firm's age, S is firm size, FL is financial leverage, K is capital intensity, G is annual economic growth, and ε is the error term.

3.2 Econometrics Techniques

Cross-Section Dependence Test
To detect cross-section correlation, we employ the CD-test for cross-sectional dependence [20, 21]. The statistics of the Pesaran's CD-test is calculated as follows:

$$CD = \sqrt{\frac{2}{N(N-1)}} \sum_{i=1}^{N-1} \sum_{j=i+1}^{N} \sqrt{T_{ij}} \hat{\rho}_{ij} \rightarrow N(0, 1)$$

Panel Unit-Root Test
The unit-root testing framework for panel data is divided on the basis of cross-section dependence. The first generation involves the studies [22–24], which does not account for cross-unit correlation. On the other hand, the second generation is robust to the correlation among panels. The latter refers to the following statistics.

$$CIPS = N^{-1} \sum_{i=1}^{N} CADF_i$$

The CADF statistics is derived from the following equation.

$$\Delta y_{it} = \alpha_i + \beta_i y_{it-1} + \theta_i \overline{y}_{t-1} + \sum_{j=0}^{p} \delta_{ij} \Delta \overline{y}_{t-j} + \sum_{j=0}^{p} \gamma_{ij} \Delta y_{it-j} + \varepsilon_{it}$$

Estimation Methodology

To align with the above empirical model, we use various estimators dedicated to the panel data model, they are pooled ordinary least squares (OLS), random effect (RE) estimator, and fixed-effect (FE) estimator. Each of them becomes the appropriate estimator, depending on the assumption of the underlying data-generating process. To quantify the effect of foreign ownership on firm performance, we use the pooled OLS, RE, and FE estimators in this study.

Next, we proceed to conduct various tests to choose the most appropriate estimator. The procedure is as follows. *First,* we employ Breusch and Pagan Lagrangian multiplier test for random effects. If the statistics is statistically significant, RE is preferable than pooled OLS. *Second,* we use Hausman's specification test for selecting FE or RE estimators. If the statistics is statistically significant, FE is preferable to RE.

It is true that the panel data model highly relates to the problem of heteroscedasticity and autocorrelation. Ignoring these issues yields inconsistent estimates [25]. To this end, we calculate the modified Wald statistic and perform Wooldridge test for serial correlation in panel-data models. In these tests, if the null hypothesis is rejected, it indicates that our data encounters heteroscedasticity and autocorrelation. To obtain robust results, we use the feasible generalised least squares.

In addition to the pooled OLS, FE, and RE estimators, we also use GMM in this study. This is because our baseline model may encounter the problem of endogeneity. In fact, various empirical studies showed that there is an effect running from corporate firm performance to foreign ownership [26]. Moreover, both FE and RE estimators are inconsistent if the true model is dynamic. Various issues relating to endogeneity, such as incorrect inference, inconsistent estimates become a stylized fact. These problems partly give rise to GMM. The GMM has two variations, difference GMM (DGMM) and system GMM (SGMM). Each of them involves two specifications. They are: one-step and two-step. Paper [27] showed that the SGMM outweighs the DGMM in providing the estimated coefficients with lower bias and standard errors. A detailed discussion on GMM and its variations can be found in the study [28].

3.3 Data

With the aim of examining the effect of foreign ownership on corporate financial performance in Vietnam, we focus on the companies listed on both two stock exchanges. By 2022, there are 410 listed companies on Ho Chi Minh Stock Exchange (HOSE), and 348 listed companies on Hanoi Stock Exchange (HNX), operating in various industries. We proposed the following criteria to select appropriate firms for investigation. They are: (1) non-financial listed firms; (2) annual financial reports are audited by the trusted third-party; and (3) the information of foreign ownership is publicly disclosed. As such, our panel data involves 73 firms, spanning from 2009 to 2018. Data is sourced from https://cafef.vn/—a reliable website for financial report at firm level in Vietnam. In relation to economic growth, we collect the data from the World Bank (Tables 1 and 2).

Table 1. Descriptive statistics

Variable	Observation	Mean	S.D	Min	Max
ROA	730	0.075	0.069	− 0.170	0.453
FO	730	0.186	0.178	0	0.769
A	730	29.089	13.456	7	92
S	730	5.592	18.030	0.064	287.974
FL	730	0.480	0.209	0.030	0.871
K	730	0.405	0.215	0.009	0.977
G	730	6.380	0.662	5.398	7.465

Table 2. Correlation table

	ROA	A	FO	S	FL	K	G
ROA	**1.000**						
A	0.109	**1.000**					
FO	0.333	− 0.041	**1.000**				
S	− 0.042	− 0.092	0.091	**1.000**			
FL	− 0.547	− 0.098	− 0.272	0.148	**1.000**		
K	− 0.031	0.032	0.122	0.128	− 0.125	**1.000**	
G	− 0.087	0.158	0.055	0.111	0.010	− 0.010	**1.000**

4 Empirical Results

This section provides the empirical results, which is derived from the techniques discussed in Sect. 3.2. Before presenting estimation results, we show findings relating to the cross-section dependence test and the unit-root test.

Table 3. The cross-sectional dependence results.

	ROA	FO	S	FL	K
CD test	15.940***	6.022***	76.897***	− 1.209	0.644
p-value	(0.000)	(0.000)	(0.000)	(0.227)	(0.519)

Table 3 demonstrates various figures relating to the cross-section dependence. According to the statistics, there is evidence of cross-unit correlation in the variables of interest, except financial leverage. These findings imply that the unit-root test should account for the problem of cross-section dependence.

Table 4. The stationary results.

Variable	Pesaran's CADF	Level of integration
ROA	– 5.067*** (0.000)	I (0)
FO	– 2.155** (0.016)	I (0)
S	– 2.545*** (0.005)	I (0)
FL	– 1.399* (0.081)	I (0)
K	– 1.992** (0.023)	I (0)

Table 4 indicates all the variables of interest are stationary at level form. This is because the null hypothesis of non-stationary is rejected at the level of 10%. This claim exhibits that ROA, FO, Size, FL, and K can join the model without differencing.

To be consistent with the econometrics techniques discussed above, we present our empirical findings in a systematic way, from pooled OLS to GMM.

Table 5 illustrates the results of estimating our baseline model, using various estimators dedicated to panel data, including pooled OLS, RE, FE, and GMM.

Before presenting the estimated coefficients, we briefly explain various statistics relating to selecting an appropriate estimator among pooled OLS, RE, and FE as well as detecting the potential problem of heteroscedasticity and autocorrelation in our dataset. According to the statistics, it is true that the RE estimator is preferable to the OLS estimator and that the FE estimator is preferable to the RE estimator due to the LM $\overline{\chi}^2 =$ 345.020 and the Hausman $\chi^2 = 40.450$, which are statistically significant. Accordingly, these claims indicate that the FE estimator is the most suitable. The figures derived from the Modified Wald test show that there is evidence of heteroscedasticity. Moreover, the Wooldridge test for autocorrelation in panel data rejects the null hypothesis of no first-order autocorrelation. All in all, these two tests suggest that our dataset encounters both heteroscedasticity and autocorrelation. In this sense, we run these estimators augmented with robust option to deal with the problem of heteroscedasticity and autocorrelation.

We report estimates based on pooled OLS (column (1)), random effect (column (2)), and fixed effect (column (3)). In general, the figures provide consistent effects of FO, FL, and K on ROA. We find that FO is positively correlated to ROA as the reported estimates of FO are 0.085; 0.074; and 0.080 with the p-value is less than 1%, implying that a higher proportion of foreign ownership improves firm performance. FL negatively affects ROA. This is because the reported estimates of FL are – 0.165; – 0.166; and – 0.153 with the p-value is less than 1%, too. The results indicate that FL hinders firm performance. The same picture can be applied to K where we find that it negatively associates with ROA. In concrete terms, the estimated coefficients of K are – 0.042; – 0.051; and – 0.066 with the p-value is less than 5%, indicating that K reduces firm performance. To other regressors, we find inconsistent impact on ROA across pooled,

Table 5. Estimation results.

Estimation method	(1) OLS	(2) RE	(3) FE	(4) One-step DGMM	(5) Two-step DGMM	(6) One-step SGMM	(7) Two-step SGMM
L.ROA	–	–	–	0.569***	0.591***	0.576***	0.582***
				(0.000)	(0.000)	(0.000)	(0.000)
FO	0.085***	0.074***	0.080***	0.197**	0.223**	0.010	– 0.003
	(0.000)	(0.001)	(0.006)	(0.010)	(0.014)	(0.626)	(0.878)
Age	0.000***	– 0.000	– 0.005***	– 0.001	– 0.001	0.000	0.000
	(0.006)	(0.656)	(0.000)	(0.183)	(0.229)	(0.772)	(0.695)
Size	0.000193*	– 0.000028	– 0.000049	– 0.000351	– 0.000379	0.000123	0.000149*
	(0.061)	(0.531)	(0.764)	(0.192)	(0.280)	(0.130)	(0.100)
FL	– 0.165***	– 0.166***	– 0.153***	– 0.111*	– 0.121*	– 0.139***	– 0.136***
	(0.000)	(0.000)	(0.000)	(0.071)	(0.081)	(0.000)	(0.000)
K	– 0.042***	– 0.051***	– 0.066**	0.086	0.062	0.008	– 0.007
	(0.000)	(0.002)	(0.011)	(0.213)	(0.509)	(0.814)	(0.838)
G	– 0.012***	– 0.009**	0.007*	0.004	0.004	0.001	0.002
	(0.000)	(0.012)	(0.066)	(0.155)	(0.235)	(0.596)	(0.534)
Observations	730	730	730	584	584	657	657
LM-P	345.020***		–	–	–	–	–
	(0.000)						
Hausman		40.450***		–	–	–	–
		(0.000)					
MW-P	34525.230***			–	–	–	–
	(0.000)						
Wooldridge	44.559***			–	–	–	–
	(0.000)						
No. IVs	–	–	–	53	53	59	59
AR(1)	–	–	–	– 3.800***	– 3.810***	– 3.670***	– 3.470***
				(0.000)	(0.000)	(0.000)	(0.001)

(continued)

Table 5. *(continued)*

Estimation method	(1) OLS	(2) RE	(3) FE	(4) One-step DGMM	(5) Two-step DGMM	(6) One-step SGMM	(7) Two-step SGMM
AR(2)	–	–	–	– 0.390	– 0.340	– 0.510	– 0.510
				(0.698)	(0.730)	(0.610)	(0.611)
Hansen statistics	–	–	–	48.940	48.940	55.69	55.69
				(0.356)	(0.356)	(0.303)	(0.303)

RE, and FE. We put our attention to the fixed effect (column (3)) to justify the effect of A, S, and G on ROA as FE is better than pooled OLS and RE in fitting data in this study. In particular, we find that A is negatively correlated with ROA while G is positively associated with ROA and that S has no effect on ROA.

Moreover, we present estimates based on one-step DGMM (column (4)), two-step DGMM (column (5)), one-step SGMM (column (6)), and two-step SGMM (column (7)). It is worth noting that DGMM performs poorly when the dependent variable is closely related to a random walk [28] and that two-step GMM outweighs one-step GMM for estimation [27]. As such, we lean on two-step SGMM to provide reliable conclusions. One of the drawbacks of the GMM estimator is the problem of instrument proliferation. Author in [28] argued that the problem of too many instruments gives rise to an implausibly good p-value of 100% in Hansen test, thus we use the sub-option *collapse* to collapse the matrix of instruments, which reduces the issue significantly.

We begin with various test statistics relating to the GMM estimator and then focus on the estimated coefficients. We are able to reject the null hypothesis of no first-order autocorrelation in first difference (row AR(1)) and unable to reject the null hypothesis of no second-order autocorrelation in first difference (row AR(2)) for all the four specifications of GMM. Similarly, the Hansen statistics (the value of χ^2 for DGMM and SGMM are 49.440, 50.470, respectively) is statistically insignificant, implying the validity of instrumental variables used in this study. We also report the number of instrumental variables (row No. IVs) for each specification. That for DGMM and SGMM are 53 and 59, respectively.

Next, we proceed to the estimated coefficients. Our estimation results show that the first lag of ROA is positively and statistically significant across the four specifications. The coefficient is slightly greater than 0.5, meaning that fluctuation in ROA can be explained by its past values. We find that FO is positively correlated to ROA as the estimated coefficients are 0.197; and 0.223 with the p-value is less than 10% in the one-step DGMM and two-step DGMM. These results reinforce our figures relating to pooled OLS, RE, and RE. Unfortunately, the impact of FO on ROA disappears in the one-step SGMM and two-step SGMM specification. Regarding FL, our figures demonstrate a consistent effect of FL that has on ROA. In concrete terms, we claim that FL hinders firm performance as the estimated coefficients are – 0.111; – 0.121; – 0.139; – 0.136 with the p-value is less than 10%. Interestingly, this relationship is also found with the pooled OLS, RE, and FE estimator. In relation to S, we find that the estimated coefficient is 0.000149 and the variable is statistically significant at the level of 10%. These finding claims that firm size negatively affects firm performance. To the other variables (A, K, and G), our figures illustrate no impact on firm performance across the four specifications.

All in all, our estimation results indicate that (1) fluctuation in ROA is partly explained by its past values; (2) FO is positively correlated with ROA; (3) A has no impact on firm performance; (4) S is negatively related to ROA, according to the two-step SGMM; (5) FL hinders firm performance. This claim is supported by the pooled OLS, RE, FE, and the four specifications of GMM; and (6) G is unrelated to firm performance in this study.

5 Discussion

This study carries out research on the effect of foreign ownership on firm performance, using a panel data of 73 listed firms in Vietnam over the period of 2009–2018 and various estimators, such as pooled OLS, RE, FE, and GMM. This section discusses the findings which emerged from the estimation results presented in the previous section.

In relation to the main interest variable - FO, the results of this study show that FO positively affects ROA within the pooled OLS, RE, FE, and DGMM setting. Our results also accord with earlier study [19], which showed that foreign ownership positively relates to corporate firm performance. However, our claims on foreign ownership-firm performance nexus are contrary to the findings of [9], who focused on the relationship over the period of 2008–2011 and used FE estimator and simultaneous equation model. As such, this discrepancy could be attributed to the time frame used, the approach of modelling, and methodologies adopted. In summary, our study supports the resource-based view theory, which posits that firms can gain competitive advantages and above-average returns through valuable and unique resources that are provided by the foreign owners.

Our estimation results illustrate that FL negatively affects ROA, which is supported by all the estimators used in this study. This finding is in accord with those of [29, 30] and differs from some published studies, such as [31–33]. Mainly, our findings reveal that firms with high financial leverage exhibit poor financial performance since it reduces financial flexibility by constraining firms from an incremental expansion and exploiting growth opportunities, which could have significant implications for their stakeholders [34].

6 Conclusion

The main goal of the current study is to examine the impact of foreign ownership on firm performance with a focus on Vietnam. To this end, we employ a model where firm performance is used as a dependent variable and measured by ROA and the independent variables are foreign ownership (FO), firm's age (A), firm's size (S), financial leverage (FL), capital intensity (K), and economic growth (G). Our panel data covers 73 listed firms over the period of 2009–2018. In this study, we use the pooled OLS, RE, FE and GMM to quantify the effect of the regressors on ROA.

Our empirical results show a consistent and positive impact of FO on ROA within the pooled OLS, RE, and FE setting. These findings are reinforced by the one-step DGMM and two-step DGMM estimators, which illustrates that FO is positively correlated with ROA. In relation to the financial leverage-firm performance, we claim that FL negatively affects ROA, which is supported by all the estimators used in this study. The capital intensity-firm performance nexus is found by the pooled, RE, and FE. In particular, our empirical findings reveal that K negatively associates with ROA. This relationship is not supported by the GMM estimator. Moreover, using the FE estimator, we claim that G is positively correlated with ROA. In relation to the firm's age and firm's size, we are unable to find their impact on firm performance in this study.

Based on these empirical findings, we suggest some policy implications for practitioners. Firstly, it sheds light on the relationship between foreign ownership and corporate financial performance in Vietnam. A thorough understanding of this relationship is crucial for Vietnamese corporations to attract foreign investment, which can alter their business strategies. In a small and open economy like Vietnam, comprehending foreign ownership is especially crucial, as global resource assessment is necessary for economic development. Secondly, stakeholders can use this paper to understand better the ongoing discussion regarding the potential expansion of foreign investor ownership. Lastly, our study examines policymakers' current trend to reduce legal barriers to cross-border capital flows and foreign direct investments. This sends a message to national governments, encouraging them to reconsider any legal restrictions on the amount of foreign investment allowed in emerging markets.

References

1. Sánchez-Ballesta, J.P., García-Meca, E.: A meta-analytic vision of the effect of ownership structure on firm performance. Corpor. Gov. Int. Rev. **15**(5), 879–892 (2007). https://doi.org/10.1111/j.1467-8683.2007.00604.x
2. Ullah, S., Akhtar, P., Zaefarian, G.: Dealing with endogeneity bias: the generalized method of moments (GMM) for panel data. Ind. Mark. Manage. **71**, 69–78 (2018). https://doi.org/10.1016/j.indmarman.2017.11.010
3. Madani, D., Davies, E., Nguyen, T.H.: Harnessing the potential of the services sector for growth (2023). Retrieved 7 Nov 2023, from The World Bank website: https://documents1.worldbank.org/curated/en/099544403132351453/pdf/IDU0343e48530e212043860bee605aae66cfb04a.pdf
4. Hoang, H.H., Bui, N.Q., Huynh, C.M.: The impact of FDI and regional factors on economic growth in Vietnamese provinces: a spatial econometric analysis. Post-Communist Econ. **35**(5), 454–474 (2023). https://doi.org/10.1080/14631377.2023.2196869
5. Nguyen, C.H.: The impact of foreign direct investment, aid and exports on economic growth in Vietnam. J. Asian Financ. Econ. Bus. **7**(10), 581–589 (2020). https://doi.org/10.13106/jafeb.2020.vol7.no10.581
6. Ning, L., Guo, R., Chen, K.: Does FDI bring knowledge externalities for host country firms to develop complex technologies? The catalytic role of overseas returnee clustering structures. Res. Policy **52**(6), 104767 (2023). https://doi.org/10.1016/j.respol.2023.104767
7. Duong, Q.N., Vo, T.P., Nguyen-Le, N.H., Nguyen, V.D.: The impact of foreign ownership on firm performance: an empirical study of listed firms in Vietnam. J. Asian Financ. Econ. Bus. **8**(6), 879–888 (2021). https://doi.org/10.13106/jafeb.2021.vol8.no6.0879

8. Nguyen, D.S., Nguyen, V.D., Tran, D.T., Dempsey, M.J.: The firm's performance in relation to capital structure and foreign ownership: evidence from Vietnam. J. Adv. Eng. Comput. **5**(1), 35 (2021). https://doi.org/10.25073/jaec.202151.315
9. Phung, D.N., Le, T.P.V.: Foreign ownership, capital structure and firm performance: empirical evidence from Vietnamese listed firms. IUP J. Corpor. Gov. **12**(2), 40–58 (2013). Retrieved from https://ssrn.com/abstract=2343066
10. Vo, X.V.: Foreign ownership and stock return volatility—evidence from Vietnam. J. Multinatl. Financ. Manag. **30**, 101–109 (2015). https://doi.org/10.1016/j.mulfin.2015.03.004
11. Faruq, M.O., Akter, T., Mizanur Rahman, M.: Does intellectual capital drive bank's performance in Bangladesh? Evidence from static and dynamic approach. Heliyon **9**(7), e17656 (2023). https://doi.org/10.1016/j.heliyon.2023.e17656
12. Raff, H., Ryan, M., Stähler, F.: Firm productivity and the foreign-market entry decision. J. Econ. Manag. Strateg. **21**(3), 849–871 (2012). https://doi.org/10.1111/j.1530-9134.2012.00346.x
13. Konings, J.: The effects of foreign direct investment on domestic firms evidence from firm-level panel data in emerging economies. Econ. Transit. **9**(3), 619–633 (2001). https://doi.org/10.1111/1468-0351.00091
14. Lichtenberg, F.R., Pushner, G.M.: Ownership structure and corporate performance in Japan. Jpn. World Econ. **6**(3), 239–261 (1994). https://doi.org/10.1016/0922-1425(94)90014-0
15. Douma, S., George, R., Kabir, R.: Foreign and domestic ownership, business groups, and firm performance: evidence from a large emerging market. Strateg. Manag. J. **27**(7), 637–657 (2006). https://doi.org/10.1002/smj.535
16. Filatotchev, I., Stephan, J., Jindra, B.: Ownership structure, strategic controls and export intensity of foreign-invested firms in transition economies. J. Int. Bus. Stud. **39**(7), 1133–1148 (2008). https://doi.org/10.1057/palgrave.jibs.8400404
17. Nguyen, T.X.H., Pham, T.H., Dao, T.N., Nguyen, T.N., Tran, T.K.N.: The impact of foreign ownership and management on firm performance in Vietnam. J. Asian Financ. Econ. Bus. **7**(9), 409–418 (2020). https://doi.org/10.13106/jafeb.2020.vol7.no9.409
18. Ferris, S.P., Park, K.: Foreign ownership and firm value: evidence from Japan (2005). https://doi.org/10.1016/S1569-3732(04)11001-3
19. Akimova, I., Schwödiauer, G.: Ownership structure, corporate governance, and enterprise performance: empirical results for Ukraine. Int. Adv. Econ. Res. **10**(1), 28–42 (2004). https://doi.org/10.1007/BF02295575
20. Pesaran, M.H.: General Diagnostic Tests for Cross Section Dependence in Panels (No. 0435). Cambridge (2004)
21. Pesaran, M.H.: Testing weak cross-sectional dependence in large panels. Economet. Rev. **34**(6–10), 1089–1117 (2015). https://doi.org/10.1080/07474938.2014.956623
22. Breitung, J.: The local power of some unit root tests for panel data (2001). https://doi.org/10.1016/S0731-9053(00)15006-6
23. Hadri, K.: Testing for stationarity in heterogeneous panel data. Economet. J. **3**(2), 148–161 (2000). https://doi.org/10.1111/1368-423X.00043
24. Levin, A., Lin, C.-F., James Chu, C.-S.: Unit root tests in panel data: asymptotic and finite-sample properties. J. Econ. **108**(1), 1–24 (2002). https://doi.org/10.1016/S0304-4076(01)00098-7
25. Gujarati, D.N.: Basic Econometrics. Prentice Hall (2009)
26. Iwasaki, I., Ma, X., Mizobata, S.: Ownership structure and firm performance in emerging markets: a comparative meta-analysis of East European EU member states, Russia and China. Econ. Syst. **46**(2), 100945 (2022). https://doi.org/10.1016/j.ecosys.2022.100945
27. Windmeijer, F.: A finite sample correction for the variance of linear efficient two-step GMM estimators. J. Econ. **126**(1), 25–51 (2005). https://doi.org/10.1016/j.jeconom.2004.02.005

28. Roodman, D.: How to do Xtabond2: an introduction to difference and system GMM in Stata. Stata J. Promot. Commun. Stat. Stata **9**(1), 86–136 (2009). https://doi.org/10.1177/1536867X0900900106
29. Goddard, J., Tavakoli, M., Wilson, J.O.S.: Determinants of profitability in European manufacturing and services: evidence from a dynamic panel model. Appl. Financ. Econ. **15**(18), 1269–1282 (2005). https://doi.org/10.1080/09603100500387139
30. Yazdanfar, D., Öhman, P.: Debt financing and firm performance: an empirical study based on Swedish data. J. Risk Financ. **16**(1), 102–118 (2015). https://doi.org/10.1108/JRF-06-2014-0085
31. Jensen, M.C.: Agency costs of free cash flow, corporate finance, and takeovers. Am. Econ. Rev. **76**(2), 323–329 (1986). Retrieved from https://www.jstor.org/stable/1818789
32. Kraus, A., Litzenberger, R.H.: A state-preference model of optimal financial leverage. J. Financ. **28**(4), 911 (1973). https://doi.org/10.2307/2978343
33. McConnell, J.J., Servaes, H.: Additional evidence on equity ownership and corporate value. J. Financ. Econ. **27**(2), 595–612 (1990). https://doi.org/10.1016/0304-405X(90)90069-C
34. Mishra, S., Modi, S.B.: Positive and negative corporate social responsibility, financial leverage, and idiosyncratic risk. J. Bus. Ethics **117**(2), 431–448 (2013). https://doi.org/10.1007/s10551-012-1526-9

Green Advertising: A Review Based on Bibliographic Coupling Analysis

Luc Phan Tan(✉)

Thu Dau Mot University, Thu Dau Mot, Binh Duong, Vietnam
lucpt@tdmu.edu.vn

Abstract. This study aims to map all scientific publications on green advertising and identify potential future research directions. The data included 204 English-language studies on green advertising that were searched in the Web of Science database. Bibliographic coupling analysis and thematic analysis methods were used to analyze the evolution of research on the topic over time. The results of the bibliometric analysis show that there are five main themes. From the thematic analysis, lines of research within each piece are identified, including 'the effectiveness of green marketing', 'green consumption intention', 'green advertising and consumer behavior', 'green message', 'method of advertising', 'strategy of green advertising', 'credibility of green advertising', 'ambiguity and skepticism in green advertising g', 'greenwashing', 'environmental cooperation', and 'platform-led green initiatives. This study offers valuable insights into the latest and most significant achievements in green advertising, explicitly focusing on state-of-the-art thematic subjects. As a result, it identifies promising areas for future research that can contribute to advancing the research community.

Keywords: Green advertising · Bibliographic coupling · Thematic analysis · Bibliometric

1 Introduction

In the face of rising environmental concerns and an increased worldwide awareness of climate change, green advertising has emerged as a crucial tactic for firms trying to demonstrate their commitment to sustainability [1]. This commitment is in direct response to a substantial shift in customer behavior [2]. Reveal that customers increasingly make purchase decisions aligned with their environmental ideals, favoring brands that actively support sustainability. Green advertising, therefore, serves a dual role in the current marketplace. It serves not just to inform consumers about eco-friendly products and services but also works as a representation of a company's corporate social responsibility.

With growing consumer consciousness about the possible ecological consequences of their purchases, there has been a surge in the popularity of items that are classified as 'eco-friendly,' 'organic,' or 'sustainable' [1]. In response to the increasing consumer

M. Pagac et al. (Eds.): SCFF 2024, LNNS 1062, pp. 438–447, 2024.
https://doi.org/10.1007/978-3-031-65656-9_43

demand for eco-friendly goods, advertising claims that promote environmental sustainability have become a crucial aspect of product advertising. Consequently, research on green advertising has garnered significant attention within the academic community [3]. Green advertising, broadly defined as any form of advertising emphasizing the relationship between a product or service and the environment, has become a crucial avenue for promoting environmentally conscious practices [4]. This form of advertising goes beyond the promotion of environmentally friendly products alone [3]. It extends its scope to supporting and advocating for a green lifestyle while creating a responsible corporate image. Despite the substantial attention green advertising research has received, it confronts several challenges. The burgeoning volume of studies necessitates a continual review of the literature to prevent redundancy in future research. Consequently, the classification and evaluation of research on green advertising become imperative to systematize both the quantity and quality of existing studies. Additionally, the vast number of studies requires diverse review methods to enable a comprehensive examination. Integrating various literature review techniques can offer intriguing perspectives on this research topic.

Bibliographic coupling analysis is a method used to analyze the similarity of scientific documents. Introduced by [5], this technique is founded on the notion that if two papers mention many of the same references, they may address related subject matter or concepts, therefore are considered to be bibliographically coupled. The more references two papers share, the stronger their bibliographic coupling strength. This notion is a cornerstone in the subject of scientometrics, which is concerned with the quantitative analysis of research, innovation, and technology. The essential idea of bibliographic coupling is intuitive: scholarly works do not exist in isolation. They are part of a complicated web of knowledge, interconnected through citations that indicate the intellectual lineage and context of the research. When two articles cite a similar third paper, it denotes a potential intellectual connection, implying that they may be "conversing" with the same scientific ideas or conclusions [6]. Bibliographic coupling analysis measures this link by calculating the number of references two papers share. The obtained bibliographic coupling strength can then be used to form a network of papers, where nodes represent documents, and edges reflect the coupling strength between them. This method offers various practical uses. Researchers can use bibliographic coupling to locate the most relevant and important papers within these clusters [7]. By evaluating these alterations, academics may monitor the dynamic process of scientific discovery and collaboration. By examining the bibliographic coupling strength of documents, such systems might recommend articles that are contextually relevant to a researcher's current reading, boosting the finding of material that might otherwise be neglected [5].

The aforementioned arguments underscore the necessity for comprehensive contributions to the current theoretical base and suggest future research directions to advance the topic of green advertising. Consequently, this study conducts a review of research on green advertising by employing two methods, including directory link analysis and thematic analysis.

2 Methods

The authors used bibliographic coupling and thematic analysis to provide insights into the knowledge structure of the literature on sustainable marketing, look at trends in research themes, and make recommendations for future study directions. The review process was conducted in three stages.

Stage 1: Data Collection

All English-language research articles related to green advertising were retrieved from the Web of Science database. The search utilized the phrase "green advertise" in the subject field (title/abstract/keywords), covering articles published before December 2023. Only peer-reviewed articles, conference papers, and book chapters in English were considered for this study. The initial search yielded 223 studies.

Stage 2: Data Review.

The author carefully read the title, abstract, and keywords of 223 papers to eliminate unrelated research and remove 19 research articles. Ultimately, 204 studies were retained for inclusion in the bibliographic coupling analysis.

Stage 3: Data Analysis

In the first step of data analysis, the author employed the VOS Viewer software for bibliographic coupling analysis. The Smart Local Movement Algorithm Method was applied to study the structure of the green advertising topic. The results obtained from directory link analysis served to identify key research topics in green advertising. Finally, in the thematic analysis step, the author thoroughly read and reviewed the content of typical research articles in each topic to identify research lines within each thematic area.

3 Bibliographic Coupling Results

Bibliographic coupling analysis resulted in the formation of five cluster (Fig. 1). Each color represents a cluster to which the belonging publications were linked. The author's surname and year of publication are labeled on each bubble. The bubble size represents the total standardized citations that the studies received. The breadth and connection between the two documents highlight the bibliographic coupling relationship. The five clusters were labeled based on the comprehension of the representative articles within each cluster. Table 1 summarizes the main sub-themes of every cluster from thematic analysis.

Cluster 1

The first cluster includes sub-themes, namely 'the effectiveness of green advertising' and 'green consumption intention'.

The first sub-theme pertains to 'the effectiveness of green advertising,' which delves into the multilayered factors that can amplify or dampen the impact of such advertising. This line of inquiry has uncovered that the environmental appropriateness of the product

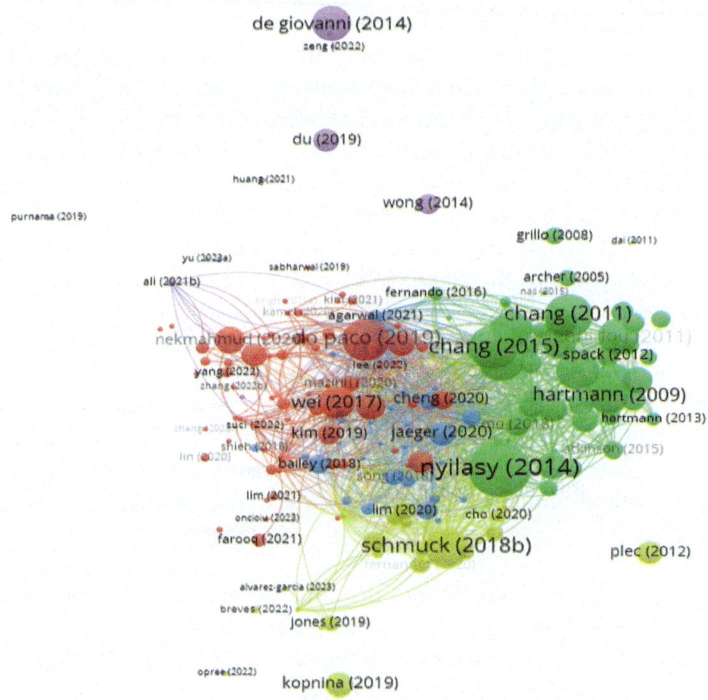

Fig. 1. Bibliographic coupling network of publications

itself is a critical determinant of advertising effectiveness. For instance, a product inherently designed with sustainability at its core is more likely to resonate with consumers who have a pre-existing pro-environmental orientation, as suggested by [8]. Moreover, the perceived corporate environmental responsibility, as explored by [9], plays a significant role in shaping consumer trust and, consequently, the persuasive power of green advertising.

The second sub-theme includes articles related to 'green consumption intention' which aims to bridge the gap between exposure to green advertising and sustainable consumption intention. Within this realm, various elements have been identified as influential. The green 4Ps marketing strategy has been examined for its capacity to mold customer values and attitudes, with findings suggesting a positive correlation between strategic green marketing applications and consumer purchase intentions [10]. Eco-labels are another facet that has garnered attention, functioning as a shorthand for consumers to identify environmentally friendly products, thereby influencing purchasing decisions [11].

Cluster 2

There are three sub-themes within this category, namely 'green advertising and consumer behavior', 'green message', and content of advertising claims and appeals.

The first sub-theme examines consumer behavior under the influence of green advertising. Tu et al. [12] delve into the nuances of green consumption patterns, attempting to decipher the complex web of consumer behavior in the face of environmental marketing. Lee et al. [13] examine consumers' attitudes toward green advertising and how these attitudes might influence their purchasing decisions. Central to this area of study is the concept of the "green gap," identified by [14], which refers to the discrepancy between consumers' expressed eco-friendly attitudes and their actual purchasing behaviors.

The second sub-theme investigates the 'green message' conveyed through advertising. This research stream aims to understand the impact such messages have on consumer perception and behavior. For instance, studies by [15, 16] scrutinize the persuasive power of green messages, assessing their effectiveness in delivering the intended environmental narrative. Ekebas-Turedi et al. [17], Kareklas et al. [18] and Li and Zhang [19] further this exploration, analyzing how the framing and presentation of green messages influence consumer attitudes and the credibility of such advertisements.

In the third sub-theme, 'method of advertising' stream delves into the tactical aspects of green advertising. Green and Peloza [20] and Matthes et al. [21] investigate the array of methods used in green advertising campaigns, from traditional media to digital platforms, and how these methods influence the effectiveness of the campaigns.

Cluster 3

The third group of topics ventures into the strategic and ethical dimensions of green advertising, dissecting how these approaches are formulated and the trust they engender among consumers. The authors found two sub-themes, namely 'strategy of green advertising' and 'credibility of green advertising'.

Within the 'strategy of green advertising' sub-theme, scholars are investigating the multifaceted approaches companies adopt to promote environmentally friendly products and practices. The role of social media as a pivotal platform for green marketing is scrutinized by [22, 23], who explore how companies leverage these digital landscapes to craft operations strategies that resonate with eco-conscious consumers. Research by [24] delves into how green advertising can be a source of competitive advantage, suggesting that companies that effectively communicate their environmental commitments can distinguish themselves in a crowded marketplace.

The second sub-theme, 'green advertising credibility', investigates the trustworthiness of the environmental claims made in green advertisements. The credibility of green advertising is a cornerstone for consumer acceptance and is critically evaluated in studies by [25, 26]. These studies examine factors such as transparency, consistency, and the substantiation of claims, which can either bolster or undermine the perceived integrity of green advertising.

Cluster 4

The fourth cluster delves into the darker side of green advertising, addressing the issues of 'ambiguity and skepticism in green advertising' along with the deceptive practice of 'greenwashing'.

The research on ambiguity and skepticism in green advertising probes the murky waters where unclear or vague claims can lead to consumer distrust. Matthes and Wonneberger [27] explore how ambiguous messages in green advertising can sow seeds of doubt in consumers' minds, which can ultimately hinder the effectiveness of these campaigns. Similarly, [28] investigate the role of ambiguity in the perceived effectiveness of ecolabels, which are intended to guide consumers to environmentally preferable choices but can sometimes contribute to confusion and skepticism when not clearly defined or standardized.

The concept of greenwashing occupies the second sub theme, representing a more insidious issue where environmental claims are not just unclear, but intentionally misleading. Greenwashing studies, such as those by [29], evaluate how organizations may manipulate their image to appear environmentally responsible when their practices do not align with their proclamations. Education is a pivotal theme in this area, with researchers like [30] investigating how educational interventions can raise awareness and critical thinking about greenwashing.

Cluster 5

The fifth cluster investigates the emerging dynamics at the intersection of environmental issues and the burgeoning platform economy, encompassing two main sub themes: 'environmental cooperation' and 'platform-led green initiatives'.

In the realm of environmental cooperation, scholars are examining the synergistic efforts between various stakeholders to enhance environmental outcomes. Bai and Meng [31] explore the factors that contribute to successful environmental cooperation, such as shared values, mutual trust, and the alignment of economic and environmental goals.

The second sub theme focuses on the influence of digital platforms in shaping consumer perceptions and behaviors towards green products and services. The 'platform-led' approach is a nod to the critical role that digital marketplaces and social media platforms play in the modern economy. Huang and Yu [32] and Du et al. [33] delve into how these platforms can harness their vast reach and influence to promote green advertising, thereby capitalizing on the increasing consumer demand for sustainable options.

Table 1. Main themes for five bibliographic coupling clusters and representative articles.

Cluster	Sub-theme	Representative articles
Cluster 1	The effectiveness of green marketing	[8, 9]
	Green consumption intention	[10, 11]
Cluster 2	Green advertising and consumer behavior	[12–14]
	Green message	[15–19]
	Method of advertising	[20, 21]
Cluster 3	Strategy of green advertising	[22–24]
	Credibility of green advertising	[25, 26]
Cluster 4	Ambiguity and skepticism in green advertising	[27, 28]
	Green washing	[29, 30]
Cluster 5	Environmental cooperation	[31]
	Platform-led green initiatives	[32, 33]

4 Future Research Directions

The first cluster concerning green advertising has illuminated various facets that influence consumer behavior and the effectiveness of eco-centric marketing. However, there remain numerous avenues for future investigation that could yield significant insights into the optimization of green advertising strategies and the reinforcement of sustainable consumer habits. Firstly, there is a need for longitudinal studies to assess the long-term impacts of green advertising on consumer attitudes and behaviors. While current research offers a snapshot of consumer responses, the durability of these attitudes over time remains inadequately explored. Secondly, with the advent of digital media, the mechanisms through which online green advertising influences consumer intention merit further exploration. As digital platforms become increasingly personalized, understanding how green advertising can be tailored to individual consumers through algorithms could be groundbreaking [34]. Finally, studies that dissect the psychological underpinnings of green advertising's impact could contribute to a deeper understanding of consumer decision-making processes.

In second cluster, the examination of green advertising and its influence on consumer behavior, the effectiveness of green messages, and the content of advertising claims and appeals has laid a substantial groundwork for understanding how environmental marketing shapes consumer choices. However, there remains a vast terrain of uncharted research possibilities that beckon for exploration to deepen our comprehension and enhance the persuasive power of green advertising. Future research could pivot towards a granular analysis of the "green gap" phenomenon by employing longitudinal studies to track changes in consumer behavior over time in response to repeated exposure to green advertising. In terms of the 'green message', an under-explored area pertains to the role of message congruence across different platforms and its impact on the consumer's perceived sincerity of the brand's environmental commitment. Different techniques could offer insights into consumers' emotional and cognitive responses to different types of green advertising content [35]. Furthermore, the impact of cultural nuances on the reception of green advertising messages is a promising avenue, especially in non-Western contexts. Comparative studies across cultures could elucidate how cultural values and norms shape the interpretation and effectiveness of green advertising [36].

The strategic and ethical dimensions of green advertising present a complex tableau, with strategy and credibility forming the nexus of consumer trust and effective environmental marketing. To navigate this landscape, future research must dissect these constructs with a multi-disciplinary lens, delving deeper into the strategic underpinnings of green campaigns and the critical nature of their credibility. One promising avenue for future research within the 'strategy of green advertising' sub-theme is to explore the cross-cultural adaptability and scalability of green advertising strategies. Regarding the 'green advertising credibility' sub-theme, future research should aim to develop a more nuanced understanding of the antecedents and consequences of consumer skepticism towards green advertising. This can encompass longitudinal studies that track changes in consumer skepticism over time, potentially informed by the rise of environmental consciousness and the prevalence of greenwashing [37]. Moreover, the interplay between green advertising credibility and regulatory frameworks deserves attention.

Within the context of green advertising, the exploration of ambiguity, skepticism, and greenwashing represents a compelling frontier for academic research. This cluster underscores the importance of transparency and genuine environmental stewardship in advertising practices. Given the rising consumer concern about sustainability, future research should prioritize understanding the cognitive mechanisms behind the perception of ambiguous green claims. Experimental designs could investigate how specific linguistic cues or lack thereof contribute to ambiguity and skepticism. Future studies could address the cognitive processing of ambiguous green claims to observe real-time consumer reactions to green advertising stimuli [38]. Another promising research direction is the analysis of the effectiveness of educational initiatives in combating greenwashing. Researchers should evaluate the long-term efficacy of educational programs highlighted by [30], in fostering critical thinking and skepticism among consumers. Such studies could involve pre-and post-intervention assessments of consumer ability to identify greenwashing in advertising.

The emerging dynamics of environmental cooperation and platform-led green initiatives present a fertile ground for future research within the context of green advertising. Building on this foundation, several research trajectories can be envisioned. Firstly, future studies could adopt a multi-level analytical approach to understand the complexities and nuances of stakeholder collaboration in environmental initiatives. Research could explore how different levels of cooperation (e.g., local, regional, international) differ in their approach and effectiveness in promoting environmental sustainability. In relation to platform-led green initiatives, there is a need to investigate the credibility and impact of digital platforms' sustainability claims. Empirical research could measure consumer responses to green marketing efforts on platforms, analyzing the trustworthiness of sustainability claims and the degree of consumer engagement [39].

5 Conclusion

This study undertook a comprehensive review of research on the topic of green advertising, drawing on data from 204 English-language studies retrieved from the Web of Science database. Employing bibliographic coupling analysis and thematic analysis methods, we delved into the diverse aspects of green advertising presented in the literature. The results of the bibliometric analysis show that there are five main themes. From the thematic analysis, lines of research within each piece are identified, including 'the effectiveness of green marketing', 'green consumption intention', 'green advertising and consumer behavior', 'green message', 'method of advertising', 'strategy of green advertising', 'credibility of green advertising', 'ambiguity and skepticism in green advertising', 'greenwashing', 'environmental cooperation', and 'platform-led green initiatives.

In light of these findings, the author has proposed future research directions for each identified line of inquiry. However, it is crucial to acknowledge the limitations of this study. Firstly, the data exclusively originated from the Web of Science source; thus, future studies could benefit from incorporating a more diverse range of database sources to offer a more comprehensive overview of the landscape of green advertising research. Secondly, this study solely employed bibliographic coupling analysis. Future research endeavors may consider employing a combination of keyword analysis and co-citation analysis to further refine and generalize the research results.

References

1. Schmuck, D., Matthes, J., Naderer, B.: Misleading consumers with green advertising? An Affect-reason-involvement account of greenwashing effects in environmental advertising. J. Advert. **47**(2), 127–145 (2018)
2. Cronin, J.J., et al.: Green marketing strategies: an examination of stakeholders and the opportunities they present. J. Acad. Mark. Sci. **39**(1), 158–174 (2011)
3. Ktisti, E., Hatzithomas, L., Boutsouki, C.J.S.: Green advertising on social media: a systematic literature review. Sustainability **14**(21), 14424 (2022)
4. Banerjee, S., Gulas, C.S., Iyer, E.: Shades of green: a multidimensional analysis of environmental advertising. J. Advert. **24**(2), 21–31 (1995)
5. Kessler, M.M.: Bibliographic coupling between scientific papers. Am. Doc. **14**(1), 10–25 (1963)
6. Jarneving, B.: A comparison of two bibliometric methods for mapping of the research front. Scientometrics **65**(2), 245–263 (2005)
7. Small, H.: Co-citation in the scientific literature: a new measure of the relationship between two documents. J. Am. Soc. Inf. Sci. **24**(4), 265–269 (1973)
8. Alniacik, U., Yilmaz, C.: The effectiveness of green advertising: influences of claim specificity, product's environmental relevance and consumers' pro-environmental orientation. Amfiteatru Econ. **14**(31), 207–222 (2012)
9. Chen, X.X., et al.: Research on the impact of corporate environmental responsibility (CER) on advertising effectiveness. J. Environ. Prot. Ecol. **23**(5), 2005–2011 (2022)
10. Ahmed, R.R., et al.: Effect of green marketing mix, green customer value, and attitude on green purchase intention: evidence from the USA. Environ. Sci. Pollut. Res. **30**(5), 11473–11495 (2023)
11. Nguyen-Viet, B.: Understanding the influence of eco-label, and green advertising on green purchase intention: the mediating role of green brand equity. J. Food Prod. Mark. **28**(2), 87–103 (2022)
12. Tu, J.C., Yang, C.H., Chen, Y.Y.: Exploring the impact of IoT and green advertising on consumer behavior. Int. J. Automat. Technol. **16**(6), 795–806 (2022)
13. Lee, J.L.M., Liu, V., Cheng, C.L.: Effect of green advertising attitude on collectivist consumer behavior: an integrative approach. Asia Pac. J. Mark. Logist. (2023)
14. Kim, Y., et al.: Closing the green gap: the impact of environmental commitment and advertising believability. Soc. Behav. Pers. **44**(2), 339–352 (2016)
15. Tu, J.C., Kao, T.F., Tu, Y.C.: Influences of framing effect and green message on advertising effect. Soc. Behav. Pers. **41**(7), 1083–1098 (2013)
16. Fernando, A.G., Sivakumaran, B., Suganthi, L.: Message involvement and attitude towards green advertisements. Mark. Intell. Plan. **34**(6), 863–882 (2016)
17. Ekebas-Turedi, C., Kordrostami, E., Benoit, I.D.: The impact of message framing and perceived consumer effectiveness on green ads. J. Consum. Mark. **38**(4), 386–396 (2021)
18. Kareklas, I., Carlson, J.R., Muehling, D.D.: The role of regulatory focus and self-view in "green" advertising message framing. J. Advert. **41**(4), 25–39 (2012)
19. Li, Z.F., Zhang, L.J.: The persuasive effects of verbal versus visual messages in green advertising. In: 6th annual China marketing international conference (CMIC). Shanghai, Peoples R. China (2018)
20. Green, T., Peloza, J.: Finding the right shade of green: the effect of advertising appeal type on environmentally friendly consumption. J. Advert. **43**(2), 128–141 (2014)
21. Matthes, J., Wonneberger, A., Schmuck, D.: Consumers' green involvement and the persuasive effects of emotional versus functional ads. J. Bus. Res. **67**(9), 1885–1893 (2014)

22. Kim, Y.K., et al.: Exploring the optimized social advertising strategy that can generate consumer engagement with green messages on social media. J. Res. Interact. Mark. **15**(1), 30–48 (2021)
23. Li, S., et al.: How does product-celebrity congruence and content sponsorship affect perceived altruism among consumers? Evidence from factorial experiments. Resour. Conserv. Recycl. **178** (2022)
24. Maziriri, E.T.: Green packaging and green advertising as precursors of competitive advantage and business performance among manufacturing small and medium enterprises in South Africa. Cogent Bus. Manag. **7**(1) (2020)
25. Kim, E., et al.: A narrative approach for overcoming the message credibility problem in green advertising. J. Bus. Res. **147**, 449–461 (2022)
26. Jäger, A.K., Weber, A.: Can you believe it? The effects of benefit type versus construal level on advertisement credibility and purchase intention for organic food. J. Clean. Prod. **257** (2020)
27. Matthes, J., Wonneberger, A.: The skeptical green consumer revisited: testing the relationship between green consumerism and skepticism toward advertising. J. Advert. **43**(2), 115–127 (2014)
28. Cho, Y.N., Taylor, C.R.: The role of ambiguity and skepticism in the effectiveness of sustainability labeling. J. Bus. Res. **120**, 379–388 (2020)
29. Abd Rahim, M.H., Shan, W.H., Lyndon, N.A.: Perception on the practices of "Greenwashing" in green product advertising among Malaysian consumer. J. Komunikasi Malays. J. Commun. **35**(1), 120–136 (2019)
30. Álvarez-García, O., Sureda-Negre, J.: Greenwashing and education: an evidence-based approach. J. Environ. Educ. **54**(4), 265–277 (2023)
31. Bai, Y., Meng, X.Y.: Environmental collaboration under various contracts in the O2O closed-loop supply chain. Discr. Dyn. Nat. Soc. (2022)
32. Huang, X., Yu, L.J.: Advertising cooperation and green technology R&D strategy in a platform-led Stackelberg game. Math. Probl. Eng. 2021 (2021)
33. Du, S.F., et al.: Platform-led green advertising: promote the best or promote by performance. Transp. Res. Part E Logist. Transp. Rev. **128**, 115–131 (2019)
34. Chang, C.: Feeling ambivalent about going green. J. Advert. **40**(4), 19–32 (2011)
35. Balaskas, S., Panagiotarou, A., Rigou, M.: Impact of environmental concern, emotional appeals, and attitude toward the advertisement on the intention to buy green products: The case of younger consumer audiences. Sustainability **15**(17), 13204 (2023)
36. Groening, C., Sarkis, J., Zhu, Q.: Green marketing consumer-level theory review: a compendium of applied theories and further research directions. J. Clean. Product. **172**, 1848–1866 (2018)
37. do Paço, A.M.F., Reis, R.: Factors affecting skepticism toward green advertising. J. Advert. **41**(4), 147–155 (2012)
38. Schmuck, D., Matthes, J., Naderer, B.: Misleading consumers with green advertising? An affect–reason–involvement account of greenwashing effects in environmental advertising. J. Advert. **47**(2), 127–145 (2018)
39. Tariq, E., et al.: The role of digital marketing, CSR policy and green marketing in brand development. Int. J. Data Netw. Sci. **6**(3), 995–1004 (2022)

Impulse Buying Behavior in Livestream on TikTok Platform: Role of Information Quality, Social Presence and Hedonic Browsing

Bang Nguyen-Viet, Quynh Huong Hoang(✉), Nu Tai Linh Truong, and Thi Bich Ngan Nguyen

University of Economics Ho Chi Minh City, 59C Nguyen Dinh Chieu, District 3, Ho Chi Minh City, Vietnam

huonghoang.31211025546@st.ueh.edu.vn

Abstract. The study employs the SOR model to investigate the influence of information quality, social presence, hedonic browsing, cognitive assimilation, and flow state on the impulse buying behavior of youth in Ho Chi Minh City. The paper follows a quantitative approach through a paper-based survey from 300 students in Ho Chi Minh City. The study used PLS-SEM and Smart PLS 4.0 to analyze the data. The findings of the study clarified that information quality, social presence has impact on cognitive assimilation, social presence, hedonic browsing has impact on flow state and hedonic browsing, cognitive assimilation and flow state has impact on impulse buying behavior. The study provides theoretical implications about impulse buying behavior and managerial implications to develop innovative marketing strategies for livestream merchants through understanding customer's impulse behavior on TikTok's livestream.

Keywords: Impulse buying behavior · Information quality · Social presence · Hedonic browsing · Cognitive assimilation · Flow state · Livestream · TikTok

1 Introduction

TikTok, beyond its social media role, has become a rapid-growing e-commerce hub, transforming consumer engagement by offering a diverse range of products swiftly; its livestream shopping feature enhances convenience and global connectivity, reshaping consumer habits (Ma 2021). According to DataReportal, statistics of TikTok indicate that the average age of TikTok users in Vietnam is between 18 and 24. The popularity of TikTok among young demographics in Vietnam is evident (Nhi 2023). Expanding the study's target audience to include individuals whose range age 16–35, encompassing Generation Z, with a collective purchasing power of $44 billion, underscores the imperative for brands to engage on platforms like TikTok. The behavior of impulsive purchases has not been thoroughly studied (Darmawan et al. 2021; Rodrigues et al. 2021). Moreover, a variety of recent research has shown various factors that impact

M. Pagac et al. (Eds.): SCFF 2024, LNNS 1062, pp. 448–459, 2024.
https://doi.org/10.1007/978-3-031-65656-9_44

consumers' impulse buying behavior on online platforms, such as: flow experience in personalized advertising (Lina et al. 2021); the relationship between internal and external stimuli: cognitive and emotional responses (Kimiagari et al. 2021) … Concerning research articles in Vietnam, numerous authors have shown a keen interest in investigating this matter within the realm of online shopping (Phan and Nguyen 2017; Nguyen et al. 2020). Research examining the variables impacting young consumers' impulsive buying behavior on the TikTok live-streaming platform in Ho Chi Minh City is still lacking. Being a swiftly evolving urban center in Vietnam, Ho Chi Minh City offers a valuable backdrop for exploring this subject to acquire a more profound comprehension of impulse buying behavior among young consumers. This study employs the SOR model to analyze impulse buying on TikTok livestreams among young individuals in Ho Chi Minh City. Drawing upon previous research conducted in Vietnam and internationally, and in alignment with the research objectives, this study is conducted to address the following research questions: (1) What factors influence the impulse buying behavior of young consumers in Ho Chi Minh City, (2) To what extent do information quality, cognitive assimilation, and flow state impact young consumers' impulse buying behavior in Ho Chi Minh City?

2 Theoretical Background and Hypotheses

2.1 SOR Model

The study uses the SOR model proposed by Mehrabian and Russell (1974) to understand impulse buying behavior. It is determined based on the interaction between three main components: (1) Stimulus (S) (Song et al. 2021), (2) Organism (O) (Wu and Li 2018), (3) Response (R) (Sherman et al. 1997). Recent research has employed the SOR model to investigate consumer behavior on the internet in e-commerce, including intentions for online repurchase and impulse buying (Zhu et al. 2020). Utilizing the SOR model, the study examines impulse buying behavior among young individuals in Ho Chi Minh City through TikTok live-streaming.

2.2 Theoretical Development

Impulse Buying Behavior

This behavior is defined as the sudden, without careful consideration and unplanned decision to buy a product, which often occurs due to stimulation from the external environment (Sengupta and Chattopadhyay 2005). It is this impulse buying behavior that can cause consumers to buy products that they do not need or want and are often made within a short period of time (Dittmar et al. 1996).

The Relationship Between Information Quality and Cognitive Assimilation

Information quality refers to product-related details, such as price, product features, and warranty policies, or the effectiveness of information content (Mollen and Wison 2010). Besides, information quality is a typical environmental signal that affects cognitive state (Xu et al. 2019). As a result, viewers adjust perceptions based on live streamer performance, influencing product understanding and satisfaction. SOR model posits information quality in live streams affects cognitive assimilation. Therefore, hypothesis is proposed:

H1: Information quality has a positive impact on cognitive assimilation.

The Relationship Between Social Presence, Cognitive Assimilation, and Flow State
Social presence refers to the extent to which a communication medium enables users to perceive others as being psychologically present (Fulk et al. 1987), the highest level is direct interaction. When viewers can feel emotionally aroused when developing a close relationship with a streamer (Shen and Khalifa 2012), this alters the viewer's level of excitement and arousal, enabling them to absorb the information conveyed by the streamer, thereby facilitating cognitive assimilation. Moreover, a number of research studies have demonstrated that social presence and flow state have a positive impact on each other in online activities (Lee and Chen 2017; Kim and Kim 2019; Chen and Wang 2020). Based on the information provided, the following hypotheses are proposed:

H2: Social presence has a positive impact on cognitive assimilation.
H3: Social presence has a positive impact on flow state.

The Relationship Between Hedonic Browsing, Flow State, and Impulse Buying Behavior
Hedonic browsing is defined as enjoyment-driven browsing on social shopping platforms without the intention to purchase, mainly for self-enjoyment (Zheng et al. 2019). Hedonic browsing that consumers receive when they surf websites is related to emotions and satisfaction, thereby affecting the flow state of shoppers (Shahpasandi et al. 2020). Previous research has identified hedonic motivation as the central factor determining consumer purchasing behavior (Kukar-Kinney and Close 2010). Furthermore, previous research determined browsing hedonism to be positively related to impulse buying (Lee and Lee 2003; Lee et al. 2009). Therefore, the hypotheses are proposed:

H4: Hedonic browsing has a positive impact on flow state.
H5: Hedonic browsing has a positive impact on impulse buying behavior.

The Relationship Between Cognitive Assimilation and Impulse Buying Behavior
Cognitive assimilation is defined as the degree to which a viewer's existing thoughts, beliefs, or attitudes are modified through internalizing, assimilating, and absorbing the streamer's influence (Piaget et al. 1989). In e-commerce, cognitive assimilation frequently results in impulse buying since customers take in recommendations while they shop, which shapes their choices (Wang et al. 2017). Cognitive assimilation has a direct impact on impulse buying behavior when examining for consumer live streaming commerce (Xu et al. 2020). The research hypothesis is proposed as follows:

H6: Cognitive assimilation has a positive impact on impulse buying behavior.

The Relationship Between Flow State and Impulse Buying Behavior
Flow state is a psychological state where participants are fully immersed in an activity, losing track of time, experiencing high satisfaction and achievement (Csikszentmihalyi 1990). The flow state induced by the online environment reflects an individual's behavior on the internet (Hyun et al. 2022) because intrinsic enjoyment can increase customer exploration behavior which in turn triggers impulse behavior (Guo and Poole 2009). Consumer attention and hedonism are frequently linked with the flow experience, leading to prolonged viewing time and influencing their buying behavior (Zhou et al. 2021). Flow state has a positive impact and increases impulse purchase behavior (Wu et al. 2020; Huo et al. 2023). The hypothesis is proposed:

H7: Flow state has a positive impact on impulse buying behavior.

3 Research Methodology

The study adapted items from the studies of Dong et al. (2014), Sun et al. (2019), Zheng et al. (2019), and Ming et al. (2021). The original English questionnaire was translated into Vietnamese. A group discussion was then conducted to adjust the item to suit the Vietnamese context and the target respondents. The constructs were measured by using a five-point Likert scale. The study used convenient sampling through a paper-based survey from January to February 2024 to students at multiple high schools and universities in Ho Chi Minh City. Of the 350 questionnaires distributed, 300 valid surveys were collected (85.71%). The demographics of the respondents are shown in Table 1.

Table 1. Demographic characteristics

Characteristic	Percentage (%)	Characteristic	Percentage (%)
Gender		Education	
Male	44.33	Upper Secondary	43.67
Female	55.67	College/University	48.0
Age		Others	8.33
16–20	42.67	Income	
21–25	32.33	Below 5 m[1] VND	45.67
26–30	14.00	5–10 m VND	25.00
31 - 35	11.00	10–15 m VND	15.00
		15 m VND or more	14.33
Total			100.00

[1] Million.

4 Research Results

4.1 Measurement Model Evaluation

The study assessed the reliability of items by Cronbach's α coefficients (Cα) and Composite Reliability (CR). In Table 2, all measurements achieved reliability with Cα > 0.6 (Nunnally and Bernstein 1994) and CR > 0.7 (Hair et al. 2010). The study assessed the convergent validity of the measures using AVE and Outer Loadings. According to the results in Table 2, the measures of study ensure convergent validity ensure convergent validity with all outer loadings of the measures ranged from [0.809; 0.909] which were above 0.7 (Hair et al. 2014) and the AVE values of all measures ranged from [0.730; 0.798] which were greater than 0.5 (Hock et al. 2010). AVE's square roots satisfying the assessment condition of Fornell and Larcker (1981) which were higher than all correlations between the variables and the HTMT values of all variable pairs are lower than 0.9 (Henseler et al. 2015) which were in the range [0.616; 0.847] (Table 3). Therefore, the measures achieve discriminant validity.

4.2 Structural Model Evaluation and Hypotheses Result

The study applied the PLS-SEM method to test the research model and hypotheses. Because PLS-SEM is a method that not only bridges the gap between explanation and prediction but also works effectively with small sample sizes (Hair et al., 2021).

Table 2. Reliability and validity for items

Constructs	OL[2]	Cα[3]	CR[4]	AVE[5]
Information quality – IQ (Dong et al. 2014)		0.908	0.931	0.730
IQ1	0.851			
IQ2	0.891			
IQ3	0.893			
IQ4	0.809			
IQ5	0.826			
Social presence – SP (Sun et al. 2019)		0.874	0.922	0.798
SP1	0.889			
SP2	0.893			

(continued)

[2] OL: Outer loadings.

[3] Cα: Cronbach's Alpha.

[4] CR: Composite Reliability.

[5] AVE: Average Variance Extracted.

Table 2. (*continued*)

Constructs	OL	Cα	CR	AVE
SP3	0.899			
Hedonic browsing – HB (Zheng et al. 2019)		0.895	0.927	0.761
HB1	0.880			
HB2	0.884			
HB3	0.864			
HB4	0.861			
Flow state – FS (Ming et al. 2021)		0.934	0.948	0.752
FS1	0.883			
FS2	0.865			
FS3	0.836			
FS4	0.870			
FS5	0.879			
FS6	0.869			
Cognitive assimilation – CA (Xu et al. 2020)		0.908	0.936	0.785
CA1	0.886			
CA2	0.897			
CA3	0.884			
CA4	0.876			
Impulse buying behavior – IPB (Ming et al. 2021)		0.915	0.940	0.798
IPB1	0.896			
IPB2	0.889			
IPB3	0.879			
IPB4	0.909			

Table 3. Discriminant validity test result

	CA	FS	HB	IPB	IQ	SP
CA	**0.866**					
FS	0.847	**0.867**				
HB	0.617	0.616	**0.872**			
IPB	0.843	0.830	0.688	**0.893**		
IQ	0.726	0.721	0.675	0.754	**0.854**	
SP	0.796	0.783	0.642	0.734	0.763	**0.893**

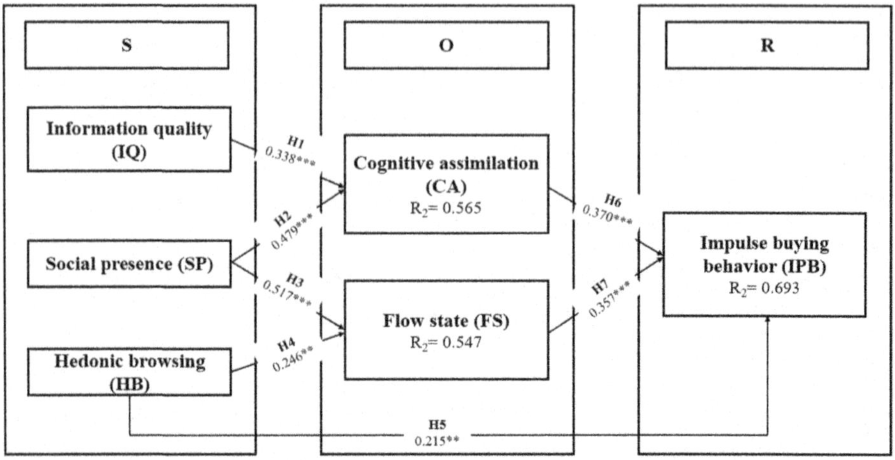

Fig. 1. Structural model test result. *Note(s)* *** Significant at 1% level, ** Significant at 5% level, * Significant at 10% level.

Figure 1 demonstrates that the R2 > 0.2 (Hair et al. 2013) which confirms the relationship between independent variables and dependent variables. Table 4 shows all standardized regression coefficients of the relationships are positive, indicating that all hypotheses have a positive impact relationship. Specifically, H1 is supported, showing that IQ has a significant impact on CA ($p = 0.000$; $\beta = 0.338$); H2 and H3 are supported, showing that SP has a strong impact on CA ($p = 0.000$; $\beta = 0.479$) and FS ($p = 0.000$; $\beta = 0.517$); H4 and H5 are supported, showing that HB has a positive impact on FS ($p = 0.001$; $\beta = 0.246$) and IPB ($p = 0.001$; $\beta = 0.215$); H6 is supported, showing that CA has a significant impact on IBP ($p = 0.000$; $\beta = 0.370$) and H7 is supported, showing that FS has a significant impact on IPB ($p = 0.000$; $\beta = 0.357$). Based on the path coefficients (β) in Table 4, the results also indicate that CA is strongest influenced by SP and then by IQ; simultaneously, FS is strongest affected by SP but weakly affected by HB; CA demonstrates the most substantial impact on IPB, with FS ranking as the second most influential - nevertheless, the disparity between CA and FS is not significant and finally, HB displaying the weakest effect on IPB.

4.3 Impulse Buying Behavior Influenced by Demographics

To determine the statistical differences between the means of two or more independent groups and then provide meaningful information for live-streaming merchants to fit each customer group, t-test and one-way anova were conducted. With T-test, impulse buying behavior is significantly different based on gender (Levene Sig = 0.006, P-value = 0.000). In that, females are more likely to make impulse buying behavior than males (M = 3.618, SD = 0.920). Meanwhile, ANOVA test results have shown that impulse buying behavior of the respondents was significantly different based on age (Levene Sig = 0.000, P-value = 0.002). In that, people from 21 to 25 (M = 3.691, SD = 1.043) are most likely to make impulse buying behavior. However, ANOVA test results have shown that impulse buying behavior of the respondents was not different based on education

Table 4. Hypothesis test result

Hypothesized path			Pathcoefficients (β)	p-values	Hypotheses	Results
IQ	→	CA	0.338	0.000	H1	Supported
SP	→	CA	0.479	0.000	H2	Supported
SP	→	FS	0.517	0.000	H3	Supported
HB	→	FS	0.246	0.001	H4	Supported
HB	→	IPB	0.215	0.001	H5	Supported
CA	→	IPB	0.370	0.000	H6	Supported
FS	→	IPB	0.357	0.000	H7	Supported

groups (Levene Sig = 0.081 > 0.05) and income levels (Levene Sig = 0.000, P-value = 0.172 > 0.05).

5 Discussion

The research explores impulse buying during livestreams among young people. Findings reveal that hedonic browsing, cognitive assimilation, and flow state positively influence impulse buying. Information quality and social presence affect cognitive assimilation. The research findings reveal that among the factors examined, social presence exhibits the strongest influence on both cognitive assimilation and flow state. Furthermore, cognitive assimilation emerges as the primary direct influencer of impulse buying behavior, closely followed by flow state. Importantly, there is minimal disparity observed in the extent of impact on impulse buying behavior between these two factors. Consequently, it can be concluded that both cognitive assimilation and flow state significantly contribute to the phenomenon of impulse buying behavior. Notably, hedonic browsing directly triggers impulse buying, suggesting emotional engagement prompts buying. Businesses should tailor content to enhance these factors, emphasizing emotions.

5.1 Limitations

The limited research time constrains this study to a short survey period, which may impact the findings on behavioral patterns and fail to assess long-term factors. Additionally, the generalizability is also restricted due to the convenience sampling method employed within a short time frame in Ho Chi Minh City. Besides the factors mentioned in the study, there are still other factors worth considering. Future research could explore alternative models and incorporate additional factors like those mentioned above.

5.2 Implications

Implications for Practitioners

The research explores factors affecting cognitive assimilation, flow state, and impulse buying behavior on TikTok livestreams among young people in Vietnam. It identifies information quality, social presence, and hedonic browsing as influential factors. Businesses should prioritize accurate information and intense interaction to enhance cognitive assimilation and maintain viewer's flow state. Curating entertaining content tailored to the relaxation needs of the audience can attract viewers and increase impulse buying likelihood. In addition to the aforementioned factors, age and gender of the viewers also directly influence impulse buying behavior, such as the female demographic and young adults aged 21–30. Understanding audience preferences and continuously monitoring performance are crucial for optimizing marketing efforts on TikTok.

Implications for Researchers

The study reveals that information quality and social presence positively affect cognitive assimilation, while social presence and hedonic browsing enhance the flow state. Furthermore, hedonic browsing, cognitive assimilation, and flow state also drive impulse buying behavior. This research also reveals that demographic variables such as gender and age also influence impulse buying behavior, warranting further investigation. Besides the factors identified in this study that influence impulse buying behavior, there are still significant factors related to viewers' perceptions and emotions (such as consumer trust, Ming et al. 2021; brand attachment emotions, Akbar et al. 2020) that could be considered for further development based on the SOR model, similar to this research.

6 Conclusions

The research investigates factors impacting cognitive assimilation, flow state, and impulse buying behavior on TikTok livestreams among young people in Vietnam, utilizing the S-O-R theory. It offers implications for both practitioners, and researchers, highlighting the significance of cognitive assimilation and flow state in retail, particularly livestream retail. Additionally, it suggests incorporating more factors into the model for a clearer understanding of their effects, enhancing insights into impulse buying behavior. This study contributes to developing models for further exploration in this field, shedding light on the dynamics of livestream shopping and its implications for retail sales strategies.

Acknowledgements. This research is funded by the University of Economics Ho Chi Minh City, Vietnam (UEH).

References

Akbar, M.I., Ahmad, B., Asif, M.H., Siddiqui, S.A.: Linking emotional brand attachment and sales promotion to post-purchase cognitive dissonance: the mediating role of impulse buying behavior. J. Asian Financ. Econ. Bus. **7**(11), 367–379 (2020). https://doi.org/10.13106/jafeb.2020.vol7.no11.367435–445. https://doi.org/10.1080/10941660802420994

Chen, Y., Wang, Q.: Live video streaming on YouTube: how social presence and community commitment influence user engagement. Comput. Hum. Behav. **104**, 106179 (2020)

Csikszentmihalyi, M.: Flow: The Psychology of Optimal Experience. Harper & Row (1990)

Darmawan, D., Gatheru, J.: Understanding impulsive buying behavior in marketplace. J. Soc. Sci. Stud. (JOS3) **1**(1), 11–18 (2021). https://doi.org/10.56348/jos3.v1i1.2

Dittmar, H., Beattie, J., Friese, S.: Objects, decision considerations and self-image in men's and women's impulse purchases. Acta Psychol. **93**(1–3), 187–206 (1996)

Dong, T.-P., Cheng, N.-C., Wu, Y.-C.J.: A study of the social networking website service in digital content industries: the Facebook case in Taiwan. Comput. Hum. Behav. **30**, 708–714 (2014). https://doi.org/10.1016/j.chb.2013.07.037

Fornell, C., Larcker, D.F.: Evaluating structural equation models with unobservable variables and measurement error. J. Mark. Res. **18**(1), 39–50 (1981)

Fulk, J., Steinfield, C.W., Schmitz, J., Power, J.G.: A social information processing model of media use in organizations. Commun. Res. **14**(5), 529–552 (1987)

Guo, Y.M., Poole, M.S.: Antecedents of Flow in online shopping: a test of alternative models. Inf. Syst. J. **19**(4), 369–39.

Hair, J.F., Black, W.C., Babin, B.J., Anderson, R.E.: Multivariate Data Analysis, 7th edn. Pearson Education, Upper Saddle River (2010)

Hair, J.F., Hult, G.T.M., Ringle, C.M., Sarstedt, M.: A Primer on Partial Least Squares Structural Equation Modeling (PLS SEM). Sage, Thousand Oaks, CA (2014)

Hair, J.F., Hult, G.T.M., Ringle, C.M., Sarstedt, M., Danks, N.P., Ray, S.: Partial Least Squares Structural Equation Modeling (PLS-SEM) Using R: A Workbook. Springer Nature (2021)

Hair, J.F., Ringle, C.M., Sarstedt, M.: Partial least squares structural equation modeling: rigorous applications, better results and higher acceptance. Long Range Plan. **46**(1–2), 1–12 (2013)

Henseler, J., Ringle, C.M., Sarstedt, M.: A new criterion for assessing discriminant validity in variance-based structural equation modeling. J. Acad. Mark. Sci. **43**(1), 115–135 (2015)

Hock, C., Ringle, C.M., Sarstedt, M.: Management of multi-purpose stadiums: importance and performance measurement of service interfaces. Int. J. Serv. Technol. Manag. **14**(2/3), 188–207 (2010)

Huo, C., Wang, X., Sadiq, M.W., Pang, M.: Exploring factors affecting consumer's impulse buying behavior in live-streaming shopping: an interactive research based upon SOR model. SAGE Open **13**(2), 21582440231172680 (2023)

Hyun, H., Thavisay, T., Lee, S.H.: Enhancing the role of flow experience in social media usage and its impact on shopping. J. Retail. Consum. Serv. **65**, 102492 (2022)

Kim, J., Kim, S.: Why do users watch live game streaming? Gratifications and relationships among content characteristics, streamer personality, user satisfaction, and continuance intention. Telematics Inform. **41**, 101253 (2019)

Kimiagari, S., Asadi Malafe, N.S.: The role of cognitive and affective responses in the relationship between internal and external stimuli on online impulse buying behavior. J. Retail. Consum. Serv. **61**, 102567 (2021). https://doi.org/10.1016/j.jretconser.2021.102567

Kukar-Kinney, M., Close, A.G.: The determinants of consumers' online shopping cart abandonment. J. Acad. Mark. Sci. **38**, 240–250 (2010)

Lee, M., Kim, Y., Fairhurst, A.: Shopping value in online auctions: their antecedents and outcomes. J. Retail. Consum. Serv. **16**(1), 75–82 (2009)

Lee, H., Lee, H.: The impacts of browsing on buying impulsiveness in Internet shopping malls. Korean Manag. Rev. **32**(5), 1235–1263 (2003)

Lee, C., Chen, L.: How do online game players experience the flow state? An empirical study of social presence and flow state in online game players. Int. J. Hum. Comput. Interact. **33**(11), 862–873 (2017)

Lina, L.F., Ahluwalia, L.: Customers' impulse buying in social commerce: the role of flow experience in personalized advertising. J. Manaj. Maranatha **21**(1), 1–8 (2021). https://doi.org/10.28932/jmm.v21i1.3837

Ma, Y.: To shop or not: understanding Chinese consumers' live-stream shopping intentions from the perspectives of uses and gratifications, perceived network size, perceptions of digital celebrities, and shopping orientations. Telemat. Inform. **59**, 101562 (2021). https://doi.org/10.1016/j.tele.2021.101562

Mehrabian, A., Russell, J.A.: An Approach to Environmental Psychology. The MIT Press (1974)

Ming, J., Jianqiu, Z., Bilal, M., Akram, U., Fan, M.: How social presence influences impulse buying behavior in live streaming commerce? The role of S-O-R theory. Int. J. Web Inf. Syst. **17**(4), 300–320 (2021). https://doi.org/10.1108/ijwis-02-2021-0012

Mollen, A., Wilson, H.: Engagement, telepresence and interactivity in online consumer experience: reconciling scholastic and managerial perspectives. J. Bus. Res. **63**, 919–925 (2010)

Nguyễn Bình Minh và nhóm tác giả: Nhân tố tác động đến hành vi mua hàng ngẫu hứng trực tuyến và hệ quả tại TP. Hà Nội. Tạp chí Tài chính Online (2020)

Nhi, N.: Thống kê độ tuổi người dùng TikTok tại Việt Nam và Thế giới. NaoTot Nhat (2023). https://www.vietnamplus.vn/infographics-viet-nam-dung-thu-6-the-gioi-ve-so-nguoi-dung-tiktok-post855934.vnp

Nunnally, J.C., Bernstein, I.H.: Psychometric Theory. McGraw-Hill, NY (1994)

Phan, Q.T., Ha, N.T.H.: Các yếu tố ảnh hưởng đến mua hàng trực tuyến của người tiêu dùng TP. Hồ Chí Minh. Open Univ. Ho Chi Minh City J. Sci. **55**(4), 38–50 (2017)

Piaget, J., Garcia, R.: Psychogenesis and the History of Science. Columbia University Press (1989)

Rodrigues, R.I., Lopes, P., Varela, M.: Factors affecting impulse buying behavior of consumers. Front. Psychol. **12**, 697080 (2021)

Sengupta, J., Chattopadhyay, A.: The influence of reference groups on impulse buying behavior. J. Consum. Psychol. **15**(2), 152–167 (2005)

Shahpasandi, F., Zarei, A., Nikabadi, M.S.: Consumers' impulse buying behavior on Instagram: examining the influence of flow experiences and hedonic browsing on impulse buying. J. Int. Commer. **19**(4), 437–465 (2020). https://doi.org/10.1080/15332861.2020.1816324

Shen, K.N., Khalifa, M.: System design effects on online impulse buying. Internet Res. **22**(4), 396–425 (2012)

Sherman, E., Mathur, A., Smith, R.B.: Store environment and consumer purchase behavior: mediating role of consumer emotions. Psychol. Mark. **14**(4), 361–378 (1997)

Song, S., Yao, X., Wen, N.: What motivates Chinese consumers to avoid Information about the COVID-19 pandemic?: The perspective of the stimulus-organism-response model. Inf. Process. Manage. **58**(1), 102407 (2021)

Sun, Y., Shao, X., Li, X., Guo, Y., Nie, K.: How live streaming influences purchase intentions in social commerce: an IT affordance perspective. Electron. Commer. Res. Appl. **37**, 100886 (2019)

Wang, S.C., Xu, X., Lin, K.C., Wu, J.H.: The effect of affordance on ubiquitous commerce consumption (2017)

Wu, Y.L., Li, E.Y.: Marketing mix, customer value, and customer loyalty in Social commerce: a stimulus-organism-response perspective. Int. Res. **28**(1) (2018)

Wu, L., Chiu, M.L., Chen, K.W.: Defining the determinants of online impulse buying through a shopping process of integrating perceived risk, expectation-confirmation model, and flow theory issues. Int. J. Inf. Manage. **52**, 102099 (2020)

Xu, X., Wu, J.H., Li, Q.: What drives consumer shopping behavior in live streaming commerce? J. Electron. Commer. Res. **21**(3), 144–167 (2020)

Xu, X., Wu, J.H., Chang, Y.T., Li, Q.: The investigation of Hedonic consumption, impulsive consumption and Social sharing in e-commerce live-streaming videos (2019)

Zheng, X., Men, J., Yang, F., Gong, X.: Understanding impulse buying in mobile commerce: an investigation into Hedonic and utilitarian browsing. Int. J. Inf. Manage. **48**, 151–160 (2019). https://doi.org/10.1016/j.ijinfomgt.2019.02.010

Zhou, M., Huang, J., Wu, K., Huang, X., Kong, N., Campy, K.S.: Characterizing Chinese consumers' intention to use live e-commerce shopping. Technol. Soc. **67**, 101767 (2021)

Zhu, B., Kowatthanakul, S., Satanasavapak, P.: Generation Y consumer online repurchase intention in Bangkok: based on stimulus-organism-response (SOR) model. Int. J. Retail Distrib. Manag. **48**(1), 53–69 (2020)

Prospects for Applying Technology to Support Digital Transformation at Vietnamese Educational Institutions: A Thematic Analysis

Thanh-Cong Truong(✉)

Faculty of Information Technology, University of Finance-Marketing, Ho Chi Minh City, Vietnam
ttcong@ufm.edu.vn

Abstract. The digital transformation of higher education institutions (HEIs) in Vietnam is a significant phenomenon, yet there exists a paucity of research specifically investigating the application of technology to support such transformations, resulting in a notable knowledge gap. This study aims to explore the potential applications of technology to facilitate digital transformation in Vietnamese HEIs and to identify key areas where technological interventions can significantly impact these institutions. Adopting a qualitative case study design, this research delves into the complexities of applying technology for digital transformation in Vietnamese HEIs. Twelve experienced professionals from the domains of education and technology participated in in-depth interviews, complemented by focus group discussions with members of the expert community involved in digital transformation initiatives. Thematic analysis was employed to analyze the collected data, ensuring rigor and consistency in data interpretation. The study revealed insights into various aspects of prospective digital transformation in Vietnamese HEIs, emphasizing infrastructure components, technology adoption, and challenges associated with integrating information and communications technology. Furthermore, the research identified the potential of technologies such as e-learning, artificial intelligence, and blockchain to enhance educational quality and foster innovation within these institutions.

Keywords: Technology · Digital transformation · Higher education institutions

1 Introduction

The trend of digital transformation has impacted every industry and sector in society. Higher education institutions (HEIs) cannot be out of this trend [1]. Technology is developing at a breakneck pace, which is leading to the constant emergence of new technologies in the field of higher education. The research and application of these new technologies will help universities improve the quality of education by enhancing student interaction and access to knowledge. In addition, the application of new technologies can also help teachers and students enhance their creativity and develop new educational solutions. This can give higher education access to modern educational methods and optimize the teaching and learning process.

M. Pagac et al. (Eds.): SCFF 2024, LNNS 1062, pp. 460–469, 2024.
https://doi.org/10.1007/978-3-031-65656-9_45

In recent years, there have been many domestic and international research projects related to the implementation of digital transformation in higher education. Author group in [2] assessed achievements and digital transformation trends in the context of Industrial Revolution 4.0. The paper also identifies infrastructure components for smart learning environments. The authors in [3] introduces 6 new technologies in higher education and analyzes impacts, challenges and solutions for technology adoption. Linh et al. in [4] Propose a framework model for digital transformation based on 3 factors: learners, teachers and technology platforms. The article also introduces technologies and applications in digital transformation.

Abad-Segura et al. asserts that ICT is key to enhancing the quality of education, fostering innovation and supporting the skills development needed for the 21st century. However, ICT also poses sustainability management challenges such as high costs, human resource shortages, integration difficulties and cybersecurity risks [5]. Mohamed Hashim et al. highlights that universities need to adapt to changes in the educational environment and take advantage of opportunities from digital technology [6]. Article by [7] provided a systematic review of relevant studies and identified benefits, challenges and recommendations for the adoption of digital transformation. The paper also proposes a framework model to support the design and implementation of digital transformation strategies.

Despite the breadth of research focusing on digital transformation within higher education, scant attention has been paid to the prospects of applying technology to support digital transformation specifically within Vietnamese HEIs. This knowledge gap is significant, given the transformative impact of technological innovation on educational access, delivery, and experience, particularly within the Vietnamese context. Consequently, the objective of this paper is to furnish an overview of the applicability of technology within Vietnamese HEIs and to propose key areas where technological interventions can effect substantial enhancements within these institutions under the expert perspectives using the thematic analysis approach.

2 The Context off Applying Information Technology at Vietnamese HEIs

The integration of technology within the educational framework of Vietnam has assumed a pivotal role, instigating a paradigm shift in pedagogical approaches and learning methodologies. The amalgamation of conventional pedagogical practices with contemporary technological advancements has engendered novel and innovative learning experiences for both learners and educators.

The discourse on digital transformation within Vietnamese higher education has gained significant traction, particularly with the proliferation of the internet and sophisticated technologies. However, the adoption of technological innovations to facilitate this transition was gradual until the onset of the Covid-19 pandemic, which served as a catalyst for digital transformation within this domain.

The Ministry of Education and Training has undertaken several pivotal measures to enhance educational quality and administrative management. A notable accomplishment is the implementation of the Higher Education Management Information System

(HEMIS), providing a robust foundation for efficient resource management. By digitizing data pertaining to 442 institutions, 152,470 educators, and 2,102,165 learners, the Ministry has established a comprehensive information system, facilitating enhanced management, evaluation, and improvement strategies [8].

Furthermore, the integration of the national population database has yielded substantial benefits. Achieving a 98% synchronization and verification rate for over 24 million learner and teacher records has streamlined educational record management [8]. Higher education institutions have made significant strides in enhancing management and teaching support mechanisms. All institutions are equipped with state-of-the-art computer facilities, LAN, wifi systems, and information portals, facilitating seamless access to technology platforms. Moreover, 90% of institutions have established portal editorial boards and regulations to safeguard information security. The adoption of Training Management and Document Management software, with a prevalence rate of up to 90%, optimizes training processes and document organization. Additionally, over 60% of institutions have implemented HR, Asset, and Equipment Management software, bolstering efficiency in personnel, asset, and equipment management. The integration of information technology has optimized processes, enhanced operational efficiency, and mitigated resource management risks [8].

The Ministry has fostered the development of digital learning repositories, a milestone in educational modernization. With over 7,500 doctoral theses archived, the digital repository serves as a valuable resource for faculty and students, fostering scientific research and development. Moreover, the implementation of online training initiatives has unlocked myriad benefits. With a 50% adoption rate, institutions have facilitated flexible and convenient access to learning content, economizing time and financial resources, and extending educational accessibility to disadvantaged demographics. Furthermore, approximately 60% of institutions have deployed digital learning materials and online assessments, contributing to a modern learning environment aligned with Industry 4.0. Moreover, the establishment of e-libraries at a 70% adoption rate has expanded avenues for accessing and utilizing learning materials, fostering deeper learning and research endeavors.

Despite these developments, challenges persist in applying technology for digital transformation. Principal challenges include lack of enthusiasm and internal communication to expedite the transformation process, absence of concrete planning and unclear assignment delineations, dearth of an overarching model and supporting architecture, infrastructure and facility inadequacies, insufficient focus on information security and access, and absence of regulations governing information system management and operation. Discrete and inflexibly integrated existing applications engender issues concerning data ownership and personal information management.

Furthermore, securing funding for digital training institutions necessitates substantial and sustained investment in construction, daily maintenance, and operations. Imperfections within the socialization mechanism for resource mobilization impede acquiring adequate funding to sustain operations.

A critical impediment is the scarcity of dedicated human resources proficient in implementing digital education models. The quantity and caliber of this workforce demand enhancement to meet burgeoning digital technology demands. Continuous knowledge updating and mastery of emerging technologies are imperative for effective application within educational contexts.

Moreover, disparities and limitations persist in digital technology utilization across individuals and entities. To ensure the efficacy of digital transformation initiatives, training and support programs are vital to cultivate digital competencies among staff, educators, and students, enabling them to leverage technology's potential in education and learning endeavors.

3 Methodology

3.1 Study Design

This research adopts a qualitative case study design, deemed appropriate for addressing the research questions regarding the applying technology to support digital transformation at Vietnamese HEIs. A case study approach allows for an in-depth exploration of a specific phenomenon within its real-life context, facilitating the examination of complex interactions and dynamics. By focusing on a single case, this design enables a thorough analysis of the intricacies and nuances inherent in the process of applying technology for digital transformation in Vietnam HEIs, offering rich and detailed insights.

3.2 Participants

The participants in this study were purposively selected to ensure relevance and diversity of perspectives. A total of 12 experienced professionals from the fields of education and technology were recruited as participants for in-depth interviews. Additionally, focus group discussions were conducted with members of the expert community, comprising educators, technologists, and policymakers involved in digital transformation initiatives in education.

3.3 Instruments

Interview protocols and focus group discussion guides were developed as primary instruments for data collection. The validity of the instruments was ensured through a thorough review by experts in the field and pilot testing with a small sample of participants. In addition, interviews and discussions were recorded in writing, and confirmed with experts to ensure data recording was accurate.

3.4 Procedures

The study followed a systematic approach to data collection, involving the following procedures, as illustrated in Fig. 1.

3.5 Data Analysis

Thematic analysis [9] was chosen as it allows for a nuanced exploration of qualitative data, enabling the identification of recurring themes and patterns. This methodological approach is well-suited for uncovering the complexities and nuances inherent in the phenomenon of digital transformation in education.

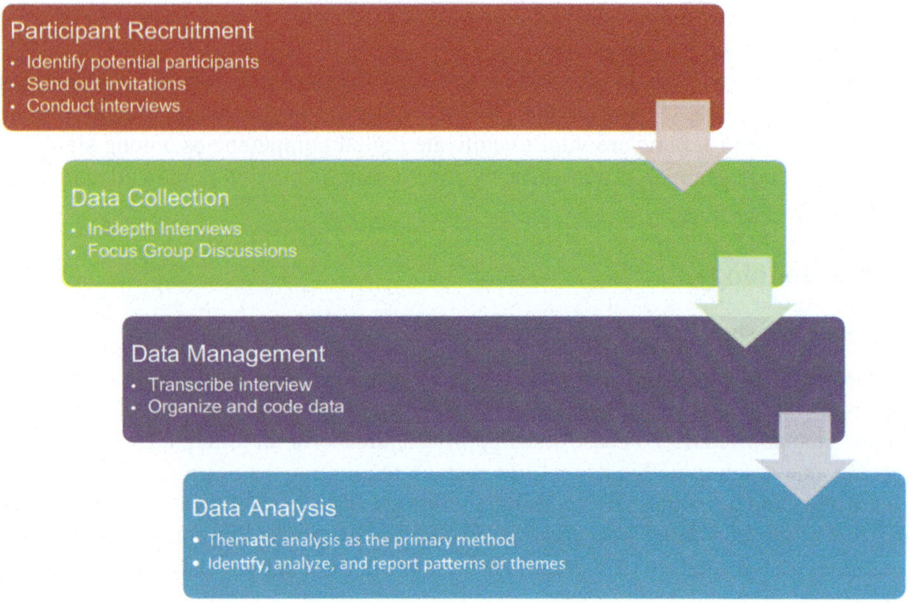

Fig. 1. The study's procedures

4 Results and Discussions

4.1 Deploy Adaptive Learning Platforms Based on AI and Data Analytics

AI-driven adaptive learning platforms utilize sophisticated algorithms to analyze learner data comprehensively, encompassing learning preferences, strengths, weaknesses, and progression. This data-centric approach facilitates the creation of tailored learning experiences attuned to the distinct requirements and learning modalities of individual learners. By delivering content and activities that resonate with each learner, adaptive learning platforms have the capacity to augment motivation, cultivate a sense of autonomy in learning, and elevate overall academic attainment.

Personalized learning pathways constitute a salient feature of adaptive learning platforms, offering customized learning trajectories based on learners' proficiency levels and mastery of concepts. Departing from conventional uniform curricula, learners navigate content at their own pace, receiving tailored assistance and resources as warranted. These adaptive pathways bolster the learning journey by furnishing supplementary hands-on exercises, remedial support, or enrichment activities tailored to each learner's competencies and learning trajectory.

Real-time feedback and assessment mechanisms are integral components of adaptive learning platforms, furnishing learners with immediate insights into their progress and grasp of concepts. Through interactive quizzes, simulations, and assessments, learners receive timely feedback on their performance, pinpointing strengths and areas necessitating further attention or practice. This prompt feedback mechanism enhances skill acquisition, affording learners the opportunity to refine their learning strategies accordingly.

Adaptive learning platforms have the potential to invigorate learning motivation by tailoring educational experiences to individual interests and aptitudes. Personalized content and activities are inherently more engaging, fostering deeper engagement with the subject matter. Moreover, the interactive features of adaptive learning platforms, including gamification elements and multimedia resources, render learning a more enjoyable and gratifying endeavor, fostering greater perseverance and academic achievement among learners.

In addition to supporting learners, adaptive learning platforms offer valuable insights and support for educators. While affording learners autonomy in their learning endeavors, these platforms furnish instructors with access to analytical dashboards and reports delineating learner progress and areas of challenge. Armed with this data-driven perspective, instructors can deploy targeted interventions, tailor instruction to meet diverse learner needs, and provide effective support to learners encountering difficulties, thus enhancing the efficacy of the teaching and learning process overall.

4.2 Online Learning Platforms and Digital Learning Materials

Access and Equality: The utilization of online platforms, digital resources, and internet connectivity can mitigate geographical barriers and facilitate educational access for individuals residing in remote or disadvantaged regions. Leveraging technology, educational institutions can extend their outreach to learners who encounter obstacles in accessing conventional modes of learning due to factors such as distance, disability, or socioeconomic circumstances. Digital learning materials ensure the universality of educational resources, irrespective of learners' geographic locations or backgrounds, thereby fostering equity and inclusivity in education.

Flexibility and Convenience: Online platforms afford flexibility and convenience to both learners and educators. Through online platforms and digital resources, learning becomes independent of time and place, enabling learners to engage in educational pursuits at their own pace and accommodate diverse schedules and commitments. Furthermore, digital learning materials offer facile updating and customization capabilities, ensuring the currency and relevance of content to evolving learner needs.

Interactive and Engaging: Digital learning materials facilitate interactive and multimedia learning experiences, stimulating learner engagement and comprehension. Incorporating features such as videos, simulations, games, and virtual laboratories, these materials enable learners to explore intricate concepts dynamically and immersively. Interactive components foster active learning, critical thinking, and problem-solving proficiencies, thereby deepening learner engagement and enhancing knowledge retention.

Collaboration and Communication: Online discussion forums, collaborative projects, and video conferencing tools facilitate meaningful interaction and knowledge exchange, engendering interpersonal learning and skill development. Additionally, the infrastructure for distance learning accommodates synchronous and asynchronous communication, catering to diverse communication preferences and learning exigencies.

4.3 Smart University with IoT

Smart Infrastructure: Internet of Things (IoT) technology has the potential to revolutionize traditional campuses by transforming them into smart, interconnected environments furnished with sensors, actuators, and network devices. Within these smart campuses, sensors embedded in school buildings, classrooms, and laboratories can capture real-time data concerning environmental conditions, occupancy rates, energy consumption, and equipment functionality. This data, upon analysis, can inform resource optimization efforts, enhance energy efficiency, bolster safety and security measures, and streamline facilities management processes.

Enhanced Teaching and Learning Experiences: IoT-enabled smart classrooms offer interactive and immersive learning environments that actively engage students and facilitate participatory learning. Through the utilization of IoT devices such as wearables and mobile applications, personalized learning experiences tailored to individual learners' interests, learning modalities, and performance metrics can be curated.

IoT Data-driven Optimization: The data generated by IoT devices furnishes valuable insights for decision-making and strategic planning within higher education institutions. By scrutinizing real-time and historical data gleaned from IoT sensors and devices, administrators can discern trends, patterns, and irregularities, thereby optimizing resource allocation and enhancing operational efficiency.

4.4 Mobile Learning

The ubiquity of mobile devices, particularly smartphones, in Vietnam presents considerable opportunities for the advancement of higher education through mobile learning.

Enhanced Personal Engagement: Mobile devices afford students the capability of accessing educational resources and participating in learning activities anytime and anywhere, thereby fostering active engagement in the teaching-learning process outside the confines of traditional classroom settings. Through mobile applications, students can access course materials, engage in discussions, collaborate on projects, and fulfill assignments, thereby cultivating a culture of continuous learning and involvement.

Supplementary Teaching Aids: Mobile devices serve as potent adjunctive teaching aids, supplementing conventional pedagogical approaches with immersive multimedia content, interactive simulations, and educational applications. Higher education institutions can capitalize on mobile technology to furnish supplementary learning materials, such as lecture recordings, electronic books, and multimedia presentations, thereby augmenting students' comprehension and retention of intricate concepts.

Flexibility and Accessibility: Mobile learning offers flexibility and accessibility, catering to the diverse needs and preferences of learners. With mobile devices, students can access educational content and resources at their convenience, enabling self-directed learning and personalized study routines. This flexibility is particularly advantageous for non-traditional learners, including working professionals, mature students, and individuals with familial or caregiving obligations, who may necessitate adaptable study options to reconcile learning with other responsibilities.

Alignment with Technological Trends: Mobile learning aligns with contemporary technological trends and student preferences. As digital natives who have grown up in the era of smartphones and tablets, present-day college students are accustomed to retrieving information and engaging with content via mobile devices. By harnessing mobile technology, higher education institutions can satisfy the expectations of modern learners and sustain relevance in an increasingly digitized milieu.

4.5 Cloud Computing

The adoption of cloud-based platforms represents a significant avenue for innovation within higher education in Vietnam, offering scalable, flexible, accessible, and cost-effective solutions to meet a variety of educational requirements. By leveraging the capabilities of cloud computing, higher education institutions can foster collaboration, facilitate knowledge dissemination, optimize resource allocation, enhance data security and privacy measures, drive innovation, and ultimately enhance the quality and accessibility of education for all learners.

Flexibility and Scalability: Cloud platforms empower higher education institutions to adjust their infrastructure and resources in accordance with fluctuating demands, enabling them to expand or contract as necessitated by evolving circumstances. Cloud computing affords the agility to swiftly and efficiently adapt to changing needs.

Accessibility: Cloud platforms ensure ubiquitous access to educational resources and services, transcending geographical constraints and enabling learning anytime and anywhere. Students and faculty can avail themselves of cloud-hosted applications, course materials, collaboration tools, and learning management systems from any internet-connected device, promoting flexibility and accessibility for learners with diverse requirements and preferences.

Cost and Resource Efficiency: Cloud computing delivers cost-effective solutions to higher education institutions by alleviating the necessity for substantial investments in on-premises infrastructure and maintenance. Through cloud platforms, organizations can curtail expenditures on hardware and software procurement, reduce energy consumption, and optimize resource utilization via a pay-per-use payment model, as well as provision computing resources on-demand.

Collaboration and Knowledge Sharing: Cloud platforms cultivate collaboration and knowledge dissemination among students, faculty, researchers, and administrators, facilitating seamless communication and cooperation across institutional boundaries. Collaboration tools hosted on the cloud, including document sharing, real-time editing, video conferencing, and virtual classrooms, underpin collaborative learning experiences, group projects, and interdisciplinary research endeavors, enriching educational experiences and fostering cross-disciplinary collaboration.

4.6 Blockchain: Potential Technology to Increase Transparency and Security

Learning Management: By storing academic records on a decentralized ledger, educational institutions can ensure the integrity of degrees and certificates, thereby mitigating issues such as fraud and misrepresentation of qualifications. This holds particular significance in Vietnam, where the verification of credentials often entails lengthy processes and complexities.

Secure Data Management: Educational institutions contend with vast amounts of sensitive data, encompassing learner records, financial details, and research data. Blockchain technology offers a secure means of storing and managing such data, safeguarding it against tampering and unauthorized access. The adoption of blockchain-based systems can augment data security and bolster compliance with privacy regulations, a paramount consideration in the digital era.

Transparent Governance: Blockchain technology facilitates transparent governance and accountability frameworks within higher education institutions. Through the utilization of smart contracts, processes such as student enrollment, course registration, and tuition payments can be automated, streamlining administrative tasks and enhancing transparency. In the context of Vietnam, blockchain holds the potential to foster transparency and accountability in educational governance.

Intellectual Property Protection: Research and innovation are integral components of educational advancement. Blockchain can aid in safeguarding intellectual property rights by timestamping and securely storing research findings, patents, and creative works. This fosters an environment conducive to innovation and knowledge dissemination while safeguarding the interests of creators and innovators, a critical element in nurturing a dynamic academic ecosystem.

5 Conclusion

The digital transformation of HEIs in Vietnam is essential for enhancing educational quality, fostering innovation, and navigating the evolving higher education landscape in the digital era. This study examines the potential applications of various technologies in supporting the digital transformation of Vietnamese HEIs. Key technologies identified include artificial intelligence, adaptive learning platforms, online learning systems, digital learning materials, IoT, mobile learning, cloud computing, and blockchain. These technologies offer solutions to challenges such as personalized learning, accessibility, flexibility, resource efficiency, collaboration, data security, and transparency. Nonetheles, successful implementation is contingent upon overcoming challenges like infrastructure inadequacies, human resource constraints, digital literacy gaps, and governance framework deficiencies. Comprehensive strategies, innovation culture cultivation, and digital competency development among stakeholders are essential. By leveraging these technologies and addressing challenges, Vietnamese HEIs can unlock transformative opportunities, revolutionize educational delivery, and cultivate a future-ready workforce. Collaboration among policymakers, educational leaders, technology experts, and industry partners is crucial for realizing this vision. A harmonized approach can enable

Vietnam to harness technology's power, shaping a progressive and inclusive educational landscape and empowering citizens to contribute to sustainable development and global competitiveness.

References

1. Truong, T.-C., Diep, Q.B.: Technological spotlights of digital transformation in tertiary education. IEEE Access **11**, 40954–40966 (2023). https://doi.org/10.1109/ACCESS.2023.3270340
2. Phong, T.C., et al.: Digital transformation in education. Vietnam J. Sci. Educ. **17**, 1–7 (2019)
3. Hang, B.T.T.: New technologies in higher education—challenges and solutions when using. Vietnam J. Sci. Educ. **28**, 13–18 (2020)
4. Linh, D.T., et al.: Digital transformation in education and training in Vietnam. Binh Duong Univ. J. Sci. Technol. **1** (2022). https://doi.org/10.56097/binhduonguniversityjournalofscienceandtechnology.vi1.14
5. Abad-Segura, E., et al.: Sustainable management of digital transformation in higher education: global research trends. Sustainability **12**(5), 2107 (2020). https://doi.org/10.3390/issue12052107
6. Mohamed Hashim, M.A., et al.: Higher education strategies in digital transformation. Educ. Inf. Technol. **27**(3), 3171–3195 (2022). https://doi.org/10.1007/s10639-021-10739-1
7. Trevisan, L.V., et al.: Digital transformation towards sustainability in higher education: state-of-the-art and future research insights. Environ. Dev. Sustain. (2023). https://doi.org/10.1007/s10668-022-02874-7
8. Vietnam Education Magazine: Digital transformation in high school and university education has achieved many impressive results. https://Giaoduc.net.vn/post-238412.gd, last accessed 21 Feb 2024
9. Braun, V., Clarke, V.: Conceptual and design thinking for thematic analysis. Qual. Psychol. **9**(1), 3–26 (2022). https://doi.org/10.1037/qup0000196

Unlocking Consumer Insights: A Comprehensive Analysis of EEG Applications in Neuro Marketing Research

Hoang-Sy Nguyen(✉)

Becamex Business School, Eastern International University, Thu Dau Mot City, Binh Duong Province, Vietnam
sy.nguyen@eiu.edu.vn

Abstract. In the ten years there has been an increase, in using neurophysiological and physiological tools to understand how consumers react to advertising. However there is a lack of in depth research on the brain processes involved in these reactions such as emotions, motivation, reward, attention, perception and memory. To address this gap this study followed the Preferred Reporting Items for Systematic Reviews and Meta Analyses (PRISMA) guidelines to find research papers. Electroencephalography (EEG) was found to be a tool among researchers due to its cost effectiveness and temporal resolution. The primary aim was to assess patterns in EEG research concerning neuromarketing by analysing freely available articles sourced from the Scopus database spanning from 2016 to 2023. It was observed that Spain and the United States were leading countries in this field while Frontiers in Psychology and Springer Proceedings in Business and Economics were journals despite publication numbers not always correlating with citation counts. This study offers an overview of EEG research in marketing highlighting the role of brain processes, in various advertising campaigns including commercial endeavors, social projects and public health initiatives. This contributes towards understanding consumer behavior and improving advertising strategies across scenarios.

Keywords: Neuromarketing · Electroencephalography · Bibliometric analysis · VOSviewer

1 Introduction

In business and advertising research, traditional methods such as surveys and interviews have been commonly used to understand consumer reactions to advertising [1]. However, these methods may be influenced by bias and often overlook the emotional and cognitive aspects of decision-making. To address these limitations, there is a growing interest in consumer neuroscience, which combines

M. Pagac et al. (Eds.): SCFF 2024, LNNS 1062, pp. 470–481, 2024.
https://doi.org/10.1007/978-3-031-65656-9_46

insights from marketing, neuroscience, and psychology to gain a deeper understanding of consumer behaviour, utilising techniques such as neuroimaging and psychophysiology [2,3]. Consumer neuroscience explores the underlying mechanisms of consumption behaviours, their psychological implications, and resultant actions, often by measuring participants' brain responses to marketing stimuli using neuroimaging and psychophysiological measures to unveil the brain processes involved in consumer decision making [4,5]. Advancements in non-invasive brain imaging techniques have fostered interdisciplinary collaboration among economists, psychologists, and neuroscientists to explore the biological underpinnings of advertising theories [6,7]. Inspired by neuromarketing investigations, our emphasis is on employing techniques that offer understanding of the brain's reactions to advertising cues, encompassing magnetoencephalography (MEG), electroencephalography (EEG), functional magnetic resonance imaging (fMRI), and functional infrared spectroscopy (fNIRS). Recent marketing studies have employed these neuroimaging tools to uncover decision-making processes related to product packaging, pricing strategies [8], various advertising formats [9], and secure payment systems [10].

Neuroimaging methods provide a non-invasive means to track brain responses to advertising stimuli, revealing subtle processes beyond self-reported data and aiding in the understanding of how ads impact us, whether through feelings of reward or personal relevance triggered by visual design, colour choices, or tone of voice. In advertising research, neuroimaging technology can (i) uncover brain activity linked to advertising concepts such as memory, attention, persuasion, and emotion; (ii) assess advertising performance by analysing brain activity and predictive factors; (iii) explore how audience demographics influence ad perception; and (iv) investigate the effectiveness of advertising strategies, including persuasion [11].

Researchers worldwide frequently employ neuroimaging methods to enhance traditional advertising research theories, aiming to understand consumer reactions and improve communication campaign effectiveness. Studies utilising fMRI, EEG, MEG, and fNIRS have explored how advertising messages influence emotions, attention, preferences, and memory, while also investigating neural factors affecting communication outcomes and media element processing [11,12]. Based on the literature, the initial fMRI investigation in neuromarketing conducted by McClure et al. [13] significantly influenced practical studies in the field [14], drawing attention from both academia and industry as a method to address the limitations of traditional marketing approaches and mitigate consumer social bias. As the use of EEG in neuromarketing is still emerging and relatively new [15], this study focuses on global research trends in studies utilizing EEG between 2015 and 2023. The paper aims to provide a comprehensive retrospective of EEG usage in academic research during this period. The main contributions of this bibliometric study include:

- Presenting an overview of cumulative research publications and journal outputs spanning from 2015 to 2023 helping newcomers comprehend the academic research trends in neuromarketing activities.
- Highlighting avenues for scholars interested in the realm of neuromarketing.

The study is organised as follows: Sect. 2 outlines the methods used in the research, while Sect. 3 examines the findings from these articles and proposes research avenues. Section 4 summarises the study's conclusions.

2 Materials and Methods

The research, adhering to the PRISMA protocol, aimed to explore neuromarketing trends, utilising analysis to examine contributions in EEG-based studies [16]. Using the Scopus database and VOSviewer software, the study analysed articles published between 2016 and 2023 that were written in English, focusing on EEG tools, to identify patterns and insights in neuromarketing research [17,18]. Data collection occurred in January 2024. Figure 1 illustrates the procedure for choosing articles for this study. The search targeted articles that integrated EEG in marketing research, utilizing searches conducted on titles, abstracts, and keywords: (TITLE-ABS-KEY ("Consumer neuroscience" OR "Neuromarketing") AND TITLE-ABS-KEY ("advertising" OR "advertisement") AND TITLE-ABS-KEY ("Neuroimaging" OR "Electroencephalography" OR "physiological" OR "EEG")) AND PUBYEAR > 2015 AND PUBYEAR < 2024 AND (LIMIT-TO (DOCTYPE, "AR") OR LIMIT-TO (DOCTYPE , "CP")) AND (LIMIT-TO (LANGUAGE, "English")).

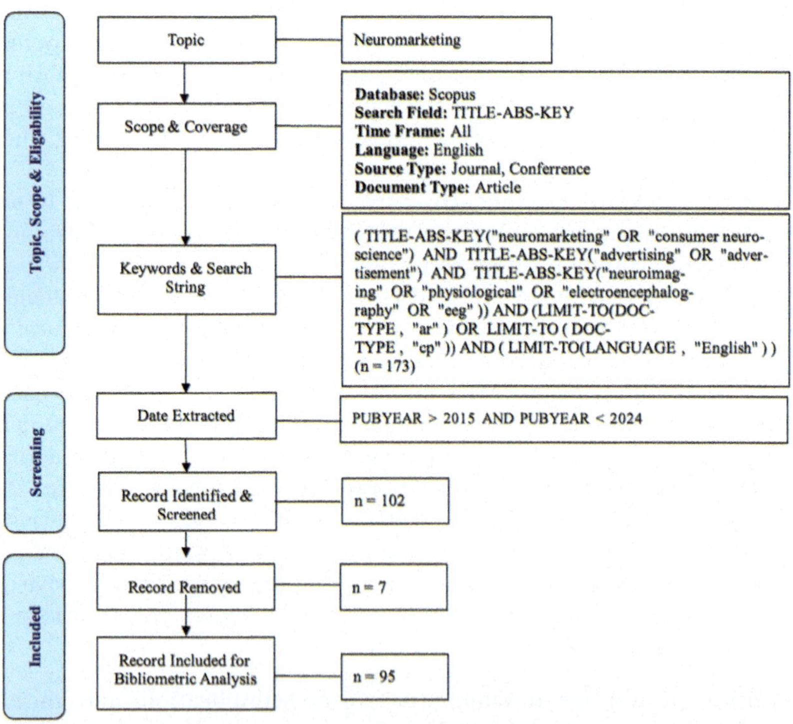

Fig. 1. The PRISMA protocol was employed to identify documents for bibliometric study.

3 Results

3.1 Descriptive Analysis

In line with the process we identified 95 articles that utilized EEG in marketing research and were published in journals Table 1. There was an uptick in publications with over 80% of all articles published between 2020 and 2023. For instance while four articles were published in 2016 this number nearly quadrupled by 2020. The year, with the number of publications was observed to be 2021 with seventeen articles being published that year.

This trend indicates a growing interest among researchers and industry professionals in neuromarketing resulting in an increase in publications and researchers utilizing EEG technology for marketing research purposes. Our findings highlight the use of neuroimaging methods, in advertising research during the five years of the study period suggesting potential advancements in this specific area of study.

Table 1. Total publications and citations by year.

	2016	2017	2018	2019	2020	2021	2022	2023
Publications	4	8	9	13	14	17	13	13
Citations	4	15	44	79	140	217	319	398

3.2 Productive Countries

A total of 95 papers were examined from the Scopus database, with the outcomes outlined in Table 2. The results reveal Spain, the United States, India, and Italy as prominent contributors to neuromarketing research in advertising, collectively constituting over 50% of all publications, emphasising their significant roles in advancing this field. Specifically, Spain led with fourteen papers (15.4% of all papers), followed by nearly eleven from the United States (12.1% of all documents), while India and Italy each contributed ten papers (11% of all documents). China and England tied for fifth and sixth place with six papers each, and Iran, Lithuania, Poland, and Malaysia also made notable contributions, with five or four publications each.

3.3 Primary Journal

In this part, we recognised nine journals that have published a minimum of two articles on advertising research related to neuromarketing, as outlined in Table 3. Frontiers in Psychology stands out with nine published articles, followed by Springer Proceedings in Business and Economics and Frontiers in Human Neuroscience, each with four articles. Additionally, Physiology and Behavior,

Table 2. Total ten productive countries.

No	Country	Total publications	Total citations	% of total articles
1	Spain	14	146	15.4
2	United States	11	118	12.1
3	India	10	220	11.0
4	Italy	10	142	11.0
5	China	6	62	6.6
6	United Kingdom	6	92	6.6
7	Iran	5	104	5.5
8	Lithuania	5	19	5.5
9	Poland	5	35	5.5
10	Malaysia	4	32	4.4

along with the Journal of Advertising Research, contributed three papers each, while Procedia Computer Science, Neuropsychological Trends, Multimedia Tools and Applications, Journal of Retailing and Consumer Services, and Frontiers in Neuroscience each presented two papers. Moreover, The citation count of a journal is regarded as an indicator of its article quality and popularity, while publication volume reflects productivity. We assessed the journals by calculating the average citation per item (ACI) using data from the Scopus database on total citations (TC) and total publications (TP). Results from Table 3 reveal that Multimedia Tools and Applications and Frontiers in Neuroscience have the ACI despite each releasing only two papers on advertising and neuromarketing, with 82.5 and 30.5 citations respectively, suggesting that publication quantity doesn't always correlate with citation count.

Table 3. Total ten journal publications.

No	Source		TC	ACI
1	Frontiers in psychology	9	79	8.8
2	Springer proceedings in business and economics	4	62	15.5
3	Frontiers in human neuroscience	4	37	0.7
4	Physiology and behavior	3	113	37.7
5	Journal of advertising research	3	66	22
6	Procedia computer science	2	15	7.5
7	Neuropsychological trends	2	13	6.5
8	Multimedia tools and applications	2	165	82.5
9	Journal of retailing and consumer services	2	25	12.5
10	Frontiers in neuroscience	2	81	40.5

3.4 Analysis of Citations

In this part, we performed a robust citation analysis to pinpoint the most widely referenced articles in the domains of advertising and neuromarketing, providing insight into their scholarly impact [19]. A total of ninety-five papers were reviewed, with outcomes detailed in Table 4, highlighting the most-cited articles in the field, including "Analysis of EEG signals and its application to neuromarketing" by M. Yadava, et al. [20], leading with 165 citations, followed by "Deep Learning for EEG-Based Preference Classification in Neuromarketing" by M. Aldayel, et al. [21], with 93 citations.

Table 4. Total ten publication articles.

No	Title	Journal	Year Pub	T.Cs
1	Analysis of EEG signals and its application to neuromarketing [20]	Multimedia Tools and Applications	2017	165
2	Deep Learning for EEG-Based Preference Classification in Neuromarketing [21]	Applied Sciences	2020	93
3	The application of EEG power for the prediction and interpretation of consumer decision-making: A neuromarketing study [22]	Physiology & Behavior	2019	92
4	Electroencephalographic, heart rate, and galvanic skin response assessment for an advertising perception study: Application to antismoking public service announcements [23]	Journal of Visualized Experiments	2017	63
5	Using Support Vector Machine on EEG for Advertisement Impact Assessment [24]	Frontiers in neuroscience	2018	57
6	Neural responses to functional and experiential ad appeals: Explaining ad effectiveness [25]	International Journal of Research in Marketing	2017	52
7	Decoding Neural Responses to Emotion in Television Commercials: An Integrative Study of Self-Reporting and fMRI Measures [26]	Journal of Advertising Research	2016	43
8	The cognitive processing of an educational app with electroencephalogram and eye tracking [27]	Comunicar	2017	40
9	EEG Spectral Dynamics of Video Commercials: Impact of the Narrative on the Branding Product Preference [28]	Scientific Report	2016	38
10	Neuromarketing, subliminal advertising, and hotel selection: An EEG study [29]	Australasian Marketing Journal	2020	36

3.5 Keywords Analysis

Bibliometric analysis quantifies keyword frequency to evaluate their relevance, while link strength measures the correlation between pairs of keywords based on their co-occurrence in papers [30]. This study performed keyword co-analysis on 56 terms extracted from 95 articles across over 30 journals, considering synonymous keywords like "neuromarketing," "electroencephalography (EEG)," and "consumer neuroscience," with each keyword required to appear in at least two papers for inclusion in the bibliometric map.

Analysing keyword co-occurrence is essential for comprehending article content and monitoring research trends, particularly in fields like neuromarketing with a focus on advertising. This approach helps identify research directions and assess topic significance, such as the use of fMRI, EEG, and fNIRS to capture brain responses to ads, revealing insights into factors like recall and recognition. By evaluating advertising campaign strengths and weaknesses, marketers can refine strategies, design impactful campaigns, and address any negative perceptions about advertising. The keyword co-occurrence map in Fig. 2 highlights EEG as a prominent focus in marketing, particularly in advertising contexts, with 56 occurrences and a total link strength of 46. "Brand" emerges as a common keyword, with 13 instances and a total link strength of 13, followed by brain functions such as attention, emotion, and memory.

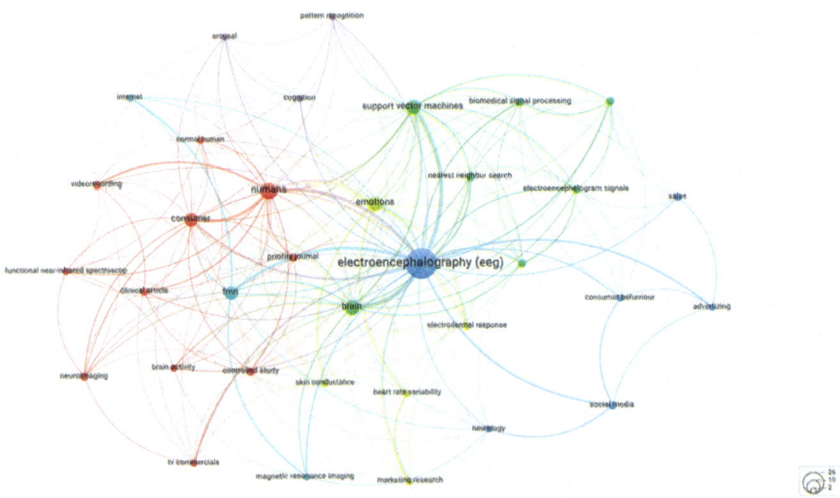

Fig. 2. The keyword co-occurrence map.

We hypothesised a strong connection between neuromarketing/consumer neuroscience and neurophysiological tools like EEG, fMRI. Additionally, we expected a significant relationship between subconscious responses like "emotions", "neuroimaging", "long short-term memory," and marketing stimuli such as "brand", "humans", "consumer". For instance, "humans" emerged as the most prevalent theme with seventeen occurrences and a total link strength (TLS) of seventeen, followed by "brain" with thirteen occurrences and thirteen TLS, "emotions" with thirteen occurrences and nine TLS, and "consumer" with eleven occurrences and eleven TLS. Notably, EEG exhibited a notable association with neuromarketing and advertising research. Table 5 provides a summary of the most frequent keywords with a minimum of four occurrences, with "electroencephalography" being the most frequently mentioned, followed by "neuromarketing."

Table 5. Top keywords with at least four occurrences.

Keyword	Occurrences	Total link strength
Electroencephalography (EEG)	56	46
Humans	17	17
Brain	13	13
Emotions	13	9
Support vector machines	12	12
Consumer	11	11
fMRI	11	9
Biomedical signal processing	4	4
Clinical article	4	4
Controlled study	4	4
Decision trees	4	4
Electroencephalogram signals	4	4
Long short-term memory	4	4
Attitudes	4	4
Neuroimaging	4	4

3.6 Future Directions

The study examined brain functions pertinent to advertising, identifying patterns of activity related to emotions, attention, memory, perception, motivation, and reward processing. Specific brain regions like the frontal and middle temporal gyri were linked to pleasure and displeasure experiences [31], while the olfactory lobe was associated with attention [32], and highlighted for its role in memory processing. Motivational components linked to hemisphere activation exhibited distinct patterns for withdrawal and approach behaviours. The involvement of the orbitofrontal cortex and ventromedial prefrontal cortex in shaping perceptions of perceived value provides valuable insights into consumer behaviour, underscoring the importance of integrating brain processes into advertising research [33]. Furthermore, proficiency in handling physiological instruments, software, and advanced technology is essential for conducting neuromarketing research, highlighting the need for expertise in utilizing tools, conducting experiments, and interpreting data [34,35].

When navigating paths in the field, addressing challenges like limited access to research techniques, along with the time-intensive nature of tasks such as data interpretation and participant recruitment, is crucial. Investing more in neuromarketing research and methods is essential to overcome these hurdles and foster advancements. Encouraging researchers from developing nations to contribute to this growing field by publishing articles and advancing knowledge is pivotal for future projects. Moving forward, it is imperative for researchers and experts to explore how advertising influences consumer persuasion, engagement, excitement. Additionally, delving into the ramifications of neuro-marketing research in domains like social sciences, stock markets, politics, public health can yield valuable insights into the influence of brain processes on consumer behaviour. This review offers an overview of academic trends in neuro-marketing research, particularly focusing on neuroimaging, physiological studies in advertising, thereby paving the way for further investigations and advancements in the field.

4 Conclusion

In conclusion, neuromarketing emerges as a promising avenue for addressing commercial challenges, optimizing advertising effectiveness, and enhancing campaigns across diverse sectors. In the competitive advertising landscape, agencies strive to uncover strategies that effectively engage consumers. The study highlights the pivotal role of neuroscientific methods in capturing consumers' cognitive and physiological responses to marketing stimuli, particularly in advertising research. This research sheds light on the intricate relationship between neural correlates and physiological responses to advertising, offering insights into consumer behavior. Neurophysiological and physiological tools facilitate the measurement of brain activity signals and bodily reactions, providing valuable data for crafting compelling advertising content. Overall, this review offers a comprehensive overview of current neuroscientific approaches in advertising research, emphasizing the importance of understanding key mental processes for achieving accurate outcomes. The findings underscore the significance of neuroscientific techniques in deciphering consumer responses to marketing stimuli, ultimately contributing to the creation of more engaging advertising campaigns across various sectors.

Acknowledgements. The authors would like to thank Eastern International University, Binh Duong Province, Vietnam for funding this research.

References

1. Carrington, M.J., Neville, B.A., Whitwell, G.J.: Lost in translation: Exploring the ethical consumer intention-behavior gap. J. Bus. Res. **67**(1), 2759–2767 (2014). https://doi.org/10.1016/j.jbusres.2012.09.022. Jan
2. Grimmer, M., Kilburn, A.P., Miles, M.P.: The effect of purchase situation on realized pro-environmental consumer behavior. J. Bus. Res. **69**(5), 1582–1586 (2016). https://doi.org/10.1016/j.jbusres.2015.10.021. May

3. Casado-Aranda, L.-A., Sánchez-Fernández, J., Montoro-Ríos, F.J.: How consumers process online privacy, financial, and performance risks: an fMRI study. Cyberpsychol. Behav. Soc. Netw. **21**(9), 556–562 (2018). https://doi.org/10.1089/cyber.2018.0196. Sep
4. Golnar-Nik, P., Farashi, S., Safari, M.-S.: The application of EEG power for the prediction and interpretation of consumer decision-making: A neuromarketing study. Physiol. Behav. **207**, 90–98 (2019). https://doi.org/10.1016/j.physbeh.2019.04.025. Aug
5. Songsamoe, S., Saengwong-ngam, R., Koomhin, P., Matan, N.: Understanding consumer physiological and emotional responses to food products using electroencephalography (EEG). Trends Food Sci. Technol. **93**, 167–173 (2019). https://doi.org/10.1016/j.tifs.2019.09.018. Nov
6. Hubert, M., Kenning, P.: A current overview of consumer neuroscience. J. Consum. Behav. **7**(4–5), 272–292 (2008). https://doi.org/10.1002/cb.251. Jul
7. Shaw, S.D., Bagozzi, R.P.: The neuropsychology of consumer behavior and marketing. Consum. Psychol. Rev. **1**(1), 22–40 (2017). https://doi.org/10.1002/arcp.1006. Dec
8. Shaw, S.D., Bagozzi, R.P.: The neuropsychology of consumer behavior and marketing. Consum. Psychol. Rev. **1**(1), 22–40 (2017). https://doi.org/10.1002/arcp.1006. Dec
9. Casado-Aranda, L.-A., et al.: Consumer processing of online trust signals: a neuroimaging study. J. Interact. Mark. **47**, 159–180 (2019). https://doi.org/10.1016/j.intmar.2019.02.006. Aug
10. Casado-Aranda, L.-A., et al.: A neuropsychological study on how consumers process risky and secure E-payments. J. Interact. Mark. **43**, 151–164 (2018). https://doi.org/10.1016/j.intmar.2018.03.001. Aug
11. Weber, R., et al.: Brain imaging in communication research: a practical guide to understanding and evaluating fMRI studies. Commun. Methods Meas. **9**(1–2), 5–29 (2015). https://doi.org/10.1080/19312458.2014.999754. Jan
12. Falk, E.B., et al.: Neural activity during health messaging predicts reductions in smoking above and beyond self-report. Health Psychol. **30**(2), 177–185 (2011). https://doi.org/10.1037/a0022259
13. McClure, S.M., et al.: Neural correlates of behavioral preference for culturally familiar drinks. Neuron **44**(2), 379–387 (2004). https://doi.org/10.1016/j.neuron.2004.09.019. Oct
14. Alsharif, A.H., et al.: Biomedical technology in studying consumers' subconscious behavior. Int. J. Online Biomed. Eng. (iJOE) **18**(08), 98–114 (2022). https://doi.org/10.3991/ijoe.v18i08.31959. Jun
15. Bazzani, A., et al.: Is EEG suitable for marketing research? A systematic review. Front. Neurosci. 14 Dec 2020. https://doi.org/10.3389/fnins.2020.594566
16. Moher, D., et al.: Preferred reporting items for systematic review and meta-analysis protocols (PRISMA-P) 2015 statement. Syst. Rev. 4(1), Jan 2015. https://doi.org/10.1186/2046-4053-4-1
17. Alsharif, A., et al.: Current trends in the application of EEG in neuromarketing: a bibliometric analysis. Sci. Ann. Econ. Bus. **69**(3), 393–415 (2022). https://doi.org/10.47743/saeb-2022-0020. Aug
18. Sánchez-Fernández, J., et al.: Consumer neuroscience techniques in advertising research: a bibliometric citation analysis. Sustainability **13**(3), 1589 (2021). https://doi.org/10.3390/su13031589. Feb

19. Nguyen, H.-S., et al.: Research trend in the field of additive manufacturing with bibliometrics study. MM Sci. J. **2023**(2) 2023. https://doi.org/10.17973/mmsj.2023_06_2023032
20. Yadava, M., et al.: Analysis of EEG signals and its application to neuromarketing. Multimedia Tools Appl. **76**(18), 19087–19111 (2017). https://doi.org/10.1007/s11042-017-4580-6. Mar
21. Aldayel, M., Ykhlef, M., Al-Nafjan, A.: Deep learning for EEG-based preference classification in neuromarketing. Appl. Sci. **10**(4), 1525 (2020). https://doi.org/10.3390/app10041525. Feb
22. Golnar-Nik, P., Farashi, S., Safari, M.-S.: The application of EEG power for the prediction and interpretation of consumer decision-making: a neuromarketing study. Physiol. Behav. **207**, 90–98 (2019). https://doi.org/10.1016/j.physbeh.2019.04.025. Aug
23. Cartocci , G., et al.: Electroencephalographic, heart rate, and galvanic skin response assessment for an advertising perception study: application to antismoking public service announcements. J. Visualized Exp. 126, Aug (2017). https://doi.org/10.3791/55872-v
24. Wei, Z., et al.: Using support vector machine on EEG for advertisement impact assessment. Front. Neurosci. **12**, Mar 2018. https://doi.org/10.3389/fnins.2018.00076
25. Couwenberg, L.E., et al.: Neural responses to functional and experiential ad appeals: explaining ad effectiveness. Int. J. Res. Mark. **34**(2), 355–366 (2017). https://doi.org/10.1016/j.ijresmar.2016.10.005. Jun
26. Shen, F., Morris, J.D.: Decoding neural responses to emotion in television commercials: an integrative study of self-reporting and fMRI measures. J. Advert. Res. **56**(2), 193 (2016). https://doi.org/10.2501/jar-2016-016
27. Cuesta-Cambra, U., et al.: The cognitive processing of an educational app with EEG and 'Eye tracking. Comunicar **25**(52), 41–50 (2017). https://doi.org/10.3916/c52-2017-04. Jul
28. Wang, R.W.Y., et al.: EEG spectral dynamics of video commercials: impact of the narrative on the branding product preference. Sci. Rep. **6**(1), Nov 2016. https://doi.org/10.1038/srep36487
29. Hsu, L., Chen, Y.-J.: Neuromarketing, subliminal advertising, and hotel selection: an EEG study. Australas. Mark. J. **28**(4), 200–208 (2020). https://doi.org/10.1016/j.ausmj.2020.04.009. Nov
30. Nguyen, H.-S., Voznak, M.: A bibliometric analysis of technology in digital health: exploring health Metaverse and visualizing emerging healthcare management trends. IEEE Access **12**, 23887–23913 (2024). https://doi.org/10.1109/access.2024.3363165
31. Shen, F., Morris, J.D.: Decoding neural responses to emotion in television commercials: an integrative study of self-reporting and fMRI measures. J. Advert. Res. **56**(2), 193 (2016). https://doi.org/10.2501/jar-2016-016
32. Smith, M.E., Gevins, A.: Attention and brain activity while watching television: components of viewer engagement. Media Psychol. **6**(3), 285–305 (2004). https://doi.org/10.1207/s1532785xmep0603_3. Aug
33. Haehl, W., Mirifar, A., Luan, M., Beckmann, J.: Dealing with failure: prefrontal asymmetry predicts affective recovery and cognitive performance. Biol. Psychol. **155**, 107927 (2020). https://doi.org/10.1016/j.biopsycho.2020.107927. Sep

34. Alsharif, A.H., et al.: Neuromarketing tools used in the marketing mix: a systematic literature and future research agenda. SAGE Open **13**(1), 215824402311565 (2023). https://doi.org/10.1177/21582440231156563. Jan
35. Alsharif, A.H., et al.: Exploring factors influencing neuromarketing implementation in Malaysian Universities: barriers and enablers. Sustainability **15**(5), 4603 (2023). https://doi.org/10.3390/su15054603. Mar

SPRINGER NATURE

GPSR Compliance

The European Union's (EU) General Product Safety Regulation (GPSR) is a set of rules that requires consumer products to be safe and our obligations to ensure this.

If you have any concerns about our products, you can contact us on ProductSafety@springernature.com

In case Publisher is established outside the EU, the EU authorized representative is:

Springer Nature Customer Service Center GmbH
Europaplatz 3
69115 Heidelberg, Germany

The manufacturer's authorised representative in the EU is Springer Nature Customer Service Centre GmbH, Europaplatz 3, 69115 Heidelberg, Germany. If you have any concerns regarding our products, please contact ProductSafety@springernature.com

Printed and bound by CPI Group (UK) Ltd, Croydon, CR0 4YY

30/09/2026

02225587-0009